Emerging Technologies in Food Preservation

Consumers worldwide are becoming increasingly aware of the significant impacts of food quality, nutrition, and food habits on their health. Demands for fresh, minimally processed foods; foods subjected to less severe preservation and processing; and foods with no synthetic chemical preservatives are expanding rapidly. Food businesses worldwide are innovating advanced processing and preservation technologies, including natural, bioactive preservatives, to fulfill these emerging consumers' demands.

Emerging Technologies in Food Preservation discusses the innovations and advancements in food processing and preservation that have emerged over the last two decades of the 21st century. Various experts in the field examine different innovative food-processing and preservation technologies, including nonthermal and minimal processing technologies, particularly those that have least destructive impact on micronutrients.

Key features of the text include

- descriptions of the hurdles and challenges in commercialization of these novel processing technologies, including their legal ramifications on food regulations and legislations;

- detailed accounts of application of each of the novel food preservation principles as they are applied in commercial food processing/food manufacturing lines worldwide;

- a complete account on recent developments and advancements in food preservation and processing technologies, including natural and biopreservatives.

Mindful of the increasing interest in alternatives to traditional thermal processing of foods and beverages, this book is a valuable reference for students, scholars, researchers, scientists, and food entrepreneurs, as well as technical professionals working in food industries and associated businesses.

Emerging Technologies in Food Preservation

Edited by Santosh Kumar, Avik Mukherjee,
Atanu Mitra, and Dipankar Halder

CRC Press
Taylor & Francis Group
Boca Raton London New York

CRC Press is an imprint of the
Taylor & Francis Group, an **informa** business

First edition published 2023
by CRC Press
6000 Broken Sound Parkway NW, Suite 300, Boca Raton, FL 33487–2742

and by CRC Press
4 Park Square, Milton Park, Abingdon, Oxon, OX14 4RN

CRC Press is an imprint of Taylor & Francis Group, LLC

Library of Congress Cataloging-in-Publication Data
Names: Kumar, Santosh, 1980– editor.
Title: Emerging technologies in food preservation / edited by Santosh Kumar, Avik
    Mukherjee, Atanu Mitra, Dipankar Halder.
Description: First edition. | Boca Raton : CRC Press, 2023. | Includes bibliographical
    references and index.
Identifiers: LCCN 2022031355 (print) | LCCN 2022031356 (ebook) | ISBN
    9780367675158 (hbk) | ISBN 9780367707835 (pbk) | ISBN 9781003147978 (ebk)
Subjects: LCSH: Food—Preservation—Technological innovations.
Classification: LCC TP371 .E435 2023 (print) | LCC TP371 (ebook) | DDC 664/
    .028—dc23/eng/20220901
LC record available at https://lccn.loc.gov/2022031355
LC ebook record available at https://lccn.loc.gov/2022031356

ISBN: 9780367675158 (hbk)
ISBN: 9780367707835 (pbk)
ISBN: 9781003147978 (ebk)

DOI: 10.1201/9781003147978

Typeset in Palatino
by Apex CoVantage, LLC

To our parents and families, friends, and all those who bestowed their encouragement and support on us, without which the completion of this endeavour wouldn't have been possible.

Contents

Preface

The book project, titled *Emerging Technologies in Food Preservation*, was born sometime in late 2020 in the context of tremendous advancements in novel food-processing technologies. It stemmed from the increasing awareness and interest among consumers and readers in food quality and safety, consumers' health, and environmental sustainability. From the perspective of a developing country, the urge to develop such a book project on this emerging area of innovative food-processing technologies was perhaps more compelling. As the world is increasingly exploring sustainable, nonthermal, novel, least invasive food-processing technologies as alternatives to traditional thermal processing of foods and beverages, this book is going to serve as a complete, systematic, comprehensive account of the contemporary developments in the area of emerging food processing technologies. It is a unique reading and resource material for educators and/or researchers, professionals in food manufacturing and handling industries, research scholars, graduate students, entrepreneurs, and people working in the area of food preservation and processing, energy savings, consumers' nutrition, and health. Also, the book describes the hurdles and challenges in commercialization of these novel processing technologies, including their legal ramifications on food regulations and legislations.

This book consists of 14 chapters covering different food processing and preservation technologies, including nonthermal, minimal processing technologies, particularly those that have least destructive impact on micronutrients. The first eight chapters detail nonthermal food processing and preservation technologies, while the next few chapters, that is, Chapters 9, 10 and 11, discuss the contemporary advancements in thermal processing of foods and beverages. The discussions on processing technologies have focused on their minimal impact on nutritional and health benefits of foods. This book also uniquely includes a couple of chapters (Chapters 12 and 13) on applications of nanotechnology in food preservation and packaging. Applications of nanomaterials as functional ingredients and/or as reinforcement(s) in food coatings and packaging have been reviewed extensively in these chapters. Finally, the last chapter of the book, Chapter 14, discusses regulations and legislations on food processing technologies, including the updates mandated by these regulations to address applications of these novel food processing technologies. In the context of globalization of food chains, this chapter has particular significance to address the complexities and the challenges that food manufacturers and businesses face to ensure safe food consistently serving the consumers' needs.

It has been a great pleasure and a privilege to collaborate with the contributing authors representing different academic institutes/universities of international repute in India and abroad. The success of this project depends on potential readers across the entire spectra of food and associated areas in academia, industries, businesses and government bodies. The readers' enthusiastic acceptance, appreciation, comments and critiques will inspire us to venture into more such collaborative projects in future. We couldn't have achieved success in our efforts without the support, cooperation and understanding of our peers, friends and families, which stems our unwavering dedication and commitment to this book project.

Santosh Kumar
Avik Mukherjee
Atanu Mitra
Dipankar Halder

Editors

Dr. Santosh Kumar graduated from the Department of Food Technology and Biochemical Engineering (FTBE), Jadavpur University, Kolkata, India with M.Tech and PhD degrees. He joined the Department of Food Engineering and Technology, Central Institute of Technology Kokrajhar, India as an assistant professor in 2011 and has been teaching undergraduate, postgraduate, and PhD students. Dr. Kumar has extensive research and academic experiences of 12-plus years in the area of food preservation and shelf-life, sustainable food packaging, and biopolymer-based functional nanocomposites. Dr. Kumar has received the "Overseas Associateship Award" from the Department of Biotechnology (DBT), Government of India, and worked as postdoctoral research fellow at KTH Royal Institute of Technology, Sweden, during November 2018–November 2019. He is a professional member of many scientific societies and institutions. He has published more than 35 peer-reviewed research papers and other articles. He also serves as an editorial board member of many reputed journals and has coedited a book, *Biopolymer-Based Food Packaging: Innovations and Technology Applications,* published by Wiley and Sons in 2022.

Dr. Avik Mukherjee graduated from the Department of Food Science and Nutrition, University of Minnesota, USA, with an MS and a PhD degree in food science. He joined the Department of Animal Sciences, Colorado State University, USA, as a postdoctoral fellow and continued his research on microbial food safety and novel food preservation technologies for about one and a half years. Dr. Mukherjee returned to India and joined the Department of Food Technology, Haldia Institute of Technology, as an assistant professor. He continued to serve the Institute for nine and half years teaching undergraduate students and was promoted first to associate professor in 2013 and then to professor position in 2015. Dr. Mukherjee joined the Department of Food Engineering and Technology, Central Institute of Technology Kokrajhar, Assam, India, in 2017 as an associate professor and has been teaching undergraduate, postgraduate, and PhD students.

Dr. Mukherjee has been actively pursuing research in microbial food safety, novel food preservation, and packaging techniques, including applications of natural preservatives and nanotechnology. One of his PhD projects also focuses on the utilization of food-processing wastes for bioethanol production through fermentation. He has received Career Awards for Young Teachers, All India Council for Technical Education (AICTE), in 2009, and has been handling extramurally funded research projects. Dr. Mukherjee has published more than 25 peer-reviewed journals, five books (one with Wiley), seven book chapters (five with Wiley), and almost 30 paper/poster presentations in numerous international and national conferences/scientific meetings. Over the last 16 years of his career, Dr. Mukherjee has served as the head of the department for 9 years and as dean for the last 4 years.

Dr. Atanu Mitra has been working as a faculty member in the Department of Chemistry (undergraduate and postgraduate), Sree Chaitanya College, Habra, West Bengal, India, since 2007. Besides teaching he is involved in active research in the diverse field of nanomaterials. He did his master's at Calcutta University and obtained a doctoral degree from Jadavpur University. Prior to his current position, Dr. Mitra worked as a researcher in different internationally recognized institutes abroad as well as within India. He also served as a visiting researcher in the University of Santiago de Compostela, Spain. Dr. Mitra is the author of several scientific papers published in internationally reputed journals.

Professor Dipankar Halder graduated from the Department of Food Technology and Biochemical Engineering, Jadavpur University, Kolkata, West Bengal, India, and later on, he joined in the same department as a faculty member. Over the last 15 years, he has been engaged in teaching food and biochemical operations for undergraduate and postgraduate students. He has been actively engaged in research as a member of his group in the area of extrusion, Ultra High Temperature (UHT) processing, bioremediation, nanoscience and -technology, and more, and what he has experienced through his journey has been translated through different journal papers and books published by frontline publishers. Compiling books that will satisfy students' desire for knowledge rooted in academic or industrial research has always been his priority, and this book is impregnated with that spirit as well.

Contributors

Maithili Adhikary
Department of Biotechnology
National Institute of Technology
  Durgapur
Durgapur, India

Uday S. Annapure
Department of Food Engineering and
  Technology
Institute of Chemical Technology
Mumbai, India
and
Institute of Chemical Technology
Marathwada Campus
Maharahtra, India

Jayita Bandyopadhyay
DSI-CSIR Centre for Nanostructures
  and Advanced Materials
Council for Scientific and Industrial
  Research
Pretoria, South Africa
and
Department of Chemical Engineering
Laval University
Quebec, Canada

Raunak Banerjee
Mycologics LLC
North Carolina, USA

Subhasis Bhattacharya
Designer and Consultant of gamma
  radiation processing plants in India
  and abroad
and
Director, Danver Hydromatics Pvt. Ltd.
Kolkata, India

Paramita Bhattacharjee Chakraborti
Department of Food Technology and
  Biochemical Engineering
Jadavpur University
Kolkata, India

Pradip Chakraborty
Food Safety and Standards Authority
  of India
New Delhi, India

Anindya Chanda
Mycologics LLC
North Carolina, USA

Abhijit Chatterjee
Department of Bio Engineering
National Institute of Technology
  Agartala
Tripura, India

Surabhi Chaudhuri
Department of Biotechnology
National Institute of Technology
  Durgapur
Durgapur, India

Jayeeta Bardhan Dasgupta
Department of Biochemistry
Vidyasagar College
Kolkata, India

Mwchangti Debbarma
Department of Food Engineering and
  Technology
School of Engineering, Tezpur University
Assam, India

Sukumar Debnath
Department of Food Engineering
CSIR-Central Food Technological
  Research Institute (CFTRI)
Mysore, India

L Susmita Devi
Department of Food Engineering and
  Technology
Central Institute of Technology
  Kokrajhar
Kokrajhar, India

Debasmita Dutta
Department of Biotechnology
National Institute of Technology
  Durgapur
Durgapur, India

Debjani Dutta
Department of Biotechnology
National Institute of Technology
  Durgapur
Durgapur, India

Madhuresh Dwivedi
Department of Food Process Engineering
NIT Rourkela
Odisha, India

Purnamsree Gogoi
Applied Microbiology Lab.,
    Department of Forestry
North Eastern Regional Institute of
    Science and Technology
Nirjuli, India

Gurbuz Gunes
Istanbul Technical University
Department of Food Engineering
Istanbul, Turkey

Akansha Gupta
Department of Dairy Science and
    Food Technology
Institute of Agricultural Sciences,
    Banaras Hindu University
Varanasi, India

Dipankar Halder
Department of Food Engineering and
    Technology
Central Institute of Technology Kokrajhar
Kokrajhar, India

Salman Hashmi
Centre of Food Technology, IPS
University of Allahabad
Uttar Pradesh, India

H. Umesh Hebbar
Department of Food Engineering
CSIR-Central Food Technological
    Research Institute (CFTRI)
Mysore, India

Rifna E J
Department of Food Process Engineering
NIT Rourkela
Odisha, India

Monica Jaiswal
Department of Food Engineering and
    Technology
School of Engineering, Tezpur University
Assam, India

Sucheta Jha
Department of Chemistry
Portland State University
Oregon, USA

Madhu Kamle
Applied Microbiology Lab.,
    Department of Forestry
North Eastern Regional Institute of
    Science and Technology
Nirjuli, India

Barjinder Pal Kaur
National Institute of Food Technology
    Entrepreneurship and Management
Kundli, India

Tatiana Koutchma
Agriculture and Agri-Food Canada/
    Government of Canada
Ontario, Canada

Pradeep Kumar
Department of Botany
University of Lucknow
Lucknow, Uttar Pradesh, India

Santosh Kumar
Department of Food Engineering and
    Technology
Central Institute of Technology
    Kokrajhar
Kokrajhar, India

Uttara Mahapatra
Department of Chemical Engineering
National Institute of Technology
    Agartala
Tripura, India

Dipendra Kumar Mahato
CASS Food Research Centre
School of Exercise and Nutrition Sciences
Deakin University
Victoria, Australia

Hilal A Makroo
Department of Food Technology
Islamic University of Science and
    Technology
Jammu & Kashmir, India

Atanu Mitra
Department of Chemistry
Sree Chaitanya College
West Bengal, India

Avik Mukherjee
Department of Food Engineering and
    Technology
Central Institute of Technology
    Kokrajhar
Kokrajhar, India

Thota Niranjan
Department of Food Process Engineering
NIT Rourkela
Odisha, India

P Srinivasa Rao
Indian Institute of Technology
    Kharagpur
Kharagpur, India

Suprakas Sinha Ray
DSI-CSIR Centre for Nanostructures
    and Advanced Materials
Council for Scientific and Industrial
    Research
Pretoria, South Africa
and
Department of Applied Chemistry
University of Johannesburg
Doornfontein, South Africa

Rahul Kumar Rout
Indian Institute of Technology
    Kharagpur
Kharagpur, India

Poulami Sarkar
Department of Food Technology and
    Biochemical Engineering
Jadavpur University
Kolkata, India

Brijesh Srivastava
Department of Food Engineering and
    Technology
School of Engineering, Tezpur University
Assam, India

Dipshikha Tamili
Department of Food Technology and
    Biochemical Engineering
Jadavpur University
Kolkata, India

Rohit Thirumdas
College of Food Science and Technology
Professor Jayashankar Telangana State
    Agricultural University
Telangana, India

1 Natural Food Preservatives

Jayeeta Bardhan Dasgupta, L Susmita Devi, Sucheta Jha, Avik Mukherjee, Santosh Kumar, Dipankar Halder, and Atanu Mitra

Corresponding authors:
Dipankar Halder, Email: dipankar.halder@jadavpuruniversity.in
Atanu Mitra, Email: mitatanu@gmail.com

CONTENTS

1.1 INTRODUCTION

Food by its nature is perishable, and with time, foods may undergo microbiological, enzymatic, physical, and chemical changes, which reduce the overall quality of food. Microbiological spoilage of food involves the contamination, growth, and reproduction of microorganisms. Fungal growth is very common in raw or processed food, and it results in significant deterioration of sensorial properties. Moreover, a few of the fungal strains have the capability to produce mycotoxin, which is highly detrimental to human health. Lipid oxidation, proteolysis, starch aging, and postharvest respiration of fruits and vegetables are the physical changes or chemical reactions that occur during the storage of food. Enzymes like polyphenol oxidase, pectinase, lipase and others inherently present in food degrade food constituents with time. However, microbial contamination and lipid oxidation are the major causes, which reduce the shelf life of food. Since ancient times, several methods, including heat treatment, reducing water activity, maintaining an inert gas environment, and adding preservatives (like salt, sugar, vinegar, etc.), have been developed and practiced for increasing the shelf life of the food materials. Among these various methods of preservation, food biopreservatives are the major focus of this current chapter.

The role of the food preservatives includes (1) maintaining the original nutritional value, (2) inhibiting the growth of any spoilage as well as pathogenic microbes, and (3) preventing any undesirable modification of physicochemical and sensory attributes during storage. The presence of preservatives increases the shelf life of the food by delaying any of the previously mentioned undesirable alterations. The preservative may be chemically synthesized or a natural product. There is an increasing demand for the application of natural products as food preservatives compared to synthetic or artificial ones as many of the artificial preservatives are harmful to consumer health (Baptista et al., 2020). Consumers prefer naturally preserved food over synthetic additives and avoid food containing synthetic additives. Preservation science principally uses antioxidants and antimicrobials to increase the shelf life of food products, in which antioxidants resist oxidative damages in the food, whereas antimicrobials prevent microbial spoilage (S. Kumar, Mukherjee, et al., 2020). The extent of oxidative damage depends on the nature of food constituents. Pigments, flavors, lipids, proteins, and vitamins in the foods are susceptible to oxidation, which causes a loss of nutritional value as well as in the shelf lives of the food materials. Such oxidative process on unsaturated, especially polyunsaturated fatty acid (PUFA) of food produces unpleasant rancid odors and harmful secondary oxidation products (Suleman et al., 2020). The extent of lipid oxidation depends on the degree of unsaturation, and even at very low-level oxidation (below 1%), food can get rancid (Gordon, 2001). Besides lipid oxidation, protein oxidation also occurs, and it is responsible for the loss of nutritional value, sensory qualities, and water-holding capacity of food products (Faustman et al., 2010).

A greater number of literature reports have appeared to explore the mechanistic details of the oxidation process that mainly occur due to free radicals. Free radicals are species with an unpaired electron, and they are highly reactive. Generally, the free radical mechanism involves chain propagation and chain termination steps after the initiation step. Light, temperature, enzymes, metals, metalloproteins, and microorganisms can act as a catalytic system (i.e., initiator) for the oxidation of lipids. The oxidation process occurs through either any one or combination of the following pathways: (1) autoxidation, (2) photooxidation, (3) thermal, and (4) enzymatic oxidation. Among the previously mentioned oxidation processes, autoxidation is the most common form in which lipid can be oxidized spontaneously by aerial oxygen and process can be accelerated with an increase in temperature. In the presence of a catalyst, the homolytic fission of C–H bond in lipid molecules generates lipid free radicals ($R^\bullet$) that react with oxygen and produces the peroxyl radicals ($ROO^\bullet$). Now the peroxyl radicals produce new free radicals ($R^\bullet$) by attacking the lipid molecules and are themselves converted to lipid hydroperoxide molecules (ROOH), a tasteless as well as odorless species that are highly unstable. The unstable hydroperoxide molecules spontaneously degrade to secondary oxidation products like alkenes, aldehydes, alcohols, and ketones, which causes a significant loss of sensory quality. The chain propagation is continued rapidly until nonradical products (R-R, ROR, ROOR etc.) are formed in the chain termination step. The bond energy of C-H bond of lipid molecules dictates the ease of generation of R$^\bullet$ radical, and accordingly presence of polyunsaturation produces free radicals more easily (Shahidi & Zhong, 2010). Photooxidation is an oxygen-, visible light–, or photosensitizer-involved oxidation process that occurs through the excitation of lipid and/or oxygen. In case of photosensitizer-involved oxidation, photosensitizer molecules are activated to an excited singlet state by absorbing energy from visible or ultraviolet light, and the excited photosensitizers transfer their energy directly to either lipid molecules or oxygen molecules. Besides, the degree of unsaturation and the oxidation of lipid also depend on the chemical structure as well as the stereochemistry of the lipid. Lipid oxidation can be prevented by the application of antioxidants, where antioxidants can act as free radical scavengers, singlet oxygen and secondary oxidation product quenchers, metal ions chelators, peroxide inactivators, inhibitors of pro-oxidative enzymes, and suppressors of oxidative initiators.

The harmful effects of artificial preservatives have forced modern food preservation techniques to lean toward the use of biopreservatives. The three major sources of bio- or natural food preservatives are (1) plants, (2) microorganisms, and (3) animals. Natural compounds, which possess either antimicrobial or antioxidant or both properties may exhibit preservation activity. Natural antimicrobials are present in food or may be incorporated into the food for delaying or preventing the proliferation as well as the growth of spoilage and pathogenic microorganisms (bacteria, yeasts, molds; Villalobos-Delgado et al., 2019). The antimicrobial effect of natural components is increasingly evident as a potent tool for the safety of food products. The increased preference for consumption of minimally processed food can be met with the use of such bioactive compounds generally recognized as safe (GRAS). The inhibition of the deleterious effects of oxidative rancidity in food spoilage demands the employment of antioxidants as a necessary constituent. This chapter elaborates on such natural preservatives or additives, which are effective in extending the shelf life of food and ensuring their quality and safety.

1.2 PLANT-DERIVED FOOD PRESERVATIVES
1.2.1 Spices and Herbs
Spices and herbs are rich in aromatic and medicinal ingredients. Spices represent the dried parts of bark, root, seeds, and flowers of the plants, while herbs are the green part of the leaves and stem of the plants. Extracts of various spices and herbs have preservative aptitudes. Useful antibacterial and antioxidant components from plant sources are essential oils (EOs), lectins, polypeptides, polyphenols, and alkaloids. Moreover, a few other categories of compounds, such as polyamines, organic acids, glucosides, and glucosinolates have also exhibited antimicrobial activity to some extent (Bensid et al., 2022). The phenolic components of polyphenolic compounds present in herbs and spices render antioxidant properties, which have an immense effect against bacteria, yeast, mold, and fungi. Several studies have provided evidence that the structure and chemical properties of such phenolic substances enable scavenging free radicals, chelating transition metals, quenching singlet

oxygen, and stimulating antioxidant enzymes (Kurutas, 2016). The antimi-crobial property is attributed to the lipophilic nature of phenolic compounds that cause microbial cell damage, thereby disrupting osmotic balance and membrane permeability of the cell. The total phenolic content is considered an indirect parameter for measuring antimicrobial activity. Recently, Gottardi et al. nicely reviewed the role of spices in food preservation as well as in food safety, in which almost 100 spices are listed along with their active ingredi-ents and effectiveness as antibacterial/antifungal/antivirus agents (Gottardi et al., 2016). Physical parameters, such as light, temperature, humidity, oxygen, and others, and biological factors, such as enzyme activity and microbial growth, irreversibly modify the food state with time rendering the food inedible.

1.2.1.1 Green Tea (Camellia sinensis)

Green tea is a well-known herb, which is rich in polyphenolic compounds like epicatechin, epigallocatechin, epicatechin gallate, epigallocatechin gallate, cat-echin, gallocatechin gallate, catechin gallate, and others (S. Kim et al., 2004). In a few studies, green tea leaf extract has been incorporated with biopolymer such as fish skin gelatin, chitosan, and synthetic polymer to fabricate food-packaging film having antimicrobial, as well as antioxidant, properties (Basumatary et al., 2022; Jessica Elizabeth et al., 2017).

1.2.1.2 Oregano (Origanum vulgare)

Oregano is one of the important members of the mint family and is popularly used as culinary herb. Oregano contains a high amount of EO in which the main bioactive components are thymol and its isomer carvacrol. Several studies demonstrated that EO of oregano exhibits significant positive inhibition against gram-positive bacteria (*Staphylococcus aureus, Bacillus cereus, Streptococcus mutans, Listeria monocytogenes, Enterococcus faecalis*) as well as against gram-negative bac-teria (*Escherichia coli, Salmonella enterica*), when the concentration of EO is higher than 5% (w/w). Many of the studies were performed with real food materials like rabbit meat, chicken meat, eggplant salad, and others. Studies also reveal that both the isomer (thymol and carvacrol) has antibacterial as well as antifun-gal activity (Veenstra & Johnson, 2019).

1.2.1.3 Rosemary (Rosmarinus officinalis)

Rosemary belongs to *Lamiacea* family and has been used in food preservation owing to its antimicrobial, as well as antioxidant, activity. However, the antioxi-dant properties of rosemary extract depend on the method of extraction as well as the type of solvent used in the extraction process. In the advanced method, an active fraction of rosemary extract is obtained through molecular distillation in which extraction carried out with ethyl ether under a refluxing condition. Today, dried rosemary leaves are generally used for the preparation of rose-mary extract. The extract of rosemary has many phenolic compounds, including carnosic acid, carnosol, rosemarinic acid hesperidine, as principal constituents, and these components exhibit high antioxidant activity. Rosemary oils can be prepared by steam distillation of fresh leaves and mainly camphor, 1,8-cineole, p-cymene, and boreneol are the major components (Bendeddouche et al., 2011). Interestingly, the antioxidant properties of rosemary are comparable or better than the chemical preservatives butylated hydroxytoluene (BHT) and butylated hydroxyanisole (BHA). Moreover, rosemary extract or its major bioactive compo-nents do not influence the organoleptic property of food, so rosemary is a green alternative to chemical antioxidants.

1.2.1.4 Turmeric (Curcuma longa L.)

Turmeric, commonly known as the 'golden spice of India', constitutes of cur-cumin and essential oils, which has wide range of antibacterial and antifungal properties. The bioactive element of *Curcuma longa* is curcumin. It has been noted that both turmeric and curcumin, along with their antioxidant properties, have high dose tolerances without any toxic effects, thus indicating a potential food preservative. Curcumin is the principal curcuminoid of turmeric that gives it its yellow color. Curcumin is reported to be hydrophobic in nature and is com-monly soluble in dismethylsulfoxide, ethanol, acetone, and oils. Experimental data support that the methoxy and hydroxyl groups of curcumin are directly involved in the antimicrobial activity. Figure 1.1 represents the chemical struc-ture of curcumene (Abd. Wahab et al., 2020; Arulkumar et al., 2017). It has been reported that turmerone and curlone components of turmeric show antibacte-rial activity against a wide variety of gram-negative and gram-positive bacteria including *Listeria monocytogenes, Vibrio parahaemolyticus, Salmonella typhimurium, S. aureus, L. innocua, Bacillus subtilis, Bacillus cereus, Bacillus coagulans, E. coli* and *Pseudomonas aeruginosa* (Y. Hu et al., 2015). In several studies, it is evident that the aqueous, n-hexane, chloroform, and methanolic extracts of turmeric exhibit excellent antifungal activities against *Candida albicans, C. glabrata,* and *C. dubliniensis* and fungal spores including *Aspergillus flavus, A. niger, Penicillium griseofulvum, P. chrysogenum,* and *Fusarium oxysporum* (Y. Hu et al., 2015). Further studies have indicated curcuminoid, a phenolic compound to be responsible for such antimicrobial activity. The extensive study on the antifungal mechanism of such naturally occurring preservatives provided an explanation based on inactivation of intracellular and extracellular enzymes and cytoplasmic granula-tion, along with membrane rupture on the action of plant lytic enzymes (Arshad et al., 2019). It has also been reported that aqueous extracts of turmeric may show zero antifungal activities due to a faulty extraction procedure, which affects the bioavailability of active components present in turmeric and their essential oils.

1.2.1.5 Clove (Eugenia caryophyllata)

Clove is a popular spice for culinary across the world and is used as a food pre-servative. Eugenol and beta caryophyllene are the main components in clove EO. Antibacterial and fungicidal activities of the major components of clove oil have been established against *E. coli, L. monocytogenes, Salmonella enterica, C. jejuni,* and *S. aureus* (Chaieb et al., 2007). The strong antioxidant property of the clove is mainly due to eugenol.

1.2.1.6 Cinnamon (Cinnamomum verum)

This spice is commonly applied in various food preparations to provide a distin-guished aroma and flavor. Besides its contribution to sensory quality, cinnamon also acts as a food preservative, as it effectively inhibits the growth of bacteria,

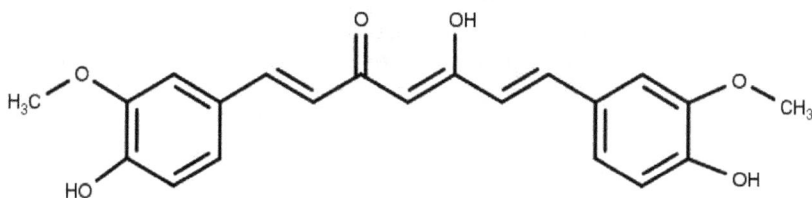

Figure 1.1 Chemical structure of curcumin.

especially gram-positive ones. This spice is obtained from the inner bark of the tree, and cinnamon oil is the main ingredient. The major compounds present in cinnamon oil are cinammyldehyde, cinammyl alcohol, and eugenol (Rao & Gan, 2014). Study reveals that cinammyldehyde is responsible for the antibacterial action of cinnamon due to its inhibition effects on (1) synthesis of cell walls, (2) function of cell membrane, and (3) synthesis of proteins as well as nucleic acids (Winias, 2015).

1.2.1.7 Black Pepper (Piper nigrum)

A spice of Indian origin, black pepper possesses good antimicrobial properties mainly due to the presence of volatile oil (Dorman & Deans, 2000). Acetone-extracted black pepper EO demonstrated an inhibitory effect on the mycelial growth of some fungi, such as *Fusarium graminearum* and *Penicillium viridicatum*. A gas chromatography–mass spectrometry (GC–MS) analysis report has shown that the principal constituents of black pepper essential oil are piperine, pierolein B, and piperamide. Study reveals that black pepper extract shows a strong antibacterial effect against both gram-positive (*S. aureus*, *Bacillus cereus*, *Streptococcus faecalis*) and gram-negative bacteria (*Pseudomonas aeruginosa*; Karsha & Lakshmi, 2010).

1.2.1.8 Curry (Murraya koenigii)

Curry is the traditional Indian spice that has become globally popular due to its characteristic aroma and flavor (Susheela, 2006). Several studies demonstrate the prominent antifungal and antibacterial properties of curry extracts (Singh et al., 2017). It has been also shown that the presence of various carbazole alkaloids and coumarins in the curry extract is responsible for its antimicrobial property and thus its use as a food preservative (Rahman & Gray, 2005).

1.2.1.9 Ginger (Zingiber officinale)

Ginger root is popular worldwide as a spice and as traditional medicine. The antioxidant and antimicrobial properties of ginger extract make it suitable as a food preservative. The antioxidant and antimicrobial properties of ginger extract have been attributed to the presence of active ingredients and polyphenolic compounds such as 6-gingerol. The amount of active ingredients in the extract depends on the extraction process. In one of the studies, it is exhibited that CO_2-extracted ginger is rich with polyphenols, and such extract has high antioxidant properties that are comparable with synthetic antioxidant BHT (Stoilova et al., 2007). Different studies demonstrate that ginger effectively inhibits the growth of several bacterial as well as fungal strains (Gupta & Ravishankar, 2005). The antibacterial or antifungal mechanism involves the rapture of microbial cells and the inactivation of enzymes through chelation of active metal ions by the polyphenolic compounds present in the ginger (Sebiomo et al., 2011).

1.2.1.10 Cumin (Cuminum cyminum)

Cumin is another popular spice that can act as a food preservative due to its good antibacterial, as well as antifungal, properties. The major chemical species in the essential oil of cumin are cuminaldehyde, b-pinene, p-cymene, and g-terpinene. Among these chemical compounds, cinammaldehyde is mainly responsible for its antimicrobial properties (Pawar & Thaker, 2006).

1.2.1.11 Basil (Ocimum basilicum L.)

Basil is a popular culinary herb and its EO has been used as a flavoring agent in different food items, including nonalcoholic beverages, salad dressings, various confectionery, and baked food products. A large variety of species of this herb have been observed worldwide, and EOs obtained from them largely vary in composition. Basil oil is natural food preservative, as it shows good antimicrobial properties. Besides conventional steam distillation technique, CO_2 extraction under liquid or supercritical condition has also been practiced for extracting oil from its leaves and flower tops. Although numerous botanical verities of basil exist, however, several studies demonstrate that eugenol, methyl eugenol, methyl chavicol, 1,8-cineole, methyl cinnamate, and α-bergamotene are the major active constituents (Suppakul et al., 2003).

1.2.2 EOs

EOs are usually a complex mixture of liquid volatile compounds, and these can be isolated from various parts of plants by either steam distillation or extraction (solvent or supercritical fluid). Various parts of the plants include leaves (basil, oregano, thyme, rosemary, sage), buds or flowers (clove), fruits (cardamom, pepper), seeds (nutmeg, parsley, fennel, caraway), and rhizomes (asafetida). EOs are mainly employed as flavoring agents in the food industry; however, they increase the shelf life of food as preservatives owing to their antimicrobial, as well as antioxidant, properties. The European Commission and the U.S. Food and Drug Administration have accepted many essential oil components as well as crude essential oil of several plants like clove, oregano, thyme, nutmeg, basil, mustard, cinnamon, and others for their use as food preservatives or additives (Mei et al., 2019). Properties of the EOs highly depend on their chemical composition and countless factors like plant organ, plant maturity degree, season of harvesting, geographical location, and the like (Dhifi et al., 2016). Although EOs contain a number of different polar and nonpolar components but the principal constituents can be divided into two structural categories (terpenoids and phenyl-propanoids) according to the hydrocarbon skeleton. Terpenoids are the derivatives of isoprene, that is, 2-methyl-1,3-butadiene (C_5H_8), first pointed out by Wallach in 1887. Mono- (C10), sesqui- (C15), di- (C20), sester- (C25), tri- (C30), tetra- (C40) nomenclature are assigned for terpene family according to the number isoprene unit present in the compound. Among these terpenes, usually mono- and sesquiterpenes are predominant in EOs and even may occupy up to 90% of the total content of EO. Other important terpenoids in the plant EOs are oxygen-containing derivatives including geraniol, nerol, linalool (acyclic alcohols), geranial, neral, citronellal (acyclic aldehydes), menthone (monocyclic ketone), menthol (monocyclic aldehyde), camphor (bicyclic terpene ketone), and others. EOs are effective against some foodborne pathogens including *E. coli* O157: *H7, L. monocytogenes, S. Typhimurium, S. aureus*, and others. Studies show that the efficiency of EOs depends on their concentration, chemical structure, the target microorganism(s), the food matrix, and the application method. The antimicrobial activity of EOs is mainly due to its inhibitory effect on membrane integrity and leakage of intracellular material resulting in cell death (Figure 1.2). The hydrophobic nature of EOs can interact with the bacterial lipid membrane, resulting in an increased permeability of the cell components (Mei et al., 2019). The structural formula of few plant-based bioactive compounds, such as polyphenols, terpenoids, phenyl-propanoids, and tocols, which act as antimicrobials, antioxidants, or both, are listed in Table 1.1.

1.2.3 Plants and Their Extract

Sahidi et al. have extensively reviewed the source and chemistry of various natural plant-based antioxidants, which can be used as food preservatives (Shahidi & Ambigaipalan, 2015). Tocols, ascorbic acid, carotenoids, and phenolic

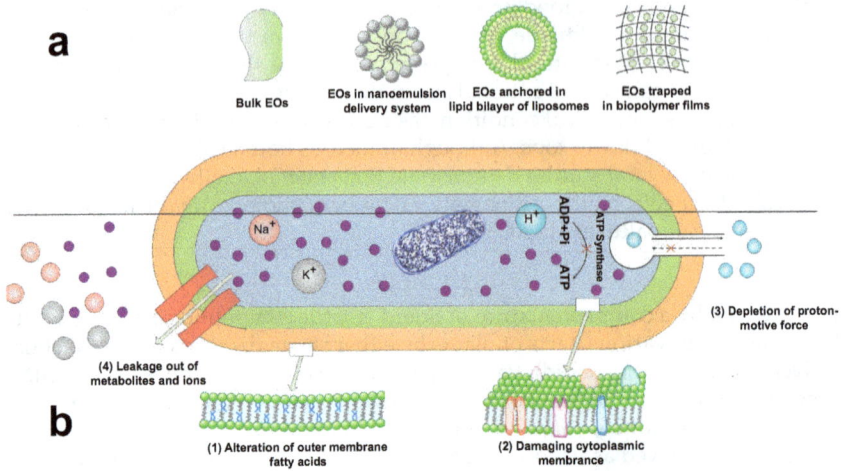

Figure 1.2 (a) Bulk EOs and different types of EO delivery systems, including nanoemulsion, liposomes, and biopolymer films. (b) Proposed common mechanisms of action and target sites of essential oils (EOs) or EO delivery systems on bacterial cell (Mei et al., 2019).

compounds are the major groups of antioxidants originating mainly from the plant kingdom (Table 1.1).

1.2.3.1 Tocopherols and Tocotrienols

Tocopherols and tocotrienols are two methylated derivatives of tocols. The high hydrophobic nature of tocols is obvious from their chemical structure, and thus, they are fat soluble. Tocotrienols possess unsaturated hydrocarbon side chain whereas tocopherol has a saturated one, and tocotrienol has better antioxidant properties than the corresponding tocopherol. Vegetable oils, legumes, cereals, and nuts, among others, are rich sources of tocols. Tocols generally terminate the oxidation chain reaction by accepting high-energy free radicals. Tocols, particularly different tocopherols are widely used as permitted food additives in different food to provide them oxidative stability and to enhance the nutritional quality. In particular, α-tocopherols (vitamin E), are commonly added as antioxidants in oils, meats, vegetable oils, and animal fats (Augustin & Sanguansri, 2010; Blom & Mørtvedt, 1991; Shahidi & Ambigaipalan, 2015).

1.2.3.2 Carotenoids

They represent a group of antioxidant compounds with polyenoic terpenoid structure, and they are widely distributed in the various parts of plants as yellow, orange, and red pigments. In the carotenoid family, carotenes (β-carotene and lycopene) are polyene hydrocarbons, and xanthophylls (lutein, zeaxanthin, capsanthin, canthaxanthin, astaxanthin, violaxanthin) contain oxygen in the form of hydroxy, oxo, or epoxy groups. Carotenes, especially β-carotene, are also found in crude palm oil, virgin olive oils, and so on. Due to their lipophilic structure, carotenoids are fat-soluble. They exert their antioxidant property by quenching singlet oxygen species and/or trapping the free radicals. The carotenoids are added to foods, such as meat, fish, fruits, cereal products, pastries,

Table 1.1 Plant-Based Bioactive Compounds, Their Structures, Sources, and Functional Activities

Plant-based bioactive compounds		Structural formula of one of the members	Food sources	Applications	References
Polyphenols	Flavonoids	Flavonols (Quercetin, kaempferol, myricetin etc.) **Kaempferol**	Fruits, vegetable, leaves, seeds, grains, capers, red onions, and kale	Antibacterial, Antioxidative properties	Barreca et al. (2021)
		Flavanones (hesperitin, naringenin, and eriodictyol) **Hesperitin**	All citrus fruits		
		Isoflavonoids (genistein and daidzein) **Genistein**	Soyabeans and other leguminous plants		
		Flavan-3-ols (epicatechin) **Epicatechin**	Bananas, apples, blueberries, peaches, and pears		

(Continued)

Table 1.1 (Continued)

Plant-based bioactive compounds		Structural formula of one of the members	Food sources	Applications	References
	Anthocyanins (cyanidin, delphinidin, malvidin, pelargonidin and peonidin)	Cyanidin	Cranberries, black currants, red grapes, merlot grapes, raspberries, strawberries, blueberries, bilberries, and blackberries	Coloring additive for beverages, antioxidant and antimicrobial activities	Albuquerque et al. (2021)
Phenolic acids	Hydroxyl benzoic (gallic acid, ellagic acid)	Gallic acid	Berries, spices, cereals, tea	Preservatives, antioxidant, antimicrobial activity	Albuquerque et al. (2021); N. Kumar & Goel (2019)
	Hydroxyl cinnamic acids (ferulic acid sinapic acid)	Ferulic acid	Coffee, yerba mate, red wine, red fruits, vegetables, whole grains	Preservatives, antioxidant, and antimicrobial activities	N. Kumar & Goel (2019)
Stilbenes	Resveratrol, pterostilbene, pinosylvin		Strawberries, grapes, peanuts, and cannabis	Antioxidant, anti-inflammatory, and cardioprotective activities	Fraga et al. (2019)

Lignans	Secoisolariciresinol, pinoresinol, lariciresinol	**Secoisolariciresinol**	Broccoli, beans, soybeans, rye, sesame seeds, pumpkin seeds, flax seeds, and some berries in very small amounts	Antioxidant, anticancer, anti-inflammatory, neuroprotective, and antimicrobial activities	Albuquerque et al. (2021)
Polyphenolic amides.	Capsaicinoids (capsaicin, dihydrocapsaicin, nordihydrocpasaicin)	**Capsaicin**	Chili peppers	Flavoring and preservative agent and used as active compound in packaging film and functional foods	Rezazadeh et al. (2021)
	Avenanthramides	**Avenanthramide a**	Oats	Antioxidant properties	Raguindin et al. (2021)
Terpenoids	Hemiterpenoids (isoprene, tiglic acid, angelic acid)	Hemiterpene **Isoprene**	Leaves of many trees (including conifers, poplars, oaks, and willows) and herbs	Natural antioxidant agent, antimicrobial properties	Gutiérrez-del-Río et al. (2021); Ludwiczuk et al. (2017)

(Continued)

Table 1.1 (Continued)

Plant-based bioactive compounds		Structural formula of one of the members	Food sources	Applications	References
Monoterpenoids	Acyclic monoterpenoids (linalool, geraniol, citronellal, citronellol)	Geraniol	Ho leaves, coriander		
	Monocyclic monoterpenoids (menthol, thymol, carvone, carvacrol)	Thymol	Citrus fruits and grapes		
	Bicyclic monoterpenoids (borneol, camphor, limonene oxide)	Borneol	Several species of *Artemisia*		
Sesquiterpenoids	Acyclic sesqiterpenoids (farnesol, β-nerolidol)	Farnesol	Citronella, cyclamen, tuberose, lemon grass, rose	Antimicrobial properties	Lyu et al. (2019)

Monocyclic sesquiterpenoids (β-bisabolene, α-zingiberene)	 **β-Bisabolene**	Ginger, cubeb		
Bicyclic sesquiterpenoids (artemisinin, β-santalol, β-caryophyllene)	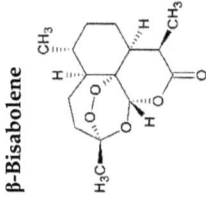 **Artemisinin**	Sweet wormwood		
Diterpenoids	Acyclic diterpenoids (phytol)	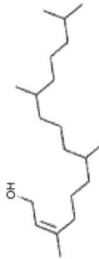 **Phytol**	Parsnips, rosemaries, red raspberries, common chokecherries, and white cabbages	Antimicrobial, antifungal, and insecticidal activity	Alihosseini (2016); Sandjo & Kuete (2013)
	Tricyclic diterpenoids (abietic acid)	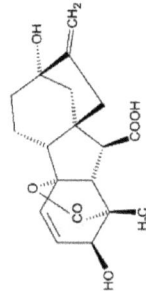 **Abietic acid**	Needles and bark of *Taxus* spp.		

(Continued)

Table 1.1 (Continued)

Plant-based bioactive compounds

		Structural formula of one of the members	Food sources	Applications	References
Triterpenoids	Squalene	Squalene	Olive oil and palm oil	Antioxidant, antimicrobial, antiviral, antiallergic, antipruritic, and antiangiogenic properties	Alihosseini (2016)
Tetraterpenoids (carotenoids)	Carotenes (γ-carotene and β-carotene)	β-Carotene	Marigold and carrots	Food colorants, antifungal and antimicrobial properties	Alihosseini (2016); Ludwiczuk et al. (2017)
	Oxygenated xanthophyll (lutein, zeaxanthin, and cryptoxanthanin)	Lutein	Spinach and paprika		
Phenyl-propanoids		Cinnamyl alcohol	Fruits, vegetables, cereal grains, beverages, spices, and herbs	Antioxidant activity	Neelam et al. (2020)
Tocols	Tocopherols and tocotrienols	α-tocopherol	Lipids, notably butter and vegetable oils (as virgin olive oil)	Prevent cancer, heart disease, and other chronic ailments, and are also used as food additives	Delgado et al. (2020); Shahidi & de Camargo (2016)

14

and dairy products (Carocho et al., 2018). In particular, lutein, zeaxanthin, and canthaxanthin are frequently used in fish products to enhance their oxidative stability, whereas lycopene, β-carotene, lutein, astaxanthin, and the like are also frequently used in the food industry.

1.2.3.3 Phenolic Compounds and Organic Acids

Phenolic compounds include phenolic acids, stilbenes, flavonoids, and lignans are another class of effective antioxidants, mainly obtained from plants. These phenolic compounds effectively inhibit lipid oxidation because of their free radical scavenging, metal chelating, and reducing properties. Moreover, such phenolic structures exhibit synergism with other antioxidants. Many organic acids that are the constituents of several natural foods are traditionally used as food preservatives, especially in homemade pickles, cakes, and so on. The acidic pH range inhibits the growth of many contaminating microbes. These acids are low cost and considered safe. For instance, ascorbic acid, which is also known as vitamin C, is a mild water-soluble reducing agent. Fruits and vegetables are rich sources of ascorbic acid. When other antioxidants are used along with ascorbic acid, ascorbic acid generally converts the inactive form (oxidized form) of the primary antioxidant into an active form (reduced form) and provides the stability to the lipid against oxidation through a synergistic interaction. Ascorbic acid exhibits significant synergistic interaction with primary antioxidants such as tocols, citric acid, and more. A few of the frequently used organic acids along with their sources and applications are listed in Table 1.2.

1.2.3.4 Algae and Mushrooms

Algae and mushrooms contain many bioactive compounds that have already found applications in food industry as antimicrobial agents and antioxidants. Several algal classes, including Phaeophyceae (brown), Rhodophyceae (red), Chlorophyceae (green), Chrysophyceae (golden), and Bacillariophyceae (diatoms), produce secondary metabolites, which contain antimicrobial compounds like fatty acids, polysaccharides, peptides, terpenes, polyacetylenes, sterols, indole alkaloids, aromatic organic acids, shikimic acid, polyketides, hydroquinones, alcohols, aldehydes, ketones, halogenated furanones, alkanes, and alkenes (Pisoschi et al., 2018). Algae are also rich with antioxidants such as α-tocopherol, carotenoids, polyphenols, phycobiliproteins and vitamin C. Studies demonstrated that edible brown algae like *Himanthalia elongata*, *Saccharina latissima*, and *Laminaria digitata* have shown significant antibacterial activity against some bacterial strains such as *L. monocytogenes*, *S, abony*, *E. faecalis*, and *P. aeruginosa*. Even higher antibacterial efficacy has been seen for the crude extract of *Himanthalia elongata* compared to chemical preservatives (sodium benzoate, sodium nitrite) against bacterial strains like *E. faecalis* and *P. aeruginosa*. Extracts of another two common brown algae species, Padina and Dictyota, can be safely added into foods prone to Gram-positive bacterial spoilage because they can effectively inhibit the growth of foodborne pathogens *L. monocytogenes*, *B. cereus*, and *S. aureus* without any toxic effect on food. The methanolic or ethanolic extracts of various edible mushrooms have shown antimicrobial activity against different foodborne pathogens and spoilage microbes. For instance, methanolic extract of *Agaricus* spp. has exhibited strong antibacterial activity against gram-positive bacteria, including *M. luteus*, *Micrococcus flavus*, *B. subtilis*, and *B. cereus* (Pisoschi et al., 2018). In another study, the extract of *F. fomentarius* mushrooms growing wild in Serbia have exhibited strong antibacterial activity.

Table 1.2 Organic Acids, Their Sources, and Their Applications

Organic Acid	Sources	Microorganism strain	Application/s	Results	References
Citric acid	Citrus and other acidic fruits such as limes, lemons, oranges, pineapples, and gooseberries	*Aspergillus niger, Candida sp, Arthrobacter paraffinens, Bacillus licheniformis,* and *Corynebacterium sp,* among others	Preservative, acidifier, antioxidant, effervescent, pH corrector.	Citric acid reduced the risk of bacterial infection and improved texture in acidified beef. Citric acid exhibits both bactericidal and bacteriostatic effects against *L. monocytogenes*	Behera et al. (2021)
Acetic acid	Produced by oxidation of ethanol or via oxidative fermentation and heterofermentative lactic acid bacteria or via anaerobic fermentation	*Acetobacter, Gluconobacter,* heterolactics, *Clostridium* species	Preservatives, Antimicrobial properties, emulsifier, stabilizer, flavor enhancer, and firming agent	Effective antimicrobial activities against *L. monocytogenes, B. cereus.* Combination of 1% acetic acid and 1% chitosan provide a larger diffusion value of between 8.36 to 12.64 mm for *E. coli, V. parahaemolyticus, S. aureus,* and *P. mirabilis*	Coban (2020)
Benzoic acid	Cinnamon, cloves, prunes, strawberries cloudberries, cranberries, and lingonberries	–	Preservatives (beverages, fruit products, bakery products, fish products, liquid eggs, margarine, mayonnaise, mustard, salad dressings, and sauces)	Extended shelf-life and prevented nutritional losses, found effective against yeasts, and molds	Kalpana & Rajeswari (2019)
Malic acid	Blackberries, blueberries cherries, apricots, peaches, mango, plums, apples, pears, quinces,	–	Flavoring agent, antibacterial activity, and pH controlling agents in foods	Combined treatment of UV and malic acid treatment on sweet lemon juice showed effectiveness against *S. typhi, S. dysenteriae, L. monocytogenes,* and *S. flexneri*	Ben Braïek & Smaoui (2021); Hashemi & Jafarpour (2020); Marques et al. (2020)

Preservative	Source	Microorganism	Function	Effect	References
Ascorbic acid (Vitamin C)	Lemons, oranges, broccoli, corn, wheat, molasses, and D-sorbitol	Black acetic acid bacteria, *Bacillus megaterium*, and *Gluconobacter oxydans*	Anti-browning agents and antioxidant	Ascorbic acid prevented oxidation and discoloration of the meat and ground meat product and showed a positive effect on the browning ratio of the fresh-cut apple	Özdemir & Gökmen, (2019); Sánchez-Escalante et al. (2001)
Lactic acid	Cow milk, sorghum, cassava bagasse, wheat, sugarcane bagasse, cheese whey, etc.	Lactic acid bacteria (*Lactobacillus* spp., *Rhizopus* spp. (*R. oryzae* and *R. arrhizus*)	Food preservative, fermentation agent, acidulant, flavor enhancer, and decontaminant	Shelf-life of spray-washed goat or sheep meat with 2% lactic acid stored at refrigeration (5–7°C) increased shelf-life to 8–11 days as against 3 days in untreated samples	Abedi & Hashemi (2020)
Adipic acid	Juice of sugar and red beets	*Candida viswanathii*	Used in bottled drinks, giving them a bubbly fizz, gives tart flavor to fruit juice and gelatin. Used in many powdered food and drink mixes to provide a sweet flavor	Adipic acid used in formulation of nanofibers improved uniformity and viscosity of the nanofibers	Hosseini et al. (2021); Karthik & Rathinamoorthy (2017); Skoog et al. (2018)
Succinic acid	Broccoli, rhubarb, sugar beets, fresh meat extracts, cheese, sauerkraut	*Anaerobiospirillum succiniciproducens*, *Actinobacillus succinogenes*, *Mannheimia succiniciproducens*	Primarily as an acidity regulator in the food and beverage industry, and also used as a flavoring agent	Unheated 1% succinic acid treatments increased shelf-life of chicken legs 1-2 days	Featherstone (2015); Saxena et al. (2017)

1.3 MICROORGANISM-DERIVED FOOD PRESERVATIVES

Interestingly, many microbes and their secondary metabolites increase the shelf life of food by inhibiting the growth of food spoilage and pathogenic microorganism. For instance, lactic acid bacteria (gram-positive bacteria) have been widely used in the food system. Although lactic acid bacteria represent a broad heterogeneous family of food-grade microorganisms, but *Lactobacillus, Lactococcus, Pediococcus, Leuconostoc,* and *Streptococcus* are the most popular for food applications. As a result of fermentation of carbohydrates, lactic acid bacteria a produce large number of by-products such as bacteriocin, extracellular polysaccharide (EPS), organic acid (lactic, acetic, propionic, pyruvic acids, etc.), acetaldehyde, diacetyl, hydrogen peroxide, and low-molecular-weight nonpolar compounds (reuterin, reutericiclin, pyroglutamic acid) that have the antimicrobial potential to be used as food preservatives (Fang et al., 2019). Bacteriocins are small peptides or proteins that can inhibit the growth of other pathogenic bacteria (Ng et al., 2020). Extensive variations of bacteriocin have been observed with respect to chain length, amino acid sequence, genetic origin, and so on. EPSs are capsular polysaccharides or slime polysaccharides that can also serve as stabilizers and emulsifiers in various food products, particularly in fermented dairy products (Zannini et al., 2016; Zhao et al., 2016). Organic acids create an adverse environment for many microorganisms by lowering the pH of food. The well-known bactericidal effect of hydrogen peroxide is due to its strong oxidizing power. The diacetyl compound exerts its antimicrobial action by blocking or modifying the catalytic sites of microbial enzymes. The extent and nature of antimicrobial compounds generated by lactic acid bacteria vary with the strain of lactic acid bacteria, nature of carbohydrate, protein of food matrix, and additives.

1.3.1 Nisin

Nisin is a bacteriocin that inhibits gram-positive bacteria such as *Bacillus cereus, Clostridium botulinum, L. monocytogenes,* and *S. aureus,* and when administered with antibiotic, nisin is also effective against gram-negative bacteria (Silva et al., 2018). Gram-positive bacteria, including *Lactobacillus* or *Streptococcus* strains, produce nisin. These microorganisms have been genetically modified, which yields bioengineered variants of nisin that exhibits enhanced inhibition activities against Gram-negative pathogens. Such examples include *L. lactis NZ9800.36,* which is genetically modified from *L. lactis NZ22186,* produce nisin Z N20K instead of natural nisin, nisin Z. Nisin inhibits *L. monocytogenes* and *S. aureus* in cheese and *B. cereus, C. botulinum,* and *Clostridium perfringens* in milk as well as milk products. Since nisin is effective against *L. monocytogenes, B. cereus,* and *C. botulinum,* so it can be used in bakery products, meat and meat products, and beverages (Choyam et al., 2019).

1.3.2 Enterocin

Enterococcus spp. produces circular bacteriocin, enterocin. The addition of both purified or semi-purified enterocin and enterocin-producing strains is in practice for food preservation. Additionally, in situ bacteriocin production in food can also be achieved by using *Enterococcus* as a starter culture (Nieto-Lozano et al., 2002). Enterocin-producing strains do exert any negative effects in the hostile food environment. Hence, addition of purified or semi-purified enterocin has been proved to be more advantageous. Enterocin inhibits *B. cereus, Bacillus macroides, Paenibacillus* spp. and *S. aureus* in fresh vegetables; *L. monocytogenes* and *Bacillus coagulans* in soybean sprouts, canned fruits, and vegetable; *L. monocytogenes* in cottage cheese, munster cheese, goat's milk, and goat milk's cheese; *B. cereus* in nonfat hard cheese; *S. aureus* and *L. monocytogenes* in the skimmed milk, yogurt, and Saint-Paulin cheese (Fu et al., 2018). Enterocin is also effective as food preservative in fish, meat, meat products, some beverages, cooked pork, dry fermented sausages, cooked ham, and Italian sausages (hadian et al., 2008). Enterocin AS-48, produced by *E. faecalis* A-48–32, inhibits *B. cereus, Alicyclobacillus acidoterrestris,* and *Bacillus licheniformis* and is used to preserve infant rice-based food, fruit juices, and apple cider (L. Y. Hu et al., 2012). Often, instead of a single application, enterocins have been found to preserve Spanish-style dry-fermented sausages, a combination of enterocin CCM 4231, enterocin 13, and sakacin K produced by *E. faecium* CCM 4231, *E. faecium* RZS C13, and *L. sakei* CTC494, respectively, is effective to inhibit *L. monocytogenes* and *L. innocua* (S. Zhang et al., 2021). *E. faecalis* BFE 1071 produced enterocin 1071 A and B, which have been found to have an inhibitory effect against *L. innocua, Staphylococcus epidermis* and *Proteus vulgaris* in fish spread (Shwaiki et al., 2019). By inhibiting *B. macroides* and *Bacillus maroccanus,* enterocin EJ97 from *E. faecalis* EJ97 can preserve vegetable (zucchini) puree (Mirzaei et al., 2020). Enterocins L50A and B from *E. faecium* L50 can inhibit *Lactobacillus brevis* and *Pediococcus damnosus* growth and hence are used in production of alcoholic and nonalcoholic beer (Mirzaei et al., 2020).

1.3.3 Pediocin

Pediococcus strains produce bacteriocin, called pediocin, which is highly stable in a wide range of temperatures and pH environments. Pediocin absorbs the amino acids in the phospholipid layer of the cytoplasmic membrane of the targeted cells, thereby killing *C. perfringens* and *L. monocytogenes* (Wang et al., 2017). Like enterocin, pediocin can also be applied in food by adding either pediocin in optimum concentration or *Pediococcus, Enterococcus,* or *Lactobacillus* strains with the optimum control. But unlike enterocin, direct pediocin addition is not beneficial as pediocin may affect its solubility and amphiphilic nature (Irwin, 1995). Pediocin PA-1, produced by *Pediococcus acidilactici* MCH14 exerts antimicrobial effects against *L. monocytogenes* and *C. perfringens,* which extend the shelf life of dried sausages and fermented meat products; against *Lactobacillus bifermentans,* which extend the shelf life of salad dressings; against *Leuconostoc mesenteroides,* which extend the shelf lives of fresh beef, vacuum-packed beef, and cottage cheese; against of *L. monocytogenes,* which extend the shelf life of fish fillets and fermented soymilk. Pediocin 34, produced from *Pediococcus pentosaceous* 34, inhibits *L. monocytogenes* in milk products and meat, thereby extending their shelf lives.

1.3.4 Other Bacteriocins

Leuconostoc spp. produce bacteriocin and leucocin. Fresh meat and sausages can be preserved by leucocin A, produced by *Leuconostoc mensenteroides* K7, which can inhibit *L. monocytogenes* and *C. divergens* UAL9 (Hornbaek et al., 2006). Similarly, milk can be preserved by leucocin K7 from *Leuconostoc mensenteroides* K7. *Debaryomyces hansenii* DSMZ70238 produces a bacteriocin, mycocin, that can inhibit *L. monocytogenes* and is used to preserve meat and meat products. Milk products and meat can also be preserved by the combined use of carno-cyclin A, carnobacteriocin BM1, and piscicolin 126 produced by *Carnobacterium maltaromaticum* UAL307. This combination of bacteriocins inhibit food pathogens like *E. coli* DH5α, *Pseudomonas aeruginosa* ATCC 14207 and *Salmonella typhimurium* ATCC 23564 (Ng et al., 2020). Both *Streptomyces natalensis* and *Streptomyces gilvo-sporeus* produce natamycin that can resist yeasts and molds in cheese, fresh dairy products, processed meat, and beverages. Fungi such as gray mold mainly cause spoilage in strawberries. During the ripening of strawberries, carbohydrates are metabolized by respiratory enzymes, producing acids and thereby lowering the pH. An important deterioration factor for postharvest fruits is respiration, on which the quality and shelf lives of postharvest fruits depend. Total soluble solid content (SSC) indicates the extent of retention of nutrients, most of which is sugar. Thus, as the storage time of strawberries increases, both pH and total SSC decrease. Lactic acid bacterial strains *Lactobacillus delbrueckii* subsp. *bulgaricus* F17 and *Leuconostoc lactis* H52 can slow this process of aging, which was evident from the higher total SSC value. This delay may be contributed by F17 and H52 metabolites or by biofilm formation (Fang et al., 2019).

1.3.5 Antimicrobial Peptides

Biocompatible and biodegradable bioactive peptides exhibit their potential as food preservatives through bacteriostasis and antioxidation. Bioactive peptides that possess antibacterial, antifungal, or antiviral properties are antimicrobial peptides (AMPs). The AMPs are amphiphilic in nature, and they interact with the polar head and nonpolar tail of microbial lipid membrane through their hydrophilic and hydrophobic parts, respectively. There are a few models that explain the mechanism of action of AMPs. When the ratio of AMP:lipid molecules reaches a certain value, AMPs vertically embed into the lipid bilayer of the membrane. This results in bending of the lipid membrane leading to membrane disruption. Normal osmotic pressure cannot be maintained leading to extravasation of the cellular contents. This mode of action is the toroidal pore model. As AMPs bind to the cell membrane, more and more AMPs tend to accumulate on the membrane, thereby forming transmembrane ion channels in the cell membrane. It resembles a barrel plate that causes the intracellular substances to overflow, leading to death (Huan et al., 2020). This mode of action is the barrel–stave model. According to the carpet model, through an electrostatic interaction, AMP molecules bind to the membrane, covering the phospholipid bilayer, and at high concentration, AMPs form micelles. This micellization leads to large-scale membrane permeation, thereby destroying the phospholipid bilayer. This causes disintegration of the cell membrane and extravasation of intracellular substances. There are four mechanisms by which AMPs can attack the intracellular components:

- AMP may directly interact with the DNA or RNA, thereby hampering the replication and transcription process that ultimately eliminates bacteria.

- AMPs may also inhibit protein translation and post-translational modification. Without protein synthesis, normal reactions cannot occur, and the bacteria die.

- AMPs may inhibit cell wall synthesis by binding to key enzymes, such as transpeptidases. Without the cell wall, the cell membrane will expand and rupture because of excessive water absorption and then lead to microbial death.

- Excessive production of reactive oxygen species (ROS) may occur due to stimulation of the fungal cells by AMPs. Ultimately, the pathogenicity of fungi is reduced because a large excess ROS causes the oxidative damage that leads to cell dysfunction and the inhibition of spore germination as well as mycelial growth (Shwaiki et al., 2019).

Antibacterial potential of AMP is the observed to weaken at high salt concentration either due to a decrease in the electrostatic interaction between AMP and negatively charged microbial cells or the salt can change the charge, structure, and activity of the AMP. Amino acids like arginine, proline, and tyrosine, however, enable the AMP to retain its antibacterial potential by replacing the salt ions around the microbial membrane. Salt concentration may alter the charge of the constituent amino acid atoms that disrupt the protein structure. Upon exposure to salt, the α-helical part may be denatured due to loss of either one or both terminal capping. Salt exposure can also damage the disulfide bonds that hold the rigid amphiphilic β-folded structure.

1.3.6 Antioxidant Peptides

Bioactive peptides that exhibit antioxidant properties are called antioxidant peptides. Antioxidant peptides scavenge free radicals, thereby terminating the chain reaction in lipid peroxidation. The antioxidant peptides disperse on the surface of the oil molecules and prevent aerial oxidation. The antioxidant peptides can either chelate the metal ions or maintain their original valence state and can reduce lipase activity. Generally, the antioxidant peptides are composed of 2–12 amino acids, but the higher the molecular weight of the antioxidant peptide, the higher the antioxidant activity (Huan et al., 2020). The antioxidant activity increases with an increase in the stability of the secondary structure and decreases in random coil content. Hydrophobic amino acids interact more strongly with unsaturated fatty acids and free radicals in the lipid matrix, which is a good solvent for peptides, resulting in increased antioxidant activity. In addition, the aromatic amino acid residues and amino acid residues with specific functions (such as His, Asp, and Glu) exhibit antioxidant property as a result of eliminating free radicals and chelating metal ions, respectively. Similar to salt concentration, high temperatures also may denature the secondary structure and lower the antioxidant activity. The presence of cysteine causes more heat-induced aggregation. In the food matrix, the resultant antioxidant effect of AMP and antioxidant peptides is dependent on the interaction between the several internal and external factors. To overcome this problem, peptides can be synthesized by directional design. Various techniques like coating, microencapsulation, and nanoencapsulation are also applied to stabilize the preservative effect of peptides (Huan et al., 2020).

1.4 PRESERVATIVES FROM ANIMAL SOURCES

Many peptides, enzymes, glycoproteins, lipids, and polysaccharides obtained from animal sources have exhibited significant antimicrobial activity and can be used as food preservatives. Lysozymes, lactoperoxidase, lactoferrin, and chitosan are frequently used as food preservatives.

1.4.1 Lysozyme

Lysozyme is an antimicrobial enzyme that exists in mammalian milk, avian eggs, fish, and others and inhibits the growth of bacteria, viruses, and protozoa. Lysozyme is a GRAS material, so it has been directly introduced in many food systems like meats, seafood, vegetables, fresh fruits, dairy products, salads, pasta, and infant formulas to enhance the shelf lives of foods (Liburdi et al., 2014). Lysozyme shows better antibacterial activity against gram-positive bacteria than gram-negative one, because the gram-negative bacterial cell wall contains less peptidoglycan (only 5 ± 10) and an outer membrane consisting of lipopolysaccharides, phospholipids, and protein. Lysozyme remains active in a pH wide range (2–10) and temperature range (4–95°C).

1.4.2 Lactoperoxidase

Lactoperoxidase is an enzyme available in cow's milk and in relatively smaller amounts in human milk. Investigation has established that this enzyme can inhibit, as well as completely destroy, Salmonella, *Shigella*, *Pseudomonas*, and coliforms. The use of lactoperoxidase for increasing the shelf life of milk is in practice, especially in countries where refrigeration is not easily available (Al-Baarri et al., 2018).

1.4.3 Lactoferrin

Lactoferrin is a globular glycoprotein component of milk, and it shows antioxidant, antimicrobial, anticarcinogenic activities (Bielecka et al., 2022). Japan and several other countries use lactoferrin as a biopreservative for yogurt and other foods. It has been reported that lactoferrin inhibits the growth of different gram-negative and gram-positive bacteria, including foodborne microbes and fungi. Studies have demonstrated that lactoferrin is less resistant against pathogenic microbes in fermented meat. Several mechanisms have been proposed to explain the antimicrobial activity of lactoferrin (Bielecka et al., 2022). One of the accepted mechanisms explains that the strong binding affinity of lactoferrin with free iron limits the availability of irons for microbial metabolism. Another view suggests that cell death may occur because of damage to the outer shell of bacteria by lactoferrin (Gruden & Poklar Ulrih, 2021). Studies have revealed that lactoferrin may inhibit the proliferation of virus and fungi. The major limitation of the application of lactoferrin in the food system as an antimicrobial agent is that the ionic environment of food can nullify the antimicrobial effect of lactoferrin.

1.4.4 Chitosan

Chitin is a long-chain linear polysaccharide of N-acetylglucosamine and is abundantly available as one of the primary constituents of the outer skeleton of arthropods (crabs, shrimps, insects; S. Kumar et al., 2019). Chitin is also a component of fungal cell wall and found in green algae, yeast, and protozoa. Chitosan is a heteropolysaccharides that produce by partial deacetylation of chitin. Chitin and chitosan both are crystalline in nature and insoluble in water at neutral pH. At acidic pH, chitosan becomes soluble in water owing to the protonation of the free amino group, whereas chitin is insoluble in water irrespective of pH. Good film formation and coating and the antimicrobial and antioxidant properties of chitosan make it suitable for food preservation applications. Several mechanisms exist to explain the antimicrobial action of chitosan. The most accepted mechanism involve interaction of positively charged amino groups (NH_3^+) of chitosan with the negatively charged carboxyl groups (COO^-), located at the cell wall of bacteria or fungi. Owing to such interactions, the permeability of the membrane may be changed, and the loss of cell components occurred. The antimicrobial efficiency of the chitosan mainly depends on the molecular weight and extent of positive charges on the polymer. The pH of the solution and the degree of deacetylation influence the extent of positive charges of the polymer. A higher degree of deacetylation and a low pH enhance the positive charge on the polymer, and as a consequence, the antimicrobial efficacy of the polymer also increases. In one of the study researchers show that chitosan losses its antibacterial activity when pH of the solution increases, and this is due to decrease of the solubility of the polymer. In fact, the solubility of the polymer is crucial for its antibacterial activity, and in another study, it was demonstrated that chitosan has no antimicrobial activity when directly used in powder form. Several studies have shown that the antimicrobial activity of chitosan may be influenced if metal ions ($Ag+$, $Cu2+$, $Zn2+$) or organic compounds like quaternary ammonium compounds, carboxymethyl, quaternized carboxymethyl, and arginine are added into the polymer solution. Enhancing the concentration of the polymer has a positive effect on antimicrobial activity possibly due to the increase in the interaction with cellular components (S. Kumar et al., 2021). On the other hand, a study revealed that with the increase in temperature, chitosan exhibits lower antimicrobial activity.

Chitosan also may act as an antioxidant. The free amino groups and the hydroxyl radicals of the chitosan are responsible for its antioxidant activity because these parts of the polymer may bind with metal ions (Fe^{2+}) and free radicals on food. Studies show that the antioxidant activity depends on several parameters, such as molecular weight, degree of deacetylation, the extent of protonation of the free amino groups, and the concentration of the polymer. More degrees of deacetylation mean more free amino groups, that is, a greater possibility to interact with metal ions. Higher-molecular-weight chitosan exists in a more compact form with strong intramolecular interaction and naturally experiences less interaction with external species. Less antioxidant activity has been observed for more protonated polymers as more protonation means more electrostatic repulsion, resulting in a linear conformation of the chain polymer. A large number of studies exist in which chitosan has been examined as a preservative in the food system. In general, an acetic acid solution of chitosan having concentration range within 0.05–4% has been employed for application in food. Studies show that acetic acid is a more preferable solvent than lactic acid as acetic acid produces more short fibers, smaller diameter, and higher load, resulting in higher viscosity. For instance, several studies report that direct incorporation of chitosan solution into seafood-based products like sausage, surimi, sausage, fish, and others increases the shelf lives of the products and maintains nutritional value and sensory quality (Basumatary et al., 2021; S. Kumar, Mudai, et al., 2020). It has been also demonstrated that nanoparticles of chitosan have higher preservative activity than larger particles.

1.5 EMULSION AND ENCAPSULATION OF NATURAL PRESERVATIVES

Many active ingredients including EOs, bioactive peptides, lipids, and others with a hydrophobic nature are either completely insoluble in water or have very low solubility. Hence, incorporating such active components into a water-based food formulation is challenging. One of the most practiced technologies to overcome this challenge is the encapsulation or lamination of active components by a hydrophilic natural polymer or by food-grade materials, which ensures more efficient delivery of the active components to food systems (I. H. Kim et al., 2013). The hydrophilic coating makes the water-insoluble hydrophobic active components, soluble in water. A further advantage is that coating materials can protect the active ingredient from environmental influences as well as undesirable chemical interactions during processing of the food. Additionally, the coatings facilitate a controlled release of the active component during storage, leading to an increased shelf life of the food. For instance, a multicomponent oil-in-water emulsion containing the EOs of ginger, cinnamon, cardamone, Tween-80, and glycerol has been prepared by mechanical stirring. An investigation revealed that decreasing the size of the emulsion droplet (EO droplet) results in better antioxidant and antimicrobial activity. Moreover, an emulsion consisting of two or three components in the oil phase has superior antioxidant and antimicrobial activities than an emulsion with a single component in the oil phase, indicating synergistic effects between combinations of EOs (Jafarizadeh-Malmiri et al., 2022). In another study, coaxial electrospinning technology has been employed to encapsulate thymol with poly(lactide-co-glycolide) as a shell material and thus obtained core-shell nanofibers (Y. Zhang et al., 2019). This study demonstrated that such core-shell nanofiber film provides better shelf life to highly perishable fruit such as strawberries in comparison to only a poly(lactide-co-glycolide) film or other common packaging materials. The shell of poly(lactide-co-glycolide) resists rapid volatilization of the core material (thymol), and the growth of bacteria, fungi, and yeast on strawberries is inhibited by the sustained release of the broad-spectrum antimicrobial thymol.

1.6 CONCLUSION

This chapter depicts in detail the microbial and oxidative damage of food. It has been shown that several parameters, such as temperature, light, metal, and others, promote the oxidative degradation of lipids or proteins, causing a significant loss of sensory and nutritional qualities of food. The application of natural ingredients with antimicrobial and /or antioxidant properties obtained from plants, algae, animals, and microorganisms in food as preservatives is safe for human health as well as the environment. The presence of several active compounds like polyphenolic compounds, small peptides, compounds present in EOs, and others are responsible for such antimicrobial and/or antioxidant activity. Extensive research has made it possible to isolate many such active compounds in pure form and utilize them as more efficient preservatives. Many studies show that applying two or more active compounds as a mixture may exhibit enhanced preservative activity owing to a synergistic effect. Most of the active compounds are hydrophobic in nature, so their solubility in an aqueous medium can be increased by employing emulsion/encapsulation technology. A direct edible coating of such active compounds, along with any natural polymers, on perishable fruits and vegetables can increase their shelf lives. However, extensive investigations are necessary to develop more efficient methods to utilize natural ingredients as preservatives.

REFERENCES

Abd. Wahab, N. A., Lajis, H. N., Abas, F., Othman, I., & Naidu, R. (2020). Mechanism of anti-cancer activity of curcumin on androgen-dependent and androgen-independent prostate cancer. *Nutrients*, 12(3), 679.

Abedi, E., & Hashemi, S. M. B. (2020). Lactic acid production – Producing micro-organisms and substrates sources-state of art. *Heliyon*, 6(10), e04974–e04974.

Al-Baarri, A. N. M., Legowo, A. M., Arum, S. K., & Hayakawa, S. (2018). Extending shelf life of Indonesian soft milk cheese (Dangke) by Lactoperoxidase system and lysozyme. *International Journal of Food Science*, 2018, 4305395.

Albuquerque, B. R., Heleno, S. A., Oliveira, M. B. P. P., Barros, L., & Ferreira, I. C. F. R. (2021). Phenolic compounds: Current industrial applications, limitations and future challenges. *Food & Function*, 12(1), 14–29.

Alihosseini, F. (2016). 10 – Plant-based compounds for antimicrobial textiles. In G. Sun (Ed.), *Antimicrobial Textiles* (pp. 155–195). Woodhead Publishing.

Arshad, M. S., Amjad, Z., Yasin, M., Saeed, F., Imran, A., Sohaib, M., Anjum, F. M., & Hussain, S. (2019). Quality and stability evaluation of chicken meat treated with gamma irradiation and turmeric powder. *International Journal of Food Properties*, 22(1), 154–172.

Arulkumar, A., Ramanchandran, K., Paramasivam, S., Palanivel, R., & Miranda, J. M. (2017). Effects of turmeric (*Curcuma longa*) on shelf life extension and biogenic amine control of cuttlefish (*Sepia brevimana*) during chilled storage. *CyTA – Journal of Food*, 15(3), 441–447.

Augustin, M. A., & Sanguansri, L. (2010). 15 – Use of encapsulation to inhibit oxidation of lipid ingredients in foods. In E. A. Decker, R. J. Elias, & D. Julian McClements (Eds.), *Oxidation in Foods and Beverages and Antioxidant Applications* (pp. 479–495). Woodhead Publishing.

Baptista, R. C., Horita, C. N., & Sant'Ana, A. S. (2020). Natural products with preservative properties for enhancing the microbiological safety and extending the shelf-life of seafood: A review. *Food Research International, 127*, 108762.

Barreca, D., Trombetta, D., Smeriglio, A., Mandalari, G., Romeo, O., Felice, M. R., Gattuso, G., & Nabavi, S. M. (2021). Food flavonols: Nutraceuticals with complex health benefits and functionalities. *Trends in Food Science & Technology, 117*, 194–204.

Basumatary, I. B., Mukherjee, A., Katiyar, V., & Kumar, S. (2022). Biopolymer-based nanocomposite films and coatings: Recent advances in shelf-life improvement of fruits and vegetables. *Critical Reviews in Food Science and Nutrition, 62*(7), 1912–1935.

Basumatary, I. B., Mukherjee, A., Katiyar, V., Kumar, S., & Dutta, J. (2021). Chitosan-based antimicrobial coating for improving postharvest shelf life of Pineapple. *Coatings, 11*(11).

Behera, B. C., Mishra, R., & Mohapatra, S. (2021). Microbial citric acid: Production, properties, application, and future perspectives. *Food Frontiers, 2*(1), 62–76.

Ben Braïek, O., & Smaoui, S. (2021). Chemistry, safety, and challenges of the use of organic acids and their derivative salts in meat preservation. *Journal of Food Quality, 2021*, 6653190.

Bendeddouche, M. S., Benhassaini, H., Hazem, Z., & Romane, A. (2011). Essential oil analysis and antibacterial activity of *Rosmarinus tournefortii* from Algeria. *Natural Product Communications, 6*(10), 1511–1514.

Bensid, A., El Abed, N., Houicher, A., Regenstein, J. M., & Özogul, F. (2022). Antioxidant and antimicrobial preservatives: Properties, mechanism of action and applications in food – A review. *Critical Reviews in Food Science and Nutrition, 62*(11), 2985–3001.

Bielecka, M., Cichosz, G., & Czeczot, H. (2022). Antioxidant, antimicrobial and anticarcinogenic activities of bovine milk proteins and their hydrolysates – A review. *International Dairy Journal, 127*, 105208.

Blom, H., & Mørtvedt, C. (1991). Anti-microbial substances produced by food associated micro-organisms. *Biochem Soc Trans, 19*(3), 694–698.

Carocho, M., Morales, P., & Ferreira, I. C. F. R. (2018). Antioxidants: Reviewing the chemistry, food applications, legislation and role as preservatives. *Trends in Food Science & Technology, 71*, 107–120.

Chaieb, K., Zmantar, T., Ksouri, R., Hajlaoui, H., Mahdouani, K., Abdelly, C., & Bakhrouf, A. (2007). Antioxidant properties of the essential oil of *Eugenia caryophyllata* and its antifungal activity against a large number of clinical Candida species. *Mycoses, 50*(5), 403–406.

Choyam, S., Srivastava, A. K., Shin, J.-H., & Kammara, R. (2019). Ocins for food safety. *Frontiers in Microbiology, 10*.

Coban, H. B. (2020). Organic acids as antimicrobial food agents: Applications and microbial productions. *Bioprocess and Biosystems Engineering, 43*(4), 569–591.

Delgado, A., Al-Hamimi, S., Ramadan, M. F., Wit, M. D., Durazzo, A., Nyam, K. L., & Issaoui, M. (2020). Contribution of tocols to food sensorial properties, stability, and overall quality. *Journal of Food Quality, 2020*, 8885865.

Dhifi, W., Bellili, S., Jazi, S., Bahloul, N., & Mnif, W. (2016). Essential oils' chemical characterization and investigation of some biological activities: A critical review. *Medicines (Basel), 3*(4).

Dorman, H. J., & Deans, S. G. (2000). Antimicrobial agents from plants: Antibacterial activity of plant volatile oils. *J Appl Microbiol, 88*(2), 308–316.

Fang, X., Li, Y., Guo, W., Ke, W., Bi, S., Guo, X., & Zhang, Y. (2019). *Lactobacillus delbrueckii* subsp. *bulgaricus* F17 and Leuconostoc lactis H52 supernatants delay the decay of strawberry fruits: A microbiome perspective. *Food & Function, 10*(12), 7767–7781.

Faustman, C., Yin, S., Tatiyaborworntham, N., & Naveena, B. M. (2010). 1 – Oxidation and protection of red meat. In E. A. Decker, R. J. Elias, & D. Julian McClements (Eds.), *Oxidation in Foods and Beverages and Antioxidant Applications* (pp. 3–49). Woodhead Publishing.

Featherstone, S. (2015). 8 – Ingredients used in the preparation of canned foods. In S. Featherstone (Ed.), *A Complete Course in Canning and Related Processes* (4th ed., pp. 147–211). Woodhead Publishing.

Fraga, C. G., Croft, K. D., Kennedy, D. O., & Tomás-Barberán, F. A. (2019). The effects of polyphenols and other bioactives on human health. *Food & Function, 10*(2), 514–528.

Fu, Y., Mu, D., Qiao, W., Zhu, D., Wang, X., Liu, F., Xu, H., Saris, P., Kuipers, O. P., & Qiao, M. (2018). Co-expression of Nisin Z and Leucocin C as a basis for effective protection against *Listeria monocytogenes* in Pasteurized Milk. *Frontiers in Microbiology, 9*.

Gordon, M. H. (2001). 2 – The development of oxidative rancidity in foods. In J. Pokorny, N. Yanishlieva & M. Gordon (Eds.), *Antioxidants in Food* (pp. 7–21). Woodhead Publishing.

Gottardi, D., Bukvicki, D., Prasad, S., & Tyagi, A. K. (2016). Beneficial effects of spices in food preservation and safety. *Frontiers in Microbiology, 7*.

Gruden, Š., & Poklar Ulrih, N. (2021). Diverse mechanisms of antimicrobial activities of lactoferrins, lactoferricins, and other lactoferrin-derived peptides. *International Journal of Molecular Sciences, 22*(20), 11264.

Gupta, S., & Ravishankar, S. (2005). A comparison of the antimicrobial activity of garlic, ginger, carrot, and turmeric pastes against *Escherichia coli* O157:H7 in laboratory buffer and ground beef. *Foodborne Pathog Dis, 2*(4), 330–340.

Gutiérrez-del-Río, I., López-Ibáñez, S., Magadán-Corpas, P., Fernández-Calleja, L., Pérez-Valero, Á., Tuñón-Granda, M., Miguélez, E. M., Villar, C. J., & Lombó, F. (2021). Terpenoids and polyphenols as natural antioxidant agents in food preservation. *Antioxidants, 10*(8).

hadian, J., Ghasemnezhad, M., Ranjbar, H., Frazane, M., & Ghorbanpour, M. (2008). Antifungal potency of some essential oils in control of postharvest decay of strawberry caused by *Botrytis Cinerea, Rhizopus Stolonifer and Aspergillus Niger. Journal of Essential Oil Bearing Plants, 11*(5), 553–562.

Hashemi, S. M. B., & Jafarpour, D. (2020). Ultrasound and malic acid treatment of sweet lemon juice: Microbial inactivation and quality changes. *Journal of Food Processing and Preservation, 44*(11), e14866.

Hornbaek, T., Brockhoff, P. B., Siegumfeldt, H., & Budde, B. B. (2006). Two subpopulations of *Listeria monocytogenes* occur at subinhibitory concentrations of leucocin 4010 and nisin. *Applied and Environmental Microbiology, 72*(2), 1631–1638.

Hosseini, A., Ramezani, S., Tabibiazar, M., Ghorbani, M., & Samadi Kafil, H. (2021). Fabrication of cumin seed oil loaded gliadin-ethyl cellulose nanofibers reinforced with adipic acid for food packaging application. *Food Packaging and Shelf Life, 30*, 100754.

Hu, L. Y., Hu, S. L., Wu, J., Li, Y. H., Zheng, J. L., Wei, Z. J., Liu, J., Wang, H. L., Liu, Y. S., & Zhang, H. (2012). Hydrogen sulfide prolongs postharvest shelf life of strawberry and plays an antioxidative role in fruits. *J Agric Food Chem, 60*(35), 8684–8693.

Hu, Y., Luo, J., Kong, W., Zhang, J., Logrieco, A. F., Wang, X., & Yang, M. (2015). Uncovering the antifungal components from turmeric (*Curcuma longa* L.) essential oil as *Aspergillus flavus* fumigants by partial least squares. *RSC Advances*, 5(52), 41967–41976.

Huan, Y., Kong, Q., Mou, H., & Yi, H. (2020). Antimicrobial peptides: Classification, design, application and research progress in multiple fields. *Frontiers in Microbiology*, 11.

Irwin, W. J. (1995). Stability and characterization of protein and peptide drugs. Case histories (Pharmaceutical biotechnology volume 5) Edited by Y. John Wang and Rodney Pearlman. *Journal of Pharmacy and Pharmacology*, 47(2), 176–176.

Jafarizadeh-Malmiri, H., Anarjan, N., & Berenjian, A. (2022). Developing three-component ginger-cinnamon-cardamom composite essential oil nano-emulsion as natural food preservatives. *Environmental Research*, 204, 112133.

Jessica Elizabeth, D. L. T., Gassara, F., Kouassi, A. P., Brar, S. K., & Belkacemi, K. (2017). Spice use in food: Properties and benefits. *Critical Reviews in Food Science and Nutrition*, 57(6), 1078–1088.

Kalpana, V. N., & Rajeswari, V. D. (2019). 1 – Preservatives in beverages: Perception and needs. In A. M. Grumezescu & A. M. Holban (Eds.), *Preservatives and Preservation Approaches in Beverages* (pp. 1–30). Academic Press.

Karsha, P. V., & Lakshmi, O. B. (2010). Antibacterial activity of black pepper (Piper nigrum Linn.) with special reference to its mode of action on bacteria. *Indian Journal of Natural Products and Resources*, 1, 213–215.

Karthik, T., & Rathinamoorthy, R. (2017). 8 – Sustainable synthetic fibre production. In S. S. Muthu (Ed.), *Sustainable Fibres and Textiles* (pp. 191–240). Woodhead Publishing.

Kim, I. H., Lee, H., Kim, J. E., Song, K. B., Lee, Y. S., Chung, D. S., & Min, S. C. (2013). Plum coatings of lemongrass oil-incorporating carnauba wax-based nanoemulsion. *J Food Sci*, 78(10), E1551–E1559.

Kim, S., Ruengwilysup, C., & Fung, D. Y. (2004). Antibacterial effect of water-soluble tea extracts on foodborne pathogens in laboratory medium and in a food model. *J Food Prot*, 67(11), 2608–2612.

Kumar, N., & Goel, N. (2019). Phenolic acids: Natural versatile molecules with promising therapeutic applications. *Biotechnology Reports*, 24, e00370.

Kumar, S., Mudai, A., Roy, B., Basumatary, I. B., Mukherjee, A., & Dutta, J. (2020). Biodegradable Hybrid Nanocomposite of Chitosan/Gelatin and Green Synthesized Zinc Oxide Nanoparticles for Food Packaging. *Foods*, 9(9), 1143.

Kumar, S., Mukherjee, A., & Dutta, J. (2020). Chitosan based nanocomposite films and coatings: Emerging antimicrobial food packaging alternatives. *Trends in Food Science & Technology*, 97, 196–209.

Kumar, S., Ye, F., Dobretsov, S., & Dutta, J. (2019). Chitosan nanocomposite coatings for food, paints, and water treatment applications. *Applied Sciences*, 9(12).

Kumar, S., Ye, F., Mazinani, B., Dobretsov, S., & Dutta, J. (2021). Chitosan nano-composite coatings containing chemically resistant ZnO – SnOx Core – Shell nanoparticles for photocatalytic antifouling. *International Journal of Molecular Sciences*, 22(9).

Kurutas, E. B. (2016). The importance of antioxidants which play the role in cellular response against oxidative/nitrosative stress: Current state. *Nutrition Journal*, 15(1), 71.

Liburdi, K., Benucci, I., & Esti, M. (2014). Lysozyme in wine: An overview of current and future applications. *Comprehensive Reviews in Food Science and Food Safety*, 13(5), 1062–1073.

Ludwiczuk, A., Skalicka-Woźniak, K., & Georgiev, M. I. (2017). Chapter 11 – Terpenoids. In S. Badal & R. Delgoda (Eds.), *Pharmacognosy* (pp. 233–266). Academic Press.

Lyu, X., Lee, J., & Chen, W. N. (2019). Potential natural food preservatives and their sustainable production in yeast: Terpenoids and polyphenols. *Journal of Agricultural and Food Chemistry, 67*(16), 4397–4417.

Marques, C., Sotiles, A. R., Farias, F. O., Oliveira, G., Mitterer-Daltoé, M. L., & Masson, M. L. (2020). Full physicochemical characterization of malic acid: Emphasis in the potential as food ingredient and application in pectin gels. *Arabian Journal of Chemistry, 13*(12), 9118–9129.

Mei, J., Ma, X., & Xie, J. (2019). Review on natural preservatives for extending fish shelf life. *Foods, 8*(10).

Mirzaei, M., Mirdamadi, S., Safavi, M., & Soleymanzadeh, N. (2020). The stability of antioxidant and ACE-inhibitory peptides as influenced by peptide sequences. *LWT, 130*, 109710.

Neelam, Khatkar, A., & Sharma, K. K. (2020). Phenylpropanoids and its derivatives: Biological activities and its role in food, pharmaceutical and cosmetic industries. *Critical Reviews in Food Science and Nutrition, 60*(16), 2655–2675.

Ng, Z. J., Zarin, M. A., Lee, C. K., & Tan, J. S. (2020). Application of bacteriocins in food preservation and infectious disease treatment for humans and livestock: A review. *RSC Advances, 10*(64), 38937–38964.

Nieto-Lozano, J. C., Reguera-Useros, J. I., Peláez-Martínez, M. C., & Hardisson de la Torre, A. (2002). Bacteriocinogenic activity from starter cultures used in Spanish meat industry. *Meat Science, 62*(2), 237–243.

Özdemir, K. S., & Gökmen, V. (2019). Effect of chitosan-ascorbic acid coatings on the refrigerated storage stability of fresh-cut apples. *Coatings, 9*(8).

Pawar, V. C., & Thaker, V. S. (2006). In vitro efficacy of 75 essential oils against *Aspergillus niger. Mycoses, 49*(4), 316–323.

Pisoschi, A. M., Pop, A., Georgescu, C., Turcuş, V., Olah, N. K., & Mathe, E. (2018). An overview of natural antimicrobials role in food. *Eur J Med Chem, 143*, 922–935.

Raguindin, P. F., Adam Itodo, O., Stoyanov, J., Dejanovic, G. M., Gamba, M., Asllanaj, E., Minder, B., Bussler, W., Metzger, B., Muka, T., Glisic, M., & Kern, H. (2021). A systematic review of phytochemicals in oat and buckwheat. *Food Chemistry, 338*, 127982.

Rahman, M. M., & Gray, A. I. (2005). A benzoisofuranone derivative and carbazole alkaloids from *Murraya koenigii* and their antimicrobial activity. *Phytochemistry, 66*(13), 1601–1606.

Rao, P. V., & Gan, S. H. (2014). Cinnamon: A multifaceted medicinal plant. *Evidence-Based Complementary and Alternative Medicine: eCAM, 2014*, 642942.

Rezazadeh, A., Hamishehkar, H., Ehsani, A., Ghasempour, Z., & Moghaddas Kia, E. (2021). Applications of capsaicin in food industry: Functionality, utilization and stabilization. *Critical Reviews in Food Science and Nutrition*, 1–17.

Sánchez-Escalante, A., Djenane, D., Torrescano, G., Beltrán, J. A., & Roncalés, P. (2001). The effects of ascorbic acid, taurine, carnosine and rosemary powder on colour and lipid stability of beef patties packaged in modified atmosphere. *Meat Science, 58*(4), 421–429.

Sandjo, L. P., & Kuete, V. (2013). 3 – Diterpenoids from the medicinal plants of Africa. In V. Kuete (Ed.), *Medicinal Plant Research in Africa* (pp. 105–133). Elsevier.

Saxena, R. K., Saran, S., Isar, J., & Kaushik, R. (2017). 27 – Production and applications of succinic acid. In A. Pandey, S. Negi, & C. R. Soccol (Eds.), *Current Developments in Biotechnology and Bioengineering* (pp. 601–630). Elsevier.

Sebiomo, A., Awofodu, A., Awosanya, A., E, A., & J, A. (2011). Comparative studies of antibacterial effect of some antibiotics and ginger (*Zingiber officinale*) on two pathogenic bacteria. *Journal of Microbiology and Antimicrobials, 3,* 18–22.

Shahidi, F., & Ambigaipalan, P. (2015). Phenolics and polyphenolics in foods, beverages and spices: Antioxidant activity and health effects – A review. *Journal of Functional Foods, 18,* 820–897.

Shahidi, F., & de Camargo, A. C. (2016). Tocopherols and tocotrienols in common and emerging dietary sources: Occurrence, applications, and health benefits. *International Journal of Molecular Sciences, 17*(10), 1745.

Shahidi, F., & Zhong, Y. (2010). Lipid oxidation and improving the oxidative stability. *Chemical Society Reviews, 39*(11), 4067–4079.

Shwaiki, L. N., Arendt, E. K., Lynch, K. M., & Thery, T. L. C. (2019). Inhibitory effect of four novel synthetic peptides on food spoilage yeasts. *International Journal of Food Microbiology, 300,* 43–52.

Silva, C. C. G., Silva, S. P. M., & Ribeiro, S. C. (2018). Application of bacteriocins and protective cultures in dairy food preservation. *Frontiers in Microbiology, 9.*

Singh, H., Charan, A., Prasad, S., Charan, A., & Charan, A. (2017). Antifungal and antibacterial activity of methanolic, ethanolic and acetonic leaf extracts of curry leaves (*Murraya koenigii*). *Journal of Pharmacognosy and Phytochemistry, 6,* 1797–1802.

Skoog, E., Shin, J. H., Saez-Jimenez, V., Mapelli, V., & Olsson, L. (2018). Biobased adipic acid – The challenge of developing the production host. *Biotechnology Advances, 36*(8), 2248–2263.

Stoilova, I., Krastanov, A., Stoyanova, A., Denev, P., & Gargova, S. (2007). Antioxidant activity of a ginger extract (*Zingiber officinale*). *Food Chemistry, 102*(3), 764–770.

Suleman, R., Wang, Z., Aadil, R. M., Hui, T., Hopkins, D. L., & Zhang, D. (2020). Effect of cooking on the nutritive quality, sensory properties and safety of lamb meat: Current challenges and future prospects. *Meat Science, 167,* 108172.

Suppakul, P., Miltz, J., Sonneveld, K., & Bigger, S. W. (2003). Antimicrobial properties of basil and its possible application in food packaging. *Journal of Agricultural and Food Chemistry, 51*(11), 3197–3207.

Susheela, R. (2006). Forms, functions, and applications of spices. In *Handbook of Spices, Seasonings, and Flavorings.* CRC Press.

Veenstra, J. P., & Johnson, J. J. (2019). Oregano (*Origanum vulgare*) extract for food preservation and improvement in gastrointestinal health. *International Journal of Nutrition, 3*(4), 43–52.

Villalobos-Delgado, L. H., Nevárez-Moorillon, G. V., Caro, I., Quinto, E. J., & Mateo, J. (2019). 4 – Natural antimicrobial agents to improve foods shelf life. In C. M. Galanakis (Ed.), *Food Quality and Shelf Life* (pp. 125–157). Academic Press.

Wang, X., Yu, H., Xing, R., Chen, X., Liu, S., & Li, P. (2017). Optimization of the extraction and stability of antioxidative peptides from mMackerel (*Pneumatophorus japonicus*) protein. *BioMed Research International, 2017,* 6837285.

Winias, S. (2015). Effect of cynammyldehyde from cinnamon extract as a natural preservative alternative to the growth of *staphylococcus aureus* bacteria. *Indonesian Journal of Tropical and Infectious Disease, 2*(1), 38–41.

Zannini, E., Waters, D. M., Coffey, A., & Arendt, E. K. (2016). Production, properties, and industrial food application of lactic acid bacteria-derived exopolysaccharides. *Applied Microbiology and Biotechnology, 100*(3), 1121–1135.

Zhang, S., Luo, L., Sun, X., & Ma, A. (2021). Bioactive peptides: A promising alternative to chemical preservatives for food preservation. *Journal of Agricultural and Food Chemistry, 69*(42), 12369–12384.

Zhang, Y., Zhang, Y., Zhu, Z., Jiao, X., Shang, Y., & Wen, Y. (2019). Encapsulation of thymol in biodegradable nanofiber via coaxial eletrospinning and applications in fruit preservation. *J Agric Food Chem, 67*(6), 1736–1741.

Zhao, X., Shi, C., Meng, R., Liu, Z., Huang, Y., Zhao, Z., & Guo, N. (2016). Effect of nisin and perilla oil combination against *Listeria monocytogenes* and *Staphylococcus aureus* in milk. *Journal of Food Science and Technology, 53*(6), 2644–2653.

2 Food Preservation by High Hydrostatic Pressure

Barjinder Pal Kaur, Rahul Kumar Rout, and P Srinivasa Rao

Corresponding author:
Prof. P Srinivasa Rao, Email: psraoiit@gmail.com

CONTENTS

2.1 INTRODUCTION

High hydrostatic pressure (HHP) processing is also known as high-pressure processing (HPP) or ultra-high-pressure processing (UHP) is a novel nonthermal food processing technology employed recently in order to meet consumer preference for fresh, natural, and chemical/additive-free foods. This new technology also meets the regulative requirements for energy-efficient and environment-friendly food production systems (Mattson et al., 2003). This upcoming technology can be employed in ways to substitute conventional food preservation techniques like thermal processing. HHP treatment can be used for the preservation of food in prepacked form to replace heat treatment as significant inactivation of microflora is achieved at pressures greater than 400 MPa, with treatment times of a few minutes even at ambient temperatures. From an economical point of view, the highest pressure employed is limited up to 600 MPa, at which higher inactivation reactions are achievable. Besides, in combination with temperatures >80 °C, even irreversible inactivation of bacterial spores can be obtained. The pressure applied is homogeneous in a food, assuming the pressure level in each section of the food is uniform, thus the heat of compression is equally distributed. The temperature rises due to heat of compression, in those cases where it is desirable, are used to improve microbial inactivation and to maximize process effectiveness. A new approach to food sterilization through HHP is possible, as the instantaneous adiabatic heating can help reach the sterilizing temperature faster, thus improving food quality significantly. However, after pressure release, the product returns to its initial temperature (or even below), which is an aspect of high interest for the processing industries, leading to high-quality foods.

In liquid foods, heat transfer can be improved by the application of heat exchangers, and thus, heat recovery rate higher than 90% is obtained, whereas the preservation of particulate/solid food requires a long time for heat transfer or exhibits a great temperature gradient. HHP has been successfully applied for various food applications such as the gelatinization of starch; tempering of chocolate; the processing of litchi (whole) and mango pulps (Kaushik et al., 2016); blanching of vegetables; the processing of aquatic products like fish, prawn, lobster, and several seafoods (Kaur et al., 2016); the tenderization of meat, fish and minced meat; and the instant freezing and thawing of meat.

2.2 PRINCIPLES, MECHANISM OF OPERATION, AND APPLICATIONS OF HHP

HHP processing is a U.S. Food and Drug Administration (FDA)–approved method of treating foods under high pressure to enhance microbial safety, and at the same time, causing minimum degradation to the food. It is based on two basic principles:

a. **Isostatic principle**

According to this principle, the application of the pressure is instantaneous and uniform throughout the sample.

b. **Le Chatelier's principle**

This principle states that any reaction involving a volume change is influenced by high-pressure applications. So a reaction with a volume decrease is accelerated, while a reaction with a volume increase is slowed.

The steps of HHP processing are shown in Figure 2.1. Generally, the food item is packaged in a flexible packaging material and kept inside the pressure vessel, which is the core of the system. The pressure vessel is sealed properly and water or any other pressure transmitting fluid is pumped inside the vessel to generate the desired pressure. Due to the isostatic principle, the pressure distribution is

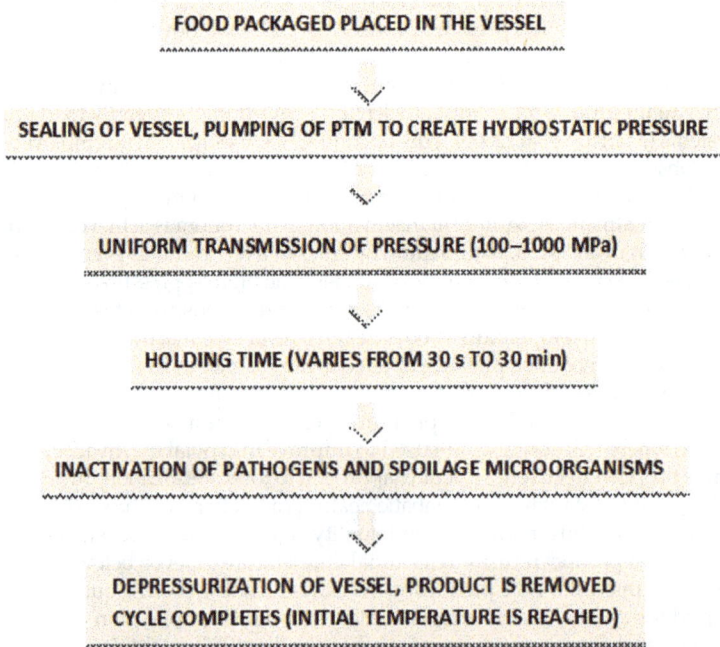

FOOD PACKAGED PLACED IN THE VESSEL

SEALING OF VESSEL, PUMPING OF PTM TO CREATE HYDROSTATIC PRESSURE

UNIFORM TRANSMISSION OF PRESSURE (100–1000 MPa)

HOLDING TIME (VARIES FROM 30 s TO 30 min)

INACTIVATION OF PATHOGENS AND SPOILAGE MICROORGANISMS

DEPRESSURIZATION OF VESSEL, PRODUCT IS REMOVED
CYCLE COMPLETES (INITIAL TEMPERATURE IS REACHED)

Figure 2.1 Flow chart showing the steps followed in HHP.

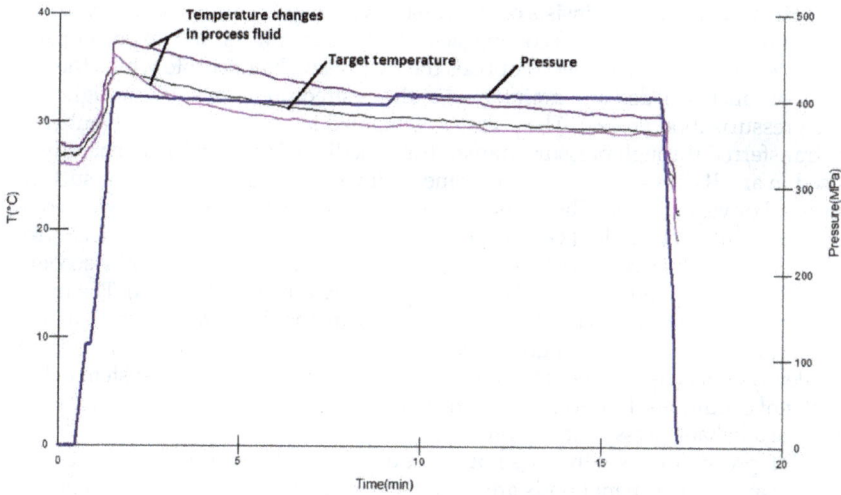

Figure 2.2 Typical single HHP cycle.

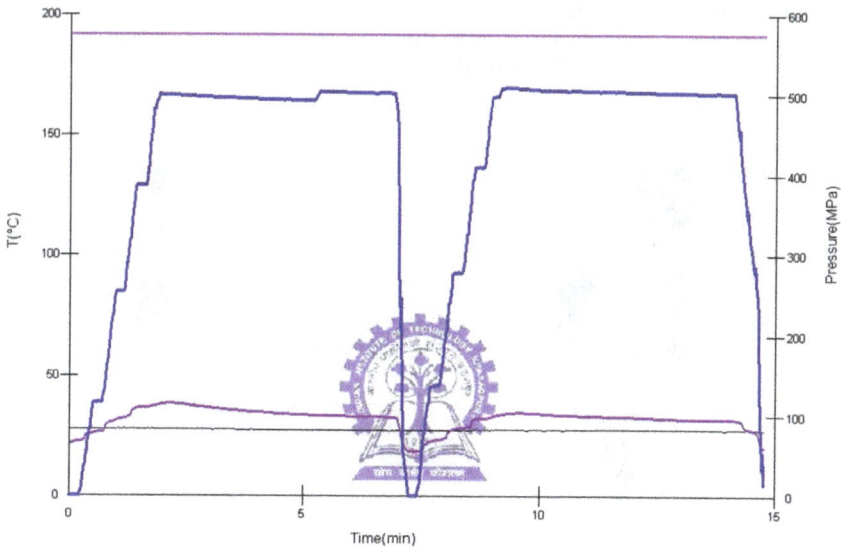

Figure 2.3 Double HHP cycle with up-ramp control.

uniform throughout the sample. This desired pressure is held for the required time, called pressure holding time, which, again, depends on the food material being processed, and varies from 30 s to 30 min (Figure 2.2). This completes the process, and then the pressure is released from the chamber and temperature comes down to the ambient. This cycle is repeated for different products. Sometimes, the pressure is applied in ramps, which results in a better effect and completes the process faster (Figure 2.3).

HHP system mainly consists of (a) pressure vessel, (b) control system, (c) intensifier/pump system, (d) cooling jackets/heat exchangers, and (e) power pack module. The pressure vessel comprises the treatment chamber into which the food product is loaded and maintained at the desired pressure level throughout the pressurization process. The pressure generated inside a treatment chamber is transferred through pressure-transmitting media (PTM). The PTM generally used in an HHP system is water combined with anticorrosive substances, such as mineral or vegetable oil. The water is preferred as PTM as it gives the lowest compression temperature rise (2.8–3°C per 100 MPa) at all pressure levels. The control system is used to give the inputs for the pressurization process, generally accomplished through a Supervisory Control and Data Acquisition (SCADA). The intensifier or a piston pump is used to apply the pressure on the PTM, which is then uniformly transmitted throughout the media, in accordance with Pascal's law. The chiller system is used optionally to maintain the temperature of the system below ambient conditions. It prevents an unnecessary increase in the temperature during the adiabatic pressurization process.

HHP processing usually does not affect the covalent bonds, whereas hydrophobic and noncovalent bonds are affected. It results in permeabilization of cell membranes and disintegration of tissue structure resulting in the death of the microorganisms (Toepfl et al., 2006). The pressure also results in reduced enzymatic activities, ensuring high-quality and shelf-stable products. Due to the high pressure, starches are swollen, resulting in structural changes. The high-pressure results in structural changes in lipids and proteins; for example, in lipids, high pressure results in the transition of cis-arrangement to trans-arrangement, and in the case of proteins, it results in the unfolding of secondary, tertiary, and quaternary structures but does not affect the primary structure (Figure 2.4).

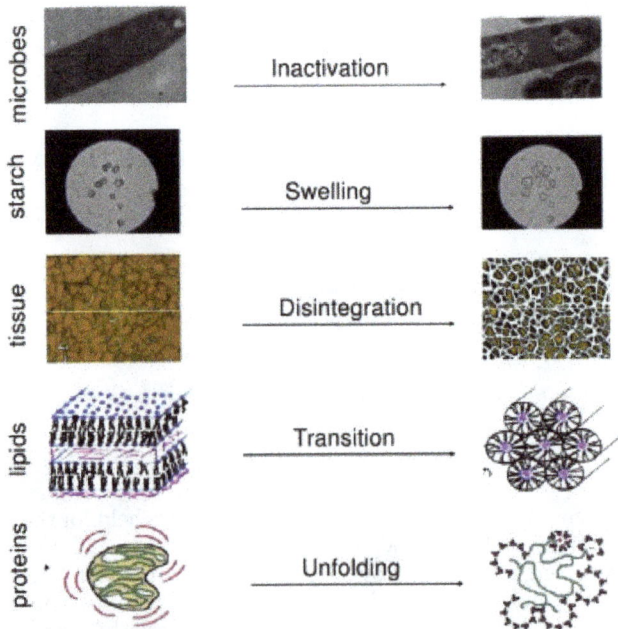

Figure 2.4 Effect of HHP on different components of food.

High-pressure pasteurization systems with a maximum pressure up to 600 MPa are employed for the treatment of high-value foods or heat-sensitive products, which may not be possible to preserve by conventional techniques. The strategic advantage of HHP is its applicability to packed foods, making efforts to prevent recontamination or providing an alternative to aseptic processing and a reduction of energy costs.

2.3 IMPACT OF HHP PROCESSING

2.3.1 Impact on Microbial Cells

HHP strongly influences the cellular structure and major functions of the microbial cell. The disastrous effect of HHP on the microorganisms is due to changes in the morphology of the cell membrane and its permeability, alteration in biochemical processes, and interference in genetic mechanisms (Yordanov & Angelova, 2010). In microorganisms, the cell membrane is generally recognized as the primary site targeted by pressurization, which is made of a fluidic structure consisting of phospholipids bilayer and proteins (Sehrawat et al., 2020). The application of HHP leads to the disruption of the microbial cell membrane, resulting in increased permeability. The greater leakages of constituents from the cell are signs of a superior degree of injury. As mentioned earlier, the membrane exhibits a fluidic nature; if the pressure is not sufficient to stimulate complete permeability (mostly in gram-negative bacteria), on the removal of pressure, the membrane recovery takes place (Yaldagard et al., 2008). It has been reported that physical damage to cell membrane occurs by pressure treatment in terms of leakage of UV-absorbing material from bacterial cells or adenosine triphosphatase (ATP) (Smelt, 1998) or increased absorption of colorants such as propidium iodide (PI), inhibition of amino acid uptake, most likely due to denaturation of membrane proteins (Benito et al., 1999).

At a pressure of 600 MPa, the alterations in the cell structures of *Saccharomyces cerevisiae* 0–39 were observed and rendered them incapable of survival (Shimada et al., 1993). The changes were related to the emanation of amino acid pools, peptides, ultraviolet (UV)–absorbing materials, and metal ions. Ritz et al. (2002) found that the application of pressure at 400 MPa for 10 min disrupted the cell membrane integrity of *Listeria monocytogenes* as the presence of bud was observed on the cell's surface using a scanning electron microscope (SEM), but cellular morphology was not affected significantly. Abe (2007) reported that hydrostatic pressure of 200 MPa can disturb the cellular functionality of the microbial population by damaging the cell membrane and the internal cell architecture. Overall, the active transportation and passive permeability of the cell membrane of microorganisms are highly impaired on applying pressure. Klotz et al. (2010) reported that increase in uptake of PI by *Escherichia coli* 8164 at pressure treatments of 100–300 MPa. Similarly, after pressure treatment of 350 MPa for 5 min, penetration of PI by *S. enterica* cell was observed, resulting in physical damage to the outer membrane (Wang et al., 2013). The fluorescent microscopic analysis clearly indicated the damage the membrane integrity of *Staphylococcus aureus* and *E. coli* O157:H7 under an HHP treatment of 200–325 MPa/1–5 min/40°C.

Pressurization leads to cell membrane detachment from the cell wall, morphological changes such as enlargement of the cell length, reduction of the gas vacuoles' size, and condensation of nuclear material (Lado & Yousef, 2002). The cell wall, being more rigid in comparison to the cell membrane, is thus less affected by HHP. Although when a light microscope was used, no physical changes were found in the cell wall of prokaryotes and lower eukaryotes, but SEM indicated intracellular damage and changes in the cell surface (Patterson, 2005). Shimada et al. (1993) observed slight modifications in shape of *Saccharomyces cerevisiae* 0–39 using SEM after pressurization. Alterations in mitochondria and cytoplasm at 400 MPa for 10 min at ambient temperature were observed using transmission electron microscopy (TEM) images, whereas 200 MPa/−20°C leads to severe disruptions in the inner architecture, and the nuclear membrane was found to have disappeared. Kalchayanand et al. (2002) found disruption in cell membrane integrity of *Leuconostoc mesenteroides* when treated at 345 MPa for 5 min at 25°C. Yang et al. (2012) reported the inactivation of different foodborne pathogens in milk at 500 MPa for 30 min and 25°C due to significant changes in their morphological structures. The most visible changes were breakdown of peptide glycan layer, a double-track bilayer structure of cell membrane, and enlargement of electron transparent ranges in the cell cytoplasm. Pilavtepe-Çelik et al. (2013) observed that *E. coli* cells get larger in size at 200 MPa and that dimples and pinches emerge when pressure intensity increased to 250 MPa. At elevated pressure (300–400 MPa), *S. aureus* exhibited a rough and cracked cell surface with no substantial effect on the cell surface at a pressurization of 200–250 MPa. The changes in the morphology induced in microbial population under HHP are presented in Table 2.1.

Table 2.1 Changes in the Morphology of Microorganisms by HHP

Target microorganism	Processing conditions	Technique used	Effects on microorganism	Reference
L. monocytogenes	400 MPa/ 10 min/ 20°C/ (citrate buffer; pH 5.6)	Scanning electron microscopy, flow cytometry	Buds' scars on the cell surface; loss of membrane integrity; reduction in membrane potential	Ritz et al. (2002)
E. coli 0157: H7 *S. aureus* 485	325 MPa/1 min/40°C 400 MPa/5 min/40°C	Scanning electron microscopy	Protein denaturation; pressure-induced phase transition of lipid membrane bilayer	Pilavtepe-Çelik et al. (2008)
S. cerevisiae	600 MPa/7 min/21 °C	Scanning electron microscopy	Disruption of cell membrane; perforations on cell wall; scars on cell surface	Marx et al. (2011)
E. coli and *S. aureus*	500 MPa/30 min/25°C	Transmission electron microscopy	Damage in cell membrane; cell wall rupture and degradation of chromosome DNA	Yang et al. (2012)
Salmonella enterica	350 MPa/5 min/25°C	Scanning electron microscopy; transmission electron microscopy	Expansion of nucleoid regions, peptidoglycan layer breakdown, rupturing of cell wall	Wang et al. (2013)
Enterobacter aerogenes	207 MPa/7 min; 267 MPa/6 min	PI fluorescence	Cell wall and membrane rupture	Maldonado et al. (2016)
L. innocua CIP80.11T, *E. coli* ATCC 7839	400 MPa/5 min/20°C 500 MPa/5 min/20 °C	Transmission electron microscopy	Compression of interior regions and the expansion of nucleoid regions	Nasilowska et al. (2021)

In general, vegetative and pathogenic microorganisms are effectually inactivated by HHP treatments in the range of 200–600 MPa. In comparison, gram-positive bacteria are more baroresistant than gram-negative due to the occurrence of teichoic acid in the cell wall (Wang et al., 2010). Yeast and mold were found to be more susceptible to pressure treatments compared to bacteria (Palou et al., 2002). Shape is another important parameter that affects the inactivation rate. Rod-shaped bacteria (*E. coli*) are more baroresistant than slim-shaped bacteria; however, extreme resistance was demonstrated by cocci-shaped bacteria (*S. aureus*; Pilavtepe-Çelik et al., 2013). Spores have been reported to show the highest resistance to pressure treatments and require higher levels of pressure in combination with moderate temperatures for the inactivation. To the contrary, nonproteolytic type B spores (spore-forming pathogens) are highly unresponsive to HHP (Balasubramaniam & Farkas, 2008). Pressures in the range of 100–300 MPa trigger the germination process in spores and reduces the volume on compression, resulting in increased solvation of spore's component (Yaldagard et al., 2008). Pressure and heat act synergistically, and when high pressure is applied in combination with temperature, the inactivation of spores took place without the germination step (Heinz & Knorr, 2001). Microorganisms' cells are more resistant to HHP in the stationery and lag phases compared to the proliferating phase (Yordanov & Angelova, 2010).

2.3.2 Impact on Food Enzymes

Enzymes are known as a special class of protein with an active site, made by the three-dimensional conformation of molecules (Rastogi et al., 2007). In enzymatic activity, the changes induced by pressure depend on the class of enzyme, nature of substrates, and, most important, the processing conditions applied. A change in the active site or protein denaturation can lead to a loss of enzyme activity. Apart from conformational changes, enzyme activity can be influenced by way of decompartmentalization induced by pressure (Hendrickx et al., 1998). When the food material is intact, enzymes and substrate are separated by compartmentalization, which gets destroyed on the application of pressure. The application of pressure damages the membrane, resulting in enzyme leakage that leads to enzyme–substrate contact. Typically, at low pressures, enzyme activity is enhanced, and at high pressure, the same is inhibited (Yaldagard et al., 2008). Pressure induced inactivation, can be classified as (1) complete irreversible inactivation, (2) complete reversibly inactivation, (3) incomplete and irreversibly inactivation, and (4) incomplete and reversibly inactivated (Miyagawa et al., 1964). Some of the works reported on effect of HHP on enzyme activity as presented in Table 2.2.

2.3.3 Impact on Food Quality and Sensory Attributes

The vital quality parameters affecting the consumer acceptance of foods are color, flavor, and texture. As discussed earlier, the covalent bonds are the least affected by HHP, and thus HPP treatment resulted in better retention of the nutritional and sensory qualities of the foods. The effect of HHP on different quality parameters of food commodities are discussed in Table 2.3.

2.4 NOVEL APPLICATIONS OF HHP IN THE FOOD INDUSTRY

2.4.1 Pasteurization and Blanching

HHP in the range of 200–600 MPa under ambient or refrigerated temperatures is suitable for pasteurizing different foods, including fruit juices, vegetable

Table 2.2 Effects of High Pressure on Enzyme Activity of Different Foods

Sample	Enzyme	Processing conditions	Inactivation rate achieved	Best conditions	Reference
Tomato puree	PG	0–700 MPa/0–2 min/20–90°C /1–2 cycles	98%	700 MPa/30 min/80°C/2 cycles	Krebbers et al. (2003)
Strawberry	PPO	400–800 MPa/5–15 min/18–22°C	100 %	800 MPa/15 min	Garcia-Palazon et al. (2004)
Fruit smoothies	PPO	450–600 MPa/5–10 min/20–37°C	83%	600 MPa/10 min	Keenan et al. (2012)
Watermelon juice	PME	300–900 MPa/5–60 min/60°C	70%	900 MPa/40 min/60°C	Zhang et al. (2011)
Litchi juice	PPO and POD	300–600 MPa/1s/ 30°C	28.4% 17%	600 MPa/1s/30°C	Kaushik et al. (2016)
Peach juice	PPO	400–600 MPa/5–25 min/25 °C	79 %	600 MPa/25 min /25°C	Rao et al. (2013)
Jackfruit bulb	PE	300–600 MPa/3–15 min/room temperature	37%	600 MPa/15 min	Ng et al. (2019)

Abbreviations: PE = pectin esterase; PG = polygalacturonase; PPO = polyphenoloxidase; POD = peroxidase; PME = pectinmethylesterase.

Table 2.3 Effect of HHP on Different Quality Aspects of Food

Food category	Commo-dity	Treatment (MPa/min/°C)	Major findings	Reference
Seafood	Albacore tuna	275–310/ 2–6/10	Pressure reduced lipid oxidation, changed color of the muscle, and resulted in increased shelf life of >22 days at 4°C.	Ramírez-Suárez and Morrissey (2006)
	Oyster	100–800/10/20	HHP resulted in increased L^* value, while decreased a^* and b^* values. High-pressure-treated oysters had higher shucking yields compared to untreated oysters.	Cruz-Romero et al. (2007)
	Red mullet	220–330/ 5–10/3–25	TMA-N and TBA values higher in pressure-treated samples than control. ΔE^* values increased significantly with increase in treatment pressure.	Erkan et al. (2010)

Food category	Commodity	Treatment (MPa/min/°C)	Major findings	Reference
	Abalone	500–550/3–8/20	TVB-N and TMA-N levels remained within the safe limits. Hardness increased and whiteness decreased after HHP. Shelf-life of 60 days at all HHP conditions applied.	Briones-Labarca et al. (2012)
	Seabass	100–400/0–30/6	With increases in pressure level and holding time, fillets became whitish and water holding capacity decreased.	Teixeira et al. (2014)
	Hilsa	250–350/5/27	ΔE^* values and hardness increased with increase in pressure. TVB-N and TMA-N levels within the safe limits. Shelf life of 25 days at 350 MPa.	Chouhan et al. (2015)
	Black tiger shrimp	361/12/46	Significant reduction inmicrobial load, TMA-N and TBA content. Shelf life of 18 days at 4°C.	Kaur et al. (2018)
Fruits, vegetables products	Raspberry puree	200–800/15/20	Anthocyanins were stable for samples processed at 200 and 800 MPa and stored at 4°C; after 9 days, 55% and 71% losses in cyaniding-3-glucoside for sample treated at 800 MPa and stored at 20 and 30°C, respectively.	Suthanthangjai et al. (2005)
	Strawberry puree	400-600/15/10-30	Nonsignificant changes in the color; no effect on the ascorbic acid and antioxidant activity; 9% increase in phenolic content at 600 MPa.	Patras et al. (2009)
	Mandarine juice	150–450/10–0 s/15, 30, 45	Carotenoid content remain unaffected; color parameters a and b decreased significantly.	Carreño et al. (2011).
	Granny Smith apple puree	400, 600/5/20	No adverse effect of pressure on nutritional quality; HP processed samples had 7 days of shelf life, 50% loss in AA after 10 days of storage.	Landl et al. (2010)

(Continued)

Table 2.3 (Continued)

Food category	Commo-dity	Treatment (MPa/min/°C)	Major findings	Reference
	Kiwi fruit	500/3/25	Microbial load reduced by 4 logs for HHP sample; 40 and 50% decrease in sugars and organic acids, respectively in HHP samples; sensory profile was maintained for HHP sample up to 21 days.	Fernandez-Sestelo et al. (2013)
	Mango pulp	100–600/0–20/30	85%, 92%, and 90% of retention of ascorbic acid, total phenolics, and antioxidant capacity, respectively.	Kaushik et al. (2014)
Dairy products	Milk	200–250/0–30/20	Control of microbial flora of milk for cheese production.	Morgan et al. (2000)
	Milk	100–600/0–30/20	Reduction in rennet coagulation time.	Huppertz et al. (2002)
	Cheese	345–483/3–7/30	Accelerated shred ability of cheddar cheese with enhanced visual acceptability and tactile handling.	Serrano et al. (2005)
	Cheese	50/3 days/25	Accelerated ripening of commercial Cheddar cheese due to degradation of αS1-casein.	O'Reilly et al. (2000)
	Cheese	50–500/20–200/30	Accelerated ripening of gouda cheese.	Messens et al. (2000)

Abbreviations: TMA-N = trimethylamine nitrogen; TBA= thiobarbituric acid.

products, salads, seafood, and deli meats (Mujica-Paz et al., 2011, Norton & Sun, 2008). Eshtiaghi and Knorr (1993) employed HHP (400 MPa for 15 min) for blanching of potato cubes at ambient temperature and observed no thermal degradation. Kingsly et al. (2009) reported the use of an HPP treatment at pressures >300 MPa/25°C for blanching of peach slices suspended in citric acid (1–1.2%), and the result showed effective inactivation of the polyphenoloxidase (PPO) enzymes.

2.4.2 Osmotic Dehydration

HHP results in permeabilization of cell structure, thereby enhancing the rate of mass transfer. It has been successfully applied for osmotic dehydration of pineapple (Rastogi & Niranjan, 1998), pears (Park et al., 2002), cherry tomatoes (Azoubel & Murr, 2004), and ginger slices (Dash et al., 2019).

2.4.3 Freezing and Thawing

Pressure-assisted freezing is carried at a constant pressure by lowering the temperature below the freezing point (LeBail et al., 2002). This technique is useful in preserving the natural quality of frozen foods. Similarly, when thawing is done at constant pressure using HHP, a significant reduction in thawing time (Massaux et al., 1999) and drip loss (Murakami et al., 1992; Rouillé et al., 2002) was observed.

2.4.4 Gelation

High pressure facilitates the gelation of protein and polysaccharides. This phenomenon can be used for altering the functional properties of foods (Rastogi et al., 2007). Keim and Hinrichs (2004) reported the formation of stable gels from whey protein isolate on the application of high pressure (600 MPa/0–30 min/30°C). The pressure-induced gel strength was higher in soya protein, with a smoother and finer appearance compared to that of heat-induced gels (An Erl King et al., 1994).

2.4.5. Extraction

The application of HHP enhanced caffeine extraction from coffee using water (Knorr, 1999). At a fixed pressure, caffeine yield was found to increase with an increase in temperature. HPP was found to be an efficient technology for anthocyanins extraction from grape skins (Corrales et al., 2009), polyphenols from green tea leaves (Xi et al., 2009, 2010), bioactive compounds from Litchi fruit pericarp (Prasad et al., 2009b) and longan fruit pericarp (Prasad et al., 2009a), and so on.

2.5 RECENT DEVELOPMENTS IN HHP TECHNIQUES

Among the nonthermal technologies, HHP is one of the most explored with different food products such as juices and jams currently marketed in different countries, namely, Japan, the United States, and European countries. HHP can destroy microorganisms and inactivate enzymes without much alteration in the organoleptic and nutrition attributes. HHP in combination with other technologies such as ultrasonication, acidification, carbon dioxide, and antimicrobial peptides can reduce the intensity of required pressure and cost of operation. HHP has promising future applications with the advancement in technology, the formulation and stipulation of food-related laws and regulations, and the enhancement of manufacturing facilities. Table 2.4 represents the recent works on HHP, with a prime focus on seafood industry.

2.6 CONCLUSION

HHP processing is an emerging nonthermal, paradigm-shifting technology that has been successfully implemented in food industries, which look for further innovation and growth in export as key tools for the improvement of competitiveness and profitability in global markets. The application of HHP can inactivate pathogenic microorganisms and enzymes and modify structures with little or no effect on the nutritional and sensory qualities of foods. It is a powerful tool for new product development, principally for the safe commercialization of natural, organic, preservative-free, ready-to-eat products, for maintaining the freshness of fruit and vegetable products, and for quality improvement in some processes of the seafood industry. HHP processing technologies have limitations related to high investment costs, absolute control of the variables associated with the process operation, and a lack of regulatory approval have delayed a wider implementation of these

Table 2.4 Recent Research in Novel Applications of HHP in Food Industry

Seafood and seafood products	Sample and storage conditions	Operating conditions	Microorganisms inhibited	Oxidation	Physicochemical changes	References
Albacore (*Thunnus alalunga*)	Locally procured Albacore and formed to steaks	High hydrostatic pressure (HHP) at 200 MPa for 0-, 2-, 4- and 6-min using water as pressure-transmitting medium	-	Oxidation of proteins and lipids increased with treatment duration	With HHP treatment, hardness was reduced in cooked samples, and lightness was increased in raw samples	Cartagena et al. (2020)
Rainbow trout fillets; fresh European catfish fillets	Trout fillets (smoked) and filleted catfish (fresh) were stored for 41 days at 5°C.	HHP applied at 200, 400 or 600 MPa at room temperature for 1 or 5 min.	*E. coli* and *Listeria monocytogenes*	-	Pressure didn't show much change; texture was improved significantly by application of HHP.	Mengden et al. (2015)
Black tiger shrimp (*Penaeus monodon*)	Black tiger shrimp (fresh), shelled, de-headed and vacuum-packed	Treated for 300, 400, 500 and 600 MPa (for 3, 6, 9, 12 and 15 min, respectively for each pressure) at room temperature	*E. coli, Staphylococcus aureus,* psychrotrophs, and aerobic mesophiles	Lipid oxidation was influenced by high pressure as well as holding time.	Lightness increased and redness decreased because of pressure treatment, loss of pigments and coagulation of proteins (globin/haem). Hardness increased due to aggregation and formation of gel.	Kaur et al. (2016)
European seabass (*Dicentrarchus labrax*)	Fillets stored for 67 days at 2°C	600 MPa/5 min at 25°C	*Pseudomonas* spp., total aerobic viable count, *Brochothrix thermosphacta,* molds and yeasts, *Enterobacteriaceae* spp., *Lactobacilli* H_2S-producing bacteria	HHP caused fractionation of sarcoplasmic proteins as confirmed with electrophoresis.	Increase in opaqueness of fillets; increase in blueness and decrease in redness values. The quality parameters were found stable during storage. Hardness and cohesiveness were influenced by HHP. Flesh discoloration, sour flavor and fishy order development were reported.	Tsironi et al. (2019)

Food	Sample	Pressure treatment	Microbial parameter	Effect on lipids/proteins	Effect on quality	Reference
Raw and sous vide cooked lobster	Fresh and shuckled raw lobster tail were used. Stored at 4°C for 28 days.	150 or 350 MPa for 10 min.	Lactic acid bacteria (LAB), total bacterial count (TBC)	–	Fresh aroma was preserved at 350 MPa for 28 days. Shear force (N) increase was reported in raw and cooked lobster at 350 MPa processing. It also increased L* and a* values.	Humaid et al. (2020)
Pacific oysters (Crassostrea gigas)	Fresh oysters (stored for 14 days at 4°C).	300 MPa for 2 min.	Aerobic plate count (APC)	Lipids oxidation was accelerated due to application of HHP.	Odor profiles of oysters were influenced due to HHP.	Rong et al. (2018)
Surimi gels of Tilapia (Orechrmis niloticus)	Surimi gels prepared from fresh tilapia.	100, 200, 300 and 400 MPa for 15 min.	–	Retention of heavy chain protein bands of actin, tropomyosin and myosin by application of HHP.	Increase in values of brightness (L*), lightness, blueness (b*) and whiteness; Increase in gel strength and hardness due to increase in pressure.	Lu et al. (2021)
Red swamp crayfish (Procambarus clarkia)	Fresh red swamp crayfish were used	100–500 MPa for 5 min.	-	Change in protein structure (like α-helix, β-turn, and β-sheets) in crayfish; disturbance of hydrogen bonding; protein oxidation due to increased pressure, increasing the carbonyl content.	Protein denaturation and aggregation resulted in increased shear force, when high-pressure treatment was given.	Shi et al. (2020)
Razor clam (Sinonovacula constricta)	Fresh razor clams.	200, 300, and 400 MPa for 1, 3, 5, and 10 min	Total viable count	Higher lipid oxidation with increased pressure treatment and holding time were reported; due to oxidation of PUFA, HHP may promote hydroperoxides formation; it also leads to oxidation of proteins, resulting in insolubility and gel formation.	Narrowing of gaps and aggregation of muscle fibers was observed at 400 MPa for 10 min treatment due to pressure-induced compaction, denaturation, aggregation, and gelation.	Xuan et al. (2018)
Salmon (Salmo salar) and cod (Gadus mrhua)	Steaks from frozen cod and salmon; stored for 14 days at 4°C.	150, 300, and 450 MPa, each for 5 min at 20°C.	Total aerobic plate count (TAPC)	HHP resulted in denaturation of actin, myosin, and sarcoplasmic proteins; carbonyl content increased at higher pressure and longer storage period; lipid oxidation was also accelerated.	Pressure increased hardness, lightness, and decreased a* value.	Arnaud et al. (2018)

technologies on commercial scale. Even the techniques described are not widely applied at an industrial scale at present, and an evaluation of the costs of investment and operation are based on results from lab-scale systems and assumptions.

REFERENCES

Abe, F. (2007). Exploration of the effects of high hydrostatic pressure on microbial growth, physiology and survival: Perspectives from piezophysiology. *Bioscience, Biotechnology, and Biochemistry, 71*(10), 2347–2357.

An Erl King, V., Yueh, P., Chang, M., & Kwo, S. J. (1994). Studies on the gelation of soy protein induced by high pressure. *Journal of the Chinese Agricultural Chemical Society, 32*(3), 309–321.

Arnaud, C., de Lamballerie, M., & Pottier, L. (2018). Effect of high pressure processing on the preservation of frozen and re-thawed sliced cod (Gadus morhua) and salmon (Salmo salar) fillets. *High Pressure Research, 38*(1), 62–79.

Azoubel, P. M., & Murr, F. E. X. (2004). Mass transfer kinetics of osmotic dehydration of cherry tomato. *Journal of Food Engineering, 61*(3), 291–295.

Balasubramaniam, V. M., & Farkas, D. (2008). High-pressure food processing. *Food Science and Technology International, 14*(5), 413–418.

Benito, A., Ventoura, G., Casadei, M., Robinson, T., & Mackey, B. (1999). Variation in resistance of natural isolates of *Escherichia coli* O157 to high hydrostatic pressure, mild heat, and other stresses. *Applied and Environment Microbiology, 65*(4), 1564–1569.

Briones-Labarca, V., Won, P. M., Zamarca, M., Radic, A. J. M., & Munizaga, T. G. (2012). Effects of high hydrostatic pressure on microstructure, texture, color and biochemical changes of red abalone (*Haliotisrufecens*) during cold storage time. *Innovative Food Science and Emerging Technologies, 13*, 42–50.

Carreño, J. M., Gurrea, M. C., Sampedro, F., & Carbonell, J. V. (2011). Effect of high hydrostatic pressure and high-pressure homogenisation on *Lactobacillus plantarum* inactivation kinetics and quality parameters of mandarin juice. *European Food Research Technology, 232*, 265–274.

Cartagena, L., Puértolas, E., & de Marañón, I. M. (2020). Evolution of quality parameters of high pressure processing (HPP) pretreated albacore (*Thunnus alalunga*) during long-term frozen storage. *Innovative Food Science & Emerging Technologies, 62*, 102–334.

Chouhan, A., Kaur, B. P., & Rao, P. S. (2015). Effect of high pressure processing and thermal treatment on quality of hilsa (*Tenualosailisha*) fillets during refrigerated storage. *Innovative Food Science and Emerging Technologies, 29*, 151–160.

Corrales, M., Avelina, F. G., Peter, B., & Bernhard, T. (2009). Extraction of anthocyanins from grape skins assisted by high hydrostatic pressure. *Journal of Food Engineering, 90*(4), 415–421.

Cruz-Romero, M., Kelly, A. L., & Kerry, J. P. (2007). Effects of high-pressure and heat treatments on physical and biochemical characteristics of oysters (*Crassostrea gigas*). *Innovative Food Science and Emerging Technologies, 8*(1), 30–38.

Dash, K. K., Balasubramaniam, V. M., & Shreya Kamat, S. (2019). High pressure assisted osmotic dehydrated ginger slices. *Journal of Food Engineering, 247*, 19–29.

Erkan, N., Üretener, G., & Alpas, H. (2010). Effect of high pressure (HP) on the quality and shelf life of red mullet (*Mullussurmelutus*). *Innovative Food Science and Emerging Technologies, 11*(2), 259–264.

Eshtiaghi, M. N., & Knorr, D. (1993). Potato cube response to water belching and high hydrostatic pressure. *Journal of Food Science, 58,* 1371–1374.

Fernández-Sestelo, A., de Saá, R. S., Pérez-Lamela, C., Torrado-Agrasar, A., Rúa, M. L., & Pastrana-Castro, L. (2013). Overall quality properties in pressurized kiwi purée: Microbial, physicochemical, nutritive and sensory tests during refrigerated storage. *Innovative Food Science and Emerging Technologies, 20,* 64–72.

Garcia-Palazon, A., Suthanthangjai, W., Kajda, P., & Zabetakis, I. (2004). The effects of high hydrostatic pressure on β-glucosidase, peroxidase and polyphenoloxidase in red raspberry (*Rubusidaeus*) and strawberry (*Fragaria×ananassa*). *Food Chemistry, 88*(1), 7–10.

Heinz, V., & Knorr, D. (2001). Effects of high pressure on spores. In M. E. G. Hendrix & D. Knorr (Eds.), *Ultra High Pressure Treatments of Foods* (pp. 77–113). Kluwer Academic/Plenum Publishers.

Hendrickx, M. L., Van den, B. I., & Weemaes, C. (1998). Effects of high pressure on enzymes related to food quality. *Trends in Food Science and Technology, 9,* 197–203.

Humaid, S., Nayyar, D., Bolton, J., Perkins, B., & Skonberg, D. I. (2020). Refrigerated shelf-life evaluation of high pressure processed, raw and sous vide cooked lobster. *High Pressure Research, 40*(3), 444–463.

Huppertz, T., Kelly, A. L., & Fox, P. F. (2002). Effects of high pressure on constituents and roperties of milk. *International Dairy Journal, 12*(7), 561–572.

Kalchayanand, N., Frethem, C., Dunne, P., Sikes, A., & Ray, B. (2002). Hydrostatic pressure and bacteriocin-triggered cell wall lysis of *Leuconostocmesenteroides*. *Innovative Food Science and Emerging Technologies, 3*(1), 33–40.

Kaur, B. P., & Rao, P. S. (2018). Effect of storage temperature and packaging on quality and shelf life of high pressure processed black tiger shrimp (*Penaeusmonodon*). *Journal of Food Processing and Preservation, 42*(1), e133–66.

Kaur, B. P., Rao, P. S., & Nema, P. K. (2016). Effect of hydrostatic pressure and holding time on physicochemical quality and microbial inactivation kinetics of black tiger shrimp (*Penaeus monodon*). *Innovative Food Science & Emerging Technologies, 33,* 47–55.

Kaushik, N., Kaur, B. P., & Rao, P. S. (2016). Inactivation of polyphenol oxidase and peroxidase enzymes during pulsed, static and cyclic pressurization of litchi (*Litchi chinensis*) juice. *Food and Bioproducts Processing, 100,* 412–423.

Kaushik, N., Kaur, B. P., Rao, P. S., & Mishra, H. N. (2014). Effect of high pressure processing on color, biochemical and microbiological characteristics of mango pulp (*Mangifera indica* cv. Amrapali). *Innovative Food Science and Emerging Technologies, 22,* 40–50.

Keenan, D. F., R¨oßle, C., Gormley, R., Butler, F., & Brunton, N. P. (2012). Effect of high hydrostatic pressure and thermal processing on the nutritional quality and enzyme activity of fruit smoothies. *LWT-Food Science and Technology, 45*(1), 50–57.

Keim, S., & Hinrichs, J. (2004). Influence of stabilizing bonds on the texture properties of high-pressure-induced whey protein gels, *International DairyJournal, 14*(4), 3rd NIZO Dairy Conference on Dynamics of Texture, Process and Perception, 355–363.

Kingsly, A. R. P., Balasubramaniam, V. M., & Rastogi, N. K. (2009). Influence of high-pressure blanching on polyphenoloxidase activity of peach fruits and its drying behavior. *International Journal of Food Properties, 3,* 671–680.

Klotz, B., Mañas, P., & Mackey, B. M. (2010). The relationship between membrane damage, release of protein and loss of viability in *Escherichia coli* exposed to high hydrostatic pressure. *International Journal of Food Microbiology, 137*(2–3), 214–220.

Knorr, D. (1999). Process assessment of high pressure processing of foods: An overview. In *Processing Foods: Quality Optimisation and Process Assessment* (pp. 249–267). CRC Press.

Krebbers, B., Matser, A. M., Hoogerwerf, S. W., Moezelaar, R., Tomassen, M. M., & van den Berg, R. W. (2003). Combined high-pressure and thermal treatments for processing of tomato puree: Evaluation of microbial inactivation and quality parameters. *Innovative Food Science and Emerging Technologies,* 4(4), 377–385.

Lado, B. H, & Yousef, A. E. (2002). Alternative food-preservation technologies: Efficacy and mechanisms. *Microbes and Infection,* 4(4), 433–440.

Landl, A., Abadias, M., Sárraga, C., Viñas, I., & Picouet, P. A. (2010). Effect of high pressure processing on the quality of acidified Granny Smith apple purée product. *Innovative Food Science and Emerging Technologies,* 11(4), 557–564.

LeBail, A., Chevalier, D., Mussa, D. M., & Ghoul, M. (2002). High pressure freezing and thawing of foods: A review. *International Journal of Refrigeration,* 25, 504–13.

Lu, W., Qin, Y., & Ruan, Z. (2021). Effects of high hydrostatic pressure on color, texture, microstructure, and proteins of the tilapia (*Orechromis niloticus*) surimi gels. *Journal of Texture Studies,* 52(2), 177–186.

Maldonado, J. A., Schaffner, D. W., Cuitiño, A. M., & Karwe, M. V. (2016). In situ studies of microbial inactivation during high pressure processing. *High Pressure Research,* 36(1), 79–89.

Marx, G., Moody, A., & Bermúdez-Aguirre, D. (2011). A comparative study on the structure of *Saccharomyces cerevisiae* under nonthermal technologies: High hydrostatic pressure, pulsed electric fields and thermo-sonication. *International Journal of Food Microbiology,* 151(3), 327–337.

Massaux, C., Be´ra, F., Steyer, B., Sindic, M., & Derqanne, C. (1999). High hydrostatic pressure effects on freezing and thawing processes of pork meat. In H. Ludwig (Ed.), *Advances in High Pressure Bioscience and Biotechnology,* Springer Verlag.

Mattson, B., & Sonesson, U. (2003). Introduction. In B. Mattson & U. Sonesson (Eds.), *Environmentally-Friendly Food Processing* (pp. 1–2). Boca Raton.

Mengden, R., Röhner, A., Sudhaus, N., & Klein, G. (2015). High-pressure processing of mild smoked rainbow trout fillets (*Oncorhynchus mykiss*) and fresh European catfish fillets (*Silurus glanis*). *Innovative Food Science & Emerging Technologies,* 32, 9–15.

Messens, W., Van de Walle, D., Arevalo, J., Dewettinck, K., & Huyghebaert, A. (2000). Rheological properties of high-pressure-treated Gouda cheese. *International Dairy Journal,* 10(5–6), 359–367.

Miyagawa, K., Sannoe, K., & Suzuki, K. (1964). Studies on taka amylase a under high pressure treatment; part II: Recovery of enzymic activity of pressure inactivated taka-amylase a and its enhancement by retreatment at moderate pressure. *Archives of Biochemistry and Biophysics,* 106, 467–474.

Morgan, S. M., Ross, R. P., Beresford, T., & Hill, C. (2000). Combination of hydrostatic pressure and lacticin 3147 causes increased killing of *Staphylococcus* and *Listeria. Journal of Applied Microbiology,* 88(3), 414–420.

Mujica-Paz, H., Valdez-Fragoso, A., Samson, C. T., Welti-Chanes, J., & Torres, J. A. (2011). High-pressure processing technologies for the pasteurization and sterilization of foods. *Food and Bioprocess Technology,* 4(6), 969–85.

Murakami, T., Kimura, I., Yamagishi, T., & Fujimoto, M. (1992). Thawing of frozen fish by hydrostatic pressure. In C. Balny, R. Hayashi, K. Heremans, & P. Masson (Eds.), *High Pressure and Biotech. Vol. 224* (pp. 329–331). John Libbey Eurotext.

Nasiłowska, J., Kocot, A., Paulina Natalia Osuchowska, P. N., & Sokołowsk, B. (2021). High-pressure-induced sublethal injuries of food pathogens – Microscopic assessment. *Foods, 10*(12), 2940.

Ng, S. K., Tan, T. B., Tan, P. F., Chong, G. H., & Tan. C. P. (2019). Effect of high pressure processing on the microbiological, physicochemical and enzymatic properties of jackfruit (*Artocarpus heterophyllus* L.) bulb. *Food Research, 3*(3), 213–220.

Norton, T., & Sun, D. W. (2008). Recent advances in the use of high pressure as an effective processing technique in the food industry. *Food and Bioprocess Technology, 1*, 2–34.

O'Reilly, C. E., O'Connor, P. M., Kelly, A. L., Beresford, T. P., & Murphy, P. M. (2000). Use of hydrostatic pressure for inactivation of microbial contaminants in cheese. *Applied and Environmental Microbiology, 66*(11), 4890–4896.

Palou, E., Lopez-Malo, J., & Welti-Chanes, J. (2002). Innovative fruit preservation methods using high pressure. In J. Welti-Chanes, G. V. Barbosa-Cánovas, & J. M. Aguilera (Eds.), *Engineering and Food for the 21st Century* (pp. 715–726). CRC Press.

Park, K. J., Bin, A., Brod, F. P. R., & Park, T. H. K. B. (2002). Osmotic dehydration kinetics of pear D'anjou (*Pyruscommunis* L.). *Journal of Food Engineering, 52*(3), 293–298.

Patras, A., Brunton, N. P., Da Pieve, S., & Butler, F. (2009). Impact of high pressure processing on total antioxidant activity, phenolic, ascorbic acid, anthocyanin content and colour of strawberry and blackberry purées. *Innovative Food Science and Emerging Technologies, 10*(3), 308–313.

Patterson, M. F. (2005). Microbiology of pressure-treated foods. *Journal of Applied Microbiology, 98*, 1400–1409.

Pilavtepe-Çelik, M., Balaban, M. O., Alpas, H., & Yousef, A. E. (2008). Image analysis based quantification of bacterial volume change with high hydrostatic pressure. *Journal of Food Science, 73*, 423–429.

Pilavtepe-Çelik, M., Yousef, A., & Alpas, H. (2013). Physiological changes of *Escherichia coli* O157:H7 and *Staphylococcus aureus* following exposure to high hydrostatic pressure. *Journal fürVerbraucherschutz und Lebensmittelsicherheit, 8*, 175–183.

Prasad, N. K., Yang, B., Zhao, M., Wei, X., Jiang, Y., & Chen, F. (2009a). High pressure extraction of corilagin from longan (*Dimocarpuslongan* Lour.) fruit pericarp. *Separation and Purification Technology, 70*, 41–45.

Prasad, N. K., Yang, B., Zhao, M. B., Wang, S., Chen, F., & Jiang, Y. (2009b). Effects of high-pressure treatment on the extraction yield, phenolic content and antioxidant activity of litchi (*Litchi chinensis* Sonn.) fruit pericarp. *International Journal of Food Science and Technology, 44*, 960–966.

Ramírez-Suárez, J. C., and Morrissey, M. T. (2006). Effect of high pressure processing (HPP) on shelf life of albacore tuna (*Thunnus alalunga*) minced muscle. *Innovative Food Science and Emerging Technologies, 7*(1–2), 19–27.

Rao, L., Guo, X., Pang, X., Tan, X., Liao, X., & Wu, J. (2013). Enzyme activity and nutritional quality of peach (*Prunus persica*) juice: Effect of high hydrostatic pressure. *International Journal of Food Properties, 17*(6), 1406–1417.

Rastogi, N. K., & Niranjan, K. (1998). Enhanced mass transfer during osmotic dehydration of high pressure treated pineapple. *Journal of Food Science, 63* (3), 508–511.

Rastogi, N. K., Raghavarao, K. S. M. S., Balasubramaniam, V. M., Niranjan, K., & Knorr, D. (2007). Opportunities and challenges in high pressure processing of foods. *Critical Reviews in Food Science and Nutrition, 47*(1), 69–112.

Ritz, M., Tholozan, J. L., Federighi, M., & Pilet, M. F. (2002). Physiological damages of *Listeria monocytogenes* treated by high hydrostatic pressure. *International Journal of Food Microbiology, 79*(1–2), 47–53.

Rong, C., Ling, Z., Huihui, S., & Qi, L. (2018). Characterization of microbial community in high-pressure treated oysters by high-throughput sequencing technology. *Innovative Food Science & Emerging Technologies, 45,* 241–248.

Rouillé, J., Le Bail, A., Ramaswamy, H. S., & Leclerc, L. (2002). High pressure thawing of fish and shellfish. *Journal of Food Engineering, 53,* 83–88.

Sehrawat, R., Kaur, B. P., Nema, P. K., Tewari, S., & Kumar, L. (2020). Microbial inactivation by high-pressure processing: Principle, mechanism and factors responsible. *Food Science and Biotechnology, 30,* 1–17.

Serrano, J., Velazquez, G., Lopetcharat, K., Ramirez, J. A., & Torres, J. A. (2005). Moderately high hydrostatic pressure processing to reduce production costs of shredded cheese: Microstructure, texture, and sensory properties of shredded milled curd cheddar. *Journal of Food Science, 70*(4), S286–S293.

Shi, L., Xiong, G., Yin, T., Ding, A., Li, X., Wu, W., & Wang, L. (2020). Effects of ultra-high pressure treatment on the protein denaturation and water properties of red swamp crayfish (*Procambarus clarkia*). *LWT, 133,* 110–124.

Shimada, S., Andou, M., Naito, N., Yamada, N., Osumi, M., & Hayashi, R. (1993). Effects of hydrostatic pressure on the ultrastructure and leakage of internal substances in the yeast *Saccharomyces cerevisiae. Applied Microbiology and Biotechnology, 40,* 123–131.

Smelt, J. P. P. M. (1998). Recent advances in the microbiology of high pressure processing. *Trends in Food Science and Technology, 9*(4), 152–158.

Suthanthangjai, W., Kajda, P., & Zabetakis, I. (2005). The effect of high hydrostatic pressure on the anthocyanins of raspberry (*Rubusidaeus*). *Food Chemistry, 90*(1), 193–197.

Teixeira, B., Fidalgo, L., Mendes, R., Costa, G., Cordeiro, C., Marques, A., Saraiva, J. A., & Nunes, M. L. (2014). Changes of enzymes activity and protein profiles caused by high-pressure processing in seabass (*Dicentrarchuslabrax*) fillets. *Journal of Agricultural and Food Chemistry, 61*(11), 2851–2860.

Toepfl, S., Mathys, A., Heinz, V., & Knorr, D. (2006). Review: Potential of high hydrostatic pressure and pulsed electric fields for energy efficient and environmentally friendly food processing. *Food Reviews International, 22*(4), 405–423.

Tsironi, T., Anjos, L., Pinto, P. I., Dimopoulos, G., Santos, S., Santa, C., & Power, D. (2019). High pressure processing of European sea bass (*Dicentrarchus labrax*) fillets and tools for flesh quality and shelf life monitoring. *Journal of Food Engineering, 262,* 83–91.

Wang, C., Chiao-Ping, H., Hsiao-Wen, H., & Yang, B. B. (2013). The relationship between inactivation and morphological damage of *Salmonella enterica* treated by high hydrostatic pressure. *Food Research International, 54,* 1482–1487.

Wang, L., Pan, J., Xie, H., Yang, Y., & Lin, C. (2010). Inactivation of *Staphylococcus aureus* and *Escherichia coli* by the synergistic action of high hydrostatic pressure and dissolved CO_2. *International Journal of Food Microbiology, 144*(1), 118–125.

Xi, J., Shen, D., Zhao, S., Lu, B., Li, Y., & Zhang, R. (2009). Characterization of polyphenols from green tea leaves using a high hydrostatic pressure extraction. *International Journal of Pharmaceutics, 382,* 139–143.

Xi, J., Zhao, S., Lu, B., Zhang, R., Li, Y., Shen, D., & Zhou, G. (2010). Separation of major catechins from green tea by ultrahigh pressure extraction. *International Journal of Pharmaceutics, 386,* 229–231.

Xuan, X. T., Cui, Y., Lin, X. D., Yu, J. F., Liao, X. J., Ling, J. G., & Shang, H. T. (2018). Impact of high hydrostatic pressure on the shelling efficacy, physicochemical properties, and microstructure of fresh razor clam (*Sinonovacula constricta*). *Journal of Food Science, 83*(2), 284–293.

Yaldagard, M., Mortazavi, S. A., & Tabatabaie, F. (2008). The principles of ultra high pressure technology and its application in food processing/preservation: A review of microbiological and quality aspects. *African Journal of Biotechnology, 7*(16), 2739–2767.

Yang, B., Shi, Y., Xia, X., Xi, M., Wang, X., Ji, B., & Meng, J. (2012). Inactivation of food borne pathogens in raw milk using high hydrostatic pressure. *Food Control, 28*(2), 273–278.

Yordanov, D. G., & Angelova, G. V. (2010). High pressure processing for foods preserving. *Biotechnology & Biotechnological Equipment, 24*(3), 1940–1945.

Zhang, C., Trierweiler, B., Li, W., Butz, P., Xu, Y., R¨ufer, C. E., Ma, Y., & Zhao, X. (2011). Comparison of thermal, ultraviolet-C and high-pressure treatments on quality parameters of watermelon juice. *Food Chemistry, 126*(1), 254–260.

3 Cold Plasma for Food Preservation

Uday S. Annapure and Rohit Thirumdas

Corresponding author:
Prof. Uday S. Annapure; Email id: us.annapure@ictmumbai.edu.in

CONTENTS

3.1 INTRODUCTION TO COLD PLASMA

In the context of emerging concern of foodborne pathogens, one of the greatest challenges faced by food processing sector is to provide safe and wholesome food to consumers. There is a significant rise in the foodborne illness outbreaks across the world that has resulted in the recall of several food products which has incurred a loss accounting for about US\$34 to 39 billion (Niemira 2012). Conventional cooking methods like heating, pasteurization, and canning that have been used for many years are accepted because of their safety. These processes require high temperatures and longer treatment times to inactivate microbes like thermophilic bacteria. However, these processing methods have some detrimental effects on the nutritional quality especially the heat-sensitive components like vitamins and sensory profiles. Moreover, many of the pathogens have acquired resistance to traditional food-processing methods. In recent years, the increased consumer demand for minimally processed foods has raised concerns about the food-processing operations employed for production and preservation. Also, concerns regarding the use of synthetic food additives like acids, antioxidants, and antibiotics for food preservation have increased. Nonthermal food-processing technologies such as cold plasma, which is a chemical-free process and has less impact on food quality and the environment, have come into existence. Nonthermal technologies are widely applied for minimally processed foods like fruits, vegetables, spices, raw meat products, and so on to preserve their physical, nutritional, and sensory qualities. Misra et al. (2017) stated that cold plasma and other nonthermal methods are designed to work at ambient temperatures to produce minimally processed foods. Some novel nonthermal technologies like high-pressure processing (HPP), ultrasound (US), pulsed electric fields (PEF), pulsed light (PL), irradiation, and nonthermal plasma have been used in food preservation.

After solid, liquid, and gaseous states, the fourth state of matter is plasma, which consists of charged and active species, electrons, ions, radicals, neutral and charged atoms, ultraviolet (UV) light, and ozone (Thirumdas et al. 2015). In 1928, Langmuir termed the word *plasma* to describe the oscillations that took place in the ionized gas. The plasma can be generated by using multiple devices

DOI: 10.1201/9781003147978-3

with different forms of energy sources at atmospheric, below-atmospheric, or partial-vacuum pressures. The excitation of gas molecules using electric discharges can produce plasma at low pressures (Zhu et al. 2020). However, the use of atmospheric air for the generation of plasma at atmospheric pressures is an important intervention for many food applications. Cold plasma has been applied in many food-processing areas like microbial and enzyme inactivation, toxins degradation, enhancing seed germination, surface functionalization of biopolymers, starch and protein modification, pesticides dissipation, and others (Thirumdas et al. 2015, 2016, 2020; Devi et al. 2017). Plasma applied for food processing depends on the mode of plasma generation, operating parameters, biological nature of food materials, surface topography, and chemical reactions induced due to the interaction of plasma species and surfaces. Active species such as oxygen and nitrogen reactive species and a few other species are capable to inactivate a wide range of microorganisms, including bacteria, yeasts, molds, and viruses (Feizollahi et al. 2021). Thirumdas et al. (2018) reported that the efficiency of microbial inactivation depends on operating parameters (i.e., power applied, treatment times, feed gases, and mode of plasma generation). Plasma active species breakdown or rupture the plasma membrane and cell wall, damaging the genetic materials, inducing the oxidative stresses by generating extra and intracellular ROS, overcoming the antioxidative mechanism of the cells, and formation of strong oxidizing agents like H_2O_2 both intra- and extracellular, and these are the primary antimicrobial mechanisms of cold plasma treatment.

3.2 METHODS OF PLASMA GENERATION

The plasma generation is governed by two well-known equations: (1) Maxwell's equation of electromagnetism and (2) Boltzmann equation of kinetic energy. The exploitation of electromagnetism and kinetic energy for plasma generation was discussed in detail by Pedrow et al. (2020) in their work. The plasma consists of positive ions, negatively charged electrons, and neutral species. However, the plasma exists as a neutral charge because of an equal number of positive and negative charge particles expect at the atomic level designating as quasi-neutral. The plasma is divided into cold/nonthermal plasma, and hot/thermal plasma with respect to overall temperature. In cold plasma, the electron temperatures (Te) are higher compared to the ion temperatures (Ti), resulting in lower temperatures (Ti >> Te) to the plasma. Since the overall temperature of the generated plasma is near or below ambient (60°C) temperature, it is designated as cold plasma (Mandal et al. 2018). This is due to the very low mass ratio of two colliding electrons ($m_2/m_1 \ll 1$), complete energy transfer does not take place between the electrons attributing to higher electron temperatures due to rapid self-heating compared to the collision of higher mass ions or neutral species, resulting in thermodynamic nonequilibrium (Turner 2016). Sharma and Singh (2020) reported that the collisions are categorized into two stages; that is, first-stage collisions are related to the electrons while second-stage collisions are heavy particle collisions. The former stage collisions resulted in ionization, dissociations, and excitation, and the reaction products of the first stage resulted in the secondary plasma collisions forming several other reactive species. Wong and Mongkolnavin (2016) reported that the collisions between the components of plasma resulted in important processes like

1. Scattering A+ e- $\rightarrow$ A + e-

2. Excitation A+ e- $\rightarrow$ A* + e

3. Ionization A + e $\rightarrow$ A+ + 2e

4. Recombination A + e $\rightarrow$ A + (hv)

The bombardment of electrons resulted in the release of secondary electrons and ions, which are capable to initiate several chemical reactions (Charoux et al. 2021). The reactive species formed from both stages were later involved in several chemical activities responsible for various applications. Plasma can be generated from different energy sources like heating, the application of electrical power and voltage, magnetic fields, radiofrequency (RF) waves, and magnetic radiations. The different forms of energies applied to a gas or gas mixture can ionize the gas generating the plasma. The different modes of generation are corona discharge, fluorescent tubes, RF plasma reactor, dielectric barrier discharges, and microwave discharges (Coutinho et al. 2021). Fridman and Friedman (2013) have postulated a few ionization mechanisms that occur during plasma generation: (1) direct ionization by electron impact, (2) step-wise ionization by electron impact, (3) heavy particles collisions, and (4) photo and surface ionization. To initiate the ionization reactions at an atmospheric pressure requires a very high-voltage electrical discharge. The well-known reactor to generate the plasma with the use of high-voltage electrodes is corona discharge. Lu et al. (2016) reported that the corona discharge can be generated by applying direct/alternating currents, and high voltages without any transition sparks. The other important plasma generator widely applied in food processing is the dielectric barrier discharge (DBD) plasma reactor, in which a dielectric layer is placed around the electrodes for safe operation without steamer-to-sparks transitions (Pedrow et al. 2020). The food materials which have dry surfaces can be processed using low-pressure plasma reactors like capacitively coupled, inductively coupled operating at low voltages. Niemira et al. (2018) reported that the formation of reactive hydroxyl radicals resulted from the breakdown of water molecules (i.e., high humidity generated due to the creation of a partial vacuum) attributed to the increase in cold plasma efficiency.

Plasma chemistry is very complex as the half-life of many reactive species formed is very short. The plasma-chemical mechanism varies from one plasma source to another, as the formation of chemical products like ions, radicals, and excited and active species will vary. Berardinelli et al. (2021) stated that O, OH, and O_3 are the important reactive oxygen species (ROS), whereas nitrogen oxide radicals and excited nitrogen molecules are important reactive nitrogen species (RNS) formed in the plasma that have shown a significant contribution to plasma efficiency. However, the half-life of ROS like hydroxyl radicals and singlet oxygen ($^1O_2$) has a shorter life but can create significant microbial destruction (Thirumdas et al. 2017). The singlet oxygen can live up to microseconds and H_2O_2, NO_2-, and NO species are long-lived and stay up to a few days (Wende et al. 2018). Several reports have shown that RNS are long-lived reactive species than ROS. Similarly, the reactive species formed from the direct treatment (direct contact of plasma with food material) are long-lived than the indirect treatment (plasma-activated liquids). The plasma chemistry of plasma-activated liquids is complex that resulted in the formation of different short-lived, long-lived, and secondary reactive species (Chen et al. 2020). Optical emission spectroscopy (OES) is the widely used analytical method for the quantification of the different ROS and RNS species formed in the plasma that are observed between 200–900 nm (Yadav et al. 2020). Presence of oxygen in feed gas results in the formation of higher concentrations of ROS like atomic oxygen and ozone possessing higher inactivation rates. The higher relative humidity percentage of the feed gas regulates the formation of more hydroxyl ions and peroxyl acid groups that resulted in higher microbial inactivation (Coutinho et al. 2021). Wende et al. (2018)

reported that reactive species like singlet oxygen radicals and one-electron oxidants like peroxynitrite generated in plasma attack the nucleobase pairs of genetic material. The viruses are more susceptible to active species compared to bacteria, molds, and yeasts (Coutinho et al. 2021).

3.3 TYPES OF COLD PLASMA SYSTEMS

3.3.1 Microwave-Pumped Cold Plasma

The use of microwaves with a frequency of more than 1 GHz can be used as a source of energy for plasma generation at low and atmospheric pressures. Electromagnetic radiations of a frequency varying between 300 MHz and 300 GHz can produce microwave discharge. The microwaves generated from the commercial magnetron can be used to generate plasma. Puligundla and Mok (2021) reported that plasma generated with microwaves at 2.45 GHz frequency attributes to higher electron densities of 1×10^{10} cm^{-3} to 1×10^{15} cm^{-3} (Wong and Mongkolnavin 2016). D'Isa et al. (2020) reported that microwave plasma can generate low-temperature electrons with high vibrational energies capable of dissociation reaction due to higher vibrational excitations. The maximum energy conversion efficiencies are high up to 90% with microwave discharges (D'Isa et al. 2020). However, the conversion efficiencies depend on the applied pressure, the higher or lower pressures have resulted in degradation. Belov et al. (2018) observed that with the increase in applied pressure, the energy conversion efficiency decreased in microwave plasma. The energy transfer is mainly influenced by the collision between the charged and uncharged species. The collision frequency increases with an increase in the gas pressures, resulting in higher energy transfer rates between the reactive species (Szabó and Schlabach 2014). The energy transfer equation for plasma generation reported by Szabó and Schlabach (2014) is

$$E \propto \frac{QZ}{m\left(f^2 + z^2\right)} \tag{3.1}$$

The energy transfer is proportional to its charge (Q), and inversely proportional to the mass of colliding particles and frequency. The addition of DC magnetic fields to the microwave plasma reactor could generate electron cyclotron resonance (ECR) plasma (Pedrow et al. 2020). The ECR plasma is a special type of microwave reactor operating at 2.45 GHz frequency coupled with the angular frequency of cyclotron motion of electrons for better performance of plasma (Mehdizadeh et al. 2015). The synchronization of ECR and microwave frequency results in enhanced electron acceleration by the electric fields. Because of synchronization, microwave oscillations match with the gyration period of free electrons increasing the kinetic energy attributing to the ionization of atoms (Mehdizadeh et al. 2015). The numerous free electrons formed in microwave plasma initiated dissociation and ionization of compounds, resulting in the formation of more chemically reactive species (Szabó and Schlabach 2014). Puligundla and Mok (2021) reported some of the advantages of ECR microwave plasma are low average electron temperature, a wide range of operating parameters, and a high degree of ionization. Important nitrogen reactive species formed in the microwave plasma, such as NO$^\bullet$, NO$_2$, NO$_2$, and NO$_3$, are mainly responsible for the microbial inactivation (Schnabel et al. 2014). Patil et al. (2016) reported that reactive species along with the UV radiations formed in microwave-driven plasma discharge have resulted in *E. coli* spore inactivation.

3.3.2 RF-Driven Cold Plasma

The gas plasma is generated by applying external radiofrequency fields with the use of electrodes or coils. Inductively coupled plasma (ICP) and capacitively coupled plasma (CCP) are widely used sources for RF- plasma generation. Puligundla and Mok (2021) reported that the generation of plasma by coupling RF fields with electrons can generate sustainable plasma by energy transferring. In ICP, the energy transfer takes place through the induction coils for the generation of the plasma torch, whereas, in CCP, the plasma is generated across the two parallel electrodes or metal plates by applying RF voltage. The generated electric field accelerates the heating of the electrons resulting in high-energy electrons, leading to ionization that forms an ion avalanche. The most common frequency for the generation of RF plasma is 13.56 MHz for industrial applications. Abdel-Fattah et al. (2011) reported that the applied frequency greatly influences the plasma active species–induced chemical reactions by modulating the physical and chemical properties of the RF plasma. Mansuroglu (2019) has stated that the exciting frequencies determine the chemical reactions taking place in the RF plasma by governing the rates of dissociation and ionization. RF plasma discharge can be generated at a controllable level due to its known ion and electron densities, making it convenient for many applications (Mansuroglu 2019). The high electron energies of the RF plasma are in the range of 10–20 eV, which is responsible for the generation of more positive ions than negative ions (Anderson 2001). Plasma diagnostics like electron densities and temperatures are important to understand the mechanism of plasma action during the processing (Mansuroglu 2019). The low-pressure RF cold plasma has an electron density up to 10^{15}–10^{18} m^{-3} and electron energy in the range of 1–4 eV (Shemakhin and Zheltukhin 2019). Chen (2007) reported that the high frequency produces plasma and low frequencies control ion distribution. The current determines the magnetic field in ICP discharge, and the electric field is directly proportional to applied frequency (Fridman and Friedman 2013). RF plasma's effectiveness depends on electrical parameters like resistance and inductance (Fridman and Friedman 2013). The generator and applicator are the two main components of the RF plasma system. The former part is used to generate the RF radiation, and latter is where the sample is being exposed. Swamy and Muthukumarappan (2021) reported that the RF plasma reactor should be carefully designed, as the loss of generated RF radiation can be absorbed by the human body, which could be hazardous to health.

3.4 IMPACT OF COLD PLASMA

3.4.1 Impact on Microbial Cells

The microbial decontamination of several fresh produce is achieved using different plasma generation systems, and the generated ROS and RNS play an important role. However, the microbial inactivation efficiency mainly depends on the type of plasma sources, operating parameters like voltage applied, power and time, distance from the produce, temperature at the product surface, and type of reactive species generated. Among the several studies conducted to decontaminate the meat and meat products, few have employed cold plasma directly on the meat surface and others have employed in meat packages. The effects of cold plasma treatment on microbial decontamination of various food products are summarized in Table 3.1.

Rød et al. (2012) observed a 50% decrease in the microbial load of *Listeria innocua* inoculated on ready-to-eat packed beef slices at 62-W power applied for 60 s. For the same treatment time, no significant difference in microbial load

Table 3.1 Effect of Cold Plasma Processing on Microbial Decontamination of Food Products

Food product	Plasma source	Processing parameters	Micro-organism	Microbial reduction ($\log_{10}$)	Reference
Lettuce	DBD cold atmospheric plasma	Feed gas: Air Exposure time: 5 min Voltage: 20 kV	*L. innocua*	2.4	Patange et al. (2019)
Strawberry	DBD cold plasma	Feed gas: Air Exposure time: 15 min Voltage: 60 kV	Bacteria	2.0	Rana et al. (2020)
Carrot	CAP	Feed gas: Air Exposure time: 15 min Voltage: 100 kV	Aerobic mesophiles	2.0	Mahnot et al. (2019)
Korean rice cake	DBD-plasma activated water	Feed gas: Air Exposure time: 20 min Voltage: 8 kV	*E. coli* *S. Typhimurium* *L. monocytogenes* *P. chrysogenum*	2.03 2.12 2.17 2.0	Han et al. (2020)
Fresh cut apple	Plasma-activated water	Feed gas: Air Exposure time: 8 min Voltage: 10 kV	Aerobic bacteria Molds Yeast Coliform bacteria	1.05 0.64 1.04 0.86	Liu et al. (2020)
Cherry tomato	DBD-Atmospheric cold plasma	Feed gas: Air Exposure time: 2 min Voltage: 70 kV	*Salmonella*	6.7	Ziuzina et al. (2014)
Black pepper	RF plasma jet	Feed gas: Air Exposure time: 15 min Power: 30 W	Mesophilic bacteria *B. subtilis* *B. atrophaeus*	0.7 0.8 1.3	Hertwig et al. (2015)
Cress seed	DBD plasma	Feed gas: Argon Exposure time: 5 min Voltage: 70 kV	*E. coli*	3.4	Butscher et al. (2016)
Almond	DBD plasma	Feed gas: Air Exposure time: 5 min Voltage: 20 kV	*Salmonella enteritidis*	5.0	Hertwig et al. (2017)
Cherry tomato	DBD plasma	Feed gas: Humid Air Exposure time: 6 min Voltage: 35 kV	*Salmonella*	0.8	Hertrich et al. (2017)
Egg	DBD plasma	Feed gas: Humid Air Exposure time: 6 min Voltage: 35 kV	*Salmonella enterica*	5.4	Georgescu et al. (2017)
Walnut	Cold plasma jet	Feed gas: Argon Exposure time: 11min Voltage: 15 kV	*Aspergillus flavus*	7.2	Amini and Ghoranneviss (2016)

(Continued)

Table 3.1 (Continued)

Food product	Plasma source	Processing parameters	Micro-organism	Microbial reduction ($\log_{10}$)	Reference
Chicken breast	DBD plasma	Feed gas: Air Exposure time: 5 min Voltage: 100 kV	Mesophilic bacteria	1.5	Moutiq et al. (2020)
Chicken breast	DBD-CAP	Feed gas: Air Exposure time: 5 min Voltage: 70 kV	Psychrophiles	0.5	Zhuang et al. (2019)
Duck egg	Arc plasma	Feed gas: Air Exposure time: 40 s Voltage: 12 kV	*Salmonella*	4.09	Gavahian et al. (2019)
Rice	DBD-CAP	Feed gas: Air Exposure time: 20 min Power: 250 Watt	Aerobic microbes	2.0	Lee et al. (2019)
Alfalfa seeds	DBD-CAP	Feed gas: Argon Exposure time: 10 min Voltage: 10 kV	*E. coli*	3.0	Butscher et al. (2016)
Cherry tomato	Corona cold plasma	Feed gas: Air Exposure time: 2 min Voltage: 8 kV	Molds	1.8	Lee et al. (2018)
Skimmed milk	Corona cold plasma	Feed gas: Air Exposure time: 20 min Voltage: 9 kV	*E. coli*	4.4	Gurol et al. (2012)
Radish sprout	Microwave driven plasma	Feed gas: Nitrogen Exposure time: 20 min Power: 900 watts	*Salmonella Typhimurium*	2.6	Oh et al. (2017)
Apple	Microwave driven plasma	Feed gas: Nitrogen Exposure time: 10 min Power: 1200 watt	*Escherichia coli*	4.6	Baier et al. (2015)
White grape juice	High voltage atmospheric plasma	Feed gas: Air Exposure time: 4 min Voltage: 80 kV	*Saccharomyces cerevisiae*	7.4	Pankaj et al. (2018)
Onion flakes	DBD-cold plasma	Feed gas: He Exposure time: 20 min Voltage: 9 kV	*S. Enteritidis*	3.1	Kim and Min (2018)
Orange juice	DBD-cold plasma	Feed gas: O_2+CO_2+N_2 Exposure time: 2 min Voltage: 90 kV	*Salmonella enterica*	5.0	Xu et al. (2017)
Red pepper powder	Microwave driven plasma	Feed gas: Helium Exposure time: 10 min Power: 900 watts	*Aspergillus flavus*	2.5	Kim et al. (2014)
Rape-seeds	Corona discharge plasma jet	Feed gas: Air Exposure time: 3 min Voltage: 20 kV	Aerobic microbes	2.2	Puligundla et al. (2017)
Rocket leaf salad	DBD-cold plasma	Feed gas: Air Exposure time: 10 min Voltage: 6 kV	*Pseduomonas* spp.	7.0	Giannoglou et al. (2020)

was observed at 15.5 W. At a higher power, the ozone disassociated due to the thermal degradation, and it might be participated in reactions with NO (Bauer 2017) and converted into NO_2, resulting in lower amounts of ozone with increasing power levels. Bauer (2017) has also stated that ozone is known for its antimicrobial activity for a short duration. The plasma generated with a feed gas containing both the nitrogen and oxygen gases were observed to have a greater decrease in the microbial load when compared to the use of nitrogen gas alone (Lee et al. 2011). Similarly, the use of oxygen gas alone resulted in the highest percentage decrease in microbial load when compared to varying O_2 gas composition on the beef sample (Gök et al. 2019). There is a decreasing trend in microbial inactivation efficiency with a decrease in oxygen composition in the feed gas. The research evidence of Lee et al. (2011) shows that ROS is more effective in microbial destruction than other reactive species, or else the synergistic effect of ROS and RNS attributes to higher efficiency rates. However, Lis et al. (2018) reported higher microbial efficiency of RNS and NO than ozone and ROS. Bauer (2017) observed a decrease in more than 2 logs of *Listeria monocytogenes*, and *Escherichia coli* when a beef loin was placed in polyamide-polythene package and exposed to atmospheric air plasma for 60 s. Ulbin-Figlewicz et al. (2015) reported a reduction of 2.09 and 0.56 logs in the total microbial count after exposure to helium and argon plasma, respectively, for a period of 10 min. The authors observed that gram-negative bacteria are more susceptible to plasma when compared to gram-positive bacteria due to the cell wall lipopolysaccharides depyrogenation. Similarly, Lee et al. (2017) reported that the reactive species of plasma attack the peptidoglycan of the cell wall of gram-negative bacteria cleaving the bonds of carbon chains, whereas in gram-positive bacteria, destruction is achieved by oxidative damage to DNA and other intracellular components (Han et al. 2016).

Rossow et al. (2018) significantly inactivated *Campylobacter jejuni* using air and argon cold plasma on chicken breast and skin. The authors have reported microbial inactivation efficiency with argon plasma is higher than the air plasma and depends on surface topography. Similarly, Dirks et al. (2012) have also observed higher microbial inactivation efficiency on chicken breast compared to chicken thigh skin. Both the previously mentioned research groups have drawn the same conclusion on the surface topography, which attributed to lower decontamination efficiencies.

Misra and Cheorun (2017) have reported that the microbes have the ability to hide in non-accessible places in the surface microstructures escaping from plasma reactive species. However, the use of plasma-activated liquids is beneficial to some extent, but the plasma chemistry of activated liquids differs from the gaseous plasma (Thirumdas et al. 2018).

In-package DBD air plasma treatment on beef loin decreased the *Listeria monocytogenes*, *Escherichia coli* and *Salmonella Typhimurium* by 1.90, 2.57, and 2.58 log CFU/g, respectively, after 10 min of treatment (Jayasena et al. 2015). The ozone, hydroxyl radicals, and NO are the important species attributed to bacterial destruction. In another study, conducted on in-package chicken breast (polypropylene-based film) exposed to DBD atmospheric plasma observed more than 90% inactivation of *Salmonella* and *Campylobacter* (Zhuang et al. 2019). In-package plasma treatment of chicken fillets was investigated by Wang et al. (2016); the bacterial count was observed less than 3 logs even after 7 days of storage. The authors further reported that the plasma-treated samples stored under modified atmospheric conditions showed 4.5 log CFU/g lesser growth than the samples stored in modified atmosphere control conditions. The synergistic effect of plasma treatment and modified atmosphere packaging improves the storage life of chicken fillets. A similar 3-log reduction in microflora growth on in-package fresh produce after DBD cold plasma treatment was reported by Misra et al. (2014). Lee et al. (2016) applied DBD plasma applied on chicken breast placed inside in a package and observed a significant decrease in *L. monocytogenes*, *E. coli*, and *S. Typhimurium* by 2.14, 2.73, and 2.71 log CFU/g, respectively, within a 10-min exposure. Yong et al. (2017) observed DBD cold plasma application to the surface of beef jerkies reduced the *Aspergillus flavus* count to 2.46 from 5.24 log CFU/g after 10 min of treatment, and they observed a greater D value for *Aspergillus flavus* compared to bacterial strains. Apart from decontamination of bacteria and molds, atmospheric cold plasma was employed to inactivate norovirus and hepatitis A virus on the surface of beef, pork, and chicken breast (Bae et al. 2015). The authors reported more than 99% and 90% reduction in norovirus and hepatitis A virus, respectively, enhancing the meat quality.

3.4.2 Impact on Food Enzymes

Food enzymes such as polyphenol oxidases (PPOs) and peroxidases (PODs), pectin methylesterase (PME), polygalacturonase, lipases, and lipoxygenase determine the food quality during the processing, ripening, transportation, and storage. Traditional heating methods like blanching, pasteurization, and sterilization are being used to inactivate the enzymes. However, these methods resulted in decreased in quality. To overcome the drawbacks of heat treatment, novel nonthermal food processing technologies have been employed to inactivate the enzymes. Enzymes like PPO, POD, and PME were significantly inactivated by cold plasma treatment. These enzymes are significantly contributed to the browning reactions resulting in the formation of melanin compounds in cut fruits and vegetables. Thirumdas and Annapure (2020) reported that the plasma active species led to a conformational change in the active site preventing substrate binding. The loss of the protein structure, mainly the secondary structure, is the main mechanism of cold plasma inactivation of enzymes. DBD cold plasma treatment altered the secondary structure of peroxidase enzymes extracted from the horseradish and litchi was reported by Wang et al. (2021). Gu et al. (2021) reported cold plasma resulted in a similar reduction in bananas' PPO and POD activity with enhanced quality (i.e., decrease in browning products and increas in color attribute). Wang et al. (2021) reported the decrease in residual enzyme activity due to the modification of hydrophobic aromatic amino acids groups in the protein structure. Similarly, Thirumdas and Annapure (2020) also reported that the arc discharge argon gas plasma inactivated the POD enzyme by destructing the enzymes heme groups. These authors have also reported that the interaction of enzymes with plasma species have led to the modifications like the oxidation of amino groups of enzymes, nitration, and nitrosylation of amino acids, sulfoxidation, and chlorination of amino acids. The previously mentioned authors have also stated that the proteins exposed to cold plasma have also resulted in crosslinking, fragmentation of backbone, folding and unfolding, modification of hydrophobicity, and so on. The other possible enzyme inactivation mechanism as stated by Misra et al. (2016) is the chemical modification of amino acid side chains by the reactive species particularly OH, superoxide anion radicals, and NO formed in the cold plasma. Any modification or a change occurring in the polypeptide chain will result in denaturation and loss of protein functionality. Sulfhydryl groups of sulfur-containing amino acids are more easily prone to the plasma reactive species compared to other amino acids (Zhou et al. 2016). Thirumdas and Annapure (2020) reported that the thiol-dependent enzymes containing the cysteine amino acid are easily prone to oxidation by ROS converting the thiol to sulfenic acid groups. Chauvin et al.

(2017) observed degradation in methionine rather than the arginine amino acid under exposure to helium jet plasma. Similarly, alanine exposed to RF argon plasma resulted in the degradation of carboxylic and amino groups. The enzyme inactivation efficiencies of cold plasma are given in Table 3.2.

It is well documented that under the influence of external (i.e., electric and magnetic) fields, the existence of dipole movement related to the secondary structure of a protein could lead to physical and chemical damage. The circular dichroism spectra can be used to analyze the protein structure modifications

Table 3.2 Effect of Cold Plasma on Food Enzymes

Enzyme	Plasma source	Process parameters	Effect on enzymatic activity	Reference
Lipase	RF cold plasma	Feed gas: Air Frequency: 13.56 MHz; Power: 60 W Exposure time: 20 min	The lipase enzyme activity is decreased by 35%	Sutar et al. (2021)
Dehy-drogenase	Co-planar surface barrier discharge plasma	Feed gas: Air Voltage: 10 kV Exposure time: 1 & 2 min	Decreased in activity by 18.5% and 27% after 1 and 2 min of treatment respectively	Henselova et al. (2012)
Peroxidase	DBD	Feed gas: Air Voltage: 50 kV Exposure time: 5 min	The enzyme residual activity is decreased	Pankaj et al. (2013)
Dehy-drogenase enzyme	RF plasma jet	Feed gas: He+O$_2$ Frequency: 13.56 MHz; Exposure time: 10 min	The enzyme activity reduced by 60%	Lackmann et al. (2013)
Alkaline phosphatase	DBD	Feed gas: Air Voltage: 60 kV Exposure time: 5 min	A 5-min treatment decreased 90% of the enzyme activity in milks	Segat et al. (2016)
Peroxidase	DBD	Feed gas: He Voltage: 10 kV Exposure time: 6 min	Enzyme activity is decreased by 95.3%	Khani et al. (2017)
Pectin methy-lesterase	HVACP	Feed gas: Air Voltage: 10 kV Exposure time: 2 min	A 74% decrease in activity	Xu et al. (2017)
POD	ACP	Feed gas: Air Voltage: 20 kV Exposure time: 15 min	A 28% in enzyme activity in coconut water	Porto et al. (2020).
PPO	Microwave driven plasma torch	Feed gas: Air Power: 1.2 kW Exposure time: 10 min	The enzyme activity is decreased by 77%	Bußler et al. (2017)
Lipo-xygenase	ACP	Feed gas: Argon Exposure time: 30 min	The enzyme activity is reduced by 87.25% in wheat germ	Tolouie et al. (2021)

Abbreviations: DBD – dielectric barrier discharge, HVACP – high voltage atmospheric cold plasma, ACP- atmospheric cold plasma.

caused by any treatment. Wang et al. (2021) analyzed the circular dichroism spectra to study the modification in α-helix% and β-sheets% after the plasma treatment. A 180-s exposure to DBD plasma resulted in a decrease of α-helix from 21.41% to 8.53% and an increase in β-sheets% to 32.1% from 23.29% in horseradish peroxidase. Surowsky et al. (2013); Zhang et al. (2015) reported a similar trend in the enzyme activity of PPO and POD and lactate dehydrogenase respectively after the cold plasma treatment. However, Ali et al. (2016) observed the opposite trend, that is, a decrease in α-helix% and an increase in β-sheets% components after the N_2 plasma treatment. Segat et al. (2016) reported a decrease in both α-helix% and β-sheets% in the alkaline phosphatase enzyme after the atmospheric cold plasma treatment. The other mechanism of inactivation could be the oxidation by UV radiations formed in the cold plasma. Manzocco and Nicoli (2015) applied UV radiations onto the apple slices as an anti-browning agent that caused photooxidation of the PPO enzyme. Similarly, the 254-nm UV radiations decreased 90% of PPO activity within 5 min of exposure to radiations. Thirumdas and Annapure (2020) also commented that the plasma treatment decreased the pH of the treating substrate/medium could be another possible reason for enzyme inactivation. Bauer et al. (2017) observed plasma treatment resulted in a 1.4 pH in the cold plasma exposed potato cubes after 10 min of exposure. The inactivation of browning enzymes mainly the PPO and POD using the cold plasma application will result in the extension of the shelf lives and quality of fruits and vegetables. However, the extent of enzyme inactivation depends on several operating parameters like feed gas composition, type of plasma source, and exposure time. From the preceding results, it can be interpreted that the cold plasma technology could be an alternate blanching method to inactivate the food enzymes. The main challenge is to understand the exact mechanism of enzyme inactivation during its interaction with the cold plasma reactive species.

3.4.3 Impact on Quality and Sensory Attributes of Food

Food quality defines all the properties and sensory attributes that determine consumer acceptance. The major challenge faced by any food-processing technology is to destroy the food spoilage microbes while preserving the quality and sensory attributes. Many food-processing operations have been designed and tailored in technology and engineering aspects to have a little effect on food quality. The conventional heating methods caused chemical changes, impaired organoleptic properties, and decreased the nutritional profile. Thus, nonthermal technologies became essential in improving the quality of a product as well as the nutritional profile. The effects of cold plasma processing on the quality of different food products are given in Table 3.3. Ganesan et al. (2021) reported that the plasma treatment preserved the sensory attributes along with the microbial destruction in multiple food products. Gavahian and Khaneghah (2020) have also commented that the cold plasma treatment destroyed the bacterial cells as well as the vegetative spores without any change in organoleptic characteristics of treated substrates. For the decontamination of red meat and its value-added products, the efficiency is emphasized based on the meat quality, particularly color, modification of surface protein structure, and meat lipids oxidation. The modified atmospheric packaging is generally employed for extending the shelf life of red meat. Chutia et al. (2020) conducted a sensory evaluation using 27 semi-trained sensory panelists of cold plasma–treated tender coconut water. The authors observed an increase in the shelf life by 7–10 times but a decrease in the sensory profile. However, the authors have enhanced the sensory profile of coconut water by blending it with fruit juices like orange.

Table 3.3 Effects of Cold Plasma Processing on Food Quality

Food item	Plasma source	Processing parameters	Key findings	Reference
Prebiotic orange juice	DBD- ACP	Feed gas: Air Exposure time: 1 min Voltage: 70 kV	• Decrease in pH • Increase in lightness of juice • Decrease in polyphenols	Almeida et al. (2015)
Pomegranate juice	Cold atmospheric gas plasma	Feed gas: Argon Exposure time: 7 min Voltage: 2.5 kV	• Increase in anthocyanin content • Positive effect on color change	Kovačević et al. (2016)
Blueberry	Cold plasma jet	Feed gas: Air Exposure time: 2 min Power: 549 watts	• Reduction in firmness • Significant change in color • Improved quality	Lacombe et al. (2017)
Strawberries	DBD-HVACP	Feed gas: Air Exposure time: 5 min Voltage: 60 kV	• Reduction in firmness • Significant change in color • Improved quality	Misra et al. (2014)
Fresh cut melon	DBD-cold plasma	Feed gas: Air Exposure time: 30 min Voltage: 15 kV	• No change in pH • Less color change • 78% decrease in PME enzyme activity	Tappi et al. (2016)
Orange juice	DBD-cold plasma	Feed gas: $O_2+CO_2+N_2$ Exposure time: 2 min Voltage: 90 kV	• No change in pH • Less color change, texture • Decrease in PME and POD enzyme activity	Xu et al. (2017)
Mandarins	Microwave – coupled plasma	Feed gas: $O_2+ N_2$ Exposure time: 10 min Power: 900 watts	• No change in pH, color, vitamin C • Increase in total phenolic content	Won et al. (2017)
Xanthan gum	RF- cold plasm	Feed gas: Air Exposure time: 30 min Power: 60 watts	• No change in proximate composition • Decrease in bulk and tapped density • Increase in Hausner's ratio and angle of repose	Bulbul et al. (2019)
Cherry tomato	Intermittent corona discharge plasma jet	Feed gas: Air Exposure time: 2 min Voltage: 8 kV	• No change in color, firmness • No change in taste, texture, flavor • Increase in shelf-life	Lee et al. (2018)
Coconut water	DBD-HVACP	Feed gas: Air Exposure time: 2 min Power: 186 watts	• Decrease in pH and acidity • No change in vitamin C	Mahnot et al. (2019)

(Continued)

Table 3.3 (Continued)

Food item	Plasma source	Processing parameters	Key findings	Reference
Wheat flour	RF-cold plasma	Feed gas: Air Exposure time: 30 min Power: 60 watts	• Enhanced wheat flour quality • Insect disinfestation • No significant change in color • Decrease in lipase activity	Sutar et al. (2021)
Corn starch	RF-cold plasma	Feed gas: Air Exposure time: 20 min Power: 60 watts	• Decrease in amylose content • Decrease in pH due to oxidation • Increase in viscosities	Banura et al. (2018)
Parboiled rice	RF-cold plasma	Feed gas: Air Exposure time: 15 min Power: 60 watts	• Increase in flour hydration properties • Increase in gel hydration properties • Crosslinking of starch	Sarangapani et al. (2015, 2016)
Rocket leaf salad	DBD-cold plasma	Feed gas: Air Exposure time: 10 min Voltage: 6 kV	• No significant change in color and texture • Shelf-life is extended to 53 days at 2°C	Giannoglou et al. (2020)

Ozen and Singh (2020) did not observe any significant change in the color of different fruit juices like grape, apple, orange, and pomegranate juices after the treatment. Similarly, the cold plasma treatment, compared to the thermal treatments, has preserved the color of blueberry juice (Huo et al. 2019). However, there is a slight increase in the pH and acidity of the processed juice. Similarly, DBD plasma treatment has decreased the pH of prebiotic orange juice after 1 min of exposure (Almeida et al. 2015). Pankaj et al. (2018) reported that the NO formed during the plasma generation has electrolyzed the waster resulting in acidification due to the formation of nitric acid attributing to a decrease in pH. Some studies haven't reported any significant change in pH due to the cold plasma treatment. Xu et al. (2017) stated that the presence of acids like citric, malic, and ascorbic acid in fruit juices resisted the change in pH. Chokeberry juice exposed to DBD plasma jet did not affect the titratable acidity and reduced the sugar percentage. Fresh strawberries when exposed to cold atmospheric plasma (CAP) did not cause any change in firmness, weight, and color (Giannoglou et al. 2021). However, atmospheric pressure DBD plasma–treated strawberries resulted in slight changes in color and an increase in firmness. Mehta and Yadav (2020) observed a positive effect on the bioactive compounds like polyphenolic compounds (gallic, epigallocatechin, naringin, phloretin), ascorbic acid in the atmospheric cold plasma (ACP)-exposed strawberry juice along with a 2-log reduction in the microbial count. Bao et al. (2020) found that the application of plasma has improved the polyphenols extraction up to 22.8% with a higher concentration of anthocyanins from the grape pomace. Ramazzina et al. (2015) observed an improvement in the kiwifruit color retention after exposure to DBD plasma for 10–20 min.

Silveira et al. (2019) improved the quality of flavored whey milk by applying N_2 plasma. The plasma treatment enhanced the creamier characteristic of milk by modulating the fat globule particle size, number, and a few rheological properties. Ribeiro et al. (2021) observed a similar sensory score of cold plasma–treated whey dairy beverages with a higher concentration in bioactive compounds. Similarly, Coutinho et al. (2019) observed the particle size increased from 87 μm to 103 μm and a higher consistency after 15 min of nitrogen plasma treatment. The particle size of chocolate milk, surface area, and consistency is increased with extending exposure time. The other mechanism of plasma treatment in enhancing the milk's quality is through the modification of milk protein functionality. Singh and Huppertz (2020) reported that the milk exposed to plasma resulted in the milk protein structural modification attributing a change in emulsifying properties. Ng et al. (2021) observed glow discharge plasma altered the secondary structure (increase in β-sheets and β-turns) of milk proteins. Chen et al. (2020) observed a similar change in the α-helix and β-sheets of secondary structure that are converted to β-turns and coils. Coutinho et al. (2019) suggested similar changes pertaining to the milk proteins like denaturation and aggregations. Ganesan et al. (2021) reported that the reactive species, particularly ROS, have resulted in the oxidation of milk proteins that resulted in a reduction of –SH (sulfhydryl) and an increase in surface hydrophobicity. The interaction of gaseous plasma species and the water content of milk resulted in the generation of plasma-activated liquid species with an increase in acidity (Coutinho et al. 2019). The application of plasma did not affect the lipid composition of the milk (Korachi et al. 2015). However, Upadhyay et al. (2020), and Yepez and Keener (2016) observed an increase in saturated fatty acids composition in cold plasma exposed to chia seed oil and soybean oil, respectively, resulting in partial hydrogenation of oils. Gavahian et al. (2019) observed an improvement in the egg quality parameters in terms of yolk color, Haugh's unit, eggshell strength, fatty acid profile, and lipid oxidation when the shell eggs are exposed to arc cold plasma to inactivate the *Salmonella* sps. However, the lipid oxidation of food substrates caused by the plasma treatment can be suppressed by decreasing the exposure time (Jadhav et al. 2021).

Abidin et al. (2018) investigated the gluten quality of the noodles after exposure to cold plasma. Wet noodles were observed to be better in gluten quality and whiter than untreated noodles. Similarly, Thirumdas et al. (2015, 2016) reported RF cold plasma–exposed basmati and brown rice resulted in an increase in the whiteness index due to the plasma etching. The atmospheric cold plasma processed hazelnut observed an improvement in the quality in terms of color, total solids, and phenolic compounds. Mahendran (2016) did not observe any significant color change in the DBD processed refined wheat flour after the treatment for 5 min. Similarly, DBD air plasma enhanced the quality of the wheat flour by changing its functional and rheological properties (Misra et al. 2015). Sutar et al. (2021) reported that RF cold plasma exposure did not affect the color of wheat flour and decreased the lipase enzyme activity by 35%, enhancing the keeping quality of flour

As reported in the earlier section 3.3.2, the cold plasma has significantly affected the activity of browning enzymes like PPO and POD. The reduction in the enzymatic activity has improved the quality of many fresh fruit cuts and fruit juices. The quality and color of the fresh-cut potato produce were improved by DBD plasma treatment without any textural change (Mahnot et al. 2019). Similarly, cold plasma application improved the color of fresh produce like pomegranates and strawberries (Rana et al. 2020). A 16% increase in residual activity of PPO within 5 min of treatment and improved the quality parameters mainly the retention of the green color of the Golden delicious cloudy apple juice (Illera et al. 2019). The inactivation of PPO and POD enzymes by the cold plasma resulted in a significant retention of color in bananas within 2 min of treatment (Gu et al. 2021). Similarly, in the atmospheric cold plasma processed acai pulp, the enzymatic activity of PPO and POD was reduced with an increase in the bio-accessibility of polyphenols (catechin, epicatechin, epigallocatechin gallate, procyanidin B1, rutin, caffeic acid, chlorogenic acid) and antioxidant property (Dantas et al. 2021).

The cold plasma has been applied to food products to enhance the drying rate while enhancing the quality. Cold plasma as a pretreatment to enhance the drying rate of wolfberry was studied by Zhou et al. (2020). The authors have found that a 50% decrease in drying time fragmentation of cell walls, facilitating and releasing moisture entrapped inside the cells. Bao et al. (2020) reported a moisture diffusion rate enhanced by 36.85% in cold plasma–treated jujube. Similarly, Zhang et al. (2019) observed the same trend in chili pepper after the treatment. Plasma etching and ablation have changed the surface functionalization and formation of micropores or micro-holes that enhanced the heat transfer rate attributing to shorter drying periods. The formation of micropores or fissures on the surface of brown rice (Thirumdas et al. 2016), green gram (Sadhu et al. 2017), and rice starch (Thirumdas et al. 2017) due to plasma etching. In several scientific studies, cold plasma has shown a positive correlation on the nutritional quality and sensory attributes along with microbial destruction.

3.5 EMERGING APPLICATIONS OF COLD PLASMA IN FOOD PRESERVATION

Tender coconut water is easily prone to spoilage and enzymatic deterioration when exposed to air. Cold plasma is applied to enhance the shelf life of coconut water by inactivation of microbes and enzymes. A 3-min plasma exposure decreased the microbial load from 150 to 50 CFU/mL while preserving the quality and enhancing the shelf life of tender coconut water (Chutia et al. 2020). DBD plasma exposure for 5 min decreased the tropomyosin allergens present in shrimp by 76%. Meinlschmidt et al. (2016) investigated the food allergens like soy allergens by analyzing the SDS-PAGE profile after the plasma treatment. The authors observed a 100% decrease in the immunoreactivity of soy allergens (glycinin and β-conglycinin). The plasma species interactions have led to formation of new proteins and modification of proteins that contributed to anti-allergenicity in food (Sarangapani et al. 2018). Ekezie et al. (2018) reported that crosslinking of proteins could be another reason for the reduction in food allergens.

The exopolysaccharides excreted by the bacteria form a film called "biofilm". Bacteria are most resistant to the external environment under protective coverings of biofilms. Zhu et al. (2020) reported that biofilms protect bacteria from several antibiotics, antibacterial chemicals, oxidizing agents, and other disinfestation chemicals. Hydrogen peroxide possessing a strong oxidation potential is widely used for the disruption of biofilms. It is assumed that the survival chance of bacteria in the biofilm is 1000 times more than for bacteria outside the films. About 80% of bacteria present on the vegetable surface exist under biofilms that are difficult to be destroyed by a conventional process. The cold plasma could be a method that can significantly affect the biofilms leading to the death of bacteria. Cold plasma can be applied on several substrates like fruits, vegetables, food-processing machinery, and more. However, the surface topography of a treated substrate is an important factor in the antimicrobial efficiency. Cui et al. (2018) applied N_2 gas plasma to investigate the anti-biofilm activity on the surface of lettuce using5 min of cold plasma exposure that completely eradicated

48-h biofilms of *L. monocytogenes* and *P. fluorescence* to undetected levels on the lettuce (Patange et al. 2019). Zhu et al. (2020) reported that the •OH and O ions formed have the ability to penetrate and disrupt the biofilm. Plasma etching is the other anti-biofilm mechanism that cleaves the chemical bonds of exopolysaccharides resulting in depolymerization. Cold plasma is also used to inactivate SARS-CoV-2. Chen et al. (2020) reported that CAP jet inactivated the coronavirus-2 within 180 s of treatment. Guo et al. (2021) activated deionized water for 5 and 10 min to generate plasma-activated water and treated the SARS-CoV-2 proteins in a pseudovirus as a model to measure antiviral activity. From the results of the above investigation, the authors have postulated that PAW activated at a lower treatment time resulted in slight morphological change, whereas the longer activated water showed virus aggregates for large complexes. The cold plasma antiviral mechanism on SARS-CoV-2 is due to surface wall disruption due to etching, damage to genetic material, and denaturation of proteins (Chen and Wriz 2020).

3.6 CONCLUSION AND PERSPECTIVES

Cold plasma is a nonthermal, emerging technology for food preservation, and it is effective in retaining the original food quality. Along with microbial inactivation, cold plasma is also applied to inactivate the food enzymes responsible for food spoilage. Different parameters such as the type of plasma source, applied voltage, treatment time, and nature of the food substrate have to be taken into consideration to achieve high inactivation rates. The reactive oxygen and nitrogen species, along with the ozone and UV radiations, are responsible for microbial and enzyme inactivation. The plasma treatment can preserve color of many cut fruits and fruit juices with the least effect on pH, acidity, and bioactive compounds. As the cold plasma research in foods is increasing day by day, there is an urgent requirement to get approval from regulatory agencies for commercialization. However, there is a need to work in the direction of standardizing the operating mechanism as the efficiency of treatment varies with plasma sources and application methods. Further research is needed for scaling up plasma treatment so that the technology is commercialized. We hope in the near future cold plasma technology will become popular as chemical-free and environment-friendly food preservation technology.

REFERENCES

Abdel-Fattah, E., M. Bazavan, and H. Sugai. 2011. "Langmuir Probe Diagnostics of Electron Energy Distributions with Optical Emission Spectroscopy in Capacitively Coupled rf Discharge in Nitrogen." *Journal of Applied Physics* 110 (11): 113303.

Abidin, Noor Shazliana Aizee, Ibni H. Rukunudin, Siti K. Zaaba, Wan A. W. Omar. 2018. "Atmospheric Pressure Cold Plasma (ACP) Treatment a New Technique to Improve Microstructure and Textural Properties of Healthy Noodles Fortified with Mango Flour." *Journal of Telecommunication, Electronic and Computer Engineering* 10: 1–17.

Ali, Anser, Zaman Ashraf, Naresh Kumar, Muhammad Rafiq, Farukh Jabeen, Ji Hoon Park, Ki Hong Choi, et al. 2016. "Influence of Plasma-Activated Compounds on Melanogenesis and Tyrosinase Activity." *Scientific Reports* 6 (1): 1–20.

Almeida, Francisca Diva Lima, Rosane Souza Cavalcante, Patrick J. Cullen, Jesus Maria Frias, Paula Bourke, Fabiano AN Fernandes, and Sueli Rodrigues. 2015. "Effects of Atmospheric Cold Plasma and Ozone on Prebiotic Orange Juice." *IFSET* 32: 127–135.

Amini, Maryam, and Mahmood Ghoranneviss. 2016. "Effects of Cold Plasma Treatment on Antioxidants Activity, Phenolic Contents and Shelf Life of Fresh and Dried Walnut (Juglans regia L.) Cultivars During Storage." *LWT* 73: 178–184.

Anderson, James M. 2001. "Biological Responses to Materials." *Annual Review of Materials Research* 31 (1): 81–110.

Bae, San-Cheong, Shin Young Park, Wonho Choe, and Sang-Do Ha. 2015. "Inactivation of Murine Norovirus-1 and Hepatitis A Virus on Fresh Meats by Atmospheric Pressure Plasma Jets." *Food Research International* 76: 342–347.

Baier, Matthias, Traute Janßen, Lothar H. Wieler, Jörg Ehlbeck, Dietrich Knorr, and Oliver Schlüter. 2015. "Inactivation of Shiga Toxin-Producing *Escherichia coli* O104: H4 Using Cold Atmospheric Pressure Plasma." *Journal of Bioscience and Bioengineering* 120 (3): 275–279.

Banura, Sidhant, Rohit Thirumdas, Amritpal Kaur, R. R. Deshmukh, and U. S. Annapure. 2018. "Modification of Starch Using Low Pressure Radio Frequency Air Plasma." *LWT* 89: 719–724.

Bao, Yiwen, Lavanya Reddivari, and Jen-Yi Huang. 2020. "Enhancement of Phenolic Compounds Extraction from Grape Pomace by High Voltage Atmospheric Cold Plasma." *LWT* 133: 109970.

Bauer, A., Y. Ni, S. Bauer, P. Paulsen, M. Modic, J. L. Walsh, and F. J. M. Smulders. 2017. "The Effects of Atmospheric Pressure Cold Plasma Treatment on Microbiological, Physical-Chemical and Sensory Characteristics of Vacuum Packaged Beef Loin." *Meat Science* 128: 77–87.

Belov, Igor, Vincent Vermeiren, Sabine Paulussen, and Annemie Bogaerts. 2018. "Carbon Dioxide Dissociation in a Microwave Plasma Reactor Operating in a Wide Pressure Range and Different Gas Inlet Configurations." *Journal of CO2 Utilization* 24: 386–397.

Berardinelli, Annachiara, Abdessalem Hamrouni, Sandra Dirè, Riccardo Ceccato, Giovanni Camera-Roda, Luigi Ragni, Leonardo Palmisano, and Francesco Parrino. "Features and Application of Coupled Cold Plasma and Photocatalysis Processes for Decontamination of Water." *Chemosphere* 262 (2021): 128336.

Bulbul, V. J., Pravin R. Bhushette, Rahul S. Zambare, R. R. Deshmukh, and Uday S. Annapure. 2019. "Effect of Cold Plasma Treatment on Xanthan Gum Properties." *Polymer Testing* 79: 106056.

Bußler, Sara, Jörg Ehlbeck, and Oliver K. Schlüter. "Pre-Drying Treatment of Plant Related Tissues Using Plasma Processed Air: Impact on Enzyme Activity and Quality Attributes of Cut Ppple and Potato." *Innovative Food Science & Emerging Technologies* 40 (2017): 78–86.

Butscher, Denis, Daniel Zimmermann, Markus Schuppler, and Philipp Rudolf von Rohr. 2016. "Plasma Inactivation of Bacterial Endospores on Wheat Grains and Polymeric Model Substrates in a Dielectric Barrier Discharge." *Food Control* 60: 636–645.

Charoux, Clémentine M. G., Apurva Patange, Sakshi Lamba, Colm P. O'Donnell, Brijesh K. Tiwari, and Amalia GM Scannell. 2021. "Applications of Nonthermal Plasma Technology on Safety and Quality of Dried Food Ingredients." *Journal of Applied Microbiology* 130 (2): 325–340.

Chauvin, Julie, Florian Judée, Mohammed Yousfi, Patricia Vicendo, and Nofel Merbahi. 2017. "Analysis of Reactive Oxygen and Nitrogen Species Generated in Three Liquid Media by Low Temperature Helium Plasma Jet." *Scientific Reports* 7 (1): 1–15.

Chen, Francis F. 2007. "Radiofrequency Plasma Sources for Semiconductor Processing." *Advanced Plasma Technology* 99–115.

Chen, Yue, Yifu Zhang, Lan Jiang, Guiyun Chen, Jiaojiao Yu, Shuhong Li, and Ye Chen. 2020. "Moisture Molecule Migration and Quality Changes of Fresh Wet Noodles Dehydrated by Cold Plasma Treatment." *Food Chemistry* 328: 127053.

Chen, Zhitong, and Richard Wirz. 2020. "Cold Atmospheric Plasma for COVID-19.", Preprint 2020 (10.20944/preprints202004.0126.v1).

Chen, Zhitong, Rong-Guang Xu, Peijian Chen, and Qiu Wang. 2020. "Potential Agricultural and Biomedical Applications of Cold Atmospheric Plasma-Activated Liquids with Self-Organized Patterns Formed at the Interface." *IEEE Transactions on Plasma Science* 48 (10): 3455–3471.

Chutia, Hemanta, Charu Lata Mahanta, Namita Ojah, and Arup Jyoti Choudhury. 2020. "Fuzzy Logic Approach for Optimization of Blended Beverage of Cold Plasma Treated TCW and Orange Juice." *Journal of Food Measurement and Characterization* 1–13.

Coutinho, N. M., M. R. Silveira, R. S. Rocha, M. Q. Freitas, M. C. K. Duarte, R. F. Quero, and A. G. da Cruz. 2021. "Cold Plasma." In *Sustainable Food Processing and Engineering Challenges*, 109–135. Academic Press.

Coutinho, Nathália M., Marcello R. Silveira, Tatiana C. Pimentel, Monica Q. Freitas, Jeremias Moraes, Leonardo M. Fernandes, Marcia C. Silva, et al. 2019. "Chocolate Milk Drink Processed by Cold Plasma Technology: Physical Characteristics, Thermal Behavior and Microstructure." *LWT* 102: 324–329.

Cui, Haiying, Mei Bai, Lu Yuan, Duraiarasan Surendhiran, and Lin Lin. 2018. "Sequential Effect of Phages and Cold Nitrogen Plasma Against *Escherichia Coli* O157: H7 Biofilms on Different Vegetables." *International Journal of Food Microbiology* 268: 1–9.

D'Isa, F. A., E. A. D. Carbone, A. Hecimovic, and U. Fantz. 2020. "Performance Analysis of a 2.45 GHz Microwave Plasma Torch for CO2 Decomposition in Gas Swirl Configuration." *Plasma Sources Science and Technology* 29 (10): 105009.

Dantas, Aline Macedo, Jéssica Dayanne Ferreira Batista, Marcos dos Santos Lima, Fabiano AN Fernandes, Sueli Rodrigues, Marciane Magnani, and Graciele da Silva Campelo Borges. 2021. "Effect of Cold Plasma on Açai Pulp: Enzymatic Activity, Color and Bioaccessibility of Phenolic Compounds." *LWT* 111883.

Devi, Yamuna, Rohit Thirumdas, C. Sarangapani, R. R. Deshmukh, and U. S. Annapure. 2017. "Influence of Cold Plasma on Fungal Growth and Aflatoxins Production on Groundnuts." *Food Control* 77: 187–191.

Dirks, B. P., D. Dobrynin, G. Fridman, et al. 2012. "Treatment of raw poultry with nonthermal dielectric barrier discharge plasma to reduce *Campylobacter jejuni* and *Salmonella enterica*." *Journal of Food Protection* 75 (1): 22–28.

Ekezie, Flora-Glad Chizoba, Jun-Hu Cheng, and Da-Wen Sun. 2018. "Effects of Nonthermal Food Processing Technologies on Food Allergens: A Review of Recent Research Advances." *Trends in Food Science & Technology* 74: 12–25.

Feizollahi, Ehsan, N. N. Misra, and M. S. Roopesh. 2021. "Factors Influencing the Antimicrobial Efficacy of Dielectric Barrier Discharge (DBD) Atmospheric Cold Plasma (ACP) in Food Processing Applications." *Critical Reviews in Food Science and Nutrition* 61 (4): 666–689.

Fridman, Alexander A., and Gary G. Friedman. 2013. *Plasma Medicine.* Chichester, UK: John Wiley & Sons.

Ganesan, Abirami R., Uma Tiwari, P. N. Ezhilarasi, and Gaurav Rajauria. 2021. "Application of Cold Plasma on Food Matrices: A Review on Current and Future Prospects." *Journal of Food Processing and Preservation* 45 (1): e15070.

Gavahian, M., Peng, HJ. & Chu, YH. Efficacy of cold plasma in producing *Salmonella*-free duck eggs: effects on physical characteristics, lipid oxidation, and fatty acid profile. *J Food Sci Technol* 56: 5271–5281 (2019). https://doi.org/10.1007/s13197-019-03996-z

Gavahian, Mohsen, and Amin Mousavi Khaneghah. 2020. "Cold Plasma as a Tool for the Elimination of Food Contaminants: Recent Advances and Future Trends." *Critical Reviews in Food Science and Nutrition* 60 (9): 1581–1592.

Georgescu, Nicolae, Livia Apostol, and Florin Gherendi. 2017. "Inactivation of *Salmonella Enterica* Serovar Typhimurium on Egg Surface, by Direct and Indirect Treatments with Cold Aatmospheric Plasma." *Food Control* 76: 52–61.

Giannoglou, Marianna, Panagiota Stergiou, Panagiotis Dimitrakellis, Evangelos Gogolides, Nikolaos G. Stoforos and George Katsaros. 2020. "Effect of Cold Atmospheric Plasma processing on quality and shelf-life of ready-to-eat rocket leafy salad." *Innovative Food Science and Emerging Technologies* 66: 102502.

Giannoglou, Marianna, Zacharoula-Maria Xanthou, Sofia Chanioti, Panagiota Stergiou, Miltiadis Christopoulos, Panagiotis Dimitrakellis, Aspasia Efthimiadou, Evangelos Gogolides, and George Katsaros. 2021. "Effect of Cold Atmospheric Plasma and Pulsed Electromagnetic Fields on Strawberry Quality and Shelf-Life." *Innovative Food Science & Emerging Technologies* 68: 102631.

Gök, Veli, Simge Aktop, Mehmet Özkan, and Oktay Tomar. 2019. "The Effects of Atmospheric Cold Plasma on Inactivation of *Listeria Monocytogenes* and *Staphylococcus Aureus* and Some Quality Characteristics of Pastırma – A Dry-Cured Beef Product." *Innovative Food Science & Emerging Technologies* 56: 102188.

Gu, Yuxiang, Wenqing Shi, Rui Liu, Yanan Xing, Xiuzhu Yu, and Hao Jiang. 2021. "Cold Plasma Enzyme Inactivation on Dielectric Properties and Freshness Quality in Bananas." *Innovative Food Science & Emerging Technologies* 69: 102649.

Guo, Li, Zhiqian Yao, Lu Yang, Hao Zhang, Yu Qi, Lu Gou, Wang Xi, et al. 2021. "Plasma-Activated Water: An Alternative Disinfectant for S Protein Inactivation to Prevent SARS-CoV-2 Infection." *Chemical Engineering Journal* 421: 127742.

Gurol, C., F. Y. Ekinci, N. Aslan, and M. Korachi. 2012. "Low Temperature Plasma for Decontamination of E. Coli in Milk." *International Journal of Food Microbiology* 157 (1): 1–5.

Han, Jin-Young, Won-Jae Song, Joo Hyun Kang, Sea C. Min, Sangheum Eom, Eun Jeong Hong, Seungmin Ryu, Seong bong Kim, Sangwoo Cho, and Dong-Hyun Kang. 2020. "Effect of Cold Atmospheric Pressure Plasma-Activated Water on the Microbial Safety of Korean Rice Cake." *LWT* 120: 108918.

Han, Lu, Sonal Patil, Daniela Boehm, V. Milosavljević, P. J. Cullen, and P. Bourke. 2016. "Mechanisms of Inactivation by High-Voltage Atmospheric Cold Plasma Differ for *Escherichia Coli* and *Staphylococcus Aureus*." *Applied and Environmental Microbiology* 82 (2): 450–458.

Henselová, Mária, Ľudmila Slováková, Michal Martinka, and Anna Zahoranová. 2012. "Growth, Anatomy and Enzyme Activity Changes in Maize Roots Induced by Treatment of Seeds with Low-Temperature Plasma." *Biologia* 67 (3): 490–497.

Hertrich, Sarah M., Glenn Boyd, Joseph Sites, and Brendan A. Niemira. 2017. "Cold Plasma Inactivation of *Salmonella* in Prepackaged, Mixed Salads is Influenced by Cross-Contamination Sequence." *Journal of Food Protection* 80 (12): 2132–2136.

Hertwig, Christian, Andriani Leslie, Nicolas Meneses, Kai Reineke, Cornelia Rauh, and Oliver Schlüter. 2017. "Inactivation of *Salmonella Enteritidis* PT30 on the Surface of Unpeeled Almonds by Cold Plasma." *Innovative Food Science & Emerging Technologies* 44: 242–248.

Hertwig, Christian, Kai Reineke, Jörg Ehlbeck, Dietrich Knorr, and Oliver Schlüter. 2015. "Decontamination of Whole Black Pepper Using Different Cold Atmospheric Pressure Plasma Applications." *Food Control* 55: 221–229.

Hou, Yanan, Ruixue Wang, Zhilin Gan, Tao Shao, Xinxue Zhang, Mohe He, and Aidong Sun. 2019. "Effect of Cold Plasma on Blueberry Juice Quality." *Food Chemistry* 290: 79–86.

Illera, A. E., S. Chaple, M. T. Sanz, S. Ng, P. Lu, J. Jones, E. Carey, and P. Bourke. 2019. "Effect of Cold Plasma on Polyphenol Oxidase Inactivation in Cloudy Apple Juice and on the Quality Parameters of the Juice During Storage." *Food Chemistry: X* 3: 100049.

Jadhav, Harsh Bhaskar, Uday S. Annapure, and Rajendra R. Deshmukh. 2021. "Non-thermal Technologies for Food Processing." *Frontiers in Nutrition* 8.

Jayasena, Dinesh D., Hyun Joo Kim, Hae In Yong, Sanghoo Park, Kijung Kim, Wonho Choe, and Cheorun Jo. 2015. "Flexible Thin-Layer Dielectric Barrier Discharge Plasma Treatment of Pork Butt and Beef Loin: Effects on Pathogen Inactivation and Meat-Quality Attributes." *Food Microbiology* 46: 51–57.

Khani, Mohammad Reza, Babak Shokri, and Khosro Khajeh. 2017. "Studying the Performance of Dielectric Barrier Discharge and Qliding Arc Plasma Reactors in Tomato Peroxidase Inactivation." *Journal of Food Engineering* 197: 107–112.

Kim, Jung Eun, Dong-Un Lee, and Sea C. Min. 2014. "Microbial Decontamination of Red Pepper Powder by Cold Plasma." *Food Microbiology* 38: 128–136.

Kim, Ji Hyeon, and Sea C. Min. 2018. "Moisture vaporization-combined helium dielectric barrier discharge-cold plasma treatment for microbial decontamination of onion flakes." *Food Control* 84: 321–329.

Korachi, May, Fatma Ozen, Necdet Aslan, Lucia Vannini, Maria Elisabetta Guerzoni, Davide Gottardi, and Fatma Yesim Ekinci. 2015. "Biochemical Changes to Milk Following Treatment by a Novel, Cold Atmospheric Plasma System." *International Dairy Journal* 42: 64–69.

Kovačević, Danijela Bursać, Predrag Putnik, Verica Dragović-Uzelac, Sandra Pedisić, Anet Režek Jambrak and Zoran Herceg. 2016. "Effects of cold atmospheric gas phase plasma on anthocyanins and color in pomegranate juice." *Food Chemistry* 190: 317–323.

Lackmann, Jan-Wilm, Simon Schneider, Eugen Edengeiser, Fabian Jarzina, Steffen Brinckmann, Elena Steinborn, Martina Havenith, Jan Benedikt, and Julia E. Bandow. 2013. "Photons and Particles Emitted from Cold Atmospheric-Pressure Plasma Inactivate Bacteria and Biomolecules Independently and Synergistically." *Journal of The Royal Society Interface* 10 (89): 20130591.

Lacombe, Alison, Brendan A. Niemira, Joshua B. Gurtler, Joseph Sites, Glenn Boyd, David H. Kingsley, Xinhui Li, and Haiqiang Chen. 2017. "Nonthermal Inactivation of Norovirus Surrogates on Blueberries Using Atmospheric Cold Plasma." *Food Microbiology* 63: 1–5.

Lee, Haelim, Hae In Yong, Hyun-Joo Kim, Wonho Choe, Suk Jae Yoo, Eun Jin Jang, and Cheorun Jo. 2016. "Evaluation of the Microbiological Safety, Quality Changes, and Qenotoxicity of Chicken Breast Treated with Flexible Thin-Layer Dielectric Barrier Discharge Plasma." *Food Science and Biotechnology* 25 (4): 1189–1195.

Lee, Hyun Jung, Heesoo Jung, Wonho Choe, Jun Sang Ham, Jun Heon Lee, and Cheorun Jo. 2011. "Inactivation of *Listeria Monocytogenes* on Agar and Processed Meat Surfaces by Atmospheric Pressure Plasma Jets." *Food Microbiology* 28 (8): 1468–1471.

Lee, Juri, Cheol Woo Lee, Hae In Yong, Hyun Jung Lee, Cheorun Jo, and Samooel Jung. 2017. "Use of Atmospheric Pressure Cold Plasma for Meat Industry." *Korean Journal for Food Science of Animal Resources* 37 (4): 477.

Lee, J. H., K. S. Woo, C. Jo, H. S. Jeong, S. K. Lee, B. W. Lee, Y. Y. Lee, B. Lee, and H. J. Kim. 2019. "Quality evaluation of rice treated by high hydrostatic pressure and atmospheric pressure plasma." *Journal of Food Quality*: 4253701.

Lee, T., P. Puligundla, and C. Mok. 2018. "Intermittent corona discharge plasma jet for improving tomato quality." *Journal of Food Engineering* 223: 168–174.

Lis, Karolina Anna, Annika Boulaaba, Sylvia Binder, Yangfang Li, Corinna Kehrenberg, Julia Louise Zimmermann, Günter Klein, and Birte Ahlfeld. 2018. "Inactivation of *Salmonella Typhimurium* and *Listeria Monocytogenes* on Ham with Nonthermal Atmospheric Pressure Plasma." *PloS One* 13 (5): e0197773.

Liu, Chenghui, Chen Chen, Aili Jiang, Xiaoyuan Sun, Qinxin Guan, and Wenzhong Hu. 2020. "Effects of Plasma-Activated Water on Microbial Growth and Storage Quality of Fresh-Cut Apple." *Innovative Food Science & Emerging Technologies* 59: 102256.

Lu, P., P. J. Cullen, and K. Ostrikov. 2016. "Atmospheric Pressure Nonthermal Plasma Sources." In *Cold Plasma in Food and Agriculture*, 83–116. Academic Press.

Mahendran, R. 2016. "Effect of Cold Plasma on Mortality of *Tribolium Castaneum* on Refined Wheat Flour." In *Proceedings of the 10th International Conference on Controlled Atmosphere and Fumigation in Stored Products (CAF 2016)*, Winnipeg, MB, Canada, pp. 7–11.

Mahnot, Nikhil Kumar, Charu Lata Mahanta, Kevin M. Keener, and N. N. Misra. 2019. "Strategy to Achieve a 5-log *Salmonella* Inactivation in Tender Coconut Water Using High Voltage Atmospheric Cold Plasma (HVACP)." *Food Chemistry* 284: 303–311.

Mandal, Ronit, Anika Singh, and Anubhav Pratap Singh. 2018. "Recent Developments in Cold Plasma Decontamination Technology in the Food Industry." *Trends in Food Science & Technology* 80: 93–103.

Mansuroglu, Dogan. 2019. "Capacitively Coupled Radio Frequency Nitrogen Plasma Generated at Two Different Exciting Frequencies of 13.56 MHz and 40 MHz Analyzed Using Langmuir Probe Along with Optical Emission Spectroscopy." *AIP Advances* 9 (5): 055205.

Manzocco, Lara, and Maria Cristina Nicoli. 2015. "Surface Processing: Existing and Potential Applications of Ultraviolet Light." *Critical Reviews in Food Science and Nutrition* 55 (4): 469–484.

Mehdizadeh, Mehrdad. 2015. *Microwave/RF Applicators and Probes: For Material Heating, Sensing, and Plasma Generation*. William Andrew.

Mehta, Deepak, and Sudesh Kumar Yadav. 2020. "Impact of Atmospheric Non-Thermal Plasma and Hydrothermal Treatment on Bioactive Compounds and Microbial Inactivation of Strawberry Juice: A Hurdle Technology Approach." *Food Science and Technology International* 26 (1): 3–10.

Meinlschmidt, P., E. Ueberham, J. Lehmann, K. Reineke, O. Schlüter, U. Schweiggert-Weisz, and P. Eisner. 2016. "The Effects of Pulsed Ultraviolet Light, Cold Atmospheric Pressure Plasma, and Gamma-irradiation on the Immunoreactivity of Soy Protein Isolate." *Innovative Food Science & Emerging Technologies* 38(Part B): 374-383. https://doi.org/10.1016/j.ifset.2016.06.007

Misra, N. N., et al. 2014. "In-package atmospheric pressure cold plasma treatment of strawberries." *Journal of Food Engineering* 125: 131–138. https://doi.org/10.1016/j.jfoodeng.2013.10.023.

Mishra, Rachna and Abhijit Kar. 2014. "Effect of storage on the physicochemical and flavour attributes of two cultivars of strawberry cultivated in Northern India." *The Scientific World Journal* 2014, Article ID 794926: 7.

Misra, N. N., and Cheorun Jo. 2017. "Applications of Cold Plasma Technology for Microbiological Safety in Meat Industry." *Trends in Food Science & Technology* 64: 74–86.

Misra, N. N., S. K. Pankaj, Annalisa Segat, and Kenji Ishikawa. 2016. "Cold Plasma Interactions with Enzymes in Foods and Model Systems." *Trends in Food Science & Technology* 55: 39–47.

Misra, N. N., Seeratpreet Kaur, Brijesh K. Tiwari, Amritpal Kaur, Narpinder Singh, and P. J. Cullen. 2015. "Atmospheric Pressure Cold Plasma (ACP) Treatment of Wheat Flour." *Food Hydrocolloids* 44: 115–121.

Moutiq, Rkia, N. N. Misra, Aubrey Mendonca, and Kevin Keener. 2020. "In-Package Decontamination of Chicken Breast Using Cold Plasma Technology: Microbial, Quality and Storage Studies." *Meat Science* 159: 107942.

Ng, Sing Wei, Peng Lu, Aleksandra Rulikowska, Daniela Boehm, Graham O'Neill, and Paula Bourke. 2021. "The Effect of Atmospheric Cold Plasma Treatment on the Antigenic Properties of Bovine Milk Casein and Whey Proteins." *Food Chemistry* 342: 128283.

Niemira, Brendan A. 2012. "Cold Plasma Decontamination of Foods." *Annual Review of Food Science and Technology* 3: 125–142.

Niemira, Brendan A. 2018. "Cold Plasma Processing to Improve Food Safety." In *Alternatives to Conventional Food Processing*, 138–152. Royal Society of Chemistry.

Oh, Yeong Ji, A. Young Song, and Sea C. Min. 2017. "Inhibition of *Salmonella Typhimurium* on Radish Sprouts Using Nitrogen-Cold Plasma." *International Journal of Food Microbiology* 249: 66–71.

Ozen, E., and R. K. Singh. 2020. "Atmospheric Cold Plasma Treatment of Fruit Juices: A Review." *Trends in Food Science & Technology* 103: 144–151.

Pankaj, S. K., N. N. Misra, and P. J. Cullen. 2013. "Kinetics of Tomato Peroxidase Inactivation by Atmospheric Pressure Cold Plasma Based on Dielectric Barrier Discharge." *Innovative Food Science & Emerging Technologies* 19: 153–157.

Pankaj, Shashi K., Zifan Wan, and Kevin M. Keener. 2018. "Effects of Cold Plasma on Food Quality: A Review." *Foods* 7 (1): 4.

Patange, Apurva, Daniela Boehm, Dana Ziuzina, P. J. Cullen, Brendan Gilmore, and Paula Bourke. 2019. "High Voltage Atmospheric Cold Air Plasma Control of Bacterial Biofilms on Fresh Produce." *International Journal of Food Microbiology* 293: 137–145.

Patil, S., P. Bourke, and P. J. Cullen. 2016. "Principles of Nonthermal Plasma Decontamination." In *Cold Plasma in Food and Agriculture*, 143–177. Academic Press.

Pedrow, Patrick, Zi Hua, Shuzheng Xie, and Mei-Jun Zhu. 2020. "Engineering Principles of Cold Plasma." In *Advances in Cold Plasma Applications for Food Safety and Preservation*, 3–48. Academic Press.

Porto, Elaine, Elenilson G. Alves Filho, Lorena Mara A. Silva, Thatyane Vidal Fonteles, Ronnyely Braz Reis do Nascimento, Fabiano AN Fernandes, Edy Sousa de Brito, and Sueli Rodrigues. 2020. "Ozone and Plasma Processing Effect on Green Coconut Water." *Food Research International* 131: 109000.

Puligundla, Pradeep, and Chulkyoon Mok. 2021. "Valorization of Sugar Beet Pulp Through Biotechnological Approaches: Recent Developments." *Biotechnology Letters*: 1–11.

Puligundla, Pradeep, Je-Wook Kim, and Chulkyoon Mok. 2017. "Effect of Corona Discharge Plasma Jet Treatment on Decontamination and Sprouting of Rapeseed (*Brassica napus* L.) Seeds." *Food Control* 71: 376–382.

Ramazzina, Ileana, Annachiara Berardinelli, Federica Rizzi, Silvia Tappi, Luigi Ragni, Giampiero Sacchetti, and Pietro Rocculi. 2015. "Effect of Cold Plasma Treatment on Physico-Chemical Parameters and Antioxidant Activity of Minimally Processed Kiwifruit." *Postharvest Biology and Technology* 107: 55–65.

Rana, Sudha, Deepak Mehta, Vasudha Bansal, U. S. Shivhare, and Sudesh
Kumar Yadav. 2020. "Atmospheric Cold Plasma (ACP) Treatment
Improved In-Package Shelf-Life of Strawberry Fruit." *Journal of Food
Science and Technology* 57 (1): 102–112.

Ribeiro, Kenny C. S., Nathalia M. Coutinho, Marcello R. Silveira, Ramon S. Rocha,
Henrique S. Arruda, Gláucia Maria Pastore, Roberto P. C. Neto, et al. 2021.
"Impact of Cold Plasma on the Techno-Functional and Sensory Properties
of Whey Dairy Beverage Added with Xylooligosaccharide." *Food Research
International* 142: 110232.

Rød, Sara Katrine, Flemming Hansen, Frank Leipold, and Susanne Knøchel.
2012. "Cold Atmospheric Pressure Plasma Treatment of Ready-to-Eat Meat:
Inactivation of *Listeria Innocua* and Changes in Product Quality." *Food
Microbiology* 30 (1): 233–238.

Rossow, Marko, Martina Ludewig, and Peggy Gabriele Braun. 2018. "Effect
of Cold Atmospheric Pressure Plasma Treatment on Inactivation of
Campylobacter Jejuni on Chicken Skin and Breast Fillet." *LWT* 91: 265–270.

Sadhu, Subham, Rohit Thirumdas, R. R. Deshmukh, and U. S. Annapure. 2017.
"Influence of Cold Plasma on the Enzymatic Activity in Germinating
Mung Beans (Vigna Radiate). *LWT* 78: 97–104.

Sarangapani, Chaitanya, Apurva Patange, Paula Bourke, Kevin Keener, and P. J.
Cullen. 2018. "Recent Advances in the Application of Cold Plasma Technology
in Foods." *Annual Review of Food Science and Technology* 9: 609–629.

Sarangapani, Chaitanya, Rohit Thirumdas, Yamuna Devi, Ajinkya Trimukhe,
Rajendra R. Deshmukh, and Uday S. Annapure. 2016. "Effect of Low-
Pressure Plasma on Physico – Chemical and Functional Properties of
Parboiled Rice Flour." *LWT-Food Science and Technology* 69: 482–489.

Sarangapani, Chaitanya, Yamuna Devi, Rohit Thirundas, Uday S. Annapure,
and Rajendra R. Deshmukh. 2015. "Effect of Low-Pressure Plasma on
Physico-Chemical Properties of Parboiled Rice." *LWT-Food Science and
Technology* 63 (1): 452–460.

Schnabel, Uta, Mathias Andrasch, Klaus-Dieter Weltmann, and Jörg Ehlbeck.
2014. "Inactivation of Vegetative Microorganisms and *Bacillus Atrophaeus*
Endospores by Reactive Nitrogen Species (RNS)." *Plasma Processes and
Polymers* 11 (2): 110–116.

Segat, Annalisa, N. N. Misra, P. J. Cullen, and Nadia Innocente. 2016. "Effect of
Atmospheric Pressure Cold Plasma (ACP) on Activity and Structure of
Alkaline Phosphatase." *Food and Bioproducts Processing* 98: 181–188.

Sharma, Shruti, and Rakesh Singh. 2020. "Cold Plasma Treatment of Dairy
Proteins in Relation to Functionality Enhancement." *Trends in Food Science
& Technology* 102: 30–36.

Shemakhin, A. Yu, and V. S. Zheltukhin. 2019. "Mathematical Modelling of RF
Plasma Flow at Low Pressures with 3D Electromagnetic Field." *Advances in
Materials Science and Engineering* 2019, Article ID 7120217: 6.

Silveira, Marcello R., Nathalia M. Coutinho, Erick A. Esmerino, Jeremias Moraes,
Leonardo M. Fernandes, Tatiana C. Pimentel, Monica Q. Freitas, et al. 2019.
"Guava-Flavored Whey Beverage Processed by Cold Plasma Technology:
Bioactive Compounds, Fatty Acid Profile and Volatile Compounds." *Food
Chemistry* 279: 120–127.

Singh, Pranav K., and Thom Huppertz. 2020. "Effect of Nonthermal Processing
on Milk Protein Interactions and Functionality." *Milk Proteins*: 293–324.

Surowsky, Bjoern, Axel Fischer, Oliver Schlueter, and Dietrich Knorr. 2013. "Cold
Plasma Effects on Enzyme Activity in a Model Food System." *Innovative
Food Science & Emerging Technologies* 19: 146–152.

Sutar, Swapnil A., Rohit Thirumdas, Bhushan B. Chaudhari, Rajendra R. Deshmukh, and Uday S. Annapure. 2021. "Effect of Cold Plasma on Insect Infestation and Keeping Quality of Stored Wheat Flour." *Journal of Stored Products Research* 92: 101774.

Swamy, Gabriela John, and Kasiviswanathan Muthukumarappan. 2021. "Microwave and Radiofrequency Processing of Plant-Related Food Products. In *Innovative Food Processing Technologies,* 731–742. Elsevier, USA.

Szabó, Dorothée Vinga, and Sabine Schlabach. 2014. "Microwave Plasma Synthesis of Materials- From Physics and Chemistry to Nanoparticles: A Materials Scientist's Viewpoint." *Inorganics* 2 (3): 468–507.

Tappi, Silvia, Giorgia Gozzi, Lucia Vannini, Annachiara Berardinelli, Santina Romani, Luigi Ragni and Pietro Rocculi. 2016. "Cold plasma treatment for fresh-cut melon stabilization." *Innovative Food Science and Emerging Technologies* 33: 225–233.

Thirumdas, Rohit, A. Trimukhe, R. R. Deshmukh, and U. S. Annapure. 2017. "Functional and Rheological Properties of Cold Plasma Treated Rice Starch." *Carbohydrate Polymers* 157: 1723–1731.

Thirumdas, Rohit, and Uday S. Annapure. 2020. "Enzyme Inactivation in Model Systems and Food Matrixes by Cold Plasma." In *Advances in Cold Plasma Applications for Food Safety and Preservation,* 229–252. Academic Press.

Thirumdas, Rohit, Anjinelyulu Kothakota, Uday Annapure, Kaliramesh Siliveru, Renald Blundell, Ruben Gatt, and Vasilis P. Valdramidis. 2018. "Plasma Activated Water (PAW): Chemistry, Physico-Chemical Properties, Applications in Food and Agriculture." *Trends in Food Science & Technology* 77: 21–31.

Thirumdas, Rohit, C. Saragapani, M. T. Ajinkya, R. R. Deshmukh, and U. S. Annapure. 2016. "Influence of Low Pressure Cold Plasma on Cooking and Textural Properties of Brown Rice." *Innovative Food Science & Emerging Technologies* 37: 53–60.

Thirumdas, Rohit, R. R. Deshmukh, and U. S. Annapure. 2015. "Effect of Low Temperature Plasma Processing on Physicochemical Properties and Cooking Quality of Basmati Rice." *Innovative Food Science & Emerging Technologies* 31: 83–90.

Tolouie, Haniye, Mohammad Amin Mohammadifar, Hamid Ghomi, and Maryam Hashemi. 2021. "Argon and Nitrogen Cold Plasma Effects on Wheat Germ Lipolytic Enzymes: Comparison to Thermal Treatment." *Food Chemistry* 346: 128974.

Turner, M. 2016. "Physics of Cold Plasma." In *Cold Plasma in Food and Agriculture,* 17–51. Academic Press.

Ulbin-Figlewicz, Natalia, Andrzej Jarmoluk, and Krzysztof Marycz. 2015. "Antimicrobial Activity of Low-Pressure Plasma Treatment Against Selected Foodborne Bacteria and Meat Microbiota." *Annals of Microbiology* 65 (3): 1537–1546.

Upadhyay, R., R. Thirumdas, R. R. Deshmukh, U. Annapure, and N. N. Misra. 2020. "An exploration of the effects of low-pressure plasma discharge on the physico-chemical properties of chia (*Salvia hispanica* L.) flour." *Journal of Engineering &Amp; Processing Management* 11 (2): 73–80.

Wang, Jiamei, Hong Zhuang, Arthur Hinton Jr, and Jianhao Zhang. 2016. "Influence of In-Package Cold Plasma Treatment on Microbiological Shelf Life and Appearance of Fresh Chicken Breast Fillets." *Food Microbiology* 60: 142–146.

Wang, Yijie, Zichong Ye, Jiahui Li, Yan Zhang, Yingxi Guo, and Jun-Hu Cheng. 2021. "Effects of Dielectric Barrier Discharge Cold Plasma on the Activity, Structure and Conformation of Horseradish Peroxidase (HRP) and on the Activity of Litchi Peroxidase (POD)." *LWT* 141: 111078.

Wende, Kristian, Thomas von Woedtke, Klaus-Dieter Weltmann, and Sander Bekeschus. 2018. "Chemistry and Biochemistry of Cold Physical Plasma Derived Reactive Species in Liquids." *Biological Chemistry* 400 (1): 19–38.

Won, M. Y., Seung Jo Lee, and Sea C. Min. 2017. "Mandarin preservation by microwave-powered cold plasma treatment." *Innovative Food Science and Emerging Technologies 39*: 25–32.

Wong, San Chiow, and Rattachat Mongkolnavin. 2016. "Methods of Plasma Generation." In *Elements of Plasma Technology*, 15–48. Springer.

Xu, Lei, Allen L. Garner, Bernard Tao, and Kevin M. Keener. 2017. "Microbial Inactivation and Quality Changes in Orange Juice Treated by High Voltage Atmospheric Cold Plasma." *Food and Bioprocess Technology* 10 (10): 1778–1791.

Yadav, Dharmendra Kumar, Manish Adhikari, Surendra Kumar, Bhagirath Ghimire, Ihn Han, Mi-Hyun Kim, and Eun-Ha Choi. 2020. "Cold Atmospheric Plasma Generated Reactive Species Aided Inhibitory Effects on Human Melanoma Cells: An in Vitro and in Silico Study." *Scientific Reports* 10 (1): 1–15.

Yepez, Ximena V., and Kevin M. Keener. 2016. "High-Voltage Atmospheric Cold Plasma (HVACP) Hydrogenation of Soybean Oil Without Trans-Fatty Acids." *Innovative Food Science & Emerging Technologies* 38: 169–174.

Yong, Hae In, Haelim Lee, Sanghoo Park, Jooyoung Park, Wonho Choe, Samooel Jung, and Cheorun Jo. 2017. "Flexible Thin-Layer Plasma Inactivation of Bacteria and Mold Survival in Beef Jerky Packaging and its Effects on the Meat's Physicochemical Properties." *Meat Science* 123: 151–156.

Zhang, Hao, Zimu Xu, Jie Shen, Xu Li, Lili Ding, Jie Ma, Yan Lan, et al. 2015. "Effects and Mechanism of Atmospheric-Pressure Dielectric Barrier Discharge Cold Plasmaon Lactate Dehydrogenase (LDH) Enzyme." *Scientific Reports* 5 (1): 1–12.

Zhang, Xiao-Lin, Chong-Shan Zhong, Arun S. Mujumdar, Xu-Hai Yang, Li-Zhen Deng, Jun Wang, and Hong-Wei Xiao. 2019. "Cold Plasma Pretreatment Enhances Drying Kinetics and Quality Attributes of Chili Pepper (Capsicum annuum L.)." *Journal of Food Engineering* 241: 51–57.

Zhou, Renwu, Rusen Zhou, Jinxing Zhuang, Zichao Zong, Xianhui Zhang, Dongping Liu, Kateryna Bazaka, and Kostya Ostrikov. 2016. "Interaction of Atmospheric-Pressure Air Microplasmas with Amino Acids as Fundamental Processes in Aqueous Solution." *PloS One* 11 (5): e0155584.

Zhou, Yu-Hao, Sriram K. Vidyarthi, Chong-Shan Zhong, Zhi-An Zheng, Yu An, Jun Wang, Qing Wei, and Hong-Wei Xiao. 2020. "Cold Plasma Enhances Drying and Color, Rehydration Ratio and Polyphenols of Wolfberry Via Microstructure and Ultrastructure Alteration." *LWT* 134: 110173.

Zhu, Yulin, Changzhu Li, Haiying Cui, and Lin Lin. 2020. "Feasibility of Cold Plasma for the Control of Biofilms in Food Industry." *Trends in Food Science & Technology* 99: 142–151.

Zhuang, Hong, Michael J. Rothrock Jr, Kelli L. Hiett, Kurt C. Lawrence, Gary R. Gamble, Brian C. Bowker, and Kevin M. Keener. 2019. "In-Package Air Cold Plasma Treatment of Chicken Breast Meat: Treatment Time Effect." *Journal of Food Quality* 2019.

Ziuzina, Dana, Sonal Patil, Patrick J. Cullen, K. M. Keener, and Paula Bourke. 2014. "Atmospheric Cold Plasma Inactivation of *Escherichia Coli*, *Salmonella Enterica* Serovar Typhimurium and *Listeria Monocytogenes* Inoculated on Fresh Produce." *Food Microbiology* 42: 109–116.

4 Ultrasound Technology for Food Preservation

Thota Niranjan, Rifna E J, Salman Hashmi, and Madhuresh Dwivedi

Corresponding author:
Madhuresh Dwivedi; Email: dwivedim@nitrkl.ac.in

CONTENTS

4.1 INTRODUCTION

Food is a biological substance originating from plants and animal sources, and it is composed of mainly macro- and micronutrients like carbohydrates, proteins, lipids, minerals, vitamins, and other bioactive compounds. Food spoils due to enzymatic and microbial activities. The physical and chemical actions on the nutrients by microorganisms result in food spoilage such as changes in color, texture, and nutrient value that make the food inedible (Amit et al., 2017). Annually, it is estimated that around the globe, about one-third of the food produced is wasted (FAO, 2015). Fulfilling the food demands of the increasing population while conserving environmental resources is a serious challenge. Decreasing food waste and increasing production and supply are the solutions to cope with the demand and to reduce the global hunger. Food preservation is the key that can be used by food processors and producers to mitigate this challenge. Food waste is generated due to a lack of preservation techniques and a delay in their processing. Therefore, the proper processing, as well as preservation techniques, is required to minimize spoilage and retain the texture, color, flavor, and nutritive value of the food. Since ancient times, people realized the importance of storing and preserving food for use in off-seasons (Amit et al., 2017). Drying, fermentation, and thermal processing were used for food preservation from primitive days to inactivate enzymes and arrest microbial growth. Thermal processing employs high temperatures and reduces microbial growth, but it also effects undesirable quality changes in the food. Food contains a high number of valuable compounds, many of which are heat-sensitive that may not sustain during thermal processing. The conventional thermal processing has low energy efficiency, with low drying rates and high drying times. The aforementioned shortcomings led to the development of novel innovative techniques in food preservation that have higher efficiency, require less time, and are environmentally friendly. New emerging technologies in preservation techniques are freezing, microwave, pulsed light, ultraviolet (UV) light, ozone, pulsed electric field, ionization radiation, high pressure, ultrasound, and cold plasma (Rifna et al., 2019). Among these, ultrasound is gaining importance in the

DOI: 10.1201/9781003147978-4

food preservation and is reported to have a lethal effect on microbes while retaining the quality attributes of food over conventional thermal technologies. When compared with other energies like heat and mechanical, acoustic energy is comparably new in food processing. Ultrasound is energy in form of sound waves that cause variation in the medium pressure and form high- and low-pressure zones. These differences in pressure are generated by the mechanical action. In ultrasound, the generated pressure-difference zones affect the biochemical and physical changes in the food. The ultrasound technology is noninvasive, as this technology does not use harmful radiation (Rifna & Dwivedi, 2021). There is no damage or loss in nutrient content as the ultrasound operates at ambient temperature. To accelerate the rate of food pasteurization and decrease processing time, ultrasound is used in combination with mild temperatures and pressures (Awad et al., 2012).

4.2 FUNDAMENTALS, PRINCIPLES, AND PROCESS DESCRIPTION

Ultrasound is mechanical waves that transmit through the medium, and their frequency range is above that perceived by human hearing (~20 kHz; Figure 4.1). Ultrasound is similar to normal sound waves in its physical properties except that it is not audible. Animals such as dolphins and bats have the ability to hear ultrasound frequency, and they communicate through ultrasound to locate prey and obstacles. Ultrasound, when transmitted through a biological material, induces mechanical variations of low-pressure zones (rare fractions) and high-pressure zones (compression). Based on the power and frequency of ultrasound, it is used for a variety of purposes in many different fields. Traditionally, ultrasound was first used in the marine industry for locating and navigation of submarines. Sonography, or ultrasound imaging, is often used in the medical field as a diagnostic technique for treating some of the diseases. In the construction field, it is used to detect cracks and flaws. In the food and chemical industries, it is used in extraction, preservation, filtration, freezing, thawing, drying, degassing, emulsification, and more. The phenomena used in food processing are fundamentally different compared with other conventional applied fields. A transducer is used for generating the sound waves in ultrasonication. There are two types of transducers: (1) piezoelectric and (2) magnetostrictive. A Langevin piezoelectric transducer is used for generating the sound vibration in ultrasonication. The transducer consists of two ceramic

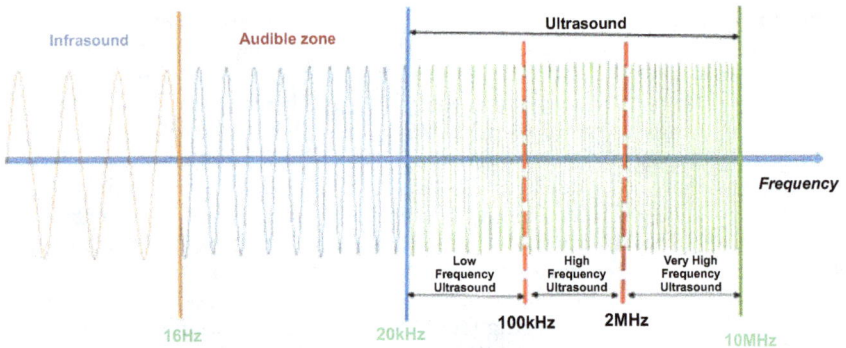

Figure 4.1 Frequency ranges of sound.

elements, which change their shape according to the applied electric field. The up-and-down motion is produced when an alternating electric field is applied. The sonication wave is primarily distinguished by four parameters: wavelength (cm), wave frequency (Hz), intensity (W/cm²), and ultrasonic power (W) (Kentish, 2017). Based on power and frequency used, ultrasound is broadly classified into two types: (1) low-intensity (P < 1 W), high-frequency (2–10 MHz) ultrasound and (2) high-intensity (P>1W), low-frequency (20–100 kHz) ultrasound. Low-intensity, high-frequency ultrasound is a nondestructive analytical technique used to monitor the process or products. The physicochemical properties of food, such as structure, physical state, and composition, are usually analyzed using high-frequency ultrasound. Low-frequency ultrasound and high-intensity ultrasound are used to enhance the process or alter the physicochemical properties of the food material (Bhargava et al., 2020). The chemical changes including oxidation occur at low frequency range due to the free radical formation. High-intensity ultrasound is used in food preservation to activate or inactivate the enzyme, and it is also used in protein isolation.

4.3 MECHANISMS OF UT AND INFLUENCING FACTORS

In ultrasound, physicochemical reactions in biological materials are predominantly due to the phenomenon known as cavitation. The cavitation referred to bubble initiation, bubble growth, and bubble implosion. When high-intensity ultrasound is applied to solid (with moisture content) or liquid food, the sound waves propagate through the medium by imparting cavitation. Cavitation is the concentration of acoustic energy in a small area and the conversion of concentrated acoustic energy into the extreme conditions of pressure and temperature. Upon sonication of food products, the sound waves propagate into the elastic medium and result in altering compression (high-pressure phase) and rarefaction (low-pressure phase) depending on the frequency. A series of high-pressure and low-pressure cycles is called a sponge effect. During this acoustic pressure difference, the molecules in the medium move from their equilibrium position. Voids are formed when the applied ultrasound power is higher than the cohesive forces between the molecules. During the compression, acoustic energy creates bubbles or voids in the medium by the gases or vapors. The formed bubbles are unstable at high pressures. In accordance with the periodic fluctuations in the pressure, the size of the bubble varies. By rectified diffusion, the bubble grows during high-pressure and low-pressure cycles. The gases or vapors are expelled, but the volume of the expelled gases is significantly lower than for diffused gases. When the bubbles grow to a critical point after many cycles, the bubble cannot absorb energy further, and they tend to collapse in the high-pressure cycle by producing high localized pressure and temperature. This phenomenon is known as "cavitation". Cavitation is associated with localized heating, shear disruption, and free radical formation. During the collapse, very high temperature of approximately 5000K and a pressure of approximately 2000 atm. are reached locally (Figure 4.2). Cavitation is considered one of the main reasons for microbial inactivation. The other reason for microbial inactivation is due to the generation of microcurrent or microstreaming. The micro-current created around the bubbles has the ability to catalyze chemical reactions and disrupt microorganisms. The collapse of the bubbles during cavitation produces high-velocity (280 m/s) liquid jets. The formation of hydrogen peroxide and free radicals results during the application of high-power ultrasound are also responsible for bactericidal properties.

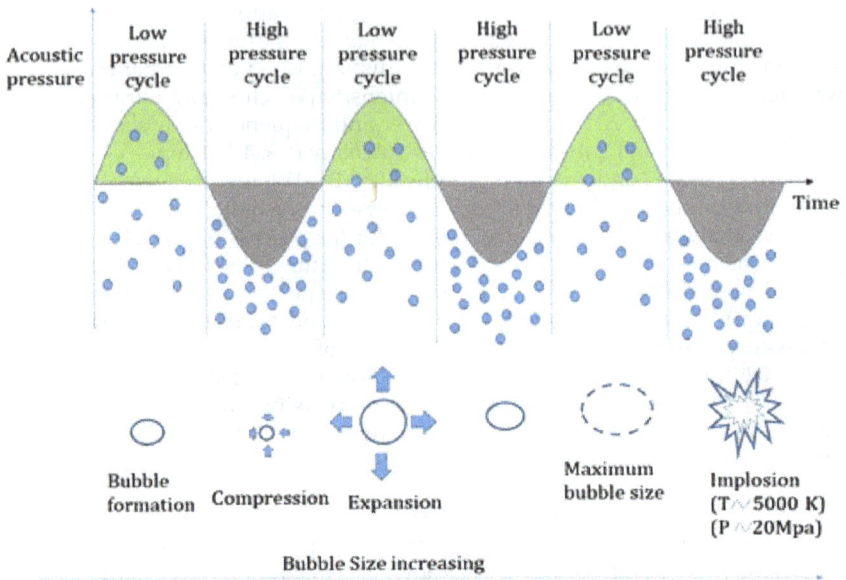

Figure 4.2 Cavitation phenomena in ultrasound technology.

4.4 EFFECTS OF POWER ULTRASOUND

4.4.1 Effects on Microorganisms

The ill effects of conventional heat treatments in food preservation like the denaturation of protein and the loss of bioactive compounds and vitamins have led to the development of novel nonthermal technologies. Ultrasound has gained attention in microbial inactivation as it has the ability to effect physiochemical changes in the food at ambient temperatures. Late in the 1920s, the sonication treatment was first used for bacterial inactivation and in 1960s' research on the mechanism of lethal effect of ultrasound on bacteria started (Harvey & Loomis, 1929, Earnshaw et al., 1995). It is considered that the cavitation is prime reason for microbial inactivation. During the high-pressure phase, the particles in the medium get compressed, resulting in an increase in pressure in the medium. In the low-pressure phase, the particles move apart, resulting in a decease in pressure in the medium. This difference in the pressure and densities in the medium creates thousands of bubbles or voids; the formed bubbles are unstable at high pressures and tend to collapse in the high-pressure cycle by producing high localized pressure and temperatures. This physical phenomenon is termed as cavitation. Cavitation is associated with localized heating, shear disruption, and free radical formation. During the collapse, very high temperatures and pressures are reached locally. In addition to cavitation, microbial inactivation occurs due to microcurrent or microstreaming, sonoluminescence, free radical formation, chemiluminescence, and heating. The microcurrent created around the bubbles has the ability to catalyze chemical reactions and disrupt microorganisms. The collapse of bubbles during cavitation results in liquid jets. The dissociation of water results in free radical formation. The lethal effect of microorganisms is due to mechanical stress and the free radical formation during cavitation. The formation of free radicals and hydrogen peroxide during the application of high-power ultrasound with bactericidal properties can damage the DNA of the microorganisms. Due to cell wall thickness, gram-negative bacteria are more susceptible than gram-positive bacteria. There are mixed results reported on the gram status and sonication treatment. Some researchers have reported that gram-negative bacteria are more susceptible to sonication treatment when compared with gram-positive bacteria (Chantapakul et al., 2019). On other hand, it has been reported that there is no remarkable relation between the sonication treatment and gram status. Cameron et al. (2008) reported that microorganisms with larger cells are more susceptible to ultrasonication than those with smaller cells, and rod-shaped bacteria are highly susceptible to sonication treatment than coccus-shaped bacteria. On the other hand, Cameron (2007) reported that there is no significant relation between the effects of ultrasound treatment and the size and shape of microbial cells. The lethal effect of ultrasound alone is less effective in destroying the microorganism.

Ultrasound is used in combination with heat and pressure to get synergistic effects that have shown to be more effective in inactivating microorganisms and enzymes. Ultrasound treatment in combination with excess pressure is called "manosonication", and in combination with excess temperature is called "thermosonication". The lethal effect sonication of microorganisms is greater when the applied temperature is more than 50°C (Tsukamoto et al., 2004). Ultrasound treatment in combination with excess pressure and temperature is called "manothermosonication", and this is more effective in killing spores (Table 4.1). Ultrasonic-assisted sterilization of milk is very effective in inactivation *Pseudomonas fluorescens*, *Listeria monocytogenes*, and *E. coli* with no adverse effect on the milk protein (Cameron et al., 2009). Various other studies on the effect of ultrasonication on pathogenic and spoilage microorganisms are summarized in Table 4.2.

4.4.2 Effects on food enzymes

Ultrasound has been investigated extensively in recent years for inhibiting the enzymatic deterioration of food. Cavitation and formation of free radicals in ultrasound treatment can lead to enzyme inactivation. At sufficiently low levels of sonication, changes in structure and metabolism can occur in cells without destructing them. Depolymerization also causes the inactivation of the enzymes in ultrasound treatment. Ultrasonication accelerates the biological activity of enzymes by altering the conformation, and this increases the contact between substrate and enzyme. It creates a suitable environment for enzymatic reactions. Ma et al. (2011) studied the effect of ultrasonic treatment on protease activity and reported that the highest alcalase activity was found at 80 W for 4 min and enzymatic activity increased by 5.8%, when compared with control. The thermodynamic parameters E_a, ΔG, ΔS, and ΔH were decreased by 70.0%, 1.3%, 34.0%, and 75.8%, respectively, after ultrasonic treatment. Wang et al. (2020) studied the

Table 4.1 Lethal Effect of Sonication with Temperature and Pressure

Inactivation by	Vegetative cells	Spores	Enzymes
Manosonication	✓	–	–
Thermosonication	✓	✓	–
Manothermosonication	✓	✓	✓

Source: Khadhraoui et al. (2019)

Table 4.2 Ultrasound Treatment in Inactivation of Microbes

Targeted micro-organism	Product/ medium	Experimental variables	Outcomes	Reference
Staphylococcus aureus	Orange juice	20 kHz 20, 30, 40, 50, and 60°C 0–90 min	• Most effective lethality of *Staphylococcus aureus*, 10.60 ± 0.13 log reduction was found at 60°C for 30 min at 20 kHz. • At 50°C for 30 min and 20 kHz a 9.44 ± 1.35 log reduction of *Staphylococcus aureus* was reported.	Tahi et al. (2021)
Escherichia coli, Staphylococcus aureus	Peptone water	45% amplitude for *E. coli;* 50% amplitude for *S. aureus.*	• Thermo-sonication treatment for 0.38 min at 60°C and 62°C reduced *E. coli and S. aureus* respectively, by 5-fold.	Baboli et al. (2020)
E. coli	Phosphate Buffer	20 kHz; 100 W; 100–500 kPa; 0.25–4 min; 40°C–61°C.	• With increase in the pressure the lethal effect on the *E. coli* increased • There is no significant effect of temperature up to 54°C. • The processing time reduced significantly.	Lee et al. (2009)
Yeast, bacteria and mold	Cherry	33 kHz, 60 W Time: 10, 20, 30, 40, 60 min	• With increase in the processing time the microbial load decreased • Ultrasound is effective in retaining the quality parameters	Muzaffar et al. (2016)
Saccharomyces cerevisiae	Apple, YPD Broth, and orange juice	100–350 bar; 31–41°C; Supercritical flow-co$_2$ assisted.	• There is no significant increase in the inactivation kinetics with application pressure and temperature. • Pressure higher than 100 bar and temperature 31°C is required to reach 7-log cycle.	Ortuno et al. (2013)
Yeast, bacteria, and mold	Peanut milk	200W, 300W and 400W	• Highest log reduction was found at 10-bar pressure. • 400W US treatment was effective for molds and yeast	Salve et al. (2019)
E. aerogenes and *Staphylococcus* spp.	Bacterial suspension	20kHz 20min 13 W	• *B. subtilis* and *E. aerogenes* and reduced up to 4.5 log cycles. • *Staphylococcus* spp. was not susceptible. • *E. aerogenes* suspensions were more sensitive. • The bacterial resistance is due to the bacterial capsule properties.	Gao et al. (2014)

Targeted micro-organism	Product/ medium	Experimental variables	Outcomes	Reference
Escherichia coli	Bacterial suspension	64, 191, 372, and 573 Wcm^{-2}; 20 kHz	• The increasing US intensity and time reduced the E. coli count by 2.16–7.10 log CFU/mL. • Bacterial cell wall became sensitive with increasing US intensity and time.	He et al. (2021)
Bacteria and fungi	Strawberry fruits	40 kHz. 3, 6 and 9 min. Ozone, flow rate ≥ 3.3 mg min^{-1}	• US treatment for 3 min reduced bacterial survival by 17-fold. • Ozone treatment for 3 min reduced bacterial survival by 44-fold. • Combined Ozone and US was very effective in reducing 98% bacterial survival	Maryam et al. (2021)
Pathogen micro-organism	Mango juice	0–600 W; 0–40 min; ultraviolet-assisted	• Total sterilization was achieved at ultrasound processing conditions of 600 W for 10 min. • The carotenoid content was enhanced for 3-fold.	Wang et al. (2020)

effect of UV-assisted ultrasound on peroxidase, pectin methylesterase, and polyphenol oxidation enzymes in mango juice and reported that there was a significant inhibition of peroxidase, pectin methylesterase, and polyphenol oxidation enzyme at the optimal ultrasound treatment at 600 W for 10 min. Sonication significantly inactivated the enzyme activity, Chantapakul et al. (2020) studied the effect of sonication, manothermosonication on soya milk enzymes and reported a synergic effect on enzyme activity by manothermosonication; the process time was significantly reduced with application of mild heat and pressure with the sonication. Li et al. (2016) evaluated the effects on enzymolysis of rice proteins (RPs) by multifrequency energy-gathered ultrasound (MFEGU), and MFEGU-assisted alkaline pretreatments. It was reported that the degree of hydrolysis and protein elution amount of rice protein were improved significantly in both the methods. The particle size and protein microstructure were destroyed in MFEGU and MFEGU-assisted alkaline pretreatments. Li et al. (2016) explored the effect of ultrasonic pretreatment using both an ultrasonic cleaning bath and a probe-type sonicator on the enzymolysis of defatted wheat germ protein (DWGP). They reported that ultrasound and temperature had a significant effect on the enzymolysis of DWGP, with temperature having detrimental role. Polypeptide concentrations showed high (231.320 and 231.019 lg/mL) in both ultrasonic cleaning bath and a probe-type sonicator under optimized conditions. There was a significant increase in the reaction rate constant (k) with the increase in the temperature. The thermodynamic parameters E_a, ΔG, ΔS, and ΔH were reduced by 68.6%

and 62.4%, 34.3% and 31.2%, 1.4% and 1.3%, and 74.1% and 67.5%, respectively. Kuldiloke et al. (2007) studied the effect of temperature and ultrasound on lemon pectinesterase and reported that a greater than 90% inactivation occurred with ultrasound when compared with the thermal process. Ultrasonic treatment alone or in combination with pressure and temperature has been reported to be more effective in deactivating enzymes like pectin, peroxidases, polyphenoloxidases, and lipooxiginase with ceasing the degradation of milk products, vegetables, and fruit juices (O'Donnell et al., 2010; Table 4.3).

Table 4.3 Effects of Ultrasound Treatment on Enzyme Activity

Targeted enzyme	Product/ medium	Experimental variables	Outcome	Reference
Peroxidase, pectin methyl esterase (PME), and polyphenol oxidation enzyme	Mango juice	0–600 W; 0–40 min; ultraviolet: assisted;	• There was a significant inhibition of peroxidase, pectin methyl esterase, and polyphenol oxidation enzyme at the optimal US treatment (10 min, 600 W).	Wang et al. (2020)
Peroxidase and lipoxygenase	Almond milk	40 kHz; 600 W; 30, 45, and 60°C; 10, 20, 30, and 40 min.	• Thermo-sonication significantly inhibited the peroxidase and lipoxygenase activity at 60°C for 30 min. • Reported 5 log cycles reduction in the yeast and 4 log cycles reduction in molds at 60 ᵒC.	Manzoor et al. (2021)
Lipoxygenase, peroxidase, and polyphenol oxidase	Soya milk	25 kHz; 25, 65, 70°C; 0, 0.5, 1, 2, 3 and 5 min; 100 and 400 kPa;	• Sonication significantly inactivated the enzyme activity. • It was reported to have a synergic effect on enzyme activity by pressure and temperature with sonication. • The process time significantly reduced with application of mild heat and pressure with the sonication.	Chantapakul et al. (2020)
PME and poly-galacturonase (PG)	Tomato juice	20 kHz; 65 µm; 50 and 75°C.	• Thermosonication increased inactivation rates of both PG and PME. • PG and PME inactivation rates were elevated by 2.3–4 times and 1.5–6 times, respectively, in 60–75°C temperature range.	Terefe et al. (2009)

Targeted enzyme	Product/ medium	Experimental variables	Outcome	Reference
Protease	Alcalase	20 kHz; 80 W; 20, 30, 40, 50°C	• The highest alcalase activity was found at 80 W for 4 min. • Enzymatic activity increased by 5.8%, when compared with control.	Ma et al. (2011)
Pectinesterase	lemon	20 kHz; 40°C-90°C.	• Reported there was a 90% inactivation occurred with ultrasound when compared with thermal process.	Kuldiloke et al. (2007)
Polyphenol oxidase	Mushroom	25 kHz 55–75 °C	• At the same temperature range the D-value of thermo-sonicated mushroom significantly reduced when compared with thermal processing. • First-order kinetic model fitted for inactivation of polyphenol oxidase.	Cheng et al. (2013)

4.4.3 Effects on food quality and sensory attributes

In ultrasonication, due to cavitation, the degradation of color pigments that takes place affects the color of the food. The degradation depends on the treatment conditions and the behavior of intercellular material during cavitation. After ultrasonic treatment, color changes have been reported in juices due to nonenzymatic browning (Rojas et al., 2016; Bi et al., 2015). The suspension of mixture of compounds like pectin, cellulose, hemicellulose, protein, lipids, and other compounds (salts, sugars, etc.) is attributed as cloud value. Due to the homogenization effect of ultrasonication, the cloud value of juices will increase. Ertugay and Baslar (2014) studied the effect of ultrasound on apple juice stability and reported an increase of 9.8 times in the cloud stability and 16.9 times in the cloud value. Cavitation and micro jets in ultrasonic treatment reduced the size of the macroparticles, reducing the sedimentation in the juices during storage. Fragmentation, increasing the relative particle surface, and structural modifications lead to the particle–particle and particle–serum interactions that improve rheological properties, reduce sedimentation, and increase turbidity. The fragmentation of macroparticles and high shear action in cavitation reduce the viscosity of juices, but in some juices, the viscosity increases due to the release of intercellular material. Bi et al. (2015) reported that the viscosity of avocado increased with ultrasonic treatment. Aadil et al. (2013) investigated the effect of sonication on the quality parameters of grape juice and reported a significant increase in ascorbic acid, 2,2-diphenyl-1-picrylhydrazyl (DPPH) free radical scavenging activity, flavonoids, total antioxidant capacity, and cloud value of treated juice. A slight difference in the color was observed, but there was no change in the acidity, Brix, and pH of the treated juice. Abid et al. (2014) investigated the effect of sonication on the quality parameters of apple juice, such as viscosity, sugars, mineral content, polyphenolic compounds, total carotenoids, electrical conductivity, and total anthocyanin. They reported a significant increase in the sugars and polyphenolic compounds, but the increase was more predominant after 30-min treatment time. No change was observed in electrical conductivity and total anthocyanin; however, P, Mg, and Cu minerals were lost due to sonication.

4.5 EMERGING APPLICATIONS OF ULTRASOUND TREATMENT IN FOOD PRESERVATION

The literature study reported that ultrasound application in food preservation and processing was limited to milk and vegetable and fruit juices. The use of ultrasound in food preservation is increasing day by day. In the food preservation, the ultrasound is used in the combination with other technologies to have a lethal effect of sonic resistance microorganisms and enzyme. Thermosonication, manosonication, and thermomanosonication are now on the table for extensive research. The effects of these technologies on the microbial cells, enzymes, and the quality attributes of wide range of products are being explored. The combination of ultrasound with other technologies like UV light, ozone, high pressure, chemical treatments (chlorine dioxide and chlorine), supercritical fluid extraction, and others is being explored to enhance the effectiveness of the treatment. Wang et al. (2020) have studied the effect of ultraviolet-assisted ultrasound (US-UV) and reported a significant reduction in pathogen bacteria. More resistant gram-positive bacteria were inactivated using ultrasound-assisted supercritical CO_2 extraction (Castillo-Zamudio et al., 2021; Gomez-Gomez et al., 2020). Lee et al. (2003) have studied the effect of ultrasound-assisted high-pressure processing and presented a significant inactivation of *E. coli* in liquid whole egg. There is a further need to explore a combination of technologies for inactivating viruses and spores responsible for foodborne diseases. Various other studies on ultrasound in food preservation are summarized in Table 4.4.

4.6 CONCLUSION AND PERSPECTIVES

Ultrasound technology is an eco-friendly and nontoxic food preservation method that offers several benefits in terms of operation time, energy input, and conservation of nutritional and heat-sensitive bioactive compounds. Even though ultrasound treatment alone has a lethal effect on microorganisms, capable of inactivating the enzymes without affecting the beneficial bioactive compounds, through the literature study, it is evident that the combination of pressure and temperature with ultrasound treatment was found to have a synergetic effect in promoting the inactivation further and reducing the processing time. Many researchers have presented the inactivation mechanism of the cell and explored combinations to enhance microbial inactivation while retaining the quality attributes. Now the new challenges for researchers are inactivating viruses and spores responsible for foodborne diseases. Furthermore, the combination of sonication technology with other novel and conventional thermal technologies will further have better results on final product quality. Research on ultrasound technologies in food preservation was conducted to a great extent, but still future research is required to optimize the process and treatment parameters on wide range of products for application on an industrial level and the commercialization of ultrasound technology as a food preservation technique.

Table 4.4 Application of Ultrasound Treatment in Food Preservation

Method	Sample	Mechanism/details	Conclusions	References
Ultrasound	Milk	Amplitude: 70 and 100% Time: 50, 100, 200, and 300 s	The optimized treatment resulted *E. coli*, *D. hansenii*, *P. fluorescens*, and *S. aureus* in 2.09, 4.61, 2.75, and 0.55 log reduction respectively.	Marchesini et al. (2015).
	Whey beverage	Sonication power:0–600 W	High intensity sonication power (600W) treatment compared with HTST (75°C for 15 s) process for in the inactivation molds and yeasts found to have 0.2- and 0.4- and 2- and 2-log reduction, respectively.	Guimarães et al. (2018)
	Peach juice	Time: 0–15 minutes Sonication power:1000 W Frequency: 20 kHz	It was reported that the stability and physical properties of peach juice improved.	Rojas et al. (2016)
	Orange juice	Sonication power: 120 and 240 W at high frequency, Sonication power: 150, 300 and 600W at low frequency Time: 15 min.	At a selected sonication treatment (500 kHz/240 W for 15 min) there it was found to have a least microbial inactivation (61.08 log CFU mL/L).	Valero et al. (2007)
Ultrasound + Temperature	Milk	Frequency: 20 kHz Time: 10–15 min Temperature: 45°C	Similar microbiological quality and physio-chemical characteristics were exhibited by the thermosonicated milk samples when compared with pasteurized milk.	Hernández-Falcón et al. (2018).
	Orange juice	Frequency: 20 kHz Temperature: 20–60°C Time: 0–90 min	Most effective lethality of *Staphylococcus aureus*, 10.60 ± 0.13 log reduction was found at 60°C for 30 min at 20 kHz. At 50°C for 30 min and 20 kHz a 9.44 ± 1.35 log reduction of *Staphylococcus aureus* was reported.	Tahi et al. (2021)
Ultrasound + Temperature and pressure	Soya milk	Frequency: 25 kHz; Temperature: 25, 65, 70°C; Time: 0–5 min; Pressure: 100 and 400 kPa	It was reported to have a synergic effect on enzyme activity and process time significantly reduced with application of mild heat and pressure with the sonication.	Chantapakul et al. (2020)

(Continued)

Table 4.4 (Continued)

Method	Sample	Mechanism/details	Conclusions	References
	Phosphate Buffer	Frequency: 20 kHz; Sonication power: 100 W; Pressure: 100–500 kPa; Time: 0.25–4 min; Temperature: 40°C–61°C.	With increase in the pressure the lethal effect on the *E. coli* increased There is no significant effect of temperature up to 54°C. Processing time also reduced significantly	Lee et al. (2009).
Ultrasound + Chemical	Cherry tomatoes	Frequency: 45 kHz Time: 10 min Dosage: sodium dichloroisocyanurate: 20 and 200 mg/L. Peracetic acid: 40 mg/L. Hydrogen peroxide: 5%.	The hybrid treatment reduced the *Salmonella Typhimurium* by 3.9 log10 cfu/g and after sanitization treatments molds and yeasts were reduced by 1.1e3.4 log10 cfu/g while contamination of Cherry tomatoes was reduced by 0.7e4.4 log10 cfu/g	Brilhante São José and Dantas Vanetti (2012)
	Strawberry juice	Frequency: 40 kHz, Sonication power: 180 W Time: 0, 15 and 30 min. Dosage: Pomegranate extract: 0, 180 and 360 mg/mL And vanillin: 0, 0.625 and 1.25 mg/mL.	The optimal conditions reduced the inoculated *E. coli* and preserved the nutritional parameters. The hurdle technology increased the shelf life of the Strawberry juice.	Tomadoni et al. (2016)
Ultrasound + Freezing	Potato	Frequency: 20, 28 and 40 kHz Time: 120 s	Ultrasound-assisted freezing improved the quality attributes and increased the drying rates.	Zhu et al. (2020)
	Apple, radish and potato	Frequency: 28 kHz Sonication powder: 0.62 W cm^{-2} Time: 2min	Ultrasound-assisted freezing promoted the freezing rate and improved the generation of ice nuclei, resulting the shortening of the phase transition time.	Zhu et al. (2018)

Ultrasound + Ultraviolet	Pomegranate Juice	Sonication power: 165, 200, 295 W. Ultraviolet: Flow rate: 1.5,3.5 L/min, UV doses: 5.1 and 10.1 mW/cm² Temperature: 40, 50, and 60°C.	The application of combined ultrasound and ultraviolet treatment significantly reduces the microbial count below the detection point and the pasteurization occurred at lower temperatures and times when compared with conventional treatment	Alabdali et al. (2020)
	Mango juice	Sonication power: 0–600 W; Time: 0–40 min; Ultraviolet: 2 UV lamps Power: 8W, wavelength: 254 nm.	The combined effect has reduced time for complete sterilization of pathogen microorganisms and higher bioactive compounds were retained.	Wang et al. (2020)
Ultrasound + Supercritical flow CO_2	Apple, YPD Broth, and orange juice	100–350 bar; 31–41°C;	• There is no significant increase in the inactivation kinetics with application pressure and temperature. • Pressure higher than 100 bar and temperature 31°C is required to reach 7-log cycle.	Ortuno et al. (2013)
Ultrasound + Ozone	Strawberry fruits	40 kHz. 3, 6 and 9 min. Ozone, flow rate ≥ 3.3 mg min^{-1}	There was 6 more days increase in shelf life of the combined ozone and ultrasound treated strawberry fruits under cold storage conditions.	Maryam et al. (2021)

REFERENCES

Aadil, R. M., Zeng, X. A., Han, Z., & Sun, D. W. (2013). Effects of ultrasound treatments on quality of grapefruit juice. *Food Chemistry, 141*(3), 3201–3206.

Abid, M., Jabbar, S., Wu, T., Hashim, M. M., Hu, B., Lei, S., & Zeng, X. (2014). Sonication enhances polyphenolic compounds, sugars, carotenoids and mineral elements of apple juice. *Ultrasonics Sonochemistry, 21*(1), 93–97.

Alabdali, T. A., Icyer, N. C., Ucak Ozkaya, G., & Durak, M. Z. (2020). Effect of stand-alone and combined ultraviolet and ultrasound treatments on physicochemical and microbial characteristics of pomegranate juice. *Applied Sciences, 10*(16), 5458.

Amit, S. K., Uddin, M. M., Rahman, R., Islam, S. R., & Khan, M. S. (2017). A review on mechanisms and commercial aspects of food preservation and processing. *Agriculture & Food Security, 6*(1), 1–22.

Awad, T. S., Moharram, H. A., Shaltout, O. E., Asker, D. Y. M. M., & Youssef, M. M. (2012). Applications of ultrasound in analysis, processing and quality control of food: A review. *Food Research International, 48*(2), 410–427.

Baboli, Z. M., Williams, L., & Chen, G. (2020). Design of a batch ultrasonic reactor for rapid pasteurization of juices. *Journal of Food Engineering, 268*, 109736.

Bhargava, N., Mor, R. S., Kumar, K., & Sharanagat, V. S. (2020). Advances in application of ultrasound in food processing: A review. *Ultrasonics Sonochemistry*, 105293.

Bi, X., Hemar, Y., Balaban, M. O., & Liao, X. (2015). The effect of ultrasound on particle size, color, viscosity and polyphenol oxidase activity of diluted avocado puree. *Ultrasonics Sonochemistry, 27*, 567e575.

Brilhante São José, J. F., & Dantas Vanetti, M. C. (2012). Effect of ultrasound and commercial sanitizers in removing natural contaminants and *Salmonella enterica* Typhimurium on cherry tomatoes. *Food Control, 24*(1–2), 95–99.

Cameron, M. (2007). *Impact of Low-Frequency High-Power Ultrasound on Spoilage and Potentially Pathogenic Dairy Microbes* (Ph.D. Dissertation). South Africa: University of Stellenbosch.

Cameron, M., McMaster, L. D., & Britz, T. J. (2008). Electron microscopic analysis of dairy microbes inactivated by ultrasound. *Ultrasonics Sonochemistry, 15* 960–964.

Cameron, M., McMaster, L. D., & Britz, T. J. (2009). Impact of ultrasound on dairy spoilage microbes and milk components. *Dairy Science and Technology, 89*(1), 83–98.

Castillo-Zamudio, R. I., Paniagua-Martínez, I., Ortuño-Cases, C., García-Alvarado, M. A., Larrea, V., & Benedito, J. (2021). Use of high-power ultrasound combined with supercritical fluids for microbial inactivation in dry-cured ham. *Innovative Food Science & Emerging Technologies, 67*, 102557.

Chantapakul, T., Lv, R., Wang, W., Chummalee, W., Ding, T., & Liu, D. (2019). Manothermosonication: Inactivation of *Escherichia coli* and *Staphylococcus aureus. Journal of Food Engineering, 246*, 16–24.

Chantapakul, T., Tao, W., Chen, W., Liao, X., Ding, T., & Liu, D. (2020). Manothermosonication: Inactivation and effects on soymilk enzymes. *Ultrasonics Sonochemistry, 64*, 104961.

Cheng, X. F., Zhang, M., & Adhikari, B. (2013). The inactivation kinetics of polyphenol oxidase in mushroom (*Agaricus bisporus*) during thermal and thermosonic treatments. *Ultrasonics Sonochemistry, 20*(2), 674–679.

Earnshaw, R. G., Appleyard, J., & Hurst, R. M. 1995. Understanding physical inactivation processes: Combined preservation opportunities using heat, ultrasound and pressure. *International Journal of Food Microbiology, 28*, 197–219.

Ertugay, M. F., & Baslar, M. (2014). The effect of ultrasonic treatments on cloudy quality-related quality parameters in apple juice. *Innovative Food Science & Emerging Technologies, 26*, 226e231.

FAO. (2015). Global initiative on food loss and waste reduction, 8, https://www.fao.org/save-food/en/.

Gao, S., Hemar, Y., Lewis, G. D., & Ashokkumar, M. (2014). Inactivation of *Enterobacter aerogenes* in reconstituted skim milk by high-and low-frequency ultrasound. *Ultrasonics Sonochemistry, 21*(6), 2099–2106.

Gomez-Gomez, A., Brito-de la Fuente, E., Gallegos, C., Garcia-Perez, J. V., & Benedito, J. (2020). Non-thermal pasteurization of lipid emulsions by combined supercritical carbon dioxide and high-power ultrasound treatment. *Ultrasonics Sonochemistry, 67*, 105138.

Guimarães, J. T., Silva, E. K., Alvarenga, V. O., Costa, A. L. R., Cunha, R. L., Sant'Ana, A. S., Freitas, M. Q., Meireles, M. A. A., & Cruz, A. G. (2018). Physicochemical changes and microbial inactivation after high-intensity ultrasound processing of prebiotic whey beverage applying different ultrasonic power levels. *Ultrasonics Sonochemistry, 44*, 251–260.

Harvey, E. N., & Loomis, A. L. (1929). The destruction of Luminous bacteria by high frequency sound waves. *Journal of Bacteriology, 17*, 373–376.

He, Q., Liu, D., Ashokkumar, M., Ye, X., Jin, T. Z., & Guo, M. (2021). Antibacterial mechanism of ultrasound against *Escherichia coli*: Alterations in membrane microstructures and properties. *Ultrasonics Sonochemistry, 73*, 105509.

Hernández-Falcón, T. A., Monter-Arciniega, A., del Socorro Cruz-Cansino, N., Alanís-García, E., Rodríguez-Serrano, G. M., Castañeda-Ovando, A., García-Garibay, M., Ramírez-Moreno, E., & Jaimez-Ordaz, J. (2018). Effect of thermoultrasound on aflatoxin M1 levels, physicochemical and microbiological properties of milk during storage. *Ultrasonics Sonochemistry, 48*, 396–403.

Kentish, S. E. (2017). Engineering principles of ultrasound technology. In *Ultrasound: Advances for Food Processing and Preservation* (pp. 1–13). Academic Press.

Khadhraoui, B., Fabiano-Tixier, A. S., Robinet, P., Imbert, R., & Chemat, F. (2019). Ultrasound technology for food processing, preservation, and extraction. In *Green Food Processing Techniques* (pp. 23–56). Academic Press.

Kuldiloke, J., Eshtiaghi, M., Zenker, M., & Knorr, D. (2007). Inactivation of lemon pectinesterase by thermosonication. *International Journal of Food Engineering, 3*(2).

Lee, D. U., Heinz, V., & Knorr, D. (2003). Effects of combination treatments of nisin and high-intensity ultrasound with high pressure on the microbial inactivation in liquid whole egg. *Innovative Food Science & Emerging Technologies, 4*(4), 387–393.

Lee, H., Zhou, B., Liang, W., Feng, H., & Martin, S. E. (2009). Inactivation of *Escherichia coli* cells with sonication, manosonication, thermosonication, and manothermosonication: Microbial responses and kinetics modeling. *Journal of Food Engineering, 93*(3), 354–364.

Li, S., Yang, X., Zhang, Y., Ma, H., Liang, Q., Qu, W., He, R., Zhou, C., & Mahunu, G. K. (2016). Effects of ultrasound and ultrasound assisted alkaline pretreatments on the enzymolysis and structural characteristics of rice protein. *Ultrasonics Sonochemistry, 31*, 20–28.

Ma, H., Huang, L., Jia, J., He, R., Luo, L., & Zhu, W. (2011). Effect of energy-gathered ultrasound on Alcalase. *Ultrasonics Sonochemistry, 18*(1), 419–424.

Manzoor, M. F., Siddique, R., Hussain, A., Ahmad, N., Rehman, A., Siddeeg, A., Alfarga, A., Alshammari, G. M., & Yahya, M. A. (2021). Thermosonication effect on bioactive compounds, enzymes activity, particle size, microbial load, and sensory properties of almond (*Prunus dulcis*) milk. *Ultrasonics Sonochemistry, 78*, 105705.

Marchesini, G., Fasolato, L., Novelli, E., Balzan, S., Contiero, B., Montemurro, F., Andrighetto, I., & Segato, S. (2015). Ultrasonic inactivation of microorganisms: A compromise between lethal capacity and sensory quality of milk. *Innovative Food Science & Emerging Technologies, 29*, 215–221.

Maryam, A., Anwar, R., Malik, A. U., Raheem, M. I. U., Khan, A. S., Hasan, M. U., Hussain, Z., & Siddique, Z. (2021). Combined aqueous ozone and ultrasound application inhibits microbial spoilage, reduces pesticide residues and maintains storage quality of strawberry fruits. *Journal of Food Measurement and Characterization, 15*(2), 1437–1451.

Muzaffar, S., Ahmad, M., Wani, S. M., Gani, A., Baba, W. N., Shah, U., Khan, A. A., Masoodi, F. A., Gani, A., & Wani, T. A. (2016). Ultrasound treatment: Effect on physicochemical, microbial and antioxidant properties of cherry (Prunus avium). *Journal of Food Science and Technology, 53*(6), 2752–2759.

O'Donnell, C. P., Tiwari, B. K., Bourke, P., & Cullen, P. J. (2010). Effect of ultrasonic processing on food enzymes of industrial importance. *Trends in Food Science & Technology, 21*(7), 358–367.

Ortuno, C., Martinez-Pastor, M. T., Mulet, A., & Benedito, J. (2013). Application of high-power ultrasound in the supercritical carbon dioxide inactivation of *Saccharomyces cerevisiae. Food Research International, 51*(2), 474–481.

Rifna, E. J., & Dwivedi, M. (2021). Emerging nondestructive technologies for quality assessment of fruits, vegetables, and cereals. In *Food Losses, Sustainable Postharvest and Food Technologies* (pp. 219–253). Academic Press.

Rifna, E. J., Singh, S. K., Chakraborty, S., & Dwivedi, M. (2019). Effect of thermal and non-thermal techniques for microbial safety in food powder: Recent advances. *Food Research International, 126*, 108654.

Rojas, M. L., Leite, T. S., Cristianini, M., Alvim, I. D., & Augusto, P. E. (2016). Peach juice processed by the ultrasound technology: Changes in its microstructure improve its physical properties and stability. *Food Research International, 82*, 22–33.

Salve, A. R., Pegu, K., & Arya, S. S. (2019). Comparative assessment of high-intensity ultrasound and hydrodynamic cavitation processing on physicochemical properties and microbial inactivation of peanut milk. *Ultrasonics Sonochemistry, 59*, 104728.

Tahi, A. A., Sousa, S., Madani, K., Silva, C. L., & Miller, F. A. (2021). Ultrasound and heat treatment effects on *Staphylococcus aureus* cell viability in orange juice. *Ultrasonics Sonochemistry, 78*, 105743.

Terefe, N. S., Gamage, M., Vilkhu, K., Simons, L., Mawson, R., & Versteeg, C. (2009). The kinetics of inactivation of pectin methylesterase and polygalacturonase in tomato juice by thermosonication. *Food Chemistry, 117*(1), 20–27.

Tomadoni, B., Cassani, L., Ponce, A., Moreira, M. D. R., & Agüero, M. V. (2016). Optimization of ultrasound, vanillin and pomegranate extract treatment for shelf-stable unpasteurized strawberry juice. *LWT-Food Science and Technology, 72*, 475–484.

Tsukamoto, I., Yim, B., Stavarache, C. E., Furuta, M., Hashiba, K., & Maeda, Y. (2004). Inactivation of *Saccharomyces cerevisiae* by ultrasonic irradiation. *Ultrasonics Sonochemistry, 11*(2), 61–65.

Valero, M., Recrosio, N., Saura, D., Muñoz, N., Martí, N., & Lizama, V. (2007). Effects of ultrasonic treatments in orange juice processing. *Journal of Food Engineering, 80*(2), 509–516.

Wang, J., Liu, Q., Xie, B., & Sun, Z. (2020). Effect of ultrasound combined with ultraviolet treatment on microbial inactivation and quality properties of mango juice. *Ultrasonics Sonochemistry, 64*, 105000.

Zhu, Z., Chen, Z., Zhou, Q., Sun, D. W., Chen, H., Zhao, Y., Zhou, W., Li, X., & Pan, H. (2018). Freezing efficiency and quality attributes as affected by voids in plant tissues during ultrasound-assisted immersion freezing. *Food and Bioprocess Technology, 11*(9), 1615–1626.

Zhu, Z., Zhang, P., & Sun, D. W. (2020). Effects of multi-frequency ultrasound on freezing rates and quality attributes of potatoes. *Ultrasonics Sonochemistry, 60*, 104733.

5 Food Preservation by Dense Phase Carbon Dioxide

Gurbuz Gunes

Corresponding author:
Prof. (Dr.) Gurbuz Gunes; Email: gunesg@itu.edu

CONTENTS

5.1 INTRODUCTION

Pasteurization and sterilization of foods are conventionally performed by thermal processes. It is well established that thermal treatments cause degradation of organoleptic and nutritional qualities of foods. Nowadays, consumers demand fresh and fresh-like foods with a minimum additives, convenience, and shelf-stability. Various nonthermal novel technologies have become a subject of interest in the food industry to meet these demands. One of these technologies is dense phase carbon dioxide (DPCO$_2$) processing. This process involves exposure of foods to pressurized CO$_2$ mainly to inactivate microorganisms. It is also referred to as high-pressure CO$_2$ processing, supercritical CO$_2$ processing in literature. CO$_2$ has the status of 'generally recognized as safe' (GRAS) because it is nontoxic. Besides, it is relatively cheap, inflammable, and nonexplosive and does not promote oxidation in foods. These characteristics make CO$_2$ widely used in food systems such as carbonated beverages, modified atmosphere packaging (MAP), and controlled atmosphere storage (CAS). The pressurization of CO$_2$ increases its density, and the resultant CO$_2$ can be in gaseous, liquid, or supercritical forms depending on the pressure and temperature. It is well established that CO$_2$ under pressure, at or above the supercritical stage, has a higher power of microbial and enzyme inactivation than CO$_2$ at ambient pressure. This chapter focuses on the fundamental aspects of DPCO$_2$ applied in food preservation. The mechanisms of this process on microbial and enzyme inactivation, as well as potential applications on liquid and solid foods, along with the treatment systems are discussed. Combinations of other treatments with DPCO$_2$ on different foods are also discussed.

DOI: 10.1201/9781003147978-5

5.2 FUNDAMENTALS AND PRINCIPLES

Carbon dioxide is widely used in food preservation systems and/or food formulations for various purposes. It is used as an ingredient in carbonated beverages. It is produced by microorganisms in fermented food products or by chemical reactions in some bakery products. In addition, CO_2 is also used in the MAP and CAS of foods to control various quality losses through the inhibition of microbial growth and physiological changes in especially fresh fruits and vegetables. Carbon dioxide has solubility in both aqueous and hydrophobic environments making it effective in a wide range of food systems. The state of CO_2 depends on its temperature and pressure. It exists as a gas under ambient pressure and temperature but can be in the form of solid, liquid, gas, or supercritical depending on temperature and pressure (Figure 5.1). CO_2 coexists as a solid, a gas, and a liquid at its triple point where pressure is 517.77 kPa and temperature is −56.6°C. CO_2 exists as a supercritical fluid above its critical point, which corresponds to 7380 kPa and 31.1°C. CO_2 has higher solvating power and high antimicrobial activity in its supercritical state. Supercritical CO_2 has a mixed property of its liquid and gaseous forms: Its density is closer to its liquid form while its diffusivity and viscosity are closer to its gaseous form. These characteristics of supercritical CO_2 improve its penetration and solvating power in food systems and improve its lethal effects on microorganisms. The use of supercritical CO_2 for microbial and enzyme inactivation in foods has been investigated extensively for more than three decades. Relatively low critical temperature (31.1°C) that is close to room temperature makes CO_2 useful as a nonthermal processing tool in food preservation. Moreover, the critical pressure of 7380 kPa is not so high and can be achievable in a processing environment with reasonable investment and operating costs. Thus, $DPCO_2$ processing is one of the attractive nonthermal processes for treating heat-sensitive food products.

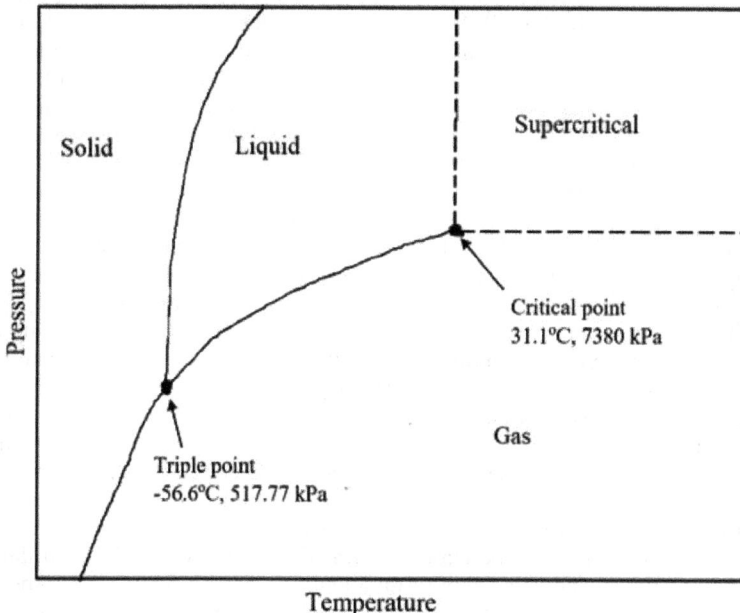

Figure 5.1 Phase diagram for carbon dioxide.

5.3 MICROBIAL INACTIVATION AND MECHANISMS

5.3.1 Inactivation Mechanisms

Mechanisms of microbial inactivation by $DPCO_2$ have been elaborated in several reviews (Yu et al., 2020; Rao et al., 2015; Garcia-Gonzalez et al., 2007; Zhang et al., 2006; Spilimbergo and Bertucco, 2003). CO_2 is dissolved in water to form carbonic acid, which dissociates into bicarbonate, carbonate, and H ions. Therefore, it causes acidification and a reduction of pH. The dissolution of CO_2 can occur not only in the external medium but also inside the cells since CO_2 can freely diffuse through the cell membrane and reach the interior of the cell. It is known that cell membranes are permeable to CO_2 so that CO_2 produced by cell metabolism can freely diffuse out without accumulation inside under normal conditions. When there is a higher concentration of CO_2 in the exterior of the cells than the interior, CO_2 diffusion from the exterior to the interior is expected according to Fick's law of diffusion. The accumulated CO_2 inside the cell can get dissolved and reduce the intracellular pH, thus disrupting the homeostasis of the cells (Giulitti et al., 2011). It has also been proposed that bicarbonate ions may interact with proteins and phospholipids at the external face of the cell membrane, and this interaction may contribute to a higher rate of CO_2 permeation into the cell (Yu et al., 2020). Disruption of the internal pH of the cell would alter the general metabolic activities of the cells by inactivating some key enzymes. It is well established that enzyme activity is highly dependent on pH and if the optimum pH is disrupted enzyme activity can be inhibited (Ballestra et al., 1996). This would result in the underproduction of some essential compound or the overproduction of some metabolites, leading to the disruption of cellular functions and eventually cell death. Other dissolved species of CO_2 may also affect cellular metabolism. For example, bicarbonate anion may interact with some enzymes and disrupts their activities (Spilimbergo and Bertucco, 2003). Carbonate ions derived from dissolved CO_2 can cause precipitation of electrolytes and this would further contribute inactivation of cells because electrolytes (like Ca^{+2}, Mg^{+2}, etc.) in the cells have important metabolic functions (like being signaling molecules or being part of enzyme systems), and they contribute to the osmotic equilibrium of the cells with their environment (Lin et al., 1993). In addition, lower extracellular pH can result in Ca^{+2} ion leakage from the cell (Hong and Pyun, 1999). Moreover, carboxylation and decarboxylation reactions in cellular metabolism can be affected by different species of dissolved CO_2 in cells (Garcia-Gonzalez et al., 2007).

Another effect of CO_2 may be associated with its dissolution power. If the concentration of CO_2 becomes high enough, it may extract phospholipids, hydrophobic compounds in cell walls and membrane, as well as some cellular constituents that may be vital for cellular function (Hong and Pyun 2001). These extracted constituents may leak out with CO_2 upon depressurization during the process, leading to permanent damage to cells. $DPCO_2$ results in damage to the cell wall, membrane, and internal cellular organization (Fraser, 1951; Foster et al., 1962). $DPCO_2$-treated cells of E. coli revealed an alteration of cellular organelles, a disruption of cytoplasm, intracellular condensation, and external membrane damage, all leading to morphological damage to the cell (Liao et al., 2010b). Moreover, $DPCO_2$ was shown to result in membrane damage and leakage of ions and nucleic acids in E. coli, and these effects were increased with increasing pressure, temperature, and exposure time (Tamburini et al. 2014). Similarly, membrane damage and leakage of ultraviolet (UV)–absorbing materials and Mg^{+2} and K^+ ions were observed in L. plantarium after being exposed to $DPCO_2$ at 7 MPa, 30°C for 10 min (Hong and Pyun 2001).

It has been suggested that $DPCO_2$ treatment caused the germination of spores, making them more sensitive to the treatment at elevated temperatures (Rao et al., 2015; Watanabe et al., 2003; Spilimbergo et al., 2003a; Furukawa et al., 2003). Heat activation of spores followed by $DPCO_2$ treatment caused significant spore inactivation (Ballestra and Cuq, 1998; Furukawa et al., 2003; Watanabe et al., 2003). However, spore inactivation by $DPCO_2$ has been explained by a damaged spore structure rather than the induction of germination (Rao et al., 2019). $DPCO_2$ treatment damages germinant receptors, leading to the blocking of spore outgrowth and spore death (Yu et al., 2020). A loss of core materials, including ions and dipicolinic acid in the inner membrane, has been detected, and this effect was associated with an increase in inner membrane permeability, resulting in reduced heat resistance of spores (Rao et al., 2020; Yu et al., 2020). There are relatively few studies on the inactivation of viruses by $DPCO_2$ in the literature. Based on the available information, enveloped viruses can be significantly inactivated by $DPCO_2$, and this was associated with the interaction of $DPCO_2$ with the virus envelop and its disruption (Perrut, 2012). This was supported by rapid inactivation of the lipid-coated *Sindbis* virus by $DPCO_2$, although it is resistant to high-pressure processing up to 700 MPa (Butz et al., 1992). Fages et al. (1998) reported more than 4-log inactivation of different viruses with or without envelop in bone allografts by $DPCO_2$ at 25 MPa and 50°C for 10 min. The overall effects of dense CO_2 on microbial inactivation can be a result of additive or synergistic effects of all these mechanisms, which are summarized in Table 5.1, since they occur simultaneously in the system. Some of these mechanisms have a theoretical basis only and their contribution to the overall inactivation needs to be verified by suitable analytical methods.

5.3.2 Factors Affecting Inactivation

$DPCO_2$ was shown to be very effective in the inactivation of microorganisms. A significant amount of information on this has been accumulated in the literature. The degree of inactivation depends on several factors as summarized in Table 5.2.

Table 5.1 Summary of Proposed Mechanisms for Microbial Inactivation by DPCO$_2$

1	CO_2 is dissolved in an aqueous medium → extracellular and intracellular pH reduction (acidification), the release of HCO_3^- and $CO_3^=$ ions
2	Disruption of the cell wall, membrane, and intracellular organization
3	Membrane permeability and fluidity increased
4	Membrane surface charge balance altered by reduced pH and bicarbonate
5	Inactivation/inhibition of enzymes due to pH reduction, denaturation, conformational changes
6	Intracellular precipitation of electrolytes (such as Ca^{2+}) and ribosomes induced by HCO_3^- and $CO_3^=$ ions leading to loss of electrolytic balance in the cytoplasm, electrolyte leakage through the cell membrane, and altered gene expression
7	Cell metabolism is affected by CO_2
8	Damage of spore structure and inhibition of spore germination, increased permeability of inner membrane causing reduced heat resistance in spores
9	Disruption of virus envelop structure

Table 5.2 Factors Affecting Microbial Inactivation by $DPCO_2$

Microbial factors	Process factors	Medium factors
Species and strains	Pressure	Characteristics of the food:
Types of cells (vegetative, spores, enveloped)	Depressurization rate	Solid/liquid
	Temperature	Particle size
Growth phase (exponential, stationary)	Exposure time	Viscosity
	The concentration of	Microstructure
Population density	CO_2 (saturation level)	pH and buffering capacity
	System design (static, continuous, well mixed vs stationary)	Chemical compositions (such as fat content)
		Hydrophobicity

Pressure is one of the most important process factors affecting the inactivation rate of microorganisms by $DPCO_2$. As the pressure increases, the physical properties of CO_2 change, making it more effective for microbial inactivation. With increased pressure, CO_2 would have higher solubility, extraction capacity, and penetration to the matrix, resulting in higher exposure of the target cells to CO_2. This effect is especially true up to about 10 MPa pressure beyond which a further increase in pressure has a minimal effect on microbial inactivation (Garcia-Gonzalez et al., 2007). This has been associated with saturation of CO_2 in the medium at about 10 MPa. Pressure applied at the typical levels in $DPCO_2$ processing does not cause any microbial inactivation by itself; it is only effective when there is CO_2 in the medium (Lin et al., 1993). Temperature is another important process factor affecting microbial inactivation by $DPCO_2$. As the temperature increase, a higher rate of inactivation can be obtained, and this can be associated with increased CO_2 diffusivity and relaxation of the cell wall which facilitates CO_2 penetration into the cell (Lin et al., 1993). On the other hand, a further increase of temperature far above the critical temperature (31.1°C) is not suggested because the solubility of CO_2 in foods/medium decrease with temperature (Garcia-Gonzalez et al., 2007). In addition, temperatures far above the critical point (especially >50°C) should be avoided because undesirable thermal effects on the quality of foods can be seen, and it would increase the energy cost during operation. Pressure and temperature effects should be considered together as they affect the physical state of CO_2. Increasing pressure and temperature to a level at or above the critical point (7.38 MPa and 31.1°C) results in supercritical CO_2 in the medium, and it is unanimously agreed that supercritical CO_2 has maximum inactivation power on microorganisms as compared to subcritical states. Further increase of pressure and temperature from the critical point has generally limited effects on inactivation rate (in the case of vegetative cells), but it would increase the cost of operation and thus is not suggested. However, if the targets are bacterial spores, an elevated pressure, with especially elevated temperatures, is needed to achieve a considerable inactivation. It has been shown that pressure cycling during treatment can increase the inactivation rate and has the potential for spore inactivation (Dillow et al., 1999). For example, *B. cereus* was inactivated by 5 log upon $DPCO_2$ applied at 20 MPa and 60°C for 2 hr with six cycles, while only 1-log inactivation was obtained when the same treatment was applied at 34°C (Dillow et al., 1999). In contrast, only a 2-log inactivation of *B. subtilis* spores was obtained by $DPCO_2$ at 15 MPa at 36–50°C for up to 20 min with 2–20 cycles (Spilimbergo et al., 2003a).

Another factor affecting the degree of microbial inactivation by $DPCO_2$ is the rate of depressurization in the process. Once the product mixed with CO_2 is exposed to the process temperature and pressure for a specific period, the pressure is released, and CO_2 and the product are naturally separated to obtain the treated product. When the applied pressure is rapidly released, the dissolved CO_2 within the microbial cell is transformed into the gaseous state, and this expansion within the cell may increase the internal pressure leading to mechanical rupture of the cells (Fraser, 1951; Lin et al., 1993; Garcia-Gonzalez et al., 2007). Although this mechanism has theoretical supports, some researchers found no effects of depressurization rate on microbial inactivation (Arreola et al., 1991b; Hong et al., 1997; Spilimbergo and Bertucco, 2003). Elucidating the effect of depressurization on microbial inactivation depends on the treatment system, analytical approaches used for assessments, and probably other process factors such as the target microorganism, the food matrix, the degree of CO_2 saturation, temperature, and pressure. The conflicting results in depressurization rate effects seem to be due to the differences in the factors stated earlier. It should be noted that a sudden depressurization may cause operational difficulty in the system as it may cause freezing of the expansion valves, which can limit the applicable rate of depressurization (Spilimbergo et al., 2003a).

Effective contact of the target cell with $DPCO_2$ is critical for process efficiency. This can be affected by the concentration of CO_2 in the product and efficient mixing in the system (Hong et al., 1997; Garcia-Gonzalez et al., 2007). The concentration of CO_2 must be determined by considering pressure and temperature as these factors would affect the degree of saturation. At the determined temperature and pressure, full saturation of CO_2 in the product would give maximum effects; thus, the concentration of CO_2 to obtain the full saturation should be determined. Increased inactivation of microorganisms by increasing the amount of CO_2 in the product during processing has been reported (Gunes et al., 2005, 2006). Besides, effective agitation in the system design would result in full saturation and efficient contact of dissolved CO_2 with the target. A microbubble system in which a high amount of CO_2 is bubbled into the feed was shown to be more effective in microbial inactivation due to the higher level of dissolved CO_2 achieved Ishikawa et al., 1997). Up to a 3-log higher inactivation of various *Bacillus* spores was achieved by $DPCO_2$ with the microbubble at 30 MPa, 40°C for 30 min as compared to the same treatment without microbubble (Ishikawa et al., 1997).

Characteristics of the medium in which the target microorganisms are present can have a significant impact on the degree of microbial inactivation by $DPCO_2$. As the acidity of the food increases, microbial inactivation increases. This is especially important for spore inactivation. For example, $DPCO_2$ at 5.4 MPa, 70°C for 120 min resulted in complete inactivation of *C. sporogenes* when the media was acidified to pH 3–2 while only 0.8-log inactivation was obtained when the pH was adjusted to 4 (Haas et al., 1989). Similarly, *A. acidoterrestris* spores suspended in apple juice with pH 3.47 were completely inactivated by $DPCO_2$ at 10 MPa, 65°C for 40 min (Bae et al., 2009). The moisture contents of the media and/or cell also affect the degree of inactivation by $DPCO_2$. A high moisture content is needed for CO_2 to be dissolved in the product and reach microbial cells to show its antimicrobial effects. For instance, *B. subtilis* spores were inactivated by 53% upon treatment at 20 MPa, 35°C for 2 h when their moisture content was 70–90%, while only 1% inactivation was achieved at the moisture content of 2–10% (Kamihira et al., 1987). An enhancement of spore inactivation by $DPCO_2$ at higher moisture content was reported by other researchers (Zhang et al., 2006; Furukawa et al., 2009). Food products are generally high in moisture content but if low moisture dehydrated products are to be treated, $DPCO_2$ would be less efficient for microbial inactivation. There was 99.6% and 99.8% inactivation of bacteria and yeast/mold in flour samples with 28% moisture, respectively, due to $DPCO_2$ at 6.2 MPa and 23°C for 2 h, while no inactivation was observed when the same treatment was applied to flour samples with 12% moisture (Haas et al., 1989). $DPCO_2$ was ineffective in microbial inactivation in dehydrated herbs and spices while significant inactivation (up to 7 log) was achieved in fresh spices (Haas et al., 1989). Higher microbial inactivation to *E. coli* in agar was observed when $DPCO_2$ was applied in wet conditions as compared to its application in dry conditions (Matsufuji et al., 2009). Physicochemical characteristics of the medium in which the cells are present also affect the extent of inactivation by $DPCO_2$. Generally, a lower rate of inactivation is achieved when the target cells are in the food matrix as compared to the cells in broth. Having a complex composition in a medium makes cells somewhat protected from the effects of $DPCO_2$. This seems to be associated with the interaction of CO_2 with medium components, making less CO_2 penetration into the microbial cells as well as protection effects, such as buffering capacity and medium components (Garcia-Gonzalez et al., 2007). Specific components of a medium can have different effects on microbial inactivation. Emulsifiers, sucrose stearate, and sodium chloride increased microbial inactivation while oil, glycerol, and sucrose decreased it (Garcia-Gonzalez et al., 2009a). Initial pH of the suspending medium, presence of antimicrobial components (natural or additives) in the medium, and viscosity are other attributes that can have an impact on the inactivation rate by $DPCO_2$.

Microorganisms have different susceptibilities to $DPCO_2$ depending on species, strains, growth stage, initial population density, forms (vegetative vs. spores), and medium. Generally, bacteria (no difference between gram-positive and gram-negative) are more sensitive than yeast and molds while spores are the most resistant (Garcia-Gonzalez et al., 2009a). Different species of *Bacillus* spores have different sensitivity to $DPCO_2$: Their sensitivities were ranked as *B. subtilis* < *G. stearothermophilus* < *B. licheniformis* < *B. coagulans* <*B. cereus* (Rao et al., 2015). Microorganisms of different strains within the same species may have different resistance to the process. For example, two strains of *A. acidoterrestris*, namely, TO-117/02 and TO-169/06, in apple juice were inactivated by 0.7 and 3.4 log, respectively, upon $DPCO_2$ treatment at 60 MPa and 75°C for 40 min (Porebska et al., 2017). Microorganisms in the log phase have been shown to be more sensitive to $DPCO_2$ than stationary phase cells. The reason behind this is unknown, but it may be due to better response/repair systems activated by stationary-phase cells to varying stress conditions like $DPCO_2$ (Ortuno et al., 2012a). The initial population density of the cells in the medium also affects the inactivation rate by $DPCO_2$. The inactivation rate seems to be lower with higher initial microbial counts. This may be associated with decreased availability of dissolved CO_2 for each cell in a dense population environment (Garcia-Gonzalez et al., 2007). Thus, a higher concentration of CO_2, higher pressure, and temperature during processing may be required for a product with high initial microbial counts. As is true for other treatments, $DPCO_2$ is less effective on spores than on vegetative cells in the same treatment conditions. This is associated with complicated spore structure; lower moisture content, which makes CO_2 difficult to penetrate and dissolve within spores; and damaged germinant receptors, leading to the prevention of outgrowth of spores (Yu et al., 2020). For this reason, more intense process conditions with combined treatments are necessary to have a significant spore inactivation.

5.4 EFFECTS ON ENZYMES AND QUALITY OF FOODS

$DPCO_2$ is a nonthermal process. Thus, thermal degradation of organoleptic quality including nutritional values of food observed in conventional thermal processes can be avoided or minimized if $DPCO_2$ is used. Moreover, the nonoxidizing characteristic of CO_2 also contributes to the retention of quality in $DPCO_2$-treated foods. Quality attributes such as flavor, texture, color, and nutritional value are important for the acceptance of foods by consumers. A comprehensive review of the effects of $DPCO_2$ on quality aspects of foods has been published (Zhou et al., 2015). Several studies reported minimal adverse effects of $DPCO_2$ on the quality of foods as compared with conventional processes. Changes in the quality of foods upon $DPCO_2$ can be due to biochemical and chemical reactions and/or physical changes in foods. $DPCO_2$ can inactivate enzymes to some extent, and this can add a further benefit of the maintenance of quality. Activities of the enzymes in foods result in alteration of color, texture, and flavor during storage. Thus, $DPCO_2$ can also contribute to the inhibition of these quality losses in foods during storage.

5.4.1 Enzymes

Enzymes such as polyphenol oxidase (PPO), lipoxygenase (LOX), pectic enzymes (PE, PME, PG), and peroxidase (POD) have been shown to be inactivated by $DPCO_2$ (Table 6.3). The degree of inactivation of an enzyme by $DPCO_2$ can be quite different depending on the source (food matrix) due to various reasons such as pH, chemical composition, and the physiological response of tissues to stress. Enzyme inactivation generally requires higher treatment pressure and temperature, and longer exposure times than microbial inactivation. It has been shown that immediate loss of enzyme activity upon $DPCO_2$ was restored to some extent during storage (Gui et al., 2006a). The degree of the restoration depends on $DPCO_2$ conditions (pressure and temperature), and enzyme. Enzyme inactivation by $DPCO_2$ has been associated with conformational changes in secondary (loss of α-helix) and tertiary structures (Liao et al., 2009). The restoration of enzyme activity has also been associated with the recovery of these conformational changes.

Polyphenol oxidase (PPO) is the enzyme catalyzing enzymatic browning in fruits and vegetables; thus, its activity would adversely affect the color of the product. $DPCO_2$ has been shown to inhibit PPO in various foods at different rates (Table 6.3). A complete inactivation of the enzymes in apple juice and apple slices by $DPCO_2$ at 20 MPa and 55°C for 20 min were reported by Murtaza et al. (2020) and Niu et al. (2010a) while only a 60% inactivation of PPO in apple juice was obtained upon $DPCO_2$ exposure at 30 MPa and 55°C for 60 min in another study (Gui et al., 2007). As the pressure and temperature increase (both within the supercritical range), PPO inactivation increases. Complete inactivation of PPO in strawberry juice upon $DPCO_2$ at 30 and 60 MPa, 45°C for 30 min was also reported (Marszalek et al., 2015). PPO in lobster, shrimp, and potato was inactivated by 98%, 78%, and 91%, respectively, using $DPCO_2$ at 5.8 MPa and 47°C for 1 min (Chen et al., 1992). A 61% and 30% inactivation of PPO in carrot juice and fresh-cut carrots, respectively, were reported (Park et al., 2002; Bi et al., 2011). On the other hand, PPO activity in fresh-cut carrots stimulated by $DPCO_2$ at low pressure (>3 MPa), 20°C, and low exposure times (<8 min) (Bi et al., 2011). D-values for PPO in carrot and celery juices were determined as 1354 min and 767 min at 10 MPa and 47°C and 60 MPa and 55°C, respectively (Marszałek et al., 2016). PPO inactivation of 80% and 75% in watermelon and melon juices, respectively, upon DPCO$_2$ treatments was also reported in other studies (Liu et al., 2012; Chen et al., 2010).

Peroxidase (POD) is another enzyme that causes color changes and is also known as the most heat-resistant enzyme in fruits and vegetables. Various degrees of inactivation of POD by DPCO$_2$ treatments in different foods have been reported (Table 5.3). D-values of POD in carrot and celery juices at 30 MPa

Table 5.3 Studies Showing the Effects of DPCO$_2$ on Various Enzymes

Enzyme	Source	Treatment conditions	Inactivation (%)	Reference
PPO	Apple juice	20 MPa, 55°C, 20 min	100	Murtaza et al., 2020
PPO	Apple slice	20 MPa 25–65°C, 20 min	100 at all temperatures	Niu et al., 2010a
PPO	Apple juice	30 MPa 55°C 60 min	60	Gui et al., 2007
PPO	Watermelon	10–30 MPa, 50°C, 15 min	20–80, from 10 to 30 MPa	Liu et al., 2012
PPO	Melon juice	35 MPa, 55°C, 60 min	75	Chen et al., 2010
PPO	Lobster	5.8 MPa, 43°C, 1 min	98	Chen et al., 1992
PPO	Shrimp	5.8 MPa, 43°C, 1 min	78	Chen et al., 1992
PPO	Potato	5.8 MPa, 43°C, 30 min	91	Chen et al. 1992
PPO	Carrot juice	2.9–4.9 MPa, 5°C, 10min	50–61	Park et al., 2002
PPO	Carrot juice	10 MPa, 47°C	D-value =1354 min	Marszalek et al., 2016
PPO	Celery juice	60 MPa, 55°C	D-value =767 min	Marszalek et al., 2016
PPO	Banana pure	20 MPa, 45–60°C, 30 min	59.3 at 45°C 88.4 at 60°C	Yu et al., 2013
PPO	Fresh-cut carrot	1.5–5 MPa, 20°C, 15 min	30 at 5 MPa for 15 min; *Activation up to 32% at low pressure and time*	Bi et al., 2011
PPO	Fresh-cut carrot	12 MPa, 22–40°C, up to 30 min	80 at 22–40°C with >15 min	Spilimbergo et al., 2013
POD	Fresh-cut carrot	1.5–5 MPa, 20°C, 15 min	24 at 5 MPa for 15 min; *Activation up to 59% at low pressure and time*	Bi et al., 2011

Enzyme	Source	Treatment conditions	Inactivation (%)	Reference
POD	Fresh-cut carrot	12 MPa, 22–40°C, up to 30 min	60 at 40°C 15 min; 55 at 40°C 30 min; 40 at 40°C 5 or 45 min; 20 at 22°C 15 min; 0 at 22°C 30 min; *200 activation at 22°C for 5 or 45 min*	Spilimbergo et al., 2013
POD	Apple juice	20 MPa, 55°C, 20 min	100	Murtaza et al., 2020
POD	Horseradish	62.1 MPa, 55oC, 15 min	100	Tedjo et al., 2000
POD	Commercial horseradish origin	62 MPa, 55°C 15 min	>90 in aq soln 80 in 20%sucrose 50 in pH8 buffer	Tedjo et al., 2000
POD	Commercial horseradish origin	8–30 MPa, 55°C, 60 min; enzyme activity measured immediately and after 7–21 days at 4°C storage	75, 60 at 8 MPa immediate and after 7–21 d at 4°C; 82, 60 at 15 MPa immediate and after 7–21 d at 4°C; 85, 80 at 22 MPa immediate and after 7–21 d at 4°C; 88, 88 at 30 MPa immediate and after 7–21 d at 4°C;	Gui et al., 2006b
POD	Watermelon	10–30 MPa, 50°C, 15 min	10–50, from 10 to 30 MPa	Liu et al., 2012
POD	Melon juice	35 MPa, 55°C, 60 min	62	Chen et al., 2010
POD	Celery juice	30 MPa, 47°C	D-value = 442 min	Marszalek et al., 2016
POD	Carrot juice	30 MPa, 47°C	D-value = 1439 min	Marszalek et al., 2016
LOX	Melon juice	35 MPa, 55°C, 60 min	95	Chen et al., 2010
LOX	Soybean	10.3 MPa, 50°C, 15 min	100	Tedjo et al., 2000
LOX	Soybean	62.1 MPa, 35°C, 15 min	95	Tedjo et al., 2000
LOX	Commercial soybean origin	35 MPa, 40°C, 15 min	95 in aq soln 40 in 20%sucrose 25 in pH7 buffer	Tedjo et al., 2000
LOX	Carrot juice	2.9–4.9 MPa, 5°C, 10 min	>70	Park et al., 2002

(Continued)

Table 5.3 (Continued)

Enzyme	Source	Treatment conditions	Inactivation (%)	Reference
PE	Orange juice	26.9 MPa, 56°C, 145 min	100	Balaban et al., 1991b
PE	Orange juice	26.9 MPa, 50°C, 4h	100	Arreola et al., 1991a
PE	Carrot juice	10 MPa, 47°C	D-value = 215 min	Marszalek et al., 2016
PE	Celery juice	30 MPa, 47°C	D-value = 200 min	Marszalek et al., 2016
PME	Watermelon juice	20–30 MPa, 50°C, 15 min	20	Liu et al., 2012
PME	Carrot juice	2.9–4.9 MPa, 5°C, 10 min	50	Park et al., 2002
PME	Fresh-cut carrot	1.5–5 MPa, 20°C, 15 min	47 at 5 MPa; 25 at 3.5 MPa; 5 at 1.5 MPa	Bi et al., 2011
PME	Fresh-cut carrot	12 MPa, 22–40°C, up to 30 min	100 at 40°C, 30 min; 50 at 22°C, 30 min	Spilimbergo et al., 2013
PME	Apple slice	20 MPa 25–65°C, 20 min	40 at 25°C; 50 at 35°C; 55 at 45°C; 75 at 55°C; 82 at 65°C	Niu et al., 2010a
PME	Orange juice	10–30 MPa, 2–21°C, 20 min	35 at 30 MPa, 2°C; 60–90 at 10–30 MPa, 21°C; increased with pressure	Briongos et al., 2016
PG	Celery juice	10 MPa, 47°C	D-value was 1645 min	Marszalek et al., 2016
PG	Carrot juice	10 MPa, 47°C	D-value was 1279 min	Marszalek et al., 2016
PG	Fresh-cut carrot	12 MPa, 22–40°C, up to 30 min	100 at 40°C, 10 min; *50 activation at 22°C, 30 min; 250 activation at 22°C 30 min*	Spilimbergo et al., 2013
ALP	Milk	8 MPa, 70°C, 30 min, 0.05 CO_2/ milk ratio	94.6	Ceni et al., 2016
ALP	Milk	20 MPa, 50°C, 50 min	100	Liao et al., 2019
Lipase	Commercial (62–68% in water	20 MPa, 35°C, 120 min	12–22	Kamihira et al., 1987

Enzyme	Source	Treatment conditions	Inactivation (%)	Reference
Lipase	Commercial (5–7% in water)	20 MPa, 35°C, 60 min	0	Taniguchi et al., 1987
Catalase	Commercial (5–7% in water)	20 MPa, 35°C, 60 min	10	Taniguchi et al., 1987
α-amylase	Commercial (62–68%)	20 MPa, 35°C, 120 min	0	Kamihira et al., 1987
Gluco-amylase	Commercial (5–7% in water)	20 MPa, 35°C, 60 min	0	Taniguchi et al., 1987
Glucose isomerase	Commercial (5–7% in water)	20 MPa, 35°C, 60 min	0	Taniguchi et al., 1987

Abbreviations: PPO: polyphenol oxidase; POD: peroxidase; LOX: lipoxygenase; PE: pectin esterase; PME: pectin methylesterase; PG: polygalacturonase; ALP: alkaline phosphatase.

and 47°C were calculated as 1439 min and 442 min, respectively (Marszałek et al., 2016). Horseradish POD was completely inactivated by $DPCO_2$ at 62 MPa and 55°C for 15 min (Tedjo et al., 2000) while a maximum of 88% inactivation by treatment at 30 MPa and 55°C for 60 min was reported in another work (Gui et al., 2006b). POD in apple juice was completely inactivated by $DPCO_2$ at 20 MPa and 55°C for 20 min (Murtaza et al., 2020). On the other hand, 24%, 50%, 62%, and 86% inactivation of POD in fresh-cut carrot, watermelon juice, melon juice, and strawberry juice, respectively, upon $DPCO_2$ treatments at different conditions were reported in other works (Bi et al., 2011; Liu et al., 2012; Chen et al., 2010; Marszalek et al., 2015). In contrast, POD activity in fresh-cut carrots stimulated by $DPCO_2$ at low pressure (>3 MPa), 20°C, and low exposure times (<8 min) (Bi et al., 2011).

LOX is the enzyme catalyzing the oxidation of lipids in foods, resulting in off-flavor and color loss. $DPCO_2$ resulted in up to 100% inactivation of commercial LOX of soybean origin in an aqueous solution (Tedjo et al., 2000). The degree of inactivation decreased to 40% or 25% when the enzyme was in a 20% sucrose solution or pH7 buffer. LOX in melon and carrot juice were inactivated by 95% and >70% upon $DPCO_2$ treatments at 35 Mpa, 55°C for 60 min, and 4.9 MPa at 5°C for 10 min, respectively (Chen et al., 2010; Park et al., 2002). The inactivation of LOX by $DPCO_2$ treatments was associated with an alteration of its molecular properties such as increased aggregation and decreased α-helix content in its structure (Liao et al., 2009).

The inactivation of various pectic enzymes by $DPCO_2$ processing has been reported in various studies (Table 6.3). $DPCO_2$ treatment at 26.9 MPa and 56°C for 145 min completely inactivated PE in orange juice (Balaban et al., 1991). Similar D-values for PE in celery and carrot juices (200–215 min) were obtained at different treatment pressures (30 and 10 MPa, respectively) delivered at 47°C (Marszałek et al., 2016). PME in orange juice was inactivated by 90% upon treatment at 30 MPa and 21°C for 20 min, while only 35% inactivation was noted when the treatment temperature was reduced to 2°C (Briongos et al., 2016). PME inactivation in apple slices increased from 40% to 82% when the treatment temperature at 20 MPa was increased from 25°C to 65°C during 20 min of exposure (Niu et al., 2010a). $DPCO_2$ in different conditions resulted in 20%, 50%, and 47% PME inactivation in watermelon juice, carrot juice, and fresh-cut carrot, respectively (Liu et al., 2012, Park et al., 2002; Bi et al., 2011). On the contrary, PME activity in fresh-cut carrots stimulated by $DPCO_2$ at low pressure (>5 MPa), 20°C, and low exposure times (<5 min) (Bi et al., 2011). $DPCO_2$ caused low levels of inactivation in polygalacturonase (PG) of celery and carrot juices: D-values at 10 MPa and 47°C were 1645 min and 1275 min, respectively (Marszałek et al., 2016).

Alkaline phosphatase (ALP) in milk was inactivated completely upon treatment at 20 MPa and 50°C for 50 min (Liao et al., 2019). A 94.6% inactivation of ALP in milk treated at 8 MPa and 70°C for 30 min was reported in another work (Ceni et al., 2016). Lipases, catalases, amylases, and glucose isomerases were reported as resistant to $DPCO_2$ treatments (Taniguchi et al., 1987; Kamihira et al., 1987). However, the advantages of enzyme-catalyzed reactions in supercritical CO_2 with various enzymes resistant to supercritical CO_2 such as lipases, cutinase, and Novozym 435 have been reported (Wimmer and Zarevucka, 2010).

5.4.2 Nutritional and Bioactive Constituents

Nutritional values and bioactive compounds are generally affected to a smaller extent by $DPCO_2$ treatments than by conventional thermal treatments. These aspects of foods as affected by $DPCO_2$ have been investigated in several studies. The changes in the bioactives of foods during $DPCO_2$ depend on the treatment conditions and type of food products. Apple juice treated by $DPCO_2$ resulted in a slight reduction in total phenolic content (Murtaza et al., 2020). Chlorogenic acid, catechin, and epicatechin levels in the juice decreased by $DPCO_2$ at 20 MPa and >45°C, but their concentrations slightly increased by treatment at 9 MPa and 30°C. The reduction of these individual phenolics was comparable with thermal treatment at similar treatment temperatures. Muscadine grape juice treated by $DPCO_2$ at 34 MPa, 8–16% CO_2, 30°C for 6.2 min had no loss of anthocyanin, phenolics, and antioxidant activity while thermal treatment at 75°C for 15 s resulted in up to 26% loss in them (Del Pozo-Insfran et al., 2006a, 2006b). Antioxidant activity and total phenolics of fresh-cut carrot were not affected by $DPCO_2$ (Spilimbergo et al., 2013) The maintenance of the phenolic contents in other $DPCO_2$-treated foods were also reported (Liu et al., 2012; Zhao et al., 2019b; Fleury et al., 2018; Ferrentino et al., 2009; Calvo et al., 2007). Anthocyanins in strawberry juice were not affected by $DPCO_2$ treatment (Marszalek et al., 2015). In contrast, some degradation of polyphenols and chlorophyll in carrot and celery juices was observed upon $DPCO_2$ at high pressures, although the reductions were lower as compared to thermal treatment (Marszałek et al., 2016). Beetroot juice treated by $DPCO_2$ had 58% and 32% loss of betacyanins and betaxanthins, respectively (Marszałek et al., 2017).

The maintenance and better retention of carotenoids in carrot juice, fresh-cut carrot, pickled carrot, melon juice by $DPCO_2$ as compared to conventional heat treatment were reported in various studies (Chen et al., 2010; Wang et al., 2019; Zhou et al., 2009; Spilimbergo et al., 2013; Marszałek et al., 2016). Better lycopene isomerization and improved bio-accessibility of lycopene in tomato juice by $DPCO_2$ as compared to heat treatment were observed (Zhao et al. 2019a, 2019b). On the other hand, $DPCO_2$ at 10–30 MPa at 50°C caused a slight reduction in the lycopene content of watermelon juice, and this was explained by its extraction and/or degradation into volatiles (Liu et al., 2012).

$DPCO_2$ treatment caused little or no loss in ascorbic acid as opposed to much higher loss in heat-treated controls in various foods (Arreola et al. 1991a; Kincal et al. 2006; Briongos et al., 2016; Chen et al., 2010; Fleury et al., 2018). Some loss of ascorbic acid in fresh-cut pears, fresh-cut carrots, and strawberry juice upon $DPCO_2$ treatment was noted (Valverde et al. 2010; Spilimbergo et al., 2013;

Marszalek et al., 2015), DPCO$_2$ at 10–20 MPa and 35°C for 10 min was shown to maintain all the functional and nutritional properties of apple juice enriched with a prebiotic (inulin) without alteration of the chemical properties of the inulin although the breakdown of inulin was seen in heat-treated samples (Silva et al., 2019). Amaral et al. (2018) found that DPCO$_2$ at the supercritical state did not adversely affect quality but improved angiotensin-converting enzyme (ACE) inhibitory activity of a whey-grape-juice-based drink.

5.4.3 Color

Apple juice from DPCO$_2$-treated apple slices had a much lower browning degree and turbidity as compared to juice made from mild heat-treated fruits Niu et al. (2010a). The maintenance of color through the inactivation of PPO in cloudy apple juice was also reported in other studies (Murtaza et al., 2020; Gui et al., 2007). DPCO$_2$ treatment resulted in inhibition of browning in other foods such as watermelon juice, melon juice, carrot juice, and orange juice (Chen et al., 2010; Liu et al., 2012; Zhou et al., 2009; Arreola et al., 1991a). The color of banana juice was better retained by DPCO$_2$ with inhibited browning as compared to mild thermal treatment, and this was associated with inactivation of PPO in DPCO$_2$-treated samples (Yu et al., 2013). The color and anthocyanin content of strawberry juice was not adversely affected by DPCO$_2$ treatment (Marszalek et al., 2015). DPCO$_2$ treatment caused an increase in L- and b-value, a decrease in the a-value of carrot and mandarin juices (Park et al., 2002; Lim et al., 2006) while the opposite trend was observed in fresh-cut pears after DPCO$_2$ treatment (Valverde et al., 2010). Tomato juice had lower L- and a-values but higher b-values upon DPCO$_2$ treatments, which result in higher total color change as compared to heat treatment (Zhao et al., 2019b). DPCO$_2$ treatment increased L- and a-values but decreased b-values in red grapefruit juice (Ferrentino et al., 2009). DPCO$_2$ caused a visual loss of green color in spinach leaves (Zhong et al., 2008).

The color of meat has been reported to be adversely affected by DPCO$_2$. A loss of the red color and cooked appearance in ground beef due to DPCO$_2$ was reported (Sirisee et al., 1998). A whitish color with the cooked and soaked appearance of chicken meat due to DPCO$_2$ was also reported (Wei et al., 1991). Similarly, an increased L-value with the cooked appearance of tilapia was observed after DPCO$_2$ treatment (Sugiharto et al., 2020). These quality defects were associated with DPCO$_2$-induced protein denaturation (Messens et al., 1997). On the other hand, no adverse effect of DPCO$_2$ on the sensory quality of marinated pork, and the color of cooked ham was reported (Choi et al. 2009a; Ferrentino et al., 2016).

5.4.4 Flavor, pH, and Sensory Quality

DPCO$_2$ is expected to decrease the pH of the food due to the dissolution of CO$_2$ in foods producing acid. However, the degree of change in pH depends on the composition of the foods treated and especially their buffering capacity. DPCO$_2$ treatments resulted in pH drop in various foods such as tomato juice, apple juice, watermelon juice, orange juice (Gui et al., 2007; Zhao et al., 2019b; Liu et al., 2012; Balaban et al., 1991). On the other hand, no change in pH after DPCO$_2$ treatment was noted in orange juice (Briongos et al., 2016; Arreola et al., 1991b) The changed pH in the foods due to DPCO$_2$ was not directly associated with sensory acceptability of the foods. The DPCO$_2$ treatment of strawberry juice caused an increase in total sugar content in which sucrose level decreased and glucose and fructose levels increased (Marszalek et al., 2015). In contrast, sucrose, glucose, fructose, and sorbitol content of apple juice were not affected by DPCO$_2$ at 10–20 MPa and 35°C for 10 min (Silva et al., 2019). The protein and lactose contents of milk were not changed by $DPCO_2$ (Liao et al., 2019). Kobayashi et al. (2016) found that free amino acid content was decreased except for phenylalanine and valine which were increased. $DPCO_2$ maintained the original flavor of tomato juice better than heat treatment (Zhao et al., 2019b). $DPCO_2$ did not adversely affect the volatile components of melon juice and maintained the aroma and flavor of beer (Chen et al., 2010; Dagan and Balaban, 2006). A lower loss of volatile compounds in apple juice was noted after $DPCO_2$ treatment at 15 MPa as compared to heat treatment (Porto et al., 2010). $DPCO_2$ treatment caused an increase in nonanal and citronellol and a decrease in ethyl butyrate and trans-2-hexenol in orange juice with better overall flavor retention than heat treatment (Niu et al., 2010b). A better sensory quality of $DPCO_2$-treated foods as compared to a heat-treated control has been reported for different foods (Arreola et al. 1991b; Kincal et al., 2006; Fabroni et al., 2010; Del Pozo-Insfran et al., 2006b; Tomic et al., 2019, 2020). $DPCO_2$ at 34 MPa, 33°C for 1 h did not adversely affect sensory attributes of orange juice, which was preferable over commercial samples (Arreola et al. 1991b). The treatment of apple juice at 10 MPa and 36°C for 10 min caused only a moderate reduction in sensory quality and volatile constituents (Gasperi et al., 2009).

5.4.5 Texture

$DPCO_2$ (6–30 MPa, 25–65°C for 10–90 min) treatment caused a loss of firmness in fresh-cut pears with increasing pressure (Valverde et al., 2010). Similarly, the firmness of the carrot slices decreased by 8% after treatment at 5 MPa for 15 min although increased firmness was observed when the exposure time was 2 min (Bi et al., 2011). This was associated with deactivation and activation of PME activity at long and short exposure times, respectively. Loss of firmness by 50–90% in fresh-cut carrots upon $DPCO_2$ at 12 MPa and 22–40°C for 15 min was also reported by Spilimbergo et al. (2013). These data show that softening in carrot tissue increased with pressure and temperature during $DPCO_2$ treatment. In contrast, no change in the hardness of fresh-cut coconuts with a slight change in microstructure upon $DPCO_2$ treatment was observed (Ferrentino et al., 2012). The changes in the texture of fresh-cut produce upon $DPCO_2$ treatment are due to the overall effects on pectic enzymes, electrolyte leakage, and structural breakdown on the cell wall. $DPCO_2$ at 8.5 MPa and 40°C did not change the texture of tilapia, but 7-MPa or 9-MPa pressure resulted in a loss of hardness (Sugiharto et al., 2020). Gel formation and induced gel strength by $DPCO_2$ in other meat products such as sausages, myosin, and shrimp surimi are other examples of textural changes related to the alteration of protein structure (Guo et al., 2017).

$DPCO_2$ resulted in cloud increase in carrot juice, orange juice, and watermelon juice, (Kincal et al., 2006; Liu et al., 2012; Zhou et al., 2009). A decrease in particle size in orange juice upon $DPCO_2$ treatment at 10–30 MPa and varying temperatures were noted in orange juice (Briongos et al., 2016). $DPCO_2$ maintained the original viscosity of tomato juice better than heat treatment (Zhao et al., 2019a, 2019b). In contrast, $DPCO_2$ treatment resulted in higher clarity, lower viscosity, lower pectin, and lower protein content than mild heat treatment, and untreated control in banana juice (Yu et al., 2013). A reduction in particle size and consistency index by $DPCO_2$ was also observed in a whey-grape juice drink, but these changes were similar to the ones induced by thermal treatment (Amaral et al., 2018). $DPCO_2$ at 20 MPa caused a reduction in the viscosity and an increase of particle size in milk, and these changes were amplified with increasing temperature and/or exposure time (Liao et al., 2019). Increased precipitation in peach juice upon $DPCO_2$ treatment at 30 MPa and 55°C was observed, and this was associated with increased particle size, decreased pH, and ζ-potential (Zhou et al., 2010). The researchers also found that $DPCO_2$ caused higher viscosity but no change in pectin and calcium content in the treated peach juice.

5.5 APPLICATIONS OF $DPCO_2$ IN FOOD PRESERVATION

5.5.1 Current Applications in Liquid Foods

$DPCO_2$ processing has been primarily studied in liquid foods. Various fruit juices, milk, and beverages have been subjected to $DPCO_2$ and microbial inactivation, as well as quality attributes, have been evaluated in several studies (Table 5.4). $DPCO_2$ can be applied to liquid foods in batch, semicontinuous, or continuous systems. The continuous treatment of liquid foods involves continuous flow of product and CO through a $DPCO_2$ system before packaging, and this approach makes the process more practical for industrial applications. The inclusion of a micropore filter in the CO_2 stream resulted in a higher rate of CO_2 dissolution in the product and make the $DPCO_2$ system more effective in microbial inactivation (Ishikawa et al., 1995; Shimoda et al., 2001).

Apple juice has been subjected to $DPCO_2$ in several studies. $DPCO_2$ treatment at an especially supercritical state resulted in significant microbial and enzyme inactivation, contributing to shelf life (Murtaza et al., 2020; Spilimbergo et al., 2007; Liao et al., 2010a; Bae et al., 2009; Gui et al., 2007). Apple juice treated by $DPCO_2$ at 30 MPa and 95°C for 90 s inactivated PPO by 60%, inhibited browning, and maintained color during 4 weeks' storage at 4°C (Gui et al., 2006a). Porto et al. (2010) treated apple juice with a continuous $DPCO_2$ at 25–25 MPa and 35°C for 15 min and compared the result with heat-treated control at 75°C for 15 s. They found that $DPCO_2$ treatment was equally effective as the heat treatment in total inactivation of natural microflora by more than 3-log. Besides, both $DPCO_2$ at 15 MPa and heat treatment resulted in similar antioxidant activity that was higher compared to the fresh untreated sample. $DPCO_2$ at 15 MPa resulted in a lower loss of volatile compounds than the treatment at 25 MPa or the heat treatment. *S. cerevisiae* in apple juice was inactivated by 4-log upon $DPCO_2$ treatment at 20 MPa and 36°C for 30 min (Spilimbergo et al., 2007). Gasperi et al. (2009) also found that $DPCO_2$ at 10 MPa and 36°C for 10 min caused a total inactivation of natural microorganisms (8-log) while a moderate reduction in sensory quality and volatile constituents in apple juice. Apple cider inoculated with *E. coli* was treated with a continuous $DPCO_2$, and it was observed that the treatment at 10 MPa and 35°C and CO_2/juice ratio greater than 70g/kg caused more than 6-log inactivation of the bacteria (Gunes et al., 2006). Apple juice enriched with a prebiotic (inulin) treated with $DPCO_2$ at 10–20 MPa and 35°C for 10 min resulted in maintenance of all the functional and nutritional properties without alteration of the chemical properties of the inulin although the breakdown of this compound was seen in heat-treated samples (Silva et al., 2019).

Orange juice has been treated with $DPCO_2$ in several studies, Arreola et al. (1991b) treated orange juice with $DPCO_2$ at 7–34 MPa and 35–60°C for 15–180 min and found that total microbial count was decreased by 2-log. Treatment at 29 MPa, 50°C for 4 h had no residual PE activity and had cloud stability during 66-day refrigerated storage (Arreola et al., 1991a). The restoration of PE activity during storage was seen on $DPCO_2$ treated juice, but this was much lower as compared to the restored activity in thermally treated control. $DPCO_2$ treated samples had higher ascorbic acid, improved color, and cloud than the thermally treated control. The authors found that treatment at 34 MPa and 33°C for 1 h did not adversely affect sensory attributes of orange juice, which was preferable over commercial samples. Kincal et al. (2005, 2006) showed that continuous $DPCO_2$ treatment of orange juice resulted in 5-log inactivation of spoilage

Table 5.4 Summary of Studies on DPCO$_2$ Applications in Liquid Foods

Medium	Microorganisms	Treatment conditions	Microbial inactivation	Effects on quality	Reference
Grape juice (muscadine)		27.6 MPa at 30°C with 7.5–15% CO$_2$		75% PPO inactivation; 80% loss of anthocyanin, total phenolic and antioxidant capacity	Del Pozo-Insfran et al. (2007)
Grape juice	Natural yeast/ mold, bacteria	34 MPa, 30°C, 8% CO$_2$, 6.2 min	7-log yeast/mold, 5-log total bacteria; Similar microbial stability and better quality during 5-wk storage as compared to heat treatment	Better retention of physicochemical and sensory quality in the samples during 5-wk storage as compared to the heat treatment at 75°C for 15 s	Del Pozo-Insfran et al. (2006b)
Grape juice	*S. cerevisiae, C. stellate,* and *K. apiculate*	6.9–48.3 MPa, 25–45°C, and 85–170 g CO$_2$/kg juice	>6-log; pressure, temperature, and CO$_2$ ratio increased microbial inactivation		Gunes et al. (2005)
Grape must	Natural yeast and bacteria (3-log); inoculated *B. subtilis* vegetative cells; inoculated *Pichia awry* yeast	8–12 MPa and 30–40°C for 30 min	3-log yeasts, 1.2-log bacteria; 6-log *B. subtilis* and *Pichia awry*		Parton et al. (2007)
Tomato juice		10–40 MPa, 55°C for 10–60 min; Heat treatment control: 95°C for 20 min		Retained viscosity and flavor profile; better consumer acceptance; better retention of phenolics; increased isomerization of lycopene; increased bioaccessibility of lycopene	Zhao et al. (2019b)
Tomato pulp	Native bacteria and yeast/molds (3-log)	10–40 MPa, 55°C for 10–40 min; Heat treatment control: 95°C for 20 min	Total inactivation of yeast/mold; 2-log bacteria	Better color and aroma as compared to heat treatment; improved consistency: increase in cis-lycopene isomerization	Zhao et al. (2019b)

Product	Microorganism	Treatment	Result	Effects	Reference
Tomato Paste	Natural lactobacillus and total bacteria	10 MPa and 50°C for 100 min	0.3-log reductions in microbial counts		Parton et al. (2007)
Apple juice (cloudy)		30 MPa and 55°C for 60 min		60% PPO inactivation, inhibited browning, color maintained during 4wk storage at 4oC	Gui et al. (2006a, 2007)
Apple juice	Natural microflora	25–25 MPa and 35°C or 15 min; heat treatment control at 75°C for 15 s	>3-log inactivation of natural microorganisms, similar microbial inactivation to heat treatment	Higher antioxidant activity than fresh control, but similar antioxidant activity to the heat treatment 15MPa treatment preserved volatiles better than the heat treatments;	Porto et al. (2010)
Apple juice	S. cerevisiae	10–20 MPa and 36°C for 30 min	4-log inactivation		Spilimbergo et al. (2007)
Apple juice	Natural microflora	10 MPa and 36°C for 10 min	Total inactivation (8-log) of natural microorganisms;	A moderate reduction in sensory quality and volatiles	Gasperi et al. (2009)
Apple cider	E. coli	10 MPa and 35°C and 70 g/kg CO_2/juice ratio	>6-log		Gunes et al. (2006)
Apple juice with inulin		10–20 MPa and 35°C for 10 min		Maintained functional and nutritional properties; preserved inulin	Silva et al. (2019)
Orange juice	Natural microflora	7–34 MPa and 35–60°C for 15–180 min	2-log	Improved color and cloudiness, sensorial acceptable	Arreola et al. (1991a, 1991b)
Orange juice	Natural microflora; L. monocytogenes, E. coli O157:H7, and S. thyphimurium		5-log inactivation of spoilage and pathogenic microorganisms;	Increased cloudiness; quality maintenance for 2 weeks	Kincal et al. (2005, 2006)

(Continued)

Table 5.4 (Continued)

Medium	Microorganisms	Treatment conditions	Microbial inactivation	Effects on quality	Reference
Orange juice		30 MPa, 40°C for 60 min		85% inactivation of PME activity; increased cloudiness; maintenance of color	Iftikhar et al. (2014)
Orange juice		10–30 MPa, 2–40°C for 3–60 min		35–90% PME inactivation; Increased cloudiness; no effect on acidity and brix; 14% reduction in ascorbic acid, small change in color	Briongos et al. (2016)
Orange juice		40 MPa, 55°C for 10–60 min		Increased cloudiness; 95% inactivation of PME; Minimal change in color; Nonanal and citronellol increased, ethyl butyrate and trans-2-hexenol decreased	Niu et al. (2010b)
Orange juice	E. coli, S. cerevisiae, total bacteria	Ultrasound+DPCO2 (10 MPa, 31–41°C) for 3.9 min	Complete inactivation of total bacteria (3-log) and E. coli (4-log), 2.6-log inactivation of S. cerevisiae Enhanced microbial inactivation;	increased cloudiness; better retention of phenolics, antioxidant activity, and ascorbic acid; higher browning index as compared to heat treatment	Paniagua-Martínez et al. (2018b)
Blood-orange juice	Native microorganisms	13–23 MPa, 36°C, 0.385–0.77 gCO$_2$/g-juice, for 15 min	Complete inactivation (3-log); microbial stability (<4.5-log) during 30-day storage at 4°C.	30–40% PME inactivation; no change in phenolics, anthocyanin, and antioxidant capacity; increased cloud	Fabroni et al. (2010)
Red grapefruit juice	Native microorganisms	13–34.5 MPa and 40°C for 5–9 min	5-log; Microbial stability for 6 weeks at 4°C.	69% PE inactivation Increased cloud, no change in acidity and brix, small change in color	Ferrentino et al. (2009)

Product	Target	Treatment conditions	Log reduction	Effects	Reference
Mandarine juice	Native microorganisms	13.8–41.4 MPa, 25–45°C, for 7–9 min	4-log	Increased cloud, L- and b-values; decreased a-values	Lim et al. (2006)
Melon juice	Native microorganisms	8–35 MPa and 35–65°C for 5–60 min	2–4 log at 45°C; complete inactivation at 65°C; microbial stability for 4 weeks at 4°C	98, 75, and 62% inactivation of LOX, PPO, and POD respectively; retention of flavor, aroma, nutrients, and shelf-life for 4 weeks at 4°C	Chen et al. (2010)
Watermelon juice		10–30 MPa at 50°C for 5–60 min		96, 85, and 60% inactivation of PPO, PME, and POD activity; Increased cloudiness, inhibited browning, retention of viscosity, very slight loss of lycopene	Liu et al. (2012)
Carrot juice		10–30 MPa, 25°C for 15–60 min		97% PPO inactivation; decreased pH and browning degree; increased cloud and viscosity; stable carotenoids and brix	Zhou et al. (2009)
Carrot juice	Native microorganisms	$DPCO_2$ (4.9 MPa, 5°C, 5 min) + HHP (300–400 MPa, 25°C for 5 min) sequential treatment	4 and 3 log by $DPCO_2$ and HHP alone; 7 log (complete inactivation) by combined treatment	Enhanced enzyme inactivation by combined treatment; decreased cloud;	Park et al. (2002)
Beetroot juice		10–60 MPa, 31–55°C, 10–30 min		Lower inactivation of enzymes as compared to thermal process; D-values for PG, PE, PPO, and POD were 256, 271, 281, and 384 min; degradation of betaxanthins (32%), betacyanins (58%), and polyphenols (30%)	Marszalek et al. (2017)

(Continued)

Table 5.4 (Continued)

Medium	Microorganisms	Treatment conditions	Microbial inactivation	Effects on quality	Reference
Liquid eggs	Native microorganisms	13 MPa and 45°C with a 50% working volume ratio and 400 min^{-1} agitation speed	Total inactivation (<3 log) of microorganisms;	Shelf-life extension to 5 weeks	Garcia-Gonzalez et al. (2009b)
Beer	Native microorganisms	26 MPa, 21°C, and 9.6% CO_2	>5-log inactivation of yeasts;	Maintenance of flavor; minor reduction in foaming	Dagan and Balaban, (2006)
Milk	*E. coli* *S. aureus*	9–15 MPa, 30°C for up to 6 h	6–7-log inactivation		Erkmen (1997, 2001)
Bovine milk		20 MPa, 50°C for 50 min		Complete inactivation of alkaline phosphatase; decreased viscosity; increased particle size	Liao et al., 2019
Milk	*E. coli*	8 MPa, 70°C, and 0.05 CO_2/milk ratio	The microbial inactivation rate was 0.09 min^{-1}	94% inactivation of alkaline phosphatase	Ceni et al. (2016)
Milk	Natural microflora	24 and 43 MPa, 40°C, and 132 gCO_2/kg milk	4.4-log bacterial inactivation		Werner and Hotchkiss (2006)
Milk	Native bacteria	2-stage CO_2 microbubble system at 4 MPa, 35–50°C for 1–10 min	3-log bacterial inactivation;	Decreased free amino acid content; casein became more sensitive to proteases	Kobayashi et al. (2016)
Milk (skimmed)		5 MPa, 35°C, and 0.30/0.33 CO_2/milk ratio for 15 min		Shelf-life extension up to 35 days; better sensory taste than thermal treatment	Di Giacomo et al. (2009)
Bovine milk	Native bacteria and yeast/molds	25 MPa, 40°C for 50 min	1.8-log bacterial; 0.65-log yeast/mold	Microbial stability maintained for 15 days at 4°C	Hongmei et al. (2014)
Whey-grape juice drink (50% V/V)		10–18 MPa, 35°C, 10 min; Heat treatment control 72°C 15 s.		Lower particle size and consistency index; better sensory quality; L, a, and b-values decreased	Amaral et al. (2018)
Dietary supplement	*E. coli, L. innocua, S. cerevisiae*	3–17 MPa, 20–55°C, 10–140 min	>6-log	Minimal loss of nutrients	Fleury et al. (2018)

microorganisms and several pathogens, such as *L. monocytogenes, E. coli* O157:H7, and *S. thyphimurium*, and an increase in cloudiness with maintained fresh juice quality during 2 weeks of refrigerated storage. Similar results on increased cloudiness in orange juice, inactivation of PME activity, and maintenance of color by $DPCO_2$ were reported in other studies (Iftikhar et al., 2014; Niu et al., 2010b). Briongos et al. (2016) observed that PME in orange juice was inactivated to a limited extent at 2°C (35% at 30 MPa), but the inactivation was more than 60% at 21°C with increasing pressure upon 20 min exposure. They found that residual activity of the enzyme in the juice treated at 30 MPa, 40°C for 40 min increased from 10 to 25% during 12-day storage at 4°C, and this brought about a decrease in cloudiness. The treatment also decreased the particle size distribution in the juice. They observed no effects of the treatments on pH, Brix, and titratable acidity but about 14% reduction in ascorbic acid and a slight reduction in L and b-value in the treated juice. Blood-orange juice treated at 13 MPa and 36°C maintained physicochemical, nutritional, microbial, and sensory quality during 20-day refrigerated storage (Fabroni et al., 2010). $DPCO_2$ 13.8–41.4 MPa, 25–45°C, for 7–9 min resulted in 4-log inactivation of bacteria, increased cloud, lightness, yellowness, and decreased redness in mandarin juice (Lim et al., 2006). Red grapefruit juice treated by a continuous $DPCO_2$ at 34.5 MPa and 40°C for 7 min had 5-log yeast, mold, and bacteria inactivated (Ferrentino et al., 2009). The treatment resulted in 6 weeks of microbial stability at 4°C, increased cloud with partial inactivation (40%) of PE enzyme, higher lightness, and redness, lower yellowness, and retention of total phenolics in the juice.

Grape juice inoculated with *S. cerevisiae, C. stellate,* and *K. apiculate* was treated with continuous $DPCO_2$ at 6.9–48.3 MPa, 25–45°C, and 85–170 gCO_2/kg juice (Gunes et al., 2005). The increased pressure and CO_2 concentration together with increased temperature resulted in increased inactivation of yeast cells (more than 6 log). Similarly, up to 7-log reduction of yeast/mold and 5-log reduction of total bacterial count in grape juice treated by $DPCO_2$ at 34.5 MPa and 30°C for 6.2 min in another work (Del Pozo-Insfran et al., 2006a). The treatment resulted in similar microbial stability but better retention of physicochemical and sensory quality in the samples during 5-wk storage as compared to the heat treatment at 75°C for 15 s (Del Pozo-Insfran et al., 2006b). $DPCO_2$ at 27.6 MPa at 30°C with 7.5% CO_2 resulted in 70% inactivation of the PPO enzyme in grape juice (Del Pozo-Insfran et al., 2007). Treatment of grape must (having 3-log natural microbial load) with a semi-continuous $DPCO_2$ at 8–12 MPa and 30°C for 30 min caused 3- and 1.2-log inactivation of natural yeast and bacteria (Parton et al., 2007). These authors also reported up to 6-log inactivation of *B. subtilis* (vegetative form) inoculated in the grape must by $DPCO_2$ at 8.5 MPa and 40°C or 11 MPa at 32°C for 5 min. Treatment of melon juice with $DPCO_2$ at 35 MPa and 55°C for 60 min resulted in shelf-life extension through microbial inactivation with better retention of aroma compounds, vitamin C, and beta-carotene in comparison with heat treatment at 90°C for 60 s (Chen et al., 2010). $DPCO_2$ at 10–30 MPa at 50°C had preserving effects on various quality attributes of watermelon such as retention of viscosity, increased cloudiness, the inhibition of browning, and limited loss of lycopene (Liu et al., 2012).

The treatment of carrot juice with $DPCO_2$ resulted in increased cloud, viscosity, and titratable acidity, inactivated PPO, inhibited browning, and maintained carotenoids and total soluble solid content (Zhou et al., 2009). The positive effects of $DPCO_2$ on the quality of carrot juice were enhanced when combined with HHP treatment (Park et al., 2002). It was found in another study that $DPCO_2$ at 10 MPa and 47°C was optimum for the inactivation of PPO, PG, and PE, while 30 MPa and 47°C were optimum for POD inactivation (Marszałek et al., 2016).

However, the reported D-values for POD, PPO, PG, and PE as 1439, 1354, 1279, and 215 min, respectively, showed that $DPCO_2$ was not sufficient for the inactivation of the enzymes. It was also noted that a further increase of pressure and temperature resulted in degradation of phenolics without further inactivation of the enzyme. The carotenoid content of carrot juice was unaffected by $DPCO_2$ while total phenolic content decreased by$DPCO_2$, although this reduction was much lower than the one due to thermal treatment. Various enzymes of celery juice required different optimum pressure and temperature for inactivation. The optimum $DPCO_2$ pressure/temperature was 30MPa/47°C for POD and PE, 60MPa/55°C for PPO, and 10MPa/47°C for PG (Marszałek et al., 2016). However, the D-values for PG, PPO, POD, and PE were 1645, 767, 442, and 200 min, respectively, and these high values showed that $DPCO_2$ was not sufficient for the inactivation of the enzymes in the juice. Further increase of the pressure and temperature resulted in degradation of chlorophyll without further increase in enzyme inactivation. $DPCO_2$ treatment resulted in a loss of chlorophyll and total phenolics in celery juice, but this was much lower as compared to the ones due to thermal treatment. Strawberry juice was treated by $DPCO_2$ at 10–60 MPa, 35–45°C for 10–30 min (Marszalek et al. 2015). Treatment at 30 MPa and 60 MPa at 45°C for 30 min exposure resulted in total inactivation of PPO, 83% and 86% inactivation in POD, respectively, and these reduced the activities of the enzymes maintained during 12 weeks of refrigerated storage. The $DPCO_2$ treatment caused an increase of total sugar content where sucrose level decreased, glucose and fructose levels increased. The treatment resulted in up to 34% loss of total vitamin C while it did not adversely affect total anthocyanin content and color.

Tomato pulp was treated with $DPCO_2$ at 10–40 MPa at 55°C for up to 40 min, which resulted in almost total inactivation of the initial 3-log microbial counts and better color, aroma, viscosity, and lycopene isomerization than the heat treatment at 95°C for 20 min (Zhao et al. 2019b). Similar positive effects of $DPCO_2$ on the quality of tomato juice with improved bio-accessibility of lycopene were also reported (Zhao et al., 2019b). On the other hand, $DPCO_2$ at 10 MPa, 50°C for 100 min caused only 0.3-log inactivation of native bacteria in tomato paste (Parton et al., 2007). A banana puree was treated with $DPCO_2$ at 20 MPa at 45–60°C for 30 min and various quality attributes for juice from the puree were evaluated and compared to thermal treatment at the same temperatures (Yu et al., 2013). The color of the juice was better retained by $DPCO_2$ with inhibited browning as compared to mild thermal treatment, and this was associated with the inactivation of PPO in $DPCO_2$-treated samples. $DPCO_2$ treatment resulted in higher clarity, lower viscosity, lower pectin, and lower protein content than mild heat treatment and untreated control. $DPCO_2$ treatment resulted in a lower degree of enzyme (PPO, PE, PG, PED) inactivation in beetroot juice than thermal treatment, and it caused significant degradations in betaxanthins (max 32%) and betacyanins (max 58%) as the temperature and pressure increased (Marszałek et al., 2017).

$DPCO_2$ showed promising results in the pasteurization of liquid eggs. The processing of whole liquid eggs with $DPCO_2$ at 13 MPa and 45°C with a 50% working volume ratio and 400 min^{-1} agitation speed resulted in effective inactivation of natural microorganisms and shelf-life extension to 5 weeks at 4°C. (Garcia-Gonzalez et al., 2009b). The treatment of beer at 26 MPa, 21°C, and 9.6% CO_2 in a continuous $DPCO_2$ system caused more than a 5-log inactivation of yeast and maintenance of aroma and flavor and a minor reduction in foaming and its stability (Dagan and Balaban, 2006).

$DPCO_2$ has been applied to dairy products in several studies. Precipitation of casein by $DPCO_2$ can be the limiting aspect of its application to liquid milk, but if the treated milk is used for other dairy products like cheese $DPCO_2$ would have potential benefits. Moreover, the process can be used to isolate and fractionate casein industrially to be used as ingredients in other products. Exposure of skimmed milk to 15 MPa 38°C for 15 min resulted in a 35-day shelf-life with better sensory taste than HTST pasteurized milk (Di Giacomo et al., 2009). A 6–7-log inactivation of *E. coli* and *S. aureus* in whole and skimmed milk was obtained upon a batch $DPCO_2$ treatment at 9–15 MPa, 30°C for up to 6 h (Erkmen, 1997, 2001). Treatment of milk by a continuous $DPCO_2$ at 24 and 43 MPa, 40°C, and 132 gCO_2/kg milk caused up to a 4.4-log inactivation of bacteria, and the inactivation was reduced as the pressure or the amount of CO_2 decreased (Werner and Hotchkiss, 2006). $DPCO_2$ at 25 MPa and 40°C for 50 min caused reductions of aerobic bacteria, yeast/mold, and coliforms to 1.8, 0.65, and 0 log, respectively, and maintained microbial stability in bovine milk for 15 days at 4°C (Hongmei et al., 2014). Kobayashi et al. (2016) treated milk with $DPCO_2$ using a two-stage low-pressure microbubble system: the first stage was 1–2 MPa at 10°C for 5 min to have saturated CO_2 in milk, and the second stage involved 4 MPa at 35–50°C for 1–10 min. They observed a 3-log reduction of native bacteria and a reduction of free amino acid content except for phenylalanine and valine, which increased. They also noted that the resistance of casein to protease decreased by the $DPCO_2$ treatment. Alkaline phosphatase activity was highly inactivated (94.6%) in milk by $DPCO_2$ at 8 MPa, 70°C, and 0.05 CO_2/milk ratio for 30 min (Ceni et al., 2016). Liao et al. (2019) obtained a complete inactivation of this enzyme by treating bovine milk at 20 MPa, 50°C for 50 min, and the inactivation rate was temperature-dependent: 40% at 20°C and 80% at 40°C. They observed a reduction of viscosity and an increase of particle size induced by $DPCO_2$ at 20 MPa, and these changes were amplified with increasing temperature and/or exposure time. They also found that the treatment increased particle size and turbidity in milk, and recommended treating bovine milk at 20 MPa, 50°C for 20 min to obtain the best physicochemical characteristics. A dietary supplement consisting of plant extracts and concentrated fruit juices was also treated with $DPCO_2$ (Fleury et al., 2018). The treatment resulted in the effective inactivation of inoculated microorganisms (*E. coli, L. innocua, S. cerevisiae*) with fully maintained phenolics and a minimum loss of ascorbic acid.

5.5.2 Emerging Applications in Solid Foods

Applications of $DPCO_2$ to solid foods in literature are relatively few as compared to liquid foods. A specific review on the application of $DPCO_2$ on solid foods has been published a decade ago (Ferrentino and Spilimbergo, 2011). Since then, several studies on various solid foods have been published. Fresh fruits and vegetable products, various meat products, dairy products, and powdered foods have been treated with $DPCO_2$ in various studies, which are briefly summarized in Table 5.5. The treatment of solid foods by $DPCO_2$ involves a batch process rather than a continuous process, and this may limit its industrial applications. Moreover, the penetration of $DPCO_2$ into the solid food matrix would be harder than into a liquid, and this would decrease its overall effects. The extraction of solid food constituents by $DPCO_2$ during treatment can happen, and this may be undesirable. Despite these limitations, $DPCO_2$ has been shown to have a lot of advantages on solid foods, which are discussed in this section.

Table 5.5 Summary of Studies on DPCO$_2$ Applications in Solid Foods

Medium	Microorganisms	Treatment conditions	Microbial inactivation	Effects on quality	Reference
Chicken breast meat	Salmonella, Listeria	13.7 MPa and 35°C for 2 h	98 and 84%, respectively	Color change to whitish; cooked appearance; loss of liquid	Wei et al. (1991)
Chicken meat	E. coli	8–14 MPa at 40°C for 15–45 min	1.4-log E. coli with 15 min 4.6-log E. coli with 45 min No difference between 8 and 14 MPa; >2.6-log native bacteria and 4-log yeast/molds with 45 min; Combination with coriander essential oil had a synergistic effect	Cooked appearance for color and texture	Gonzales-Alonso et al. (2020)
Chicken breast meat	Salmonella Native bacteria	10 MPa and 40°C for up to 420 min for drying process with or without ultrasound (US)	Salmonella: 1.5 and 5.3-log with DPCO$_2$ alone and combined with US for 15 min; >7 log with 45 min alone or combined Native bacteria: 3 and 5 log with DPCO$_2$ alone and combined with US for 45 min	Combined treatment had a similar drying rate to oven drying at 75°C; better rehydration obtained with combined treatment; better retention of vitamins; color changed similar to steamed or boiled samples	Morbiato et al. (2019)
Ground beef	E. coli and S. aureus	31.03 MPa, 42.5°C, 180 min	1 and 3-log, respectively	Cooked color/appearance in the treated meat	Sirisee et al. (1998)
Ground beef	Total bacteria, E. coli O157:H7, E. coli, Salmonella spp.	10.3 MPa, 36°C, 15 min	0.78, 0.94, 0.94, and 1.23-log, respectively	Higher tenderness; No significant changes in juiciness, flavor intensity, and off-flavor intensity	Meurehg (2006)

Food	Microorganism	Conditions	Log reduction	Effects	Reference
Pork loins marinated	E. coli, E. coli O157:H7, S. thyphimurium, L. monocytogenes	14 MPa and 45°C for 40 min	2–2.5 log; slightly higher (10%) inactivation when the meat was in soy sauce-marinates than in hot pepper-marinates		Choi et al. (2009a)
Fresh pork	L. monocytogenes, E. coli O157:H7, E. coli, and S. thyphimurium	12 MPa CO$_2$ at 35°C for 30 min.	1.5-log with DPCO$_2$ alone; 2–2.5-log with DPCO$_2$ combined with organic acid		Choi et al. (2009b)
Meat: porcine longissimus dorsi muscle		7.4 and 15.2 MPa at 31.1°C for 10 min.		No effect on pH, tenderness, and water-holding capacity; Increased lightness and sarcoplasmic protein denaturation	Choi et al. (2008)
Dry-cured ham	L. monocytogenes, Native microorganism	8–12 MPa, 35–50°C	2–7.5 log in Listeria; high effects of temperature but no pressure effect; 3 log in native microbial population at 50°C	No major effects on color, pH, and sensory quality	Ferrentino et al. (2014); Ferrentino et al. (2013)
Sausages		10 MPa at 50 and 60°C for 15 min		Positive contribution to gel formation; cooked appearance; higher weight loss	Rao et al. (2016)
Tilapia	Native microorganisms	8–8.5 MPa and 40°C for 15 min	2–3 log	60% inactivation of protease; color loss, cooked appearance	Sugiharto et al., 2020
Oyster	Total aerobic bacteria	17.2 MPa, 60°C for 60 min	3-log	Retention of the overall acceptability	Meujo et al. (2010)
Fresh-cut carrot	E. coli	6 and 8 MPa, 22°C, 30 min; 10–12 MPa, 22°C, 10 min; 10–12 MPa, 35°C, 10 min; 8 MPa, 35°C, 15 min; 6 MPa, 35°C, 20 min	8-log at 35°C and 12, 10, 8, and 6 MPa for 8, 10, 15, and 20 min, respectively; 7.5 log, 22°C, and 12, 10, 8, and 6 MPa for 10, 10, 30, and 30 min, respectively; due to presence of viable but not culturable cells detected by flow cytometry, the actual inactivation was 2–2.5 log		Tamburini et al. (2014)

(Continued)

Table 5.5 (Continued)

Medium	Microorganisms	Treatment conditions	Microbial inactivation	Effects on quality	Reference
Fresh-cut carrot	E. coli	6–12 MPa, 26–35°C	7 log at 12 MPa, 26–35°C for 7 min or at 10 MPa, 26–35°C for 12 min; 4–4.5 log at 8 MPa, 26–35°C for 15 min; 2.5–3.0 log at 6 MPa, 26–35°C for 15 min		Galvanin et al. (2014)
Fresh-cut carrot	Native microorganism	8–12 MPa, 22–45°C for 5–45 min		Bioactive compounds, antioxidant capacity was maintained during 4-wk storage; loss of texture (90%) and ascorbic acid (40%)	Spilimbergo et al. (2013)
Pickled carrot	Native microorganisms	20 MPa, 20°C for 30 min	2.5 log total bacteria; 5 log lactic acid bacteria; Lower rate of microbial growth during storage as compared to thermal treatment (95°C for 5 min)	Increased electrolyte leakage and MDA content; brittleness, hardness, and pectin content decreased; higher beta-carotene than control and thermally treated samples; retention of color	Wang et al. (2019)
Carrot		20 MPa at 40–60°C for 150 min for drying process		Better retention of microstructure as compared to air drying; paler color of dehydrated samples but become similar to raw carrot after rehydration; reversible reduction in transparency of the cell wall	Brown et al. (2008)

Sample	Microorganism	Treatment conditions	Microbial result	Quality result	Reference
Fresh coriander leaves		10 MPa at 40–50°C for 150 min for drying process	Total inactivation of yeast/molds (5 log); 4 log inactivation of mesophilic bacteria; <1-log inactivation of spores	Similar phenolic constituent to air-dried samples; 82% moisture loss, aw reduced to 0.42	Zambon et al. (2018)
Fresh coriander leaves	*E. coli* O157:H7 *Salmonella* spp. *L. monocytogenes*	10 MPa at 40°C for pathogen tests; 8 MPa at 35°C for 150 min for drying process	7.4, 4.7, and 4.9-log inactivation of the pathogens, respectively; 2.8-, 5.0-, and 4.6-log reduction in native aerobic bacteria, yeast/molds, and *Enterobacteriaceae*, respectively, in dried samples; >5-log inactivation of the pathogens after drying treatment	Similar visual quality and smell as compared to freeze-drying	Bourdoux et al. (2018)
Fresh basil		8–10 MPa, 40°C for 2–4 h for drying process		Moisture content decreased to <10%; reduction in L-value; chlorophyll degradation by 41%; better taste, appearance, and color than air-drying; longer exposure (4 h) caused loss of phenolics, antioxidant capacity, and ascorbic acid	Busic et al. (2014)
Various fresh-cut vegetables with (wet) or without (dry) being in solution		6 MPa, 35°C for 10 min (G-wet); 10 MPa, 35°C for 10 min (S-wet)	G-wet: up to 2 log; inactivation was similar to NaOCl treatment	Soften structure; G-wet: total phenolics, ascorbic acid, and antioxidant activity decreased greater than G-dry and NaOCl, loss of green color; S-wet: caused freezing of vegetables during decompression	Matsufuji et al. (2009)

(Continued)

Table 5.5 (Continued)

Medium	Microorganisms	Treatment conditions	Microbial inactivation	Effects on quality	Reference
Fresh-cut pear	S. cerevisiae	10 MPa, 50°C for 10 min	4 log	Consistency loss; softer aspect; loss of liquid; transition to brown coloring	Valverde et al. (2010)
Fresh-cut apples	E. coli O157:H7 Salmonella spp. L. monocytogenes	10 MPa and 40°C for 10 min for pathogen test; 12.5 MPa and 50°C for 16 h for drying process	5-log inactivation in all pathogens	Dehydration study: a_w reduced to 0.19, lower native microbial counts during 12-month storage as compared to air drying at 60°C for 8 h. Similar and better sensory quality as compared to freeze-drying and air-drying, respectively, after 6–12-month storage	Zambon et al. (2021); Djekic et al. (2018); Tomic et al. (2019)
Fresh-cut coconut	S. enterica	12 MPa and 40–50°C	1 and 2 log at 40 and 50°C, respectively, for 15 min; 3 and 4 log at 40 and 50°C, respectively, for 60 min		Galvanin et al. (2014)
Spinach	E. coli	5–10 MPa and 40°C 10 min	5 log at 7.5–10 MPa 2.5 log at 5 MPa	Discoloration; decreased leaf firmness	Zhong et al. (2008)
Cheese	Native bacteria	1.4–6.2 MPa and 23°C for 16 h	50–99%		Haas et al. (1989)
Cheese	Native bacteria	10–40 MPa and 35°C for 50 min	2.5 log in aerobic bacteria; 2–4 log lactic acid bacteria	48–57% fat reduction, maintenance of phospholipids	Sanchez-Macias et al. (2013)

Food	Target	Conditions	Results	Observations	Reference
Cheese	E. coli, L. innocua, G. stearothermophilus spores, Native microorganisms	9.9 MPa and 35°C for 30 min with or without peracetic acid (PAA, 50–100 ppm)	Synergistic effects of PAA and DPCO$_2$; with combined treatment: >7 log in E. coli, 2.9–4.6 log in Listeria, 3.8 log G. Stearothermophilus, 6–7 log in native microorganisms	pH was not affected by DPCO$_2$ alone; PAA decreased pH from 5.9 to 5.1; combined DPCO$_2$/PAA (100) further decreased pH to 4.6	Sikin et al. (2016)
Dry pepper with 30% moisture added	Native bacteria	5.5 MPa and 23°C for 2 h	90%		Haas et al. (1989)
Paprika powder	Total aerobic bacteria	30 MPa, 90°C for 45 min	5.5 log	A slight reduction of color intensity; decrease of water content	Calvo and Torres (2010)
Cacao powder	Total bacteria and spores	30.0 MPa, 65°C, 40 min	Total inactivation	decreased water content No effect on the physical aspect	Calvo et al. (2007)
Ginseng powder	Total aerobic bacteria	10 MPa, 60°C for 15 h	2.7 log; synergistic effects of additives (water, ethanol, H$_2$O$_2$) observed		Dehghani et al. (2008)
Alfalfa seeds	L. monocytogenes, S. thyphimurium, E. coli O157:H7	10–15 MPa and 35–45°C for 5–15 min	2.65 and 2.48 log for Listeria and Salmonella at 10MPa and 45°C for 5 min; 3.5 log for E. coli at 15 MPa and 35°C for 15 min	No effect on germination rate	Jung et al. (2009)

DPCO$_2$ at 12 MPa and 22°C resulted in 5-log reductions in total coliforms and total yeast/mold counts in fresh-cut carrot after 5- and 10-min exposures, respectively (Spilimbergo et al., 2013). On the other hand, only 3- and 2-log reductions on mesophilic and lactic acid bacterial count, respectively, were noted upon 10-min exposure. Total inactivation of pectolytic enzymes required treatment at higher treatment temperature (40°C) for 30 min. The hardness of the samples decreased by 50% and 90% by the treatments at 22°C and 40°C, respectively. The initial loss of hardness by DPCO$_2$ treatment was not correlated with pectolytic enzyme activities, so the applied pressure seems to be the cause of this reduction. The inactivation of PPO was 80% after 15-min exposure at both temperatures while a maximum of 60% inactivation of POD was achieved by the treatment at 40°C for 15 min. The DPCO$_2$ treatment caused no adverse effects on phenolics, carotenoids, and antioxidant capacity but decreased ascorbic acid and hardness of the carrot samples significantly. The DPCO$_2$ treatment at 12 MPa and 40°C resulted in a 4-week shelf life at 4°C whereas the treatment at 22°C maintained quality for only 2 weeks during storage. Fresh-cut carrots treated by subcritical CO$_2$ at 5 MPa at 20°C for 15 min resulted in the inactivation of PPO, POD, and PME moderately in another study (Bi et al., 2011). However, lower exposure times (up to 8 min) especially at lower pressures caused increased activities of these enzymes. The DPCO$_2$ treatments resulted in increased electrolyte leakage, especially at higher pressures. The firmness of the carrot slices decreased by 8% after treatment at 5 MPa for 15 min, although increased firmness was observed when the exposure time was 2 min. Wang et al. (2019) showed that DPCO$_2$ treatment at 20 MPa and 20°C for 30 min resulted in better sterilization effects, retention of color, and β-carotene content in pickled carrot than conventional thermal treatment at 95°C for 5 min.

Ferrentino and Spilimbergo (2017) treated apple slices in syrup with DPCO$_2$ at 8–12 MPa, 35–55°C for 15 min and evaluated quality and shelf life. They found that 10 MPa at 55°C caused complete inactivation of PPO enzyme together with total inactivation of initial contamination levels of 3.4, 2, and 2.1 log of mesophilic bacteria, coliforms, and yeast/molds, respectively. The microbial counts remained below 1 log, and color, pH, and ascorbic acid levels were maintained after 60 days of storage at 25°C. In another work, DPCO$_2$ treatment at 10 MPa, 40°C for 10 min resulted in total inactivation (>5 log) of E. coli O157:H7, Salmonella spp., and L. monocytogenes in apple slices (Zambon et al., 2021). Niu et al. (2010a) treated apple slices with DPCO$_2$ at 20 MPa at temperatures from 25–65°C for 20 min and processed the fruit into cloudy apple juice. They found no residual PPO activity in the juice made from the fruits treated at all temperatures. PME inactivation by the DPCO$_2$ treatment increased with temperature: 40% at 25°C, 60% at 45°C and 82% at 65°C. They also found that apple juice from DPCO$_2$-treated fruits had a much lower browning degree and turbidity as compared to juice made from mild heat-treated fruits.

Fresh strawberries exposed to a gas mixture of 90% N$_2$+10% CO$_2$ at 5.5 MPa and 45°C for 2 h caused effective inhibition of mold growth for 1 week, but extensive tissue damage was observed (Haas et al., 1989). The extent of the tissue damage was much higher when the treatment was delivered with pure CO$_2$. Fresh-cut coconuts treated by DPCO$_2$ at 12 MPa and 45°C for 15 min brought about 4-log reductions in background microflora, a slight modification of microstructure, but no effects on hardness (Ferrentino et al., 2012). Spinach leaves were treated by DPCO$_2$ at 5–10 MPa at 40°C for 10 min (Zhong et al., 2008). The treatment at supercritical pressures resulted in 5-log inactivation of E. coli while the subcritical pressure caused only 2-log reduction. Background microbial count, on the other hand, was inactivated by only 1 log by the $DPCO_2$ treatments. A visual loss of green color was noted due to $DPCO_2$ treatment.

Ground beef inoculated with *E. coli* and *S. aureus* was exposed to $DPCO_2$ at 31 MPa and 42.5°C, and it was found that 180 min exposure caused a 1-log inactivation of *E. coli* and 120 min exposure caused a 3-log inactivation of *S. aureus* (Sirisee et al., 1998). The microorganisms were protected by the ground beef matrix from the process since only 1.7 min at the same treatment conditions was required to achieve the same inactivation level in phosphate buffer. The authors reported a cooked color/appearance in the treated meat, which could limit its practical use. Similar protecting effects of the food matrix on microorganisms were observed by Erkmen (2000): A 5-log inactivation of *B. thermosphacta* was observed upon treating skinned beef meat with $DPCO_2$ at 6.1 MPa and 45°C for 150 min while only 1-log inactivation was noted in minced beef at the same treatment. $DPCO_2$ treatment of chicken breast meat with no skin at 13.7 MPa and 35°C for 2 h resulted in a 98% reduction in *Salmonella* and 84% reduction in *Listeria* (Wei et al., 1991). However, the treatment resulted in color disruption in the samples. The disruption of color and cooked appearance in chicken meat upon $DPCO_2$ treatment was also noted by Gonzales-Alonso et al. (2020). Pork marinated in soy sauce treated at 14 MPa CO_2 at 45°C for 40 min had 2.5-log inactivated pathogens such as *L. monocytogenes*, *E. coli* O157:H7, and *S. thyphimurium* (Choi et al., 2009a). The inactivation levels with the same treatment were up to 10% lower when the pork was marinated in a hot-pepper paste. The combination of $DPCO_2$ with organic acids (lactate and acetate) resulted in increased pathogen inactivations in fresh pork (Choi et al., 2009b). Treatment of porcine meat with $DPCO_2$ 7–15 MPa and 30°C for 15 min caused no change in pH, water holding capacity, and tenderness, but a loss of color and protein denaturation was noted (Choi et al., 2008). *L. monocytogenes* on dry-cured ham was effectively inactivated (7 logs) by $DPCO_2$ at 12 MPa and 50°C for 15 min with minimal effects on the quality of the product (Ferrentino et al., 2013). The same treatment at 35°C caused a 5-log reduction while treatment at 8 MPa at these temperatures caused 2–3-log reductions in the pathogen. These authors found in another study that treatment of cubed cooked ham at 12 MPa, 50°C for 15 min resulted in up to 3-log inactivation of background bacteria with an undetectable level of yeast and molds, and the treatment had no adverse effects on quality attributes (Ferrentino et al., 2014).

Rao et al. (2017) treated sausages with $DPCO_2$ at 10 MPa at 50°C and 60°C and evaluated gel properties as compared to corresponding mild heat treatment. They found that $DPCO_2$-treated sausages for 15 min had higher hardness, springiness, and chewiness than the mild heat-treated samples at the same temperatures, and this suggested a positive contribution of the $DPCO_2$ treatments to gel formation. The $DPCO_2$ treatment resulted in higher weight loss which was associated with the extraction of fat and reduction of protein solubility. The treatment caused a higher L- and b-values and a lower a-value, which results in a cooked appearance as compared to the mild heat treatment. $DPCO_2$-induced gel formation in myosin and increased gel strength in shrimp surimi, and this was associated with the transformation of α-helix to β-sheet, resulting in increased interactions and cross-links in protein structure (Guo et al., 2017). Sugiharto et al. (2020) found that exposure of tilapia to $DPCO_2$ at 8–8.5 MPa and 40°C for 15 min caused a 2–3-log reduction in microbial counts and 50–60% inactivation of protease with no significant change in texture. However, the treatment resulted in a significant loss of color (increase in L-value), resulting in cooked fish appearance. They also found that as the pressure increased further increase of protease inactivation was observed, but this resulted in unacceptable texture in the sample.

Powdered foods have also been treated with $DPCO_2$ in literature. $DPCO_2$ applied at 30 MPa, 65°C for 40 min resulted in total inactivation of background bacterial spores (>3 log) and mesophilic bacteria (>4 log) in cacao powder with 10% added water, while no significant inactivation was achieved on non-humidified samples (Calvo et al., 2007). The authors found that the moisture content of the humidified sample returned to the initial level and that total phenolic and flavanol monomers and dimers were not affected by the treatment. A 2.7-log reduction of total aerobic count in ginseng powder was achieved by treatment at 10 MPa, 60°C for 15 h (Dehghani et al., 2008). The treatment of paprika powder humidified to 35% moisture with $DPCO_2$ at 30 MPa, 90°C for 45 min caused a 4.5-log inactivation (1 log higher than thermal treatment) of total aerobic microorganism while the inactivation was limited to 1.5 log at 30 MPa, 80°C for 30 min (Calvo and Torres, 2010). The intensity of treatments required to achieve a significant microbial inactivation in powdered foods with low moisture would to be higher as compared to the one for high moisture foods. Higher moisture content (28%) in wheat flour resulted in improved microbial inactivation by $DPCO_2$ at 6.2 MPa, 25–50°C for 2 h while no effect was seen at 12% moisture content (Haas et al., 1989). The authors also noted that $DPCO_2$ was not effective for microbial inactivation on dry herbs while it significantly inactivated microorganisms on fresh herbs. Without adversely affecting the germination capacity of alfalfa seeds, the maximum inactivation of *L. monocytogenes* and *S. thyphimurium* was 2.65 and 2.48 log, respectively, obtained by $DPCO_2$ treatment at 10 MPa and 45°C for 5 min, and a maximum *E. coli* O157:H7 inactivation was 3.5 log at 15 MPa at 35°C for 15 min (Jung et al., 2009). These results showed that $DPCO_2$ can be a potential alternative to commercial treatments such as ethylene oxide, irradiation, or steam treatments used for sterilization of some powdered foods with relatively high moisture content.

The potential use of $DPCO_2$ to produce low-fat cheese with improved texture and microbial quality, meat with low fat and cholesterol, and low-cholesterol milk powder the improved properties of whey has been proposed by several researchers (Chitra et al., 2015; Chao et al., 1991; Sanchez-Macías et al., 2013; Yee et al., 2008; Yee et al., 2007; Zhong and Jin, 2008). $DPCO_2$ processing has also been evaluated for drying of various foods such as apples, coriander, basil, carrot, and beet slices in several studies (Zambon et al., 2021; Tomic et al., 2019; Djekic et al., 2018; Bourdoux et al., 2018; Brown et al., 2008; Busic et al., 2014; Tomic et al., 2020). The drying process by $DPCO_2$ requires a much longer time than for microbial inactivation and is based on the extraction of water from tissues by supercritical CO_2. Zambon et al. (2021) applied $DPCO_2$ at 12.5 MPa and 50°C for 15 h to apple slices for the drying process. The researchers found that the $DPCO_2$ drying process caused a reduction of in the water activity of the samples to 0.18, which was comparable to air drying at 60°C for 8 h. Besides, the $DPCO_2$ dried apple slices had much lower natural microbial counts than the air and freeze-dried samples during 12 months' storage. The authors pointed out that $DPCO_2$ treatment has great potential for drying fruits with improved microbial quality and safety. $DPCO_2$ drying resulted in dehydrated apple slices stable during 6–12 months in protective packages with better retention of chemical compositions and sensory acceptance than air-drying (Tomic et al., 2019). Tomic et al. (2020) found that $DPCO_2$ drying can be a good alternative drying method for making beetroot snack with very good sensory quality. Drying of basil by $DPCO_2$ at 8–10 MPa and 40°C for 2–3 h resulted in products with similar quality to freeze-drying (Busic et al., 2014). $DPCO_2$ at 10 MPa and 40°C resulted in a 7-log inactivation in

E. coli O157:H7, 4.7-log inactivations of *Salmonella* and *L. monocytogenes*, and more than 4-log inactivation of total bacteria and yeast/mold counts in fresh coriander leaves (Bourdoux et al., 2018; Zambon et al., 2018).

5.6 COMBINATION OF DPCO$_2$ WITH OTHER TREATMENTS

A combination of other treatments with DPCO$_2$ in a hurdle approach can have synergistic effects on microbial and enzyme inactivation. The first patent claiming inactivation of microorganisms by pressurized CO$_2$ was a combined treatment with a low dose of ionizing radiation (Urbain et al., 1969). Several other treatments have been studied in combination with DPCO$_2$, and they are briefly discussed in this section. Selected methods must be compatible with DPCO$_2$ treatment from a practical and economical point of view. One of the common treatments combined with DPCO$_2$ has been mild heat treatments. This combined treatment is relatively easy to apply since mild heat treatment can be simultaneously applied in DPCO$_2$ systems. Combined DPCO$_2$ and mild heat treatments have been shown to effectively inactivate microorganisms and enzymes in foods. Apple juice treated with DPCO$_2$ at 20 MPa for 30 min combined with mild heat at 52°C simultaneously resulted in effective inactivation of aerobic bacteria, yeast, and molds and extended shelf life during storage while 62°C thermal treatment alone was required to achieve similar results (Liao et al., 2010a). Combined DPCO$_2$ at 20 MPa with mild heat up to 60°C resulted in higher inactivation of PPO and better color retention, higher clarity, and lower viscosity in banana juice as compared to mild heat or DPCO$_2$ at lower temperature alone (Yu et al., 2013). Combined heat and DPCO$_2$ is especially required for the inactivation of bacterial spores as DPCO$_2$ alone is not sufficient for their inactivation. *G. stearothermophilus* was inactivated by 5 log upon DPCO$_2$ at 30 MPa for 120 min when combined with 95°C while mild heat at this temperature alone or DPCO$_2$ at lower temperatures caused a very little effect on this microorganism (Watanaba et al., 2003). The enhanced inactivation of bacterial spores by DPCO$_2$ combined with mild heat has also been shown in various studies (Haas et al., 1989; Enomoto el al., 1997; Ballestra and Cuq, 1998; Spilimbergo et al., 2002; Shimoda et al., 2002).

DPCO$_2$ can be combined with antimicrobial and/or antioxidants for appropriate foods. Ground pork was exposed to DPCO$_2$ at 13.8 MPa at 35°C for 2 h with or without rosemary, and its quality during storage was evaluated (Huang et al., 2017). It was found that DPCO$_2$ combined with 5% rosemary had synergistic effects on the inactivation of background microflora and caused a 2-log-lower count on day 1 and a lower rate of growth during further refrigerated storage as compared to DPCO$_2$ treatment alone. The combined treatment inhibited fat oxidation, increased L- and b-values, but had no effects on the a-value, pH, and total volatile basic nitrogen (TVB-N) during storage as compared to DPCO$_2$ alone. Bi et al. (2014) found that the combination of nisin with DPCO$_2$ decreased the time required for complete inactivation of the *E. coli* O157:H7 in carrot juice. Similarly, DPCO$_2$ at 30 MPa and 60°C for 30 min combined with nisin at 0.01% resulted in more than 7- and 5-log inactivation on *B. Subtilis* and *G. stearothermophilus* spores, respectively (Silva et al., 2016). Coriander essential oil (0.5% v/w) in raw chicken meat had a synergistic effect on the inactivation of *E. coli* by DPCO$_2$ at 14 MPa and 40°C for 45 min, although fresh coriander had no effects on it (Gonzales-Alonso et al., 2020). Synergistic effects of peracetic acid (50–100 ppm) and DPCO$_2$ at 9.9 MPa and 35°C for 30 min were obtained on the inactivation of native and inoculated microorganisms such as *E. coli*, *L. innocua*, and *G. stearothermophilus* in shredded cheese (Sikin et al., 2016).

A combination of high hydrostatic pressure (HHP) with CO$_2$ has been studied with different foods. There have been different systems used in the combined HHP and CO_2 treatments. For example, a liquid product can be carbonated by bubbling it with CO_2 at ambient pressure and then exposed to an HHP treatment or packaged product with its headspace flushed with CO_2 can be treated by HHP in especially solid foods. Salmon filet was treated by HHP at 150 MPa for 5 min at room temperature in a plastic bag with varying headspace CO_2 levels from 0–100% (Perez-Won et al., 2020). Lipase activity was enhanced by HHP combined with CO_2 while the high enzyme activity after 10 days of storage was only retained in the samples treated by 150 MPa in 100% CO_2. There was an initial reduction in protease activity after treatment at 150 MPa plus 100% CO_2, but the enzyme activity was higher in this sample after 10 days of storage. Collagenase activity increased during storage in the control samples, but this increase was inhibited in samples treated with HHP plus CO_2. On the other hand, no synergistic effects of HHP on microbial quality and shelf life were observed over CO_2 packaging alone. Ortuno et al. (2013a) studied the effects of HHP combined with CO_2 on feijoa puree in a different system. They carbonated the puree in a bag by bubbling it with CO_2 1.28 L/min for 5 min at refrigerated temperature and atmospheric pressure, and the sealed bags of samples with or without further CO_2 flush at 8.5 mLCO_2/g in their headspace were treated by HHP at 300–600 MPa for 5 min. They found increased inactivation of POD, PPO, and PME enzymes when HHP was combined with CO_2 through carbonation and headspace flushing. However, all the treatments except the carbonated samples treated by 450 MPa resulted in significant color change in the puree. Park et al. (2002) studied the combined effects of $DPCO_2$ and HHP applied sequentially on carrot juice. Carrot juice was first treated by $DPCO_2$ at up to 4.9 MPa and 5°C for 5 min, and then the treated juice was packaged in plastic bags and exposed to HHP treatment for 5 min at 25°C and up to 600 MPa. Synergistic effects of $DPCO_2$ and HHP on microbial inactivation were observed: 8-log inactivation was achieved by $DPCO_2$ at 4.9 MPa followed by HHP at 300 MPa. The combined treatment at the highest pressures inactivated LOX, PPO, and PME enzymes synergistically by up to 90%. The combined treatment did not cause any color change, but the $DPCO_2$ treatment caused cloud loss in the samples.

Another treatment tested in combination with $DPCO_2$ is ultrasound. Integration of the ultrasound process with $DPCO_2$ equipment has been developed, and this allows both treatments applied simultaneously (Benedito et al., 2011). Ultrasound creates cavitation in the process medium resulting in mechanical disruption of cells, increased extraction, higher penetration, and dissolution of CO_2 resulting in synergistic effects on microbial inactivation (Ortuno et al., 2014). It has been shown that a combination of ultrasound treatment with $DPCO_2$ shortened the processing time to achieve at least 7-log inactivation of E. coli and S. cerevisiae, and prevented protective effects of solute in fruit juices on microbial inactivation by $DPCO_2$ treatment (Ortuno et al., 2012b, 2013b, 2014). Ultrasound-assisted $DPCO_2$ treatment resulted in 6.8-log inactivation of S. cerevisiae in apple juice while only 4.3-log inactivation was achieved by $DPCO_2$ alone at the same condition (Paniagua-Martínez et al., 2016). Ultrasound-assisted $DPCO_2$ treatment of orange juice resulted in highly significant reductions in microbial counts, increased cloudiness, better retention of phenolics and ascorbic acid but a higher browning index compared to heat treatment (Paniagua-Martínex et al., 2018b; Ortuno et al., 2013b). The combined use of ultrasound (10 W) with $DPCO_2$ at 12 MPa and 45°C for 15 min resulted in complete inactivation of lactic acid, bacteria, and yeast/molds, and a 4-log inactivation of mesophilic bacteria on cooked ham, although the same degree of inactivation required 60 min of exposure when the $DPCO_2$ treatment was used alone (Ferrentino and Spilimbergo, 2016).

Pineapple juice treated with ultrasound-assisted $DPCO_2$ at 10 MPa and 31°C for 4.6 min caused complete inactivation (up to 4 log) of mesophilic bacteria, yeast, and *E. coli* with minimal effects on juice quality attributes such as pH, Brix, and vitamin C (Paniagua-Martínez et al., 2018a). Ultrasound combined with $DPCO_2$ has also been tested for drying of foods such as coriander and chicken breast where improvements in the process and product quality were noted (Michelino et al., 2018; Morbiato et al., 2019). The combined use of pulsed electric field with $DPCO_2$ on the inactivation of several microorganisms such as *E. coli, S. aureus,* and *B. cereus* was studied by Spilimbergo et al. (2003b). They found that pretreatment with PEF had a synergistic effect on microbial inactivation by $DPCO_2$, where a complete inactivation of the bacteria and a minimum 3-log inactivation of the spore were observed.

5.7 CONCLUSION AND PERSPECTIVES

$DPCO_2$ has great capacity in the inactivation of various microorganisms in foods, and this makes it a potential nonthermal technology for the production of a wide range of foods. The inactivation of bacterial spores and enzymes is also possible by $DPCO_2$ but requires a more intense $DPCO_2$ process and in combination with other treatments. There has been a large number of scientific publications on $DPCO_2$ in batch or continuous systems for different food products. However, these studies are conducted on small-scale systems (mostly lab-scale). Commercialization of $DPCO_2$ processing still requires equipment that is designed and manufactured at industrial scales with reasonable investment costs and treatment capacity. The existing scientific results obtained in small-scale systems have to be verified and scaled up to an industrial scale before commercially applicable. $DPCO_2$ process seems to be more appropriate for fluid foods that can be pumped through a continuous treatment system to obtain a high rate of production that is required for industrial applications. $DPCO_2$ process has also been shown to be applicable for solid foods, but they can be treated in batch systems and usually requires longer exposure times that would limit its practical applications. Thus, the development of large-scale $DPCO_2$ processing systems with sufficiently high capacities and reasonable investment costs is a high priority for the commercialization of this process for various foods.

Further understanding of the mechanistic effects of $DPCO_2$ on microorganisms, enzymes, and various components and quality attributes of foods is needed. This requires advanced analytical tools and approaches to characterize the effects of $DPCO_2$ on these constituents in foods. A deep understanding of the mechanisms of $DPCO_2$ would certainly contribute to a better process and system design for different applications. Comprehensive mathematical models for predicting microbial and enzyme inactivation as functions of process parameters and food matrix properties should be developed for better process design. Further studies on the effects of $DPCO_2$ on organoleptic quality, nutritional and bioactive components, and the shelf lives of foods should be conducted to determine its advantages over conventional processes. More comprehensive studies on the combination of $DPCO_2$ with other compatible technologies or treatments would expand the potential of $DPCO_2$ in food preservation.

REFERENCES

Amaral, G.V., Silva, E.K., Costa, A.L.R., Alvarenga, V.O., Cavalcanti, R.N., Esmerino, E.A., et al. 2018. Whey-grape juice drink processed by supercritical carbon dioxide technology: Physical properties and sensory acceptance. *LWT-Food Science Technology*, 92, 80–86.

Arreola, A.G., Balaban, M.O., Marshall, M.R., Peplow, A.J., Wei, C.I., and Cornell, J.A. 1991a. Supercritical carbon dioxide effects on some quality attributes of single strength orange juice. *Journal of Food Science*, 56, 1030–1033.

Arreola, A.G., Balaban, M.O., Wei, C.I., Peplow, A., Marshall, M., and Cornell, J. 1991b. Effect of supercritical carbon dioxide on microbial populations in single strength juice. *Journal of Food Quality*, 14, 275–284.

Bae, Y.Y., Lee, H.J., Kim, S.A., and Rhee, M.S. 2009. Inactivation of *Alicyclobacillus acidoterrestris* spores in apple juice by supercritical carbon dioxide. *International Journal of Food Microbiology*, 136, 95–100.

Balaban, M.O., Arreola, A.G., Marshall, M.R., Peplow, A., Wei, C.I., and Cornell, J. 1991. Inactivation of pectinesterase in orange juice by supercritical carbon dioxide. *Journal of Food Science*, 56, 743–6, 750.

Ballestra, P., and Cuq, J.L. 1998. Influence of pressurized carbon dioxide on the thermal inactivation of bacterial and fungal spores. *LWT-Food Science and Technology*, 31, 84–88.

Ballestra, P., Dasilva, A.A., and Cuq, J.-L. 1996. Inactivation of *Escherichia coli* by carbon dioxide under pressure. *Journal of Food Science*, 61, 829.

Benedito, J.J., Martínez-Pastor, M.T., Mulet, A., Ortuño, C., and Peña, R. 2011. Procedure of inactivation microorganisms by combination of supercritical fluids and ultrasound. *Spain*. Patent No. P201131099.

Bi, X., Wang, Y., Zhao, F., Zhang, Y., et al. 2014. Inactivation of *Escherichia coli* O157:H7 by high pressure carbon dioxide combined with nisin in physiological saline, phosphate buffered saline and carrot juice. *Food Control*, 41, 139–146.

Bi, X.F., Wu, J.H., Zhang, Y., Xu, Z.H., and Liao, X.J. 2011. High pressure carbon dioxide treatment for fresh – cut carrot slices. *Innovative Food Science and Emerging Technologies*, 12, 298–304.

Bourdoux, S., Rajkovic, A., De Sutter, S., Vermeulen, A., Spilimbergo, S., Zambon, A., Hofland, G., Uyttendaele, M., and Devlieghere, F. 2018. Inactivation of *Salmonella, Listeria monocytogenes* and *Escherichia coli* O157:H7 inoculated on coriander by freeze-drying and supercritical CO2 drying. *Innovative Food Science and Emerging Technologies*, 47, 180–186.

Briongos, H., Illera, A.E., Sanz, M.T., Melgosa, R., Beltrán, S., Solaesa, A.G. 2016. Effect of high pressure carbon dioxide processing on pectin methylesterase activity and other orange juice properties. *LWT-Food Science and Technology*, 74, 411–419.

Brown, Z., Fryer, P., Norton, I., Bakalis, S., and Bridson, R. 2008. Drying of foods using supercritical carbon dioxide – investigations with carrot. *Innovative Food Science and Emerging Technologies*, 9, 280–289.

Busic, A., Vojvodić, A., Komes, D., Akkermans, C., Belščak-Cvitanović, A., Stolk, M., and Hofland, G. 2014. Comparative evaluation of CO2 drying as an alternative drying technique of basil (*Ocimum basilicum* L.) – the effect on bioactive and sensory properties. *Food Research International*, 64 No. Supplement C, 34–42

Butz, P., Habison, G., and Ludwig, H. 1992. Influence of high pressure on a lipid coated virus. In: C. Balny et al. (Ed.), *High Pressure and Biotechnology*, 224, John Libbey Eurotext Ltd., pp. 61–64.

Calvo, L., Muguerza, B., and Cienfuegos-Jovellanos, E. 2007. Microbial inactivation and butter extraction in a cocoa derivative using high pressure CO2. *Journal of Supercritical Fluids*, 42, 80–87.

Calvo, L., and Torres, E. 2010. Microbial inactivation of paprika using high pressure CO2. *Journal of Supercritical Fluids*, 52, 134–141.

Ceni, G., Fernandes Silva, M., Valério, C. Jr, Cansian, R.L., Oliveira,. J.V., Dalla Rosa, C., and Mazutti, M.A. 2016. Continuous inactivation of alkaline phosphatase and *Escherichia coli* in milk using compressed carbon dioxide as inactivating agent. *Journal of CO2 Utilization*, 13, 24–28.

Chao, R.R., Mulvaney, S.J., Bailey, M.E., and Fernando, L.N. 1991. Supercritical CO2 conditions affecting extraction of lipid and cholesterol from ground beef. *Journal Food Science*, 56, 183–187.

Chen, J.L., Zhang, J., Song, L.J., Jiang, Y., Wu, J.H., Hu, X.S. 2010. Changes in microorganism, enzyme, aroma of hami melon (*Cucumis melo* L.) juice treated with dense phase carbon dioxide and stored at 4°C. *Innovative Food Science and Emerging Technologies*, 11, 623–629.

Chen, J.S., Balaban, M., Wei, C.I., Marshall, M.R., and Hsu, W.Y. 1992. Inactivation of polyphenol oxidase by high-pressure carbon dioxide. *Journal of Agricultural and Food Chemistry*, 40, 2345–2349.

Chitra, J., Deb, S., and Hari, N.M. 2015. Selective fractionation of cholesterol from whole milk powder: Optimisation of supercritical process conditions. *International Journal of Food Science and Technology*, 50, 2467–2474.

Choi, Y.M., Bae, Y.Y., Kim, K.H., Kim, B.C., and Rhee, M.S. 2009a. Effects of super-critical carbon dioxide treatment against generic *Escherichia coli*, *Listeria monocytogenes*, *Salmonella typhimurium*, and *E. coli* O157:H7 in marinades and marinated pork. *Meat Science*, 82, 419–424.

Choi, Y.M., Kim, O.Y., Kim, K.H., Kim, B.C., and Rhee, M.S. 2009b. Combined effect of organic acids and supercritical carbon dioxide treatments against nonpathogenic *Escherichia coli*, *Listeria monocytogenes*, *Salmonella typhimurium* and *E. coli* O157:H7 in fresh pork. *Letters in Applied Microbiology*, 49, 510–515.

Choi, Y.M., Ryu, Y.C., Lee, S.H., Go, G.W., Shin, H.G., Kim, K.H., et al. 2008. Effects of supercritical carbon dioxide treatment for sterilization purpose on meat quality of porcine longissimus dorsi muscle. *LWT-Food Science and Technology*, 41, 317–322.

Dagan, G.F., and Balaban, M.O. 2006. Pasteurization of beer by a continuous dense-phase CO2 system. *Journal of Food Science*, 71, E164–E169.

Dehghani, F., Annabi, N., Titus, M., Valtchev, P., and Tumilar, A. 2008. Sterilization of Ginseng using a high pressure CO2 at moderate tempera-tures. *Biotechnology and Bioengineering*, 102, 569–576.

Del Pozo-Insfran, D., Balaban, M.O., and Talcott, S.T. 2006a. Microbial stability, phytochemical retention, and organoleptic attributes of dense phase CO2 processed muscadine grape juice. *Journal of Agricultural and Food Chemistry*, 54, 5468–5473.

Del Pozo-Insfran, D., Balaban, M.O., and Talcott, S.T. 2006b. Enhancing the reten-tion of phytochemicals and organoleptic attributes in muscadine grape juice through a combined approach between dense phase CO2 processing and copigmentation. *Journal of Agricultural and Food Chemistry*, 54, 6705–6712.

Del Pozo-Insfran, D., Balaban, M.O., and Talcott, S.T. 2007. Inactivation of poly-phenol oxidase in muscadine grape juice by dense phase-CO2 processing. *Food Research International*, 40, 894–899.

Di Giacomo, G., Taglieri, L., and Carozza, P. 2009. Pasteurization and steriliza-tion of milk by supercritical carbon dioxide treatment. In: *Proceeding of ISSF 2009 New Trends in Supercritical Fluids: Energy, Materials*, Processing, Bordeaux (France).

Dillow, A.K., Dehghani, F., Hrkach, J.S., Foster, N.R., and Langer, R. 1999. Bacterial inactivation by using near- and supercritical carbon dioxide. *Proceedings of the National Academy of Sciences of the United States of America*, 96, 10344–10348

Djekic, I., Tomic, N., Bourdoux, S., Spilimbergo, S., Smigic, N., Udovicki, B., Hofland, G., Devlieghere, F., and Rajkovic, A. 2018. Comparison of three types of drying (supercritical CO2, air and freeze) on the quality of dried apple – quality index approach. *LWT-Food Science and Technology*, 94, 64–72.

Enomoto, A., Nakamura, K., Hakoda, M., and Amaya, N. 1997. Lethal effect of high-pressure carbon dioxide on a bacterial spore. *Journal of Fermentation and Bioengineering*, 83, 305–307.

Erkmen, O. 1997. Antimicrobial effect of pressurized carbon dioxide on *Staphylococcus aureus* in broth and milk. *LWT-Food Science and Technology*, 30, 826–829.

Erkmen, O. 2000. Antimicrobial effects of pressurized carbon dioxide on *Brochothrix thermosphacta* in broth and foods. *Journal of the Science of Food and Agriculture*, 80, 1365–1370.

Erkmen, O. 2001. Effects of high-pressure carbon dioxide on *Escherichia coli* in nutrient broth and milk. *International Journal of Food Microbiology*, 65, 131–135.

Fabroni, S., Amenta, M., Timpanaro, N., and Rapisarda, P. 2010. Supercritical carbon dioxide-treated blood orange juice as a new product in the fresh fruit juice market. *Innovative Food Science and Emerging Technologies*, 11, 477–484.

Fages, J., Jean, E., Frayssinet, P., Mathon, D., Poirier, B., Autefage, A., and Larzul, D. 1998. Bone allografts and supercritical processing: Effects on osteointegration and viral safety. *Journal of Supercritical Fluids*, 13, 351–356.

Ferrentino, G., Balzan, B., Dorigato, A., et al. 2012. Effect of supercritical carbon dioxide pasteurization on natural microbiota, texture, and microstructure of fresh-cut coconut. *Journal of Food Science*, 77, 137–143.

Ferrentino, G., Balzan, S., and Spilimbergo, S. 2013. Supercritical carbon dioxide processing of dry cured ham spiked with *Listeria monocytogenes*: Inactivation kinetics, color, and sensory evaluations. *Food and Bioprocess Technology*, 6, 1164–1174.

Ferrentino, G., Balzan, S., and Spilimbergo, S. 2014. Optimization of supercritical carbon dioxide treatment for the inactivation of the natural microbial flora in cubed cooked ham. *International Journal of Food Microbiology*, 161, 189–196.

Ferrentino, G., Plaza, M., Ramirez, R.M., Ferrari, G., and Balaban, M. 2009. Effects of dense phase carbon dioxide pasteurization on the physical and quality attributes of a red grapefruit juice. *Journal of Food Science*, 74, E333–E341.

Ferrentino, G., and Spilimbergo, S. 2011. High pressure carbon dioxide pasteurization of solid foods: Current knowledge and future outlooks. *Trends in Food Science & Technology*, 22, 427–441.

Ferrentino, G., and Spilimbergo, S. 2016. A combined high pressure carbon dioxide and high power ultrasound treatment for the microbial stabilization of cooked ham. *Journal of Food Engineering*, 174, 47–55.

Ferrentino, G., and Spilimbergo, S. 2017. Non-thermal pasteurization of apples in syrup with dense phase carbon dioxide. *Journal of Food Engineering*, 207, 18–23.

Fleury, C., Savoire, R., Harscoat-Schiavo, C., Hadj-Sassi, A., and Subra-Paternault, P. 2018. Optimization of supercritical CO2 process to pasteurize dietary supplement: Influencing factors and CO2 transfer approach. *Journal of Supercritical Fluids*, 141, 240–251.

Foster, J.W., Cowan, R.M., and Maag, T.A. 1962. Rupture of bacteria by explosive decompression. *Journal of Bacteriology*, 83, 330–334.

Fraser, D. 1951. Bursting bacteria by release of gas pressure. *Nature*, 167, 33–34.

Furukawa, S., Watanabe, T., Koyama, T., Hirata, J., Narisawa, N., Ogihara, H., and Yamasaki, M. 2009. Inactivation of food poisoning bacteria and *Geobacillus stearothermophilus* spores by high pressure carbon dioxide treatment. *Food Control*, 20, 53–58.

Furukawa, S., Watanabe, T., Tai, T., Hirata, J., Ogihara, H., and Yamasaki, M. 2003. Effect of high pressure gaseous and supercritical carbon dioxide treatments on bacterial spores. *Biocontrol Science*, 8, 97–100.

Galvanin, F., Luca, R.D., Ferrentino, G., Barolo, M., Spilimbergo, S., and Bezzo, F. 2014. Bacterial inactivation on solid food matrices through supercritical CO2: A correlative study. *Journal of Food Engineering*, 120, 146–157.

Garcia-Gonzalez, L., Geeraerd, A.H., Elst, K., Van Ginneken, L., Van Impe, J.F., and Devlieghere, F. 2009a. Influence of type of microorganism, Food ingredients and food properties on high-pressure carbon dioxide inactivation of microorganisms. *International Journal of Food Microbiology*, 129, 253–263.

Garcia-Gonzalez, L., Geeraerd, A.H., Elst, K., Van Ginneken, L., Van Impe, J.F., and Devlieghere, F. 2009b. Inactivation of naturally occurring microorganisms in liquid whole egg using high pressure carbon dioxide processing as an alternative to heat pasteurization. *Journal of Supercritical Fluids*, 51, 74–82.

Garcia-Gonzalez, L., Geeraerd, A.H., Spilimbergo, S., Elst, K., Van Ginneken, L., Debevere, J., Van Impe, J.F., and Devlieghere, F. 2007. High pressure carbon dioxide inactivation of microorganisms in foods: The past, the present and the future. *International Journal of Food Microbiology*, 117, 1–28.

Gasperi, F., Aprea, E., Biasioli, F., Carlin, S., Endrizzi, I., Pirretti, G., and Spilimbergo, S. 2009. Effects of supercritical CO2 and N2O pasteurisation on the quality of fresh apple juice. *Food Chemistry*, 115, 129–136.

Giulitti, S., Cinquemani, C., and Spilimbergo, S. 2011. High pressure gases: Role of dynamic intracellular pH in pasteurization. *Biotechnology and Bioengineering*, 108, 1211–1214.

Gonzales-Alonso, V., Cappelletti, M., Bertolini, F.M., Lomolino, G., Zambon, A., and Spilimbergo, S. 2020. Microbial inactivation of raw chicken meat by supercritical carbon dioxide treatment alone and in combination with fresh culinary herbs. *Poulty Science*, 99, 536–545.

Gui, F., Wu, J., Chen, F., Liao, X., Hu, X., Zhang, Z., and Wang, Z. 2006a. Change of polyphenol oxidase activtiy, color, and browning degree during storage of cloudy apple juice treated by supercritical carbon dioxide. *European Food Research and Technology*, 223, 427–432.

Gui, F.Q., Chen, F., Wu, J., Wang, Z., Liao, X., and Hu, X. 2006b. Inactivation and structural change of horseradish peroxidase treated with supercritical carbon dioxide. *Food Chemistry*, 97, 480–489.

Gui, F.Q., Wu, J., Chen, F., Liao, X., Hu, X., Zhang, Z., et al. 2007. Inactivation of polyphenol oxidases in cloudy apple juice exposed to supercritical carbon dioxide. *Food Chemistry*, 100, 1678–1685.

Gunes, G., Blum, L.K., and Hotchkiss, J.H. 2005. Inactivation of yeasts in grape juice using a continuous dense phase carbon dioxide processing system. *Journal of the Science of Food and Agriculture*, 85, 2362–2368.

Gunes, G., Blum, L.K., and Hotchkiss, J.H. 2006. Inactivation of *Escherichia coli* (ATCC 4157) in diluted apple cider by dense-phase carbon dioxide. *Journal of Food Protection*, 69, 12–16.

Guo, M., Liu, S., Ismail, M., et al. 2017. Changes in the myosin secondary structure and shrimp surimi gel strength induced by dense phase carbon dioxide. *Food Chemisty*, 227, 219–226.

Haas, G.J., Prescott, J.R., Dudley, R.D., Hintlian, C., and Keane, L. 1989. Inactivation of microorganism by carbon dioxide under pressure. *Journal of Food Safety*, 9, 253–265.

Hong, S.I., Park, W.S., and Pyun, Y.R. 1997. Inactivation of *Lactobacillus* sp. from kimchi by high pressure carbon dioxide. *LWT-Food Science and Technology*, 30, 681–685.

Hong, S.I., and Pyun, Y.R. 1999. Inactivation kinetics of *Lactobacillus plantarum* by high pressure carbon dioxide. *Journal of Food Science*, 64, 728–733.

Hong, S.I., and Pyun, Y.R. 2001. Membrane damage and enzyme inactivation of *Lactobacillus plantarum* by high pressure CO2 treatment. *International Journal of Food Microbiology*, 63, 19–28.

Hongmei, L., Zhong, K., Liao, X., and Hu, X. 2014. Inactivation of microorganisms naturally present in raw bovine milk by high-pressure carbon dioxide. *International Journal of Food Science and Technology*, 49, 696–702.

Huang, S., B. Liu, D. Ge, et al. 2017. Effect of combined treatment with supercritical CO2 and rosemary on microbiological and physicochemical properties of ground pork stored at 4 °C. *Meat Science*, 125, 114–120.

Iftikhar, T., Wagner, M.E., and Rizvi, S.S.H. 2014. Enhanced inactivation of pectin methyl esterase in orange juice using modified supercritical carbon dioxide treatment. *International Journal of Food Science and Technology*, 49, 804–810.

Ishikawa, H., Shimoda, M., Shiratsuchi, H., and Osajima, Y. 1995. Sterilization of microorganisms by the supercritical carbon dioxide micro-bubble method. *Bioscience, Biotechnology and Biochemistry*, 59, 1949–1950.

Ishikawa, H., Shimoda, M., Tamaya, K., Yonekura, A., and Kawano, I. 1997. Inactivation of *Bacillus* spores by the supercritical carbon dioxide micro-bubble method. *Bioscience, Biotechnology and Biochemistry*, 61, 1022–1023.

Jung, W.Y., Choi, Y.M., and Rhee, M.S. 2009. Potential use of supercritical carbon dioxide to decontaminate *Escherichia coli* O157:H7, *Listeria monocytogenes*, and *Salmonella typhimurium* in alfalfa sprouted seeds. *International Journal of Food Microbiology*, 136, 66–70.

Kamihira, M., Taniguchi, M., and Kobayashi, T. 1987. Sterilization of microorganisms with supercritical carbon dioxide. *Agricultural and Biological Chemistry*, 51, 407–412.

Kincal, D., Hill, W.S., Balaban, M., Portier, K.M., Sims, C.A., Wei, C.I., and Marshall, M.R. 2006. A continuous high-pressure carbon dioxide system for cloud and quality retention in orange juice. *Journal of Food Science*, 71, 338–344.

Kincal, D., Hill, W.S., Balaban, M.O., Portier, K.M., Wei, C.I., and Marshall, M.R. 2005. A continuous high pressure carbon dioxide system for microbial reduction in orange juice. *Journal of Food Science*, 70, M249–M254.

Kobayashi, F., Odake, S., Kobayashi, K., and Sakurai, H. 2016. Effect of pressure on the inactivation of enzymes and hiochi bacteria in unpasteurized sake by low-pressure carbon dioxide microbubbles. *Journal of Food Engineering*, 171, 52–56.

Liao, H., Zhang, F., Liao, X., Hu, X., Chen, Y., and Deng, L. 2010b. Analysis of *Escherichia coli* cell damage induced by HPCD using microscopies and fluorescent staining. *International Journal of Food Microbiology*, 144, 169–176.

Liao, H., Zhang, L., Hu, X., and Liao, X. 2010a. Effect of high pressure CO2 and mild heat processing on natural microorganisms in apple juice. *International Journal of Food Microbiology*, 137, 81–87

Liao, H., Zhong, K., Hu, X., and Liao, X. 2019. Effect of high pressure carbon dioxide on alkaline phosphatase activity and quality characteristics of raw bovine milk. *Innovative Food Science and Emerging Technologies*, 52, 457–462

Liao, X., Zhang, Y., Bei, J., Hu, X., and Wu, J. 2009. Alterations of molecular properties of lipoxygenase induced by dense phase carbon dioxide. *Innovative Food Science and Emerging Technologies*, 10, 47–53.

Lim, S., Yagiz, Y., and Balaban, M.O. 2006. Continuous high pressure carbon dioxide processing of mandarin juice. *Food Science and Biotechnology*, 15, 13–18.

Lin, H.M., Yang, Z., and Chen, L.F. 1993. Inactivation of *Leuconostoc dextranicum* with carbon dioxide under pressure. *The Chemical Engineering Journal,* 52, B29–B34.

Liu, Y., Hu, X., Zhao, X., and Song, H. 2012. Combined effect of high pressure carbon dioxide and mild heat treatment on overall quality parameters of watermelon juice. *Innovative Food Science and Emerging Technologies,* 13, 112–119.

Marszalek, K., Skąpska, S., Woźniak, L., and Sokołowska, B. 2015. Application of supercritical carbon dioxide for the preservation of strawberry juice: Microbial and physicochemical quality, enzymatic activity and the degradation kinetics of anthocyanins during storage. *Innovative Food Science and Emerging Technologies,* 32, 101–109.

Marszałek, K., Krzyżanowska, J., Woźniak, Ł., and Skąpska, S. 2016. Kinetic modelling of tissue enzymes inactivation and degradation of pigments and polyphenols in cloudy carrot and celery juices under supercritical carbon dioxide. *Journal of Supercritical Fluids,* 117, 26–32.

Marszałek, K., Krzyżanowska, J., Woźniak, Ł., and Skąpska, S. 2017. Kinetic modelling of polyphenol oxidase, peroxidase, pectin esterase, polygalacturonase, degradation of the main pigments and polyphenols in beetroot juice during high pressure carbon dioxide treatment. *LWT-Food Science and Technology,* 85, 412–417.

Matsufuji, H., Furukawa, S., Teranishi, K., Kawaharada, K., et al. 2009. Effects of nonthermal processes on the inactivation of microorganisms and Antioxidants in minimally processed vegetables. *Food Science and Technology Research,* 15, 153–162.

Messens, W., Van Camp, J., and Huyghebaert, A. 1997. The use of high pressure to modify the functionality of food proteins. *Trends in Food Science & Technology,* 8, 107–112.

Meujo, D.A.F., Kevin, D.A., Peng, J., Bowling, J.J., Liu, J., and Hamann, M.T. 2010. Reducing oyster-associated bacteria levels using supercritical fluid CO2 as an agent of warm pasteurization. *International Journal of Food Microbiology,* 138, 63–70.

Meurehg, T.C.A. 2006. Control of *Escherichia coli* O157:H7, generic *Escherichia coli,* and *Salmonella* spp. On beef trimmings prior to grinding using a controlled phase carbon dioxide (cpCO2) system. PhD Thesis, Manhattan, Kansas: Kansas State University.

Michelino, F., Zambon, A., Vizzotto, M.T., Cozzi, S., and Spilimbergo, S. 2018. High power ultrasound combined with supercritical carbon dioxide for the drying and microbial inactivation of coriander. *Journal of CO2 Utilization,* 24, 516–521.

Morbiato, G., Zambon, A., Toffoletto, M., Poloniato, G., Dall'Acqua, S., de Bernard, M., and Spilimbergo, S. 2019. Supercritical carbon dioxide combined with high power ultrasound as innovate drying process for chicken breast. *Journal of Supercritical Fluids,* 147, 24–32.

Murtaza, A., Iqbal, A., Marszałek, K., Iqbal, M.A., Ali, S.W., Xu, X., Pan, S., and Hu, W. 2020. Enzymatic, phyto-, and physicochemical evaluation of apple juice under high-pressure carbon dioxide and thermal processing. *Foods,* 9, 243.

Niu, L.Y., Hu, X.S., Wu, J.H., Liao, X.J., Chen, F., Zhao, G.H., and Wang, Z.F. 2010b. Effect of dense phase carbon dioxide process of physicochemical properties and flavor compounds of orange juice. *Journal of Food Processing and Preservation,* 34, 530–548.

Niu, S., Xu, Z., Fang, Y., Zhang, L., Yang, Y., Liao, X., and Hu, X. 2010a. Comparative study on cloudy apple juice qualities from apple slices treated by high pressure carbon dioxide and mild heat. *Innovative Food Science and Emerging Technologies,* 11, 91–97.

Ortuno, C., Duong, T., Balaban, M.O., and Benedito, J. 2013a. Combined high hydrostatic pressure and carbon dioxide inactivation of pectin methylesterase, polyphenol oxidase and peroxidase in feijoa puree. *Journal of Supercritical Fluids*, 82, 56–62.

Ortuno, C., Martínez-Pastor, M.T., Mulet, A., and Benedito, J. 2013b. Application of high power ultrasound in the supercritical carbon dioxide inactivation of *Saccharomyces cerevisiae*. *Food Research International*, 51, 474–481.

Ortuno, C., Martínez-Pastor, M.T., Mulet, A., and Benedito, J. 2012a. Supercritical carbon dioxide inactivation of *Escherichia coli* and *Saccharomyces cerevisiae* in different growth stages. *Journal of Supercritical Fluids*, 63, 8–15.

Ortuno, C., Martínez-Pastor, M.T., Mulet, A., and Benedito, J. 2012b. An ultrasound-enhanced system for microbial inactivation using supercritical carbon dioxide. *Innovative Food Science and Emerging Technologies*, 15, 31–37.

Ortuno, C., Quiles, A., and Benedito, J. 2014. Inactivation kinetics and cell morphology of *E. coli* and *S. cerevisiae* treated with ultrasound-assisted supercritical CO2. *Food Research International*, 62, 955–964.

Paniagua-Martínez, I., Mulet, A., García-Alvarado, M.A., and Benedito, J. 2016. Ultrasound- assisted supercritical CO2 treatment in continuous regime: Appliaction in *Sacharomyces cerevisiae* inactivation. *Journal of Food Engineering*, 181, 42–49.

Paniagua-Martínez, I., Mulet, A., García-Alvarado, M.A., and Benedito, J. 2018a. Inactivation of the microbiota and effect on the quality attributes of pineapple juice using a continuous flow ultrasound assisted supercritical carbon dioxide system. *Food Science and Technology International*, 24, 547–554.

Paniagua-Martínez, I., Mulet, A., García-Alvarado, M.A., and Benedito, J. 2018b. Orange juice processing using a continuous flow ultrasound-assisted supercritical CO2 system: Microbiota inactivation and product quality. *Innovative Food Science and Emerging Technologies*, 47, 362–370.

Park, S.J., Lee, J.I., and Park, J. 2002. Effects of a combined process of high-pressure carbon dioxide and high hydrostatic pressure on the quality of carrot juice. *Journal of Food Science*, 67, 1827–1834.

Parton, T., Bertucco, A., and Bertoloni, G. 2007. Pasteurisation of grape must and tomato paste by dense-phase CO_2. *Italian Journal of Food Science*, 19, 425–437.

Perez-Won, M., Lemus-Mondaca, R., Herrera-Lavados, C.E., Reyes, J., Roco, T., Palma-Acevedo, A., Tabilo-Munizaga, G.P., and Aubourg, S. 2020.Combined treatments of high hydrostatic pressure and CO2 in Coho Salmon (*Oncorhynchus kisutch*): Effects on enzyme inactivation, physicochemical properties, and microbial shelf life. *Foods*, 9, 273.

Perrut, M. 2012. Sterilization and virus inactivation by supercritical fluids (a review). *Journal of Supercritical Fluids*, 66, 359–371.

Porebska, I., Sokolowska, B., Skapska, S., and Rzoska, S.J. 2017. Treatment with high hydrostatic pressure and supercritical carbon dioxide to control *Alicyclobacillus acidoterrestris* spores in apple juice. *Food Control*, 73, 24–30.

Porto, C.D., Decorti, D., and Tubaro, F. 2010. Effects of continuous dense-phase CO2 system on antioxidant capacity and volatile compounds of apple juice. *International Journal of Food Science and Technology*, 45, 1821–1827.

Rao, L., Bi, X.F., Zhao, F., Wu, J.H., Hu, X.S., and Liao, X.J. 2015 Effect of high-pressure CO2 processing on bacterial spores. *Critical Reviews in Food Science and Nutrition*, 56, 1808–1825.

Rao, L., Wang, Y., Chen, F., Hu, X., Liao, X., and Zhao, L. 2020. High pressure CO2 reduces the wet heat resistance of *Bacillus subtilis* spores by perturbing the inner membrane. *Innovative Food Science and Emerging Technologies*, 60, 102291.

Rao, L., Zhao, L., Wang, Y., Chen, F., Hu, X., and Liao, X. 2016. Investigating the inactivation mechanism of *Bacillus Subtilis* spores by high pressure CO2. *Frontiers in Microbiology*, 7, 1411.

Rao, L., Zhao, L., Wang, Y., Chen, F., Hu, X., Setlow, P., and Liao, X. 2019. Mechanism of inactivation of *Bacillus subtilis* spores by high pressure CO2 at high temperature. *Food Microbiology*, 82, 36–45.

Rao, W. Li, X., Wang, Z., Yang, Y., Qu, Y., Gao, Y., Chen, L., and Zhang, D. 2017. Dense phase carbon dioxide combined with mild heating induced myosin denaturation, texture improvement and gel properties of sausage. *Journal of Food Process Engineering*, 40, e12404.

Sanchez-Macías, D., Laubscher, A., Castro, N., Argüello, A., and Jiménez-Flores, R. 2013. Effects of supercritical fluid extraction pressure on chemical composition, microbial population, polar lipid profile, and microstructure of goat cheese. *Journal of Dairy Science*, 96, 1325–1334.

Shimoda, M., Cocunubo-Castellanos, J., Kago, H., Miyake, M., Osajima, Y., and Hayakawa, I. 2001. The influence of dissolved CO2 concentration on the death kinetics of *Saccharomyces cerevisiae*. *Journal of Applied Microbiology*, 91, 306–311

Shimoda, M., Kago, N.H., Kojima, Miyake, M., Osajima, Y., Hayakawa, I. 2002. Accelerated death kinetics of *Aspergillus niger* spores under high pressure carbonation. *Applied and Environmental Microbiology*, 68, 4162–4167

Sikin, A.M., Walkling-Ribeiro, M., and Rizvi, S.S.H. 2016. Synergistic effect of supercritical carbon dioxide and peracetic acid on microbial inactivation in shredded Mozzarella-type cheese and its storage stability at ambient temperature. *Food Control*, 70, 174–182.

Silva, E.K., Arruda, H.S., Eberlin, M.N., Pastore, G.M., and Meireles, M.A.A. 2019. Effects of supercritical carbon dioxide and thermal treatment on the inulin chemical stability and functional properties of prebiotic-enriched apple juice. *Food Research International*, 125, 108561.

Silva, M.A., Araujo, A.P., Ferreira, J.S., and Kieckbusch, T.G. 2016. Inactivation of *Bacillus subtilis* and *Geobacillus stearothermophilus* inoculated over metal surfaces using supercritical CO2 process and nisin. *Journal of Supercritical Fluids*, 109, 87–94.

Sirisee, U., Hsieh, F., and Huff, H.E. 1998. Microbial safety of supercritical carbon dioxide processes. *Journal of Food Processing and Preservation*, 22, 387–403.

Spilimbergo, S., and Bertucco, A. 2003. Non-Thermal bacteria inactivation with dense CO2. *Biotechnology and Bioengineering*, 84, 627–638.

Spilimbergo, S., Bertucco, A., Lauro, F.M., and Bertoloni, G. 2003a. Inactivation of *Bacillus subtilis* spores by supercritical CO2 treatment. *Innovative Food Science and Emerging Technologies*, 4, 161–165.

Spilimbergo, S., Dehghani, F., Bertucco, A., and Foster, N.R. 2003b. Inactivation of bacteria and spores by pulse electric field and high pressure CO2 at low temperature. *Biotechnology and Bioengineering*, 1, 118–125.

Spilimbergo, S., Elvassore, N., and Bartucco, A. 2002. Microbial inactivation by high-pressure. *Journal of Supercritical Fluids*, 22, 55–63.

Spilimbergo, S., Komes, D., Vojvodic, A., et al. 2013. High pressure carbon dioxide pasteurization of fresh-cut carrot. Journal of Supercritical Fluids, 79, 92–100.

Spilimbergo, S., Mantoan, D., and Dalser, A. 2007. Supercritical gases pasteurization of apple juice. *Journal of Supercritical Fluids*, 40, 485–489.

Sugiharto, S.P., Bintoro, N., Karyadi, J.N.W., and Pranoto, Y. 2020. Supercritical carbon dioxide pasteurization to reduce the activity of muscle protease and its impact on physicochemical properties of Nile tilapia. *Research Ideas and Outcomes*, 6, e56887

Tamburini, S., Anesi, A., Ferrentino, G., Spilimbergo, S., Guella, G., and Jousson, O. 2014. Supercritical CO2 induces marked changes in membrane phospholipids composition in *Escherichia coli* K12. *Journal of Membrane Biology*, 247, 469–477.

Taniguchi, M., Kamihira, M., and Kobayashi, T. 1987. Effect of treatment with supercritical CO2 on enzyme activity. *Agricultural and Biological Chemistry*, 51, 593–4.

Tedjo, W., Eshtiaghi, M.N., and Knorr, D. 2000. Impact of supercritical carbon dioxide and high pressure on lipoxygenase and peroxidase activity. *Journal of Food Science*, 65, 1284–1287.

Tomic, N., Djekic, I., Hofland, G., Smigic, N., Udovicki, B., and Rajkovic, A. 2020. Comparison of supercritical CO2-drying, freeze-drying and frying on sensory properties of beetroot. *Foods*, 9, 1201.

Tomic, N., Djekic, I., Zambon, A., Spilimbergo, S., Bourdoux, S., Holtze, E., Hofland, G., Sut, S., Dall'Acqua, S., Smigic, N., et al. 2019. Challenging chemical and quality changes of supercritical CO2 dried apple during long-term storage. *LWT-Food Science and Technology*, 110, 132–141.

Urbain, W.M., Shank, J.L., and Kauffman, F.L. 1969. Irradiation with CO2 under pressure. US Patent: 3,483,005.

Valverde, M.T., Marin-Iniesta, F., and Calvo, L. 2010. Inactivation of *Saccharomyces cerevisiae* in conference pear with high pressure carbon dioxide and effects on pear quality. *Journal of Food Engineering*, 98, 421–428.

Wang, D., Ma, Y., Suna, X., Zhang, M., Zha, Y., and Zhao, X. 2019. Effect of dense phase carbon dioxide treatment on physicochemical and textural properties of pickled carrot. *CYTA – Journal of Food*, 17, 988–996.

Watanabe, T., Furukawa, S., Hiratak, J., Koyama, T., Ogihara, H., and Yamasaki, M. 2003. Inactivation of *Geobacillus stearothermophilus* spores by high-pressure carbon dioxide treatment. *Applied and Environmental Microbiology*, 69, 7124–7129.

Wei, C.I., Balaban, M.O., Fernando, S.Y., and Peplow, A.J. 1991. Bacterial effect of high pressure CO2 treatment on foods spiked with *Listeria* or *Salmonella*. *Journal of Food Protection*, 54, 189–193.

Werner, B.G., and Hotchkiss, J.H. 2006. Continuous flow nonthermal CO2 processing: The lethal effects of subcritical and supercritical CO2 on total microbial populations and bacterial spores in raw milk. *Journal of Dairy Science*, 89, 872–881.

Wimmer, Z., and Zarevucka, M. 2010. A Review on the effects of supercritical carbon dioxide on enzyme activity. *International Journal of Molecular Sciences*, 11, 233–253;

Yee, J.L., Khalil, H., and Jiménez-Flores, R. 2007. Flavor partition and fat reduction in cheese by supercritical fluid extraction: Processing variables. *Lait*, 87, 269–285.

Yee, J.L., Walker, J., Khalil, H., and Jiménez-Flores, R. 2008. Effect of variety and maturation of cheese on supercritical fluid extraction efficiency. *Journal of Agricultural and Food Chemistry*, 56, 5153–5157.

Yu, T., Niu, L., and Iwahashi, H. 2020. High-pressure carbon dioxide used for pasteurization in food industry. *Food Engineering Reviews*, 12, 364–380.

Yu, Y., Xiao, G., Wu, J., Xu, Y., Tang, D., Chen, Y., Wen, J., and Fu, M. 2013. Comparing characteristic of banana juices from banana pulp treated by high pressure carbon dioxide and mild heat. *Innovative Food Science and Emerging Technologies*, 18, 95–100.

Zambon, A., Bourdoux, S., Pantano, M.F., MariaPugno, N., Boldrin, F., Hofland, G., Rajkovic, A., Devlieghere, F., and Spilimbergo, S. 2021. Supercritical CO2 for the drying and microbial inactivation of apple's slices. *Drying Technology*, 39, 259–267.

Zambon, A., Michelino, F., Bourdoux, S., Devlieghere, F., Sut, S., Dall'Acqua, S., Rajkovic, A., Spilimbergo, S. 2018. Microbial inactivation efficiency of supercritical CO2 drying process. *Drying Technology*, 36, 2016–2021

Zhang, J., Davis, T.A., Matthews, M.A., Drews, M.J., LaBerge, M., and An, Y.H. 2006. Sterilization using high-pressure carbon dioxide. *Journal of Supercritical Fluids*, 38, 354–372.

Zhao, W., Sun, Y., Cheng, Y., Ma, Y., and Zhao, X. 2019a. Effect of high-pressure carbon dioxide on the quality of cold- and hot-break tomato pulps. *Journal of Food Processing and Preservation*, 43, e13959.

Zhao, W., Sun, Y., Ma, Y., and Zhao, X. 2019b. Dense phase carbon dioxide treatment of tomato juice: Effect on physico-chemical properties, phenolic composition, lycopene isomerisation and in vitro bioaccessibility. *International Journal of Food Science and Technology*, 54, 1658–1669.

Zhong, Q., Black, D.G., Davidson, P.M., et al. 2008. Nonthermal inactivation of *Escherichia coli* K-12 on spinach leaves, using dense phase carbon dioxide. *Journal of Food Protection*, 71, 1015–1017.

Zhong, Q., and Jin, M. 2008. Enhanced functionalities of whey proteins treated with supercritical carbon dioxide. *Journal of Dairy Science*, 91, 490–499.

Zhou, L., Bi, X., Xu, Z., Yang, Y., and Liao, X. 2015. Effects of high-pressure co2 processing on flavor, texture, and color of foods. *Critical Reviews in Food Science and Nutrition*, 55, 750–768.

Zhou, L., Wang, Y., Hu, X.S., Wu, J.H., and Liao, X.J. 2009. Effect of high pressure carbon dioxide on the quality of carrot juice. *Innovative Food Science and Emerging Technologies*, 10, 321–327.

Zhou, L.Y., Zhang, Y., Leng, X.J., Liao, X.J., and Hu, X.S. 2010. Acceleration of precipitation formation in peach juice induced by high-pressure carbon dioxide. *Journal of Agricultural and Food Chemistry*, 58, 9605–9610.

6 Food Preservation by Ozonation

Maithili Adhikary, Raunak Banerjee, Surabhi Chaudhuri, and Anindya Chanda

Corresponding authors:
Anindya Chanda; chanda@mycologics.net
Surabhi Chaudhuri; surabhi.chaudhuri@bt.nitdgp.ac.in

CONTENTS

6.1 INTRODUCTION

Foodborne illness and risks associated with the consumption of contaminated foods are increasing steadily. The world health organization estimates that 1 in 10 people fall ill after eating contaminated food, leading to 600 million foodborne illnesses worldwide and 420,000 deaths every year [1, 2]. Children younger than 5 years carry 40% of this illness burden, resulting in 125,000 deaths per year [2]. Hence, food safety is critical for the food industry, and thus, vast resources are invested in bringing advanced technologies to the market to deliver clean and safe food products. Production of safe food products involves examining materials entering the food supply chain, controlling microbial growth, and minimizing or eliminating foodborne pathogens either during

DOI: 10.1201/9781003147978-6

postharvest or/and postprocessing. While thermal treatments such as pasteurization or sterilization are traditionally applied to pasteurize or sterilize food, such treatments remain most common for heat-stable foods; however, thermal treatments deteriorate sensory and nutritional qualities [3, 4]. Furthermore, consumers prefer nonthermally treated foods over thermal sterilized foods because of an underlying notion that thermal treatments make foods 'less fresh' because it loses nutrients and sensory quality upon heat treatment [5, 6]. Hence the industry has consistently searched for nonthermal technologies that could retain most of the fresh intrinsic attributes while keeping the food safe and stable for longer duration [7–12].

Among the nonthermal technologies used for food preservation, antimicrobials are the most primitive and common. Antimicrobials have been commonly used to extend the shelf lives of vegetables and fruits. Some common examples of antimicrobials include chlorine, peracetic acid, electrolyzed water, and hydrogen peroxide [13–17]. Sodium hypochlorite (that leads to chlorine formation during treatment) is widely used by the food industry for decontamination, especially during aqueous formulations under various conditions (washing, spraying, etc.) [18–20]. Chlorine is a very effective antimicrobial agent and can efficiently inactivate foodborne pathogens without degrading the overall quality of processed food during its shelf life [21–28]. However, recent studies suggest that chlorine reacts with organic matters like iodine and bromine components and forms hazardous chemicals such as iodinated and brominated disinfection by-products, such as monochloramine, organochlorinated byproducts, halo acetic acids, and trihalomethanes [29–31]. These by-products are cytotoxic, mutagenic, and persistent in the environment [32–37]. Hence, chlorine's use for disinfection has been discontinued in several European countries, and there has been a continuous and increasing demand for natural preservatives and environment-friendly technologies, especially for preserving fresh fruits and vegetables. The application of ozone in food preservation emerged as an up-and-coming technology that can address the need for "green" and nonthermal preservation technologies and has therefore received wide attention in the food industry.

Forty-eight years after ozone was discovered in 1840 by Christian Schonbein, Fewson invented a generator to make ozone in the United States [38]. In 1904 the first commercial ozone-based disinfection machine using ozone to treat drinking water was invented [39]. Europe has used ozone for water purification since 1906 [40]. The U.S. Food and Drug Administration (FDA) has approved ozone as generally recognized as safe (GRAS) for food applications in 1995 [41]. Furthermore, in 2001 the FDA has approved ozone as an antimicrobial supplement in both gaseous and aqueous phases for direct contact with foods [42]. Ozone treatment of natural mineral water is now permitted by the European Council of Ministers [43]. The French Food Safety Authority (AFSSA, now called ANSES) extended the use of ozone as a complementary technology to treat wheat grains before milling, and as a technological aid for washing ready-to-use salads [44].

Ozone has powerful oxidizing property with an oxidation potential (2.07 V) higher than chlorine (1.36 V) and oxygen (1.23 V) [45]. Hence, it is used as a preservative against viruses, bacteria, mold, and yeast. It inactivates microbes in a short contact time and serves as an excellent substitute alternative to the existing food preservation techniques [46]. The rapid decomposition of ozone to oxygen after treatment, with minimal toxic residues in the substrate, makes ozone an attractive eco-friendly antimicrobial. The large domain of ozone's application in the food industry is illustrated in Figure 6.1. This chapter reviews ozone's

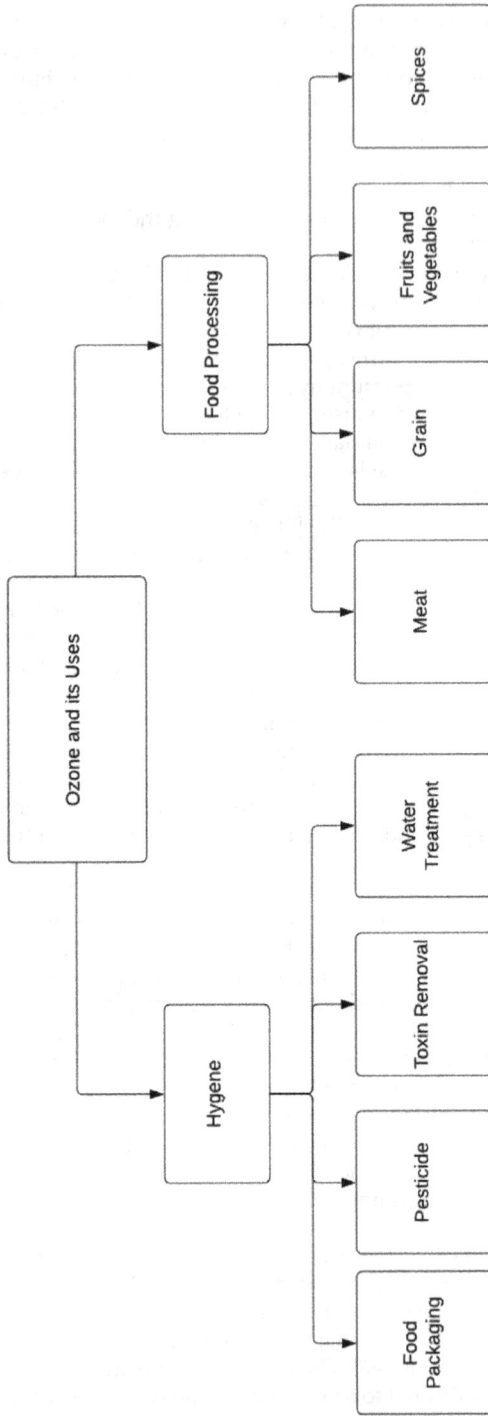

Figure 6.1 The application areas of ozonation in the food industry.

physicochemical and antimicrobial properties and the risks associated with its usage in the food industry.

6.2 FUNDAMENTALS AND PRINCIPLES

6.2.1 Physiochemical properties of ozone

Ozone (O_3) is an allotropic form of oxygen (O_2) and has three oxygen atoms instead of two atoms in the 'common oxygen'. Its physical properties are shown in Table 6.1.

Ozone gas is naturally present in the atmosphere in gaseous form and formed due to lightning or high-energy ultraviolet (UV) radiation [47]. During the formation of ozone, a diatomic oxygen molecule splits and creates free radicals of oxygen, which then react with other oxygen molecules to form a triatomic ozone molecule. The lightning or the UV provides the enormous high energy that is needed to break the O–O bond. The three oxygen atoms in the ozone contain two unpaired electrons in its valence shell, making it a highly reactive molecule [45] and very unstable. Hence, it reacts with most surfaces and leaves no residues after treatment [45]. Human exposure to a high ozone concentration can lead to lung toxicity, so it would be handled carefully during its application. It also caused symptoms like headaches, nosebleeds, eye irritation, dry throat, and respiratory irritation it is wise to avoid worker exposure [48].

6.2.2 Antimicrobial Properties

Due to its highly reactive (electrophilic) property, ozone is one of the most powerful oxidizing agents. It can destroy up to 99.9% of microorganisms and pesticides usually found in food. Both gaseous ozone and water-dissolved ozone are used in the food-processing industry for treating food products and contact surfaces [26, 49–53]. According to the types and states of microorganisms, the susceptibility to ozone may vary. For example, bacterial spores are the most resistant, and bacterial vegetative cells are the most sensitive to ozone [54–56]. In addition, the physiological conditions of the treated microorganism may affect its vulnerability to ozone [57]. For example, cells in the exponential phase are more susceptible to ozone than those in the stationary phase. Also, microorganisms show more resistance to ozone in their natural microflora on food than in the laboratory [57]. For example, in previously reported product inoculation studies, ozone treatments with a concentration of 3 ppm for 5 min were most effective and decreased the microbial population by 5 log units [58]. In contrast, in naturally occurring populations, the same treatment generally reduces the microbial population by 3 log units.

Several studies indicated that ozone reacts with microbial cell membranes [59–61] and oxidizes the lipids on bacteria's cell envelope, leading to intracellular cell contents leakage [59]. Additionally, it oxidizes intracellular enzymes and damages their genetic material, finally leading to the death of the cell [59]. Electron micrographs of ozone-damaged microbial cells have also been shown in prior studies [56, 57, 62] that showed damage was more evident in

Table 6.1 Physical Properties of Pure Ozone

Boiling point:	$-111.97 \pm 0.3°C$
Critical temperature:	$-12.1°C$
Melting point:	$-192.57 \pm 0.4°C$
Critical pressure:	54.6 atm

gram-negative bacteria (such as *Pseudomonas fluorescens* and *Escherichia coli* O157:H7) than in gram-positive bacteria (such as *Leuconostoc mesenteroides* and *Listeria monocytogenes*) [62]. Gram-positive bacteria treated with ozone appeared to lose only a few mucoid materials outside the cell wall. However, gram-negative bacteria tend to collapse and lose cellular fluids and components. Gram-positive bacteria are affected differently on treatment with ozone, leading to intracellular damage that efficiently inactivates these bacterial cells [62]. In summary, the oxidizing mechanism of ozone includes direct reactions of ozone molecules and reactivating the free radicals. In addition, it forms bonds with sulfhydryl groups and unsaturated lipids and causes the distraction of the cellular activity [63]. Finally, leakage of intracellular components destroys the genetic material and causes cell death due to oxidation [64]. Ozone can also destroy foodborne viruses such as norovirus and hepatitis A [65–67]. The microbial killing by ozone depends on the type of microorganisms, the intensity of ozone, and contact time with the microbial cell [64].

6.3 TECHNOLOGY OF OZONE PROCESSING

6.3.1 Ozone Production Technologies

The term *technical ozone* is a general term used for the ozone produced artificially as an oxidant. Since the gas is very unstable, it has to be produced *in situ* at the place of use [53] and cannot be stored or transported like other industrial gases. Ozone production is endothermic, and the following methods commonly obtain the required energy needed to generate technical ozone.

6.3.1.1 Corona Discharge Method

The corona discharge (CD) method is most widely used for the production of technical ozone in food-processing applications. The technique uses a high-tension electrode and a low-tension electrode (also known as a ground electrode), which are separated by a ceramic dielectric medium [68]. In the presence of oxygen, a high-voltage alternating current is applied, which induces the breakdown of oxygen molecules into oxygen atoms and then combines with the other oxygen molecules to produce ozone (Figure 6.2). The preparation of proper air is crucial to CD ozone systems. The ozone production varies depending on current frequency, voltage, dielectric material thickness and property, discharge gap, and absolute pressure within the discharge gap difference [56]. Therefore, the preparation of proper air is also crucial to CD ozone systems. For example, 1–4% ozone can be produced if the air is passed through the generator as the feed gas. However, it can reach up to 6–14% ozone if pure oxygen is used. In this method, electrical energy is used to split oxygen molecules to free oxygen atoms, which, in turn, binds with an unsplit oxygen molecule to form ozone.

6.3.1.2 UV, Irradiation, and Chemonuclear Methods

Ozone is also made by splitting oxygen molecules with other energy sources. For example, UV-based ozone generation is popular in the industry. In the UV method, atmospheric oxygen is split into free oxygen atoms using a 188-nm UV light [69]. These oxygen atoms collide with other oxygen atoms and produce ozone molecules. However, this method produces ozone in low intensity. Other energy sources include chemical, nuclear, thermal, and electrolytic processes [69].

6.3.1.3 Electrolysis of Water

Finally, a newer method of producing ozone involves using an electrochemical procedure in which water is broken down into hydrogen and oxygen atoms by

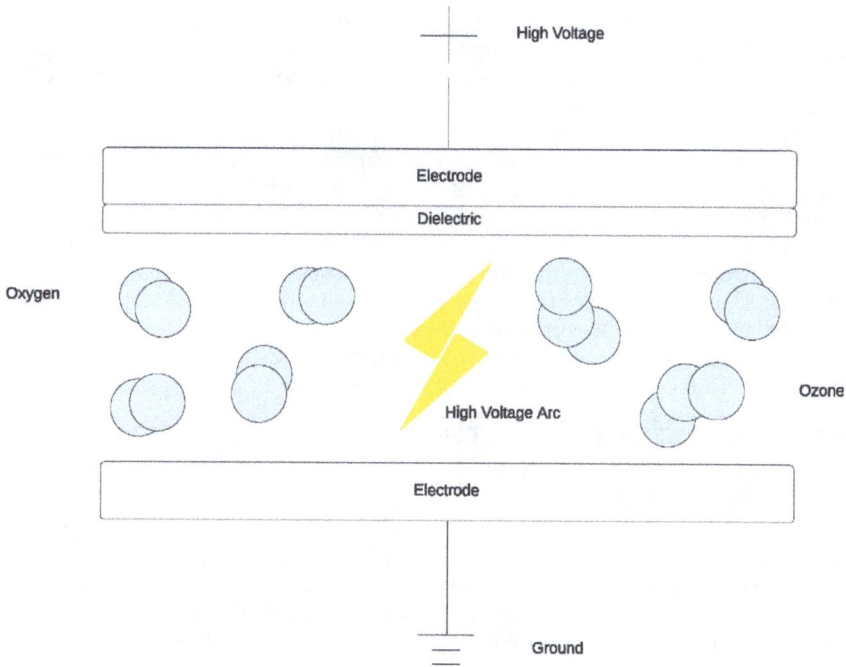

Figure 6.2   The overall concept of ozone generation using the CD method.

electrolysis. The hydrogen molecules are separated from the gas–water mixture, and the free oxygen atoms combine to form ozone and diatomic oxygen. This method produces ozone at concentrations three to four times higher (10–18%) than the CD method [56].

6.3.2 Ozone Monitoring after Production

The key factors that regulate ozone treatment are ozone concentration in the carrier gas and ozone application pressure [70]. If applied in solution, additional factors include size and rise rate of bubbles, hydrodynamics at the gas–liquid interface, pH, and temperature of the solution [70]. If applied in the gaseous form on a solid surface, then additional factors include solid-phase transfer, the structure of the solid, surface/volume ratio (particle size), the physical structure of solid surface periphery available to gas, the chemical configuration of the solid (reactivity), and water activity of the solid [70]. Both applications have been frequently used in the preservation of fresh vegetables. No matter what form of ozone is used, reasonable control of ozone contact with the target medium is crucial because it will determine the ozone consumption, reaction kinetics, and disinfecting power.

UV photometry (at 253.7 nm) is primarily used for quantifying gaseous ozone and is dependable for field operations. For quantification of ozone in solution, two ways are commonly used: volumetry and/or spectrophotometry. The volumetric method is more laborious and includes iodometric titration [71], but it is the most common method used to analyze ozone. Simultaneously, a different variety of spectrophotometric methods also exists. The fundamental principle is using a specific compound, such as potassium iodide, indigo trisulfonate, and methyl orange, that reacts with ozone which transforms into a product that can be quantified using UV-vis spectrophotometry [71]. In addition, amperometry is also common for the continuous measurement of dissolved ozone. The method involves measuring cells comprised of gas-permeable membranes stretched tightly over a gold cathode that completes a circuit together with a silver anode and an electrolyte solution [72]. The current between the cathode and the anode is created by applying a polarizing voltage and is directly proportional to the rate of diffusion of ozone through the membrane into the sensor [72]. Additionally, it is essential to ensure maximum efficiency and the lowest residual ozone mass due to the high expenses associated with ozone production. In conclusion, regardless of the ozone application methodology that is chosen, it is vital to be able to measure and supervise the concentration of ozone used during the process to be able to define the quantities strictly necessary to ensure adequate efficacy for the disinfection of a solution and/or a solid medium.

6.4 FACTORS AFFECTING THE EFFICACY OF OZONE TREATMENT
6.4.1 Extrinsic Parameters

Extrinsic factors influencing the efficacy of ozone treatment include the ozone concentration used, the residual ozone on the food matrix after the treatment, and a variety of environmental factors that influence the degradation of ozone, such as pH, organic matter in the food matrix, temperature, the application method (whether gaseous form or aqueous form is used), the concentration of ozone used and the treatment time [73]. To efficiently kill foodborne pathogens, the ozone decomposition rate should be minimal. The pH of the treatment surface is one significant extrinsic factor that affects ozone decomposition. Previous studies have shown that ozone remains relatively stable within pH 3.0–4.0 [74]; however, as pH increases toward neutral, the ozone degradation rate increases due to the formation of hydroxyl radicals. With further increases in pH toward an alkaline pH > 7.0, peroxy-radicals formed that further reduce ozone stability. Hence, the lower the pH of the treatment environment, the better the ozone stability, and therefore, the more efficient is the ozone treatment. For example, killing two *E. coli* strains in apple juice could be achieved faster at pH 3.0 than at pH 5.0 [75]. Similar studies suggesting better ozone efficacy at lower pH values have been reported in studies with various water samples [76] and other microbes in different treatment media [77].

The presence of organic matter surrounding the treatment medium also leads to rapid ozone degradation, resulting in decreased efficacy of ozone treatment [78]. Ozone-based *E. coli* inactivation, for example, was significantly reduced in the presence of humic acid [79]. The inactivation of *E. coli* in a model orange juice was faster with lower pulp content (1 min) than higher pulp content (6 min). However, if the juice was utterly unfiltered (containing all its pulp content), a 15–18 min of treatment was needed [80]. Also, the type of organic matter influences the treatment more than the quantity of organic matter [81]. For example, the killing of test bacteria *Staphylococcus aureus*, *Listeria monocytogenes*, *E. coli*, and *Salmonella typhimurium* in water was better if the water contained starch as organic matter compared to water containing an equal concentration of bovine serum albumin [81]. Temperature is another extrinsic factor that affects the efficiency of ozone treatments. Lower temperatures increase ozone solubility and stability in an aqueous medium, increasing availability and antimicrobial efficiency [72]. Conversely, ozone becomes more unstable and less soluble with temperature increase, and its effectivity decreases significantly.

Ozone efficacy can also depend on whether it is applied in a gaseous or aqueous form. Previous reports suggest that ozone is more active in higher relative humidity (RH) and is more effective in the aqueous form [72], which is why the ozone treatments of fruits and vegetables are done at higher RH (>80%) [82]. Reducing RH from 90% to 50% can make an ozone treatment completely ineffective [83]. Also, dynamic and static applications have entirely different effects [57]. The antimicrobial effect is higher when aqueous ozone is applied in a dynamic state than static [84]. For example, apples with *E. coli* were sterilized better when ozone was bubbled during washing than dipping those apples in ozonated water [85], and the sterilization of shredded lettuce with ozone was better when bubbled during a higher-speed stirring than a lower-speed stirring [86]. The size of the ozone bubble in water also influences the treatment efficacy. Smaller-sized bubbles lead to more effective ozone contact with the treatment medium and work significantly better than bigger bubbles, so the ozone dosage needed for the same treatment can be reduced in ultrafine bubbles [87].

Finally, the ozone concentration (C) and the processing times (T) determine ozone treatment's efficacy, which is typically described using the product of C and T, where C denotes the residual concentration of ozone and T denotes the contact time. Hence, a low ozone concentration applied for a long time can be equivalent to the treatment using a high concentration of ozone applied over a short time if CT for both treatments is the same [53].

6.4.2 Intrinsic Parameters

The physical properties of the food matrix appear to have an enormous influence on ozone treatment efficacy. Higher porosity, for example, impacts the diffusion of gases through a food material that allows microbes to attach and remain embedded within the food matrix and escape ozone treatments [53, 57]. This explains why it is more challenging to treat carrots with a rough and porous matrix [53, 88] than apples, tomatoes, and green pepper with smooth surfaces that allow easy exposure to ozone [89–91]. Also, intact leafy vegetables are more accessible to decontaminate than cut leaf tissues, which would enable microbes to attach to the trichrome, stomata, and cut edges of damaged leaves [53, 57]. The other intrinsic factors influencing the treatment is the resident microbiota and whether they reside as biofilms. Biofilms produce exopolysaccharides that resist ozone treatments [92, 93]. Also, the type and concentrations of microorganisms, their age, and their physiology collectively determine their susceptibility to ozone treatments [94, 95]. Younger colonies are more susceptible than older colonies. Sometimes refrigeration increases a microbe's resilience to ozone, as reported previously in the case of *Pseudomonas* strains [96]. Gram-negative bacteria appear to be more susceptible to ozone than Gram-positive bacteria [97] and for extended treatments, the efficacy depends on the resident microbial strains [88]. Because of their spore coat, spores are more tolerant to ozone than vegetative cells, which vary from strain to strain [54, 55, 98, 99].

6.5 EFFECTS OF OZONE ON MICROBIAL CELLS

Ozone kills a microorganism through progressive oxidation of its vital cellular components, eventually leading to its death. The cellular constituents damaged by ozone include the microbial cell envelope made up of unsaturated lipids, the lipopolysaccharide membrane (specific to gram-negative bacteria), the intracellular enzymes, and the genetic materials of microbes [100]. Upon contact with a bacterium, ozone first oxidizes the lipids on the cell envelope resulting in leakage of intracellular cell contents. The oxidation process generates free radicals that form chemical bonds with sulfhydryl groups in intracellular enzymes, unsaturated lipids, and genetic material in the cells leading to disruption of the cellular activity and eventual death of the cell [63]. Due to a thinner cellular covering in gram-negative bacteria, damage in these bacteria is more drastic than in gram-positive bacteria (such as *Leuconostoc mesenteroides* and *Listeria monocytogenes*) [62]. In addition to the bacteria, the gaseous and aqueous forms of ozone have also been very effective against various foodborne fungi, yeasts, parasites, and viruses [57].

6.6 EFFECTS OF OZONE ON PRODUCT QUALITY AND NUTRITION

Previous studies collectively support that ozone treatments do not damage the quality characteristics of food after treatment if used in doses deemed appropriate for that food [73]. However, if used at higher doses, ozone results in harmful effects on food's physical and sensory qualities. Hence, it is critical to determine an effective and safe treatment condition for the foods to be treated. Vegetables are commonly treated with ozone (both in gaseous and aqueous forms), especially during handling. For root vegetables like carrots, ozone did not affect the firmness of carrots, although some studies have suggested an adverse effect of ozone on fresh carrots [73]. Such negative effects changes in cellulose, lignin, and hemicellulose content, which was due to the reduced lignification of cell walls and is associated with a delay of tissue toughening and reduction of firmness [67, 70, 76]. In addition, during carrot storage, gaseous ozone usage in postharvest treatment is associated with undesirable physiological disruptions, including increased respiration rates, electrolyte leakage, and sucrose concentration [68, 70]. But on the other hand, lipid oxidation increases terpenes and hexanal, enhancing carrot flavor [70]. Hence, the key to a successful ozone treatment is applying a safe dose that protects disease and does not result in physiological damage. For carrots, a 15 μL L^{-1} application for 28 days (8 h a day) provided some disease protection with a minimum of physical and physiological damage [68]. The philosophy remains the same for other foods. In grains, for example, ozonation at higher than recommended doses may promote oxidation degradation of chemical constituents, leading to undesirable odors. However, if used in lower doses, even the amino acid and fatty acid profiles remain unaltered [101].

6.7 COMMERCIAL APPLICATION OF OZONE IN FOOD PROCESSING

The commercial use of ozone for food processing in the United States was initiated for reconditioning recycled poultry chilling water in 1997 [69]. In 2001, the FDA officially approved ozone for applications in the food industry and direct contact with food products, including use as an antimicrobial and surface sanitizer in CIP (clean-in-place) systems and fish, meat, and poultry processing plants [93]. In European countries, ozone has been used for food processing for a long time [69]. Also, food industries in Japan have used ozonated water treatments to increase the shelf life of vegetables and fruits, including carrot, cabbage, lettuce, apple, orange, grape, Chinese gooseberry, Japanese persimmon, Japanese radish, cucumber, spinach, bean sprout, parsley, melon, burdock, and potato. Currently, ozone is used frequently in the food industry due to its advantages over traditional food preservation techniques [93]. Its use as a fumigant is also a viable alternative from environmental and economic perspectives. The precursor for industrial ozone production is that O_2 is abundant and inexhaustible. Therefore, ozone is not very expensive to manufacture, and most byproducts of ozone can easily integrate themselves into the environment. Ozone shows antimicrobial activity against bacteria, fungi, viruses, protozoa, and bacterial and fungal spores [56]. In addition, ozone can also be used against storage pests and can degrade mycotoxins. One of the biggest advantages of using ozone as a preservative is that it generally leaves no residue in food since

146

Table 6.2 Inactivation of Different Foodborne Pathogens by Ozone in Different Food Products

Food products	Target pathogen	Ozone concentration	Treatment time	Outcome	Reference
Fresh carrots	E. coli O157:H7	8.2 g m^{-3}	5 and 15 min	Complete inactivation	88
Full ripe tomatoes	Botrytis cinera	0.005–1.0 µmol/mol	10 days	Reduction in spore production or viability	129
Cherry tomatoes	E. coli O157:H7, Listeria innocua, Salmonella typhimurium,	1000ppm	1–3 min	6 and 3 log CFU unit^{-1} reductions in Listeria, 2.7 log and 2.1 in Salmonella, and 1.8 to 2.6 in E. coli	51
Strawberry	Salmonella enterica	14–80 mg/L	5–30 min	2.1–1.6 log reduction	137
Bell pepper	S. enterica	0–1.0 mg/L	0.5–24 h	96.4% reduction	138
Grape tomatoes	S. enterica and S. typhimurium	1.71, 3.43, 6.85 mg/L	2–4 h	2 log reduction	139
Sweet basil	S. typhimurium	1.0 mg/L	7 min	2.6 log reduction	140
Alfalfa sprouts	S. enterica	5.0 mg/L	10–20 min	3.5 log v	141
Chicken drumstick	S. typhimurium and Salmonella choleraesuis	8 mg/L	4 min x 10 washes	Complete reduction	142
Hen eggs	S. typhimurium, Salmonella enteritidis, and Salmonella senftenberg	600 mg/h	120 min	4.79 log reduction	143
Tambaqui fish	S. typhimurium	0.35–9.1 mg/L	3 min	Complete reduction	144
Almonds, pistachios	Salmonella enterica	160 mg/L	30 min	2.3–2.9 log reduction	145
Grapes	Grey molds	0.3 ppm	7 weeks	Almost complete prevention of nesting	109
Apple juice	E. coli	3 ppm	5 min	Reduction to an undetectable level	58

excess ozone auto-decomposes rapidly to produce oxygen, thus making it safe to use with consumable products. Table 6.2 summarizes the ozone application in the inactivation of pathogens on select fruits and vegetables.

6.7.1 Fruit Processing

6.7.1.1 Apple

Ozone has been applied to inactivate microbes for maintaining quality and enhancing the shelf life of apple juice. A reduction in E. coli by 3.7 or 2.6 log

CFU/g was observed after apples are washed or dipped in ozone water, respectively, containing 25 mg/L of ozone [89]. It was observed that when apples were washed in 3 ppm ozone for 5 min, which resulted in a reduction of pathogens (*E. coli* and *Listeria monocytogenes*) to an undetectable level that remained below the detection limit during the period of 9 days' storage at 4°C [58]. A group of scientists showed that the treatment of apple juice with ozone can be used as an effective method for reducing bacteria, but the pH of the juice plays a major role in the inactivation of *E. coli* [102]. Ozone concentration and treatment time notably affect the color and total phenol content of apple juice. Ozone treatment of *E. coli* in apple juice has been shown to achieve the desired 5-log reduction within 5 min [75]. In the ozone treatment, it was observed that efficacy of ozone in the reduction of foodborne pathogenic bacteria up to 5 log units (*E. coli*, *Salmonella typhimurium*, *L. monocytogenes*) was mainly dependent on the solids content of the juice and increased with an increase in apple juice concentration from 18 to 72°Bx [103].

6.7.1.2 Berries

The shelf life of blackberries is lower due to fungal decay, and it can be increased by a continuous supply of 0.3 ppm gaseous ozone [104], and the red color of the blackberries can be maintained up to 12 days when stored at 2°C, after the ozone treatment. During the ozone treatment of fruits, methanol, ethanol, and 2-nonanone were synthesized, which implied that ozone resulted in stress in the fruits. When blueberries were sprayed with water containing ozone of 1 ppm concentration, it showed that it was effective in reducing microbial counts on fruit inoculated with *Pseudomonas fluorescens* and *Escherichia agglomerans*. When ozone was used in a combination with hydrogen peroxide for a longer time, the experiment did not result in any improvement in the antimicrobial effectiveness. While studying the effect of ozone treatment (0–7.8% w/w, 0–10 min) on the anthocyanin content and color of fresh blackberry juice, it was observed that ozone notably influences both [105].

6.7.1.3 Grapes

Continuous exposure to ozone (0.3 ppm) was employed to inhibit the growth of gray mold that grows on grapes during its storage for 7 weeks at 5°C [106]. A significant decrease in the decay percentage of grapes in their stored condition was observed when ozone was applied continuously or intermittently (2 ppm, 12 h/day) [107]. It was demonstrated that during conventional grape harvesting, ozone treatment can be used instead of sulfur dioxide [108]. The amount of pesticide residue such as fenhexamid, cyprodinil, pyrimethanil, and pyraclostrobin can also be reduced by using ozone [109]. At different relative humidity, ozone was studied to control the growth of conidia of *Penicillium digitatum*, *Penicillium italicum*, and *Botrytis cinerea* in table grapes. It was observed that conidia died more rapidly at higher humidity, and *B. cinerea* is more susceptible to ozone whereas *P. digitatum* and *P. italicum* were more resistant to ozone [110]. Postharvest pathogenic fungi could be controlled by fumigation with ozone and can also be employed to disinfect processing equipment and storage rooms. There is a significant reduction in color as well as anthocyanin content in grape juice due to the ozone treatment.

6.7.1.4 Orange

The growth of green and blue mold on inoculated oranges got delayed when it was treated continuously with 0.3 ppm of ozone for about 1 week compared to ambient air. Due to the treatment with ozone, conidia development of *P. italicum* and external mycelial growth of *P. digitatum* were also prevented or reduced [106, 111]. Furthermore, the surface growth of mold on packages, walls, and floors was also reduced by using ozone in the concentration of 0.25 ppm in the storage atmosphere. Later, it was observed that after exposure to 0.72 ppm of ozone for 14 days, there was no sign of spores of either *P. digitatum* or *P. italicum* on inoculated 'Lanelate' oranges [106]. However, the ethylene level within the containers can be effectively reduced by using ozone treatments, retarding the ripening process of the oranges. When freshly squeezed orange juice were treated with ozone, it produced no significant changes in pH, degrees Brix, titratable acidity, cloud value, or nonenzymatic browning [112]. However, in such exposure of ozone, the color of the juice was significantly affected. A complete reduction of *E. coli* was achieved in model orange juice in 1 min and in juice with low pulp content in 6 min, but in unfiltered juice, inactivation was only achieved after 15–18 min. Furthermore, it was demonstrated that ozone treatment of orange juice inoculated with mild acid-stressed and mild acid stress–habituated (pH 5.5) cells of *L. monocytogenes* resulted in higher inactivation time compared to control non–acid stressed cells [102].

6.7.2 Vegetable Processing

6.7.2.1 Cabbage

Upon treatment with aqueous ozone with a concentration of 3 ppm resulted in decreased PPO activity, total bacteria count, *E. coli*, yeast, and mold counts of fresh cabbage. After the treatment with ozone, this effect was observed immediately and lasted for 3 days in cold storage, but there was no significant difference between the control and the ozone-treated samples after 4 days onward [113].

6.7.2.2 Capsicum

It was observed in precut capsicum, ozonated water was more effective in reducing microbial load than non-ozonated water [114], and a log reduction of 0.66 was observed after treatment with non-ozonated water. Washing with ozone water (0.300.36, 0.380.45, and 3.853.95 mg/L) showed a reduction of up to 0.72 log compared to the untreated control [115]. 2.3 log unit^{-1} reductions were observed when ozone treatment for 3 min at a concentration of 2.0 ppm was done [116]. It was also observed that chlorination for 15 min at a concentration of 20 ppm or ozonation for 15 min at a concentration of 0.5 mg/L resulted in 1- and 0.5-log reduction, respectively [117]. Whereas continuous ozonation treatment resulted in a 2-log reduction in the first 15 min, and a 3.5-log reduction after 30 min exposure.

6.7.2.3 Tomato

Tomatoes were kept in an ozone-enriched atmosphere (0.05–5 µmol/mol) for up to 13 days and inhibition of 94–99%, 55–80%, and 13–74% were obtained in spore formation by *B. cinerea*, *A. alternata*, and *C. coccodes*, respectively [118, 119]. Cherry tomatoes inoculated with *Salmonella enteritidis* in a low dose resulted in complete inactivation on ozone treatment for 5 min in a concentration of 10 mg/L [120]. Tomatoes, after being treated with ozone, had a good appearance and overall quality after 15 days of storage [121]. It was observed that gaseous ozone treatment for a short time, that is, for 10 min using 10 µl/L of ozone, was useful for reducing damage and excessive softening of tomatoes without negatively affecting other quality attributes such as fruit color, sugar content, acidity, or antioxidant capacity [122]. However, the ozone-treated fruits resulted in delayed softening. For inactivating *S. enterica* and *E. coli* on pre-inoculated tomatoes, three nonthermal antimicrobial technologies were evaluated and those were, chlorine dioxide gas (10 mg/L, 3 min), ozone gas (4.3 mg/L, 5 min), and electron beam irradiation (7 kGy, 1 min) [123]. A reduction of 4 log CFU/g was observed after tomatoes are treated with ozone gas as compared to the initial load of pathogenic bacteria, while ClO_2 and e-beam treatments resulted in 5.3 and 4.4 log CFU/g reductions, respectively [124]. Coliform and *E. coli* were reduced by 1.9 and 2.0 logs, respectively [125]. Treatment with ozone enhanced protection against fungal infection as well as retained the firmness of the fruits [126].

6.7.2.4 *Carrots*

It was observed that when aqueous chlorine dioxide (ClO_2, 10.0 mg/L for 10 min), ozonated water (9.7 mg/L for 10 min), or a thyme oil suspension (1.0 mL/L for 5 min) was used on baby carrots, a log reduction of 1.48–1.97 log_{10} CFU/g in E. coli was obtained. Changing the order of sequence of the washing steps like thyme oil followed by aqueous ClO_2/ozonated water or ozonated water/ aqueous ClO_2 resulted in a reduction of 3.99 and 4.34 log per units, respectively [127]. In another study, the effect of ozone treatment (300 and 1000 nL/L for 1, 2, and 4 days) of carrots stored at 10°C and high relative humidity on saprophytic mold growth, which was effective for about 8 weeks after the treatment [128]. It was seen that ozone treatment (300 or 1000 nL/L, 2 or 4 days) induced spoilage resistance against *B. cinerea*, however, no such effect was observed against *Sclerotinia sclerotiorum* [129]. In a separate study, it was found that when carrots were washed with tap water, the maximum reduction in the bacterial population obtained was 0.5 log units; on the other hand, washing with water containing 4 ppm ozone resulted in a 1.5-log-unit microbial reduction, and no further reduction in bacteria was observed by extending the washing time up to 10 min [130]. Carrots with ozone treatment (50 ± 10 nL/L) resulted in reduced spoilage caused by *S. sclerotiorum* and *B. cinerea* [131]. On disinfection with chlorine, a final inactivation of 3.5 log units resulted after 15 min, whereas there was no inactivation after citric acid and ultraviolet light treatments. However, disinfection of carrot with ozone showed a low degree of inactivation even after 15 min [132].

6.8 PROGRESS IN OZONE PROCESSING

There are good advantages of ozone applications in the food industry, such as food surface hygiene, sanitation of food plant equipment, the reuse of wastewater, treatment, and the lowering of food and plant waste [52, 57, 59, 72, 133]. The bactericidal effects of ozone have been documented on various organisms, including gram-positive and gram-negative bacteria, spores, and vegetative cells. Hence, it has been used as an effective, potent sanitizer in the food industry for vegetables, fruits, and fish and for processing marine products, cereal and cereal products, confectionery products, meat and meat products, and processing plants. Ozone not only is capable of inactivation of a variety of organisms, including bacteria, fungi, yeast, parasites, and viruses but also can oxidize natural organic compounds as well as synthetic ingredients, such as detergents, herbicides, and composite pesticides [134–136]. Hence, ozone gradually replaces traditional sterilization technologies such as chlorine, steam, or hot water. It is a cost-effective, environment-friendly, and healthy approach to food safety management and, therefore, gaining momentum in the food-processing industry.

6.9 CONCLUSION AND PERSPECTIVES

Ozone can be used in both gaseous and water-soluble forms as a preservative on various vegetables and fruits. It has great antimicrobial activity against foodborne pathogens due to its strong oxidative capacity against microorganisms' lipid constituents, enzymes, nucleic acids, membranes, and other cellular constituents. Ozone-based techniques are considered eco-friendly techniques and are preferred over conventional chemical-based sanitizers such as chlorine, acids, and more. Ozone treatment has been shown to enhance safety and increase the shelf lives of foods with a limited impact on food product quality. The shelf lives of fruits and vegetables such as carrots, cabbages, lettuce, apples, oranges, grapes, cucumbers, spinach, and potatoes have significantly increased after ozone treatment. Ozone is also effective for reducing the effect of the harmful pesticide residues present on fruits and vegetables. An important advantage of using ozone is that it degrades rapidly after treatment resulting in minimal residue and it has no toxic effect on food items. This is in contrast to harmful pesticides that leave residues and lead to serious health issues for the consumer. However, it is critical to ensure appropriate safety precautions during applications that can restrict human exposure to ozone because ozone inhalation can be toxic to the lungs and result in severe lung conditions after exposure. It is also essential that the applied ozone concentration is varied depending on material treated, the microorganisms, and the environmental conditions. Current research efforts are aimed to produce ozone efficiently that can result in the best effects with minimal dosage. Side by side, ensuring safety is a priority, and efforts to develop new systems to minimize human exposure risks are ongoing. Ozone treatment can be combined with other preservation techniques for better effectiveness such as ozone and ultraviolet-C, ozone, and modified atmosphere packaging, among others.

REFERENCES

1. Lee, H. and Y. Yoon, *Etiological Agents Implicated in Foodborne Illness World Wide.* Food Sci Anim Resour, 2021. **41**(1): p. 1–7.
2. Organization, W.H., *Estimating the burden of foodborne diseases: A practical handbook for countries: A guide for planning, implementing and reporting country-level burden of foodborne disease.* 2021.
3. Balasubramaniam, V., et al., *Kinder, gentler food processing.* Food Technology, 2016. **12**: p. 20–28.
4. Jadhav, H.B., U.S. Annapure, and R.R. Deshmukh, *Non-thermal technologies for food processing.* Frontiers in Nutrition, 2021. **8**.
5. Chakka, A.K., M. Sriraksha, and C. Ravishankar, *Sustainability of emerging green non-thermal technologies in the food industry with food safety perspective: A review.* LWT, 2021. **151**: p. 112140.
6. Perrea, T., K.G. Grunert, and A. Krystallis, *Consumer value perceptions of food products from emerging processing technologies: A cross-cultural exploration.* Food Quality and Preference, 2015. **39**: p. 95–108.
7. Artes-Hernandez, F., et al., *Phytochemical Fortification in Fruit and Vegetable Beverages with Green Technologies.* Foods, 2021. **10**(11).
8. Barbhuiya, R.I., P. Singha, and S.K. Singh, *A comprehensive review on impact of non-thermal processing on the structural changes of food components.* Food Res Int, 2021. **149**: p. 110647.
9. De Aguiar Saldanha Pinheiro, A.C., et al., *Innovative Non-Thermal Technologies for Recovery and Valorization of Value-Added Products from Crustacean Processing By-Products-An Opportunity for a Circular Economy Approach.* Foods, 2021. **10**(9).
10. Masotti, F., et al., *Current insights into non-thermal preservation technologies alternative to conventional high-temperature short-time pasteurization of drinking milk.* Crit Rev Food Sci Nutr, 2021: p. 1–18.
11. Soni, A., et al., *Applications of novel processing technologies to enhance the safety and bioactivity of milk.* Compr Rev Food Sci Food Saf, 2021. **20**(5): p. 4652–4677.

12. Speranza, B., et al., *Innovative Preservation Methods Improving the Quality and Safety of Fish Products: Beneficial Effects and Limits*. Foods, 2021. **10**(11).

13. Abnavi, M.D., et al., *Chlorine inactivation of Escherichia coli O157:H7 in fresh produce wash process: Effectiveness and modeling*. Int J Food Microbiol, 2021. **356**: p. 109364.

14. Botta, C., et al., *Impact of Electrolyzed Water on the Microbial Spoilage Profile of Piedmontese Steak Tartare*. Microbiol Spectr, 2021. **9**(3): p. e0175121.

15. Han, J., et al., *Chlorine dioxide oxidation of hemicellulose from alkaline hydrolysate bagasse to remove lignin unit in lignin-carbohydrate complex*. Carbohydr Polym, 2022. **277**: p. 118817.

16. Truchado, P., M.I. Gil, and A. Allende, *Peroxyacetic acid and chlorine dioxide unlike chlorine induce viable but non-culturable (VBNC) stage of Listeria monocytogenes and Escherichia coli O157:H7 in wash water*. Food Microbiol, 2021. **100**: p. 103866.

17. Wang, H., et al., *Pre-oxidation of spent lettuce wash water by continuous Advanced Oxidation Process to reduce chlorine demand and cross-contamination of pathogens during post-harvest washing*. Food Microbiol, 2022. **103**: p. 103937.

18. Dai, X., et al., *Efficacy of different sanitizing agents and their combination on microbe population and quality of fresh-cut Chinese chives*. J Food Sci, 2012. **77**(7): p. M348–353.

19. Durak, M.Z., J.J. Churey, and R.W. Worobo, *Efficacy of UV, acidified sodium hypochlorite, and mild heat for decontamination of surface and infiltrated Escherichia coli O157:H7 on green onions and baby spinach*. J Food Prot, 2012. **75**(7): p. 1198–206.

20. Yang, G., et al., *Comparison of Inactivation Effect of Slightly Acidic Electrolyzed Water and Sodium Hypochlorite on Bacillus cereus Spores*. Foodborne Pathog Dis, 2021. **18**(3): p. 192–201.

21. Annous, B.A., D. Buckley, and A. Burke, *Evaluation of Chlorine Dioxide Gas against Four Salmonella enterica Serovars Artificially Contaminated on Whole Blueberries*. J Food Prot, 2020. **83**(3): p. 412–417.

22. Gu, G., et al., *Salmonella inactivation and sponge/microfiber mediated cross-contamination during papaya wash with chlorine or peracetic acid as sanitizer*. Food Microbiol, 2021. **95**: p. 103677.

23. Ogunniyi, A.D., et al., *Comparative antibacterial activities of neutral electrolyzed oxidizing water and other chlorine-based sanitizers*. Sci Rep, 2019. **9**(1): p. 19955.

24. Ortiz-Suarez, L.E., et al., *Optimization of the In Vitro Bactericidal Effect of a Mixture of Chlorine and Sodium Gallate against Campylobacter spp. and Arcobacter butzleri*. J Food Prot, 2021. **84**(7): p. 1127–1135.

25. Park, S.H., S.S. Kim, and D.H. Kang, *Development of sustained release formulations of chlorine dioxide gas for inactivation of foodborne pathogens on produce*. Food Sci Technol Int, 2021. **27**(8): p. 726–733.

26. Sheng, L. and M.J. Zhu, *Practical in-storage interventions to control foodborne pathogens on fresh produce*. Compr Rev Food Sci Food Saf, 2021. **20**(5): p. 4584–4611.

27. Wason, S., T. Verma, and J. Subbiah, *Validation of process technologies for enhancing the safety of low-moisture foods: A review*. Compr Rev Food Sci Food Saf, 2021. **20**(5): p. 4950–4992.

28. Yeom, W., et al., *Inactivation of Escherichia coli O157:H7 on radish and cabbage seeds by combined treatments with gaseous chlorine dioxide and heat at high relative humidity*. Food Microbiol, 2021. **99**: p. 103805.

29. Liu, X., et al., *Characterization of carbonyl disinfection by-products during ozonation, chlorination, and chloramination of dissolved organic matters*. Environmental Science & Technology, 2020. **54**(4): p. 2218–2227.

30. Richardson, S.D. and C. Postigo, *Drinking water disinfection by-products*, in *Emerging organic contaminants and human health*. 2011, Springer. p. 93–137.
31. Chaukura, N., et al., *Contemporary issues on the occurrence and removal of disinfection byproducts in drinking water-A review*. Journal of Environmental Chemical Engineering, 2020. **8**(2): p. 103659.
32. Henao, L.D., A. Turolla, and M. Antonelli, *Disinfection by-products formation and ecotoxicological effects of effluents treated with peracetic acid: A review*. Chemosphere, 2018. **213**: p. 25–40.
33. Boorman, G.A., *Drinking water disinfection byproducts: Review and approach to toxicity evaluation*. Environ Health Perspect, 1999. **107 Suppl 1**: p. 207–17.
34. Chowdhury, S., K. Alhooshani, and T. Karanfil, *Disinfection byproducts in swimming pool: Occurrences, implications and future needs*. Water Res, 2014. **53**: p. 68–109.
35. Oller-Arlandis, V. and J. Sanz-Valero, *[Cancer in the population under 19 years of age caused by chemical contamination in drinking water: A systematic review]*. Rev Panam Salud Publica, 2012. **32**(6): p. 435–43.
36. Sharma, V.K., R. Zboril, and T.J. McDonald, *Formation and toxicity of brominated disinfection byproducts during chlorination and chloramination of water: A review*. J Environ Sci Health B, 2014. **49**(3): p. 212–28.
37. Weber, L.W., M. Boll, and A. Stampfl, *Hepatotoxicity and mechanism of action of haloalkanes: Carbon tetrachloride as a toxicological model*. Crit Rev Toxicol, 2003. **33**(2): p. 105–36.
38. Massoud, R., et al., *Ozone Technology in Food Preservation*.
39. LEBOUT, H., *Fifty Years of Ozonation at Nice*. 1959, ACS Publications.
40. Glaze, W.H., *Drinking-water treatment with ozone*. Environmental science & technology, 1987. **21**(3): p. 224–230.
41. Hampson, B.C. and S.R. Fiori. *Applications of ozone in food processing operations*. in *IOA PAG Conf., Lake Tahoe*. 1997. Citeseer.
42. Perry, J.J. and A.E. Yousef, *Decontamination of raw foods using ozone-based sanitization techniques*. Annual review of food science and technology, 2011. **2**: p. 281–298.
43. Tiwari, B. and R.G. Rice, *Regulatory and legislative issues*. Ozone in food processing, 2012: p. 7–17.
44. Sarron, E., P. Gadonna-Widehem, and T. Aussenac, *Ozone treatments for preserving fresh vegetables quality: A critical review*. Foods, 2021. **10**(3): p. 605.
45. Greene, A.K., Z. Guzel-Seydim, and A.C. Seydim, *Chemical and Physical Properties of Ozone*. 2012, John Wiley and Sons, Oxford, UK. p. 19–31.
46. Chacha, J.S., et al., *Revisiting Non-Thermal Food Processing and Preservation Methods – Action Mechanisms, Pros and Cons: A Technological Update (2016–2021)*. Foods, 2021. **10**(6): p. 1430.
47. Pawar, E.K.R., et al., *Ozone Treatment: The Green Technology in Food Industry*.
48. Holm, S.M. and J.R. Balmes, *Systematic Review of Ozone Effects on Human Lung Function, 2013 Through 2020*. Chest, 2022. **161**(1): p. 190–201.
49. Afonso, R.B., R.H.R. Moreira, and P.L.R. de Almeida, *Can ozone be used as antimicrobial in the dairy industry? A systematic review*. J Dairy Sci, 2022. **105**(2): p. 1051–1057.
50. Afsah-Hejri, L., P. Hajeb, and R.J. Ehsani, *Application of ozone for degradation of mycotoxins in food: A review*. Compr Rev Food Sci Food Saf, 2020. **19**(4): p. 1777–1808.
51. Fan, X. and Y. Song, *Advanced Oxidation Process as a Postharvest Decontamination Technology To Improve Microbial Safety of Fresh Produce*. J Agric Food Chem, 2020. **68**(46): p. 12916–12926.

52. Panebianco, F., S. Rubiola, and P.A. Di Ciccio, *The Use of Ozone as an Eco-Friendly Strategy against Microbial Biofilm in Dairy Manufacturing Plants: A Review.* Microorganisms, 2022. **10**(1).

53. Sarron, E., P. Gadonna-Widehem, and T. Aussenac, *Ozone Treatments for Preserving Fresh Vegetables Quality: A Critical Review.* Foods, 2021. **10**(3).

54. Foegeding, P., *Ozone inactivation of Bacillus and Clostridium spore populations and the importance of the spore coat to resistance.* Food Microbiology, 1985. **2**(2): p. 123–134.

55. Güzel-Seydim, Z., P.I. Bever Jr, and A.K. Greene, *Efficacy of ozone to reduce bacterial populations in the presence of food components.* Food Microbiology, 2004. **21**(4): p. 475–479.

56. Khadre, M., A. Yousef, and J.G. Kim, *Microbiological aspects of ozone applications in food: A review.* Journal of food science, 2001. **66**(9): p. 1242–1252.

57. Kim, J.-G., A.E. Yousef, and S. Dave, *Application of ozone for enhancing the microbiological safety and quality of foods: A review.* Journal of Food Protection, 1999. **62**(9): p. 1071–1087.

58. Rodgers, S.L., et al., *A comparison of different chemical sanitizers for inactivating Escherichia coli O157: H7 and Listeria monocytogenes in solution and on apples, lettuce, strawberries, and cantaloupe.* Journal of Food Protection, 2004. **67**(4): p. 721–731.

59. RG, S.M., D. Singla, and A. Singh, *Ozone Revisited.* Journal of Advanced Oral Research, 2015. **6**(2): p. 5–9.

60. Scott, D.M. and E.C. Lesher, *Effect of ozone on survival and permeability of Escherichia coli.* Journal of bacteriology, 1963. **85**(3): p. 567–576.

61. Thanomsub, B., et al., *Effects of ozone treatment on cell growth and ultrastructural changes in bacteria.* The Journal of general and applied microbiology, 2002. **48**(4): p. 193–199.

62. Kim, J.-G., *Ozone as an antimicrobial agent in minimally processed foods.* 1998: The Ohio State University.

63. Fielding, L. and R. Bailey, *Ozone decontamination in hygiene management,* in *Handbook of hygiene control in the food industry.* 2005, Elsevier. p. 507–515.

64. Kim, J.G., *Environmental friendly sanitation to improve quality and microbial safety of fresh-cut vegetables.* Sammour R, Biotechnology – Molecular Studies and Novel Applications for Improved Quality of Human Life, 2012: p. 173–196.

65. Brie, A., et al., *Inactivation of murine norovirus and hepatitis A virus on fresh raspberries by gaseous ozone treatment.* Food Microbiol, 2018. **70**: p. 1–6.

66. Hirneisen, K.A., S.M. Markland, and K.E. Kniel, *Ozone inactivation of norovirus surrogates on fresh produce.* J Food Prot, 2011. **74**(5): p. 836–839.

67. Predmore, A., et al., *Control of human norovirus surrogates in fresh foods by gaseous ozone and a proposed mechanism of inactivation.* Food Microbiol, 2015. **50**: p. 118–25.

68. Gonçalves, A.A., *Ozone: An emerging technology for the seafood industry.* Brazilian archives of Biology and Technology, 2009. **52**(6): p. 1527–1539.

69. Guzel-Seydim, Z.B., A.K. Greene, and A. Seydim, *Use of ozone in the food industry.* LWT-Food Science and Technology, 2004. **37**(4): p. 453–460.

70. Rakness, K., et al., *Guideline for measurement of ozone concentration in the process gas from an ozone generator.* 1996.

71. Tjahjanto, R.T. and S. Wardhani, *Ozone determination: A comparison of quantitative analysis methods.* The Journal of Pure and Applied Chemistry Research, 2012. **1**(1): p. 18–25.

72. O'Donnell, C., et al., *Ozone in Food Processing.* 2012: John Wiley & Sons.

73. Brodowska, A.J., A. Nowak, and K. Śmigielski, *Ozone in the food industry: Principles of ozone treatment, mechanisms of action, and applications: An overview.* Critical reviews in food science and nutrition, 2018. **58**(13): p. 2176–2201.

74. Gurol, M.D. and P.C. Singer, *Kinetics of ozone decomposition: A dynamic approach.* Environ Sci Technol, 1982. **16**(7): p. 377–83.
75. Patil, S., et al., *Inactivation of Escherichia coli by ozone treatment of apple juice at different pH levels.* Food microbiology, 2010. **27**(6): p. 835–840.
76. Jamil, A., S. Farooq, and I. Hashmi, *Ozone disinfection efficiency for indicator microorganisms at different pH values and temperatures.* Ozone: Science & Engineering, 2017. **39**(6): p. 407–416.
77. Britton, H.C., M. Draper, and J.E. Talmadge, *Antimicrobial efficacy of aqueous ozone in combination with short chain fatty acid buffers.* Infection Prevention in Practice, 2020. **2**(1): p. 100032.
78. Cho, M., H. Chung, and J. Yoon, *Disinfection of water containing natural organic matter by using ozone-initiated radical reactions.* Appl Environ Microbiol, 2003. **69**(4): p. 2284–91.
79. Hunt, N.K. and B.J. Mariñas, *Inactivation of Escherichia coli with ozone: Chemical and inactivation kinetics.* Water Research, 1999. **33**(11): p. 2633–2641.
80. Patil, S., et al., *Inactivation of Escherichia coli in orange juice using ozone.* Innovative Food Science & Emerging Technologies, 2009. **10**(4): p. 551–557.
81. Restaino, L., et al., *Efficacy of ozonated water against various food-related microorganisms.* Appl Environ Microbiol, 1995. **61**(9): p. 3471–5.
82. Glowacz, M. and D. Rees, *The practicality of using ozone with fruit and vegetables.* Journal of the Science of Food and Agriculture, 2016. **96**(14): p. 4637–4643.
83. Ishizaki, K., N. Shinriki, and H. Matsuyama, *Inactivation of Bacillus spores by gaseous ozone.* Journal of Applied Bacteriology, 1986. **60**(1): p. 67–72.
84. Marino, M., et al., *Inactivation of foodborne bacteria biofilms by aqueous and gaseous ozone.* Frontiers in Microbiology, 2018: p. 2024.
85. And, M.A. and A. Yousef, *Efficacy of ozone against Escherichia coli O157: H7 on apples.* Journal of Food Science, 2001. **66**(9): p. 1380–1384.
86. KIM, J.G., A.E. Yousef, and G.W. Chism, *Use of ozone to inactivate microorganisms on lettuce.* Journal of Food Safety, 1999. **19**(1): p. 17–34.
87. Hashimoto, K., et al., *Reduction of ozone dosage by using ozone in ultrafine bubbles to reduce sludge volume.* Chemosphere, 2021. **274**: p. 129922.
88. Evrendilek, G.A. and P. Ozdemir, *Effect of various forms of non-thermal treatment of the quality and safety in carrots.* LWT, 2019. **105**: p. 344–354.
89. Achen, M. and A. Yousefa, *Efficacy of Ozone Against Escherichia Coli O157: H7 on Apples. J. 961.* 2001, Van Nostrand Reinhold Co., New York.
90. Alexopoulos, A., et al., *Evaluation of ozone efficacy on the reduction of microbial population of fresh cut lettuce (Lactuca sativa) and green bell pepper (Capsicum annuum).* Food Control, 2013. **30**(2): p. 491–496.
91. Bermúdez-Aguirre, D. and G.V. Barbosa-Cánovas, *Disinfection of selected vegetables under nonthermal treatments: Chlorine, acid citric, ultraviolet light and ozone.* Food Control, 2013. **29**(1): p. 82–90.
92. Panebianco, F., S. Rubiola, and P.A. Di Ciccio, *The Use of Ozone as an Eco-Friendly Strategy against Microbial Biofilm in Dairy Manufacturing Plants: A Review.* Microorganisms, 2022. **10**(1): p. 162.
93. Ziyaina, M. and B. Rasco, *Inactivation of microbes by ozone in the food industry: A review.* African Journal of Food Science, 2021. **15**(3): p. 113–120.
94. Gibson, K.E., et al., *Inactivation of bacteria on fresh produce by batch wash ozone sanitation.* Food Control, 2019. **106**: p. 106747.
95. Wani, S., J. Barnes, and I. Singleton, *Investigation of potential reasons for bacterial survival on 'ready-to-eat'leafy produce during exposure to gaseous ozone.* Postharvest Biology and Technology, 2016. **111**: p. 185–190.
96. Ripamonti, B., et al., *Use of gaseous ozone as a disinfectant in meat industry.* Italian Journal of Food Safety, 2009: p. 29–32.

97. Moore, G., C. Griffith, and A. Peters, *Bactericidal properties of ozone and its potential application as a terminal disinfectant.* J Food Prot, 2000. **63**(8): p. 1100–1106.

98. Akbas, M.Y. and M. Ozdemir, *Application of gaseous ozone to control populations of Escherichia coli, Bacillus cereus and Bacillus cereus spores in dried figs.* Food Microbiology, 2008. **25**(2): p. 386–391.

99. Broadwater, W.T., R. Hoehn, and P. King, *Sensitivity of three selected bacterial species to ozone.* Applied Microbiology, 1973. **26**(3): p. 391–393.

100. Giese, A.C. and E. Christensen, *Effects of ozone on organisms.* Physiological zoology, 1954. **27**(2): p. 101–115.

101. Mendez, F., et al., *Penetration of ozone into columns of stored grains and effects on chemical composition and processing performance.* Journal of Stored Products Research, 2003. **39**(1): p. 33–44.

102. Patil, S., V.P. Valdramidis, P.J. Cullen, J.M. Frias, P. Bourke, *Ozone inactivation of acid stressed Listeria monocytogenes and Listeria innocua in orange juice using a bubble column.* Food Control 2010a. **21**: p. 1723–1730.

103. Choi, M.R., Q. Liu, S.Y. Lee, J.H. Jin, S. Ryu, D.H. Kang, *Inactivation of Escherichia coli O157:H7, Salmonella typhimurium and Listeria monocytogenes in apple juice with gaseous ozone.* Food Microbiology 2012. **32**: p. 191–195.

104. Barth, M.M., C. Zhou, J. Mercier, F.A. Payne, *Ozone storage effects on anthocyanin content and fungal growth in blackberries.* Journal of Food Science. 1995. **60**: p. 1286–1288.

105. Song, J., L. Fan, C.F. Forney, M.A. Jordan, P.D. Hildebrand, W. Kalt, D.A.J. Ryan, *Effect of ozone treatment and controlled atmosphere storage on quality and phytochemicals in high bush blueberries.* Acta Horticulturae. 2003a. **600**: p. 417–423.

106. Palou, L., J.L. Smilanick, C.H. Crisosto, M. Mansour, P. Plaza, *Ozone gas penetration and control of sporulation of Penicillium digitatum and Penicillium italicum within commercial packages of oranges during cold storage.* Crop Protection. 2003. **22**: p. 1131–1134.

107. Cayuela, J.A., A. Va´zquez, A.G. Pe´rez, J.M. Garcı´a, *Control of table grapes postharvest decay by ozone treatment and resveratrol induction.* Food Science and Technology International. 2009. **15**: p. 495–502.

108. Gabler, F.M., J.L. Smilanick, M.F. Mansour, H. Karaca, *Influence of fumigation with high concentrations of ozone gas on postharvest graymold and fungicide residues on table grapes.* Postharvest Biology and Technology. 2010a. **55**: p. 85–90.

109. Gabler, F.M., J. Mercier, J.I. Jimenez., J.L. Smilanick, *Integration of continuous biofumigation with Muscodor albus with pre-cooling fumigation with ozone or sulfur dioxide to control postharvest graymold of table grapes.* Postharvest Biology and Technology. 2010b. **55**: p. 78–84.

110. Ozkan, R., J.L. Smilanick, O.A. Karabulut, *Toxicity of ozone gas to conidia of Penicillium digitatum, Penicillium italicum, and Botrytis cinerea and control of graymold on table grapes.* Postharvest Biology and Technology. 2011. **60**: p. 47e51.

111. Renzo, G.C.D., G. Altieri, L. D'Erchia, G. Lanza, M.C. Strano, *Effects of gaseous ozone exposure on cold stored orange fruit.* Acta Horticulturae. 2005. **682**: p. 1605–1610.

112. Tiwari, B.K., K. Muthukumarappan, C.P. O'donnell, C.P. Cullen, *Kinetics of freshly squeezed orange juice quality changes during ozone processing.* Journal Agriculture Food Chemistry. 2008a. **56**: p. 6416–6422.

113. Youm, H., J. Jang, K. Kim, H. Kim, E. Jeon, E. Park, M. Kim, K.B. Song, *Effect of chemical treatment with citric acid or ozonated water on microbial growth and polyphenoloxidase activity in lettuce and cabbage.* Journal of Food Science Nutrition. 2004. **9**: p. 121–125.

114. Ketteringham, L., R. Gausseres, S.J. James, C. James, *Application of aqueous ozone for treating pre-cut green peppers (Capsicum annuum L.).* Journal of Food Engineering. 2006. **76**: p. 10.

115. Han, Y., J.D. Floros, R.H. Linton, S.S. Nielsen, P.E. Nelson, *Response surface modelling for the inactivation of Escherichia coli O157:H7 on green peppers (Capsicum annuum) by ozone gas treatment.* Journal of Food Science. 2002. **67**: p. 1188–1193.

116. Elisabete, M.C.A., M.S.P. Dora, R.S. Teresa, C. Brandao, L.M. Silva, *Influence of aqueous ozone, blanching and combined treatments on microbial load of red bell peppers, strawberries and watercress.* Journal of Food Engineering. 2011. **105**: p. 277–282.

117. Alexopoulos, A., S. Plessas, S. Ceciu, V. Lazar, I. Mantzourani, C. Voidarou, E. Stavropoulou, E. Bezirtzoglou, *Evaluation of ozone efficacy on the reduction of microbial population of fresh cut lettuce (Lactuca sativa) and green bell pepper (Capsicum annuum).* Food Control. 2013. **30**: p. 491–496.

118. Tzortzakis, N., I. Singleton, J. Barnes, *Deployment of low-level ozone-enrichment for the preservation of chilled fresh produce.* Postharvest Biology and Technology. 2007a. **43**: p. 261–270.

119. Tzortzakis, N., A. Borland, B.J. Singleton, *Impact of atmospheric ozone-enrichment on quality-related attributes of tomato fruit.* Postharvest Biology and Technology. 2007b. **45**: p. 317–325.

120. Das, E., G.C. Gu'rakan, A. Bayindirli, *Effect of controlled atmosphere storage, modified atmosphere packaging and gaseous ozone treatment on the survival of Salmonella enteritidis on cherry tomatoes.* Food Microbiology. 2006. **23**: p. 430–438.

121. Tzortzakis, N., A. Borland, B.J. Singleton, *Impact of atmospheric ozone-enrichment on quality-related attributes of tomato fruit.* Postharvest Biology and Technology. 2007b. **45**: p. 317–325.

122. Rodoni, L., N. Casadei, A. Concellon, A.R. Chaves Alicia, A.R. Vicente, *Effect of short-term ozone treatments on tomato (Solanum lycopersicum L.) fruit quality and cell wall degradation.* Journal of Agricultural and Food Chemistry. 2010. **58**: p. 594–599.

123. Trinetta, V., N. Vaidya, R. Linton, M.A. Morgan, *Comparative study on the effectiveness of chlorine dioxide gas, ozone gas and e-beam irradiation treatments for inactivation of pathogens inoculated onto tomato, cantaloupe and lettuce seeds.* International Journal of Food Microbiology. 2011. **146**: p. 203e206.

124. Long, W., S. Pao, P. Inserra, E. Westbrook, S. Ahn, *Efficacy of ozone produce washers in reducing natural and artificially inoculated microorganisms on Roma tomatoes and green onions.* Journal of Food Safety. 2011. **31**: p. 268–275.

125. Tzortzakis, N., T. Taybi, R. Roberts, I. Singleton, A. Borland, J. Barnes, *Low-level atmospheric ozone exposure induces protection against Botrytis cinerea with down-regulation of ethylene, jasmonate and pathogenesis-related genes in tomato fruit.* Postharvest Biology and Technology. 2011. **61**: p. 152–159.

126. Tuffi, R., R. Lovino, S. Canese, L.M. Cafiero, F. Vitali, *Effects of exposure to gaseous ozone and negative air ions on control of epiphytic flora and the development of Botrytis cinerea and Penicillium expansum during cold storage of strawberries and tomatoes.* Italian Journal of Food Science. 2012. **24**: p. 102–114.

127. Singh, N., R.K. Singh, A.K. Bhunia, R.L. Stroshine, *Efficacy of chlorine dioxide, ozone, and thyme essential oil or a sequential washing in killing Escherichia coli O157:H7 on lettuce and baby carrots.* LWT e Food Science and Technology. 2002. **35**: p. 720–729.

128. Song, J., L. Fan, C.F. Forney, M.A. Jordan, P.D. Hildebrand, W. Kalt, D.A.J. Ryan, *Effect of ozone treatment and controlled atmosphere storage on quality and phytochemicals in high bush blueberries.* Acta Horticulturae. 2003a. **600**: p. 417–423.

129. Forney, C.F., J. Song, P.D. Hildebrand, L. Fan, K.B. McRae, *Interactive effects of ozone and 1-methylcyclopropene on decay resistance and quality of stored carrots.* Postharvest Biology and Technology. 2007. **45**: p. 341–348.

130. Hassenberg, K., A. Frohling, M. Geyer, O. Schluter, W.B. Herppich, *Ozonated wash water for inhibition of Pectobacterium carotovorum on carrots and the effect*

on the physiological behaviour of produce. European Journal of Horticulture Science. 2008. **73**: p. 37–42.

131. Hildebrand, P.D., C.F. Forney, J. Song, L. Fan, K.B. McRae, *Effect of a continuous low ozone exposure (50 nL/L) on decay and quality of stored carrots.* Postharvest Biology and Technology. 2008. **49**: p. 397–402.

132. Bermudez, A.D., and G.V.C. Barbosa, *Disinfection of selected vegetables under nonthermal treatments: Chlorine, acid citric, ultraviolet light and ozone.* Food Control. 2013. **29**: p. 82–90.

133. Kim, J.-G., A.E. Yousef, and M.A. Khadre, *Ozone and its Current and Future Application in the Food Industry.* Advances in Food and Nutrition Research, 2003. **45**: p. 167–218.

134. Brunet, R., M. Bourbigot, and M. Dore, *Oxidation of Organic Compounds Through the Combination Ozone –Hydrogen Peroxide.* The Journal of the International Ozone Association, 1984. **6**(3): p. 163–183.

135. Bailey, P.S., *The reactions of ozone with organic compounds.* Chemical Reviews, 1958. **58**(5): p. 925–1010.

136. Siddiqui, M.S., G.L. Amy, and B.D. Murphy, *Ozone enhanced removal of natural organic matter from drinking water sources.* Water Research, 1997. **31**(12): p. 3098–3106.

137. Zhou, L., W. Liu, R. Stockmann, and N.S. Terefe, *Effect of citric acid and high pressure thermal processing on enzyme activity and related quality attributes of pear puree.* Innovative Food Science & Emerging Technologies. 2018. **45**: p. 196–207.

138. Yousuf, B., V. Deshi, B. Ozturk, and M.W. Siddiqui, *Fresh-cut fruits and vegetables: Quality issues and safety concerns.* Fresh-Cut Fruits and Vegetables, 1–15 (Elsevier). 2020.

139. Aadil, R.M., X.-A. Zeng, A. Ali, et al., *Influence of different pulsed electric field strengths on the quality of the grapefruit juice.* International Journal of Food Science and Technology. 2015. **50**: p. 2290–2296.

140. Phaephiphat, A., and W. Mahakarnchanakul, *Surface decontamination of Salmonella Typhimurium and Escherichia coli on sweet basil by ozone microbubbles.* Cogent Food & Agriculture. 2018. **4**.

141. Mohammad, Z., A. Kalbasi-Ashtari, G. Riskowski, and A. Castillo, *Reduction of Salmonella and Shiga toxin-producing Escherichia coli on alfalfa seeds and sprouts using an ozone generating system.* International Journal of Food Microbiology. 2019. **289**: p. 57–63.

142. Megahed, A., B. Aldridge, and J. Lowe, *Antimicrobial efficacy of aqueous ozone and ozone – lactic acid blend on salmonella-contaminated chicken drumsticks using multiple sequential soaking and spraying approaches.* Frontiers in Microbiology. 2020. **11**(3121).

143. Mattioli, S., R. Ortenzi, S. Scuota, A.C. Mancinelli, A. Dal Bosco, E. Cotozzolo, et al., *Impact of ozone and UV irradiation sanitation treatments on the survival of Salmonella and the physical – chemical characteristics of hen eggs.* The Journal of Applied Poultry Research. 2020. **29**(2): p. 409–419.

144. Luiz, D.d. B., S.R. Campelo, V.R.V.d. Santos, L. K. F.d. Lima, P.C.M.S. Chicrala, and M.K.P. Iwashita, *Evaluation of the effectiveness of ozone as a sanitizer for fish experimentally contaminated with Salmonella sp.* Brazilian Journal of Food Technology. 2017. **20**: p. e2016150.

145. Perry, J.J., M. Pena-Melendez, and A.E. Yousef, *Ozone-based treatments for inactivation of Salmonella enterica in tree nuts: Inoculation protocol and surrogate suitability considerations.* International Journal of Food Microbiology. 2019. **297**: p. 21–26.

7 Food Preservation Using Ultraviolet Light

Monica Jaiswal, Mwchangti Debbarma, Hilal A Makroo,
Tatiana Koutchma, and Brijesh Srivastava

Monica Jaiswal and Mwchangti Debbarma contributed
equally to the chapter as first authors.

*Corresponding author:
Prof. Brijesh Srivastava, Email: brijesh@tezu.ernet.in

CONTENTS

DOI: 10.1201/9781003147978-7

7.1 INTRODUCTION

The conventional food processing methods primarily include thermal process-ing and the addition of chemical preservatives. Thermal processing can lead to the damage of heat-sensitive nutrients such as polyphenols, vitamins, and other bioactive compounds. In contrast, most chemical preservatives used in food have been found to cause health issues, including death-causing cancer (Vanga et al., 2017). With the passage of time and the application of modern technolo-gies in health and medical fields, consumers' cognizance of healthy food and what they consume have risen. Consumers today ask for food products that must offer other things besides nutrition, such as accessibility, variety, extended storage life, low in calories, lower pricing, and eco-friendliness (Jermann et al., 2015). Important attributes determining food quality like appearance, texture, taste, and nutritional content are greatly affected by the methods of food pro-cessing (Priyadarshini et al., 2019). Considering the limitations associated with the conventional food processing methods in the last few decades, many new processing techniques (including micro/nanofiltration, ultrasonication, high-pressure processing, field-based technique, ionizing radiation, cold atmospheric plasma, ultraviolet [UV] light, and pulsed light) as alternative food preservation processes have been introduced to enhance food quality attributes and function-ality by reducing processing impact to help improve the safety and acceptability of food (Hernández-Hernández et al., 2019). The objective of developing these technologies is to increase the processing efficiency with few or no modifications to food nutritional characteristics, lower energy requirements, and minimize food losses due to spoilage by extending storage life. The contact-based emerg-ing and modern processing technologies do have specific limitations, such as ultrasonication failing to achieve complete inactivation of polyphenol oxidase (PPO) in fruit juice due to juice sonication, fouling, and clogging in the mem-brane at the highest temperature. This issue is rectified with the tangential microfiltration process, which makes the thermo-sonication complex multiple-step process for industrial application, whereas high-pressure processing is time-consuming and expensive (Mala et al., 2021; Vollmer et al., 2021). However, such limitations can be overcome by contactless and light-based food processing technologies such as UV light processing.

UV light is nonionizing radiation with wavelengths ranging from 100 to 400 nm. It has a broad antimicrobial effect, inactivating viruses (Eischeid et al., 2009), bacteria (both vegetative and spores), fungi, conidia (Gómez-López et al., 2005), and parasites (Hijnen et al., 2006). UV light food treatment is a nonthermal, chemical-free approach to inactivate microorganisms, making it eco-friendly without generating any by-products or irradiation. UV light is safe to use for food disinfection and can help food businesses to save energy and money. The term *disinfection* is defined as the reduction by means of biological or chemical agents and/or physical methods in the number of viable microorganisms on surfaces, in water, or air to a level that does not compromise food safety and/or suitability. UV light is also known as surface germicidal technology commonly utilized for water and surface disinfection in the food sector (Guerrero-Beltrán and Barbosa-Cánovas, 2004). UV light disinfection of food has also been shown to be an excellent physical preservation method with favorable consumer perception and commercial appeal. UV light technology can be applied to food by using either continuous sources or pulsed UV sources. Continuous UV light is emitted by the source that consists of an inert-gas flash lamp that converts high-power electricity to high-power radiation like mercury lamps or UV light-emitting diodes (LEDs) that continuously generate UV photons in monochromatic and polychromatic modes. On the other hand, pulsed-light (PL) treatment is an advanced technology involving the use of intense, short-duration pulses of broad-spectrum light, higher in UV-C range emitted by inert-gas flash lamps like xenon, which rapidly and repeatedly release light pulses on the surface of the sample. UV-C wavelengths are capable of destroying the DNA and RNA of pathogens, thus preventing them from dividing and multiplying, effectively rendering them inactive. Thus, this alternative nonthermal technique has the potential for inactivating spoilage as well as pathogenic microorganisms in foods and, as a result, extending their shelf lives (Elmnasser et al., 2007).

The present chapter is composed of the fundamental principle of UV-based food processing technique, UV-light sources, and the mechanism of inactivation of microbes in food by UV light. Furthermore, this chapter also covers different ways of applying UV light and the effects on food quality and characteristics. Finally, the future scope of this technology in food processing has been discussed briefly.

7.2 FUNDAMENTALS AND PRINCIPLES OF UV LIGHT TECHNOLOGY

Ultraviolet radiation/illumination/light is a form of electromagnetic spectra with wavelengths ranging from 100 to 400 nm (Figure 7.1). It is further divided into UV-A (315–400 nm) generally accountable for skin tanning, UV-B (280–315 nm) causing sunburn and ultimately increases the risk rate of skin cancer, UV-C (200–280 nm) is known as the bacterial range as it adequately inhibits pathogens, and vacuum UV (100–200 nm) is a wavelength at which a vacuum apparatus is solely used because the UV light gets strongly absorbed in the air. The energy exchange throughout the collision of UV-C photons of wavelength <240 nm

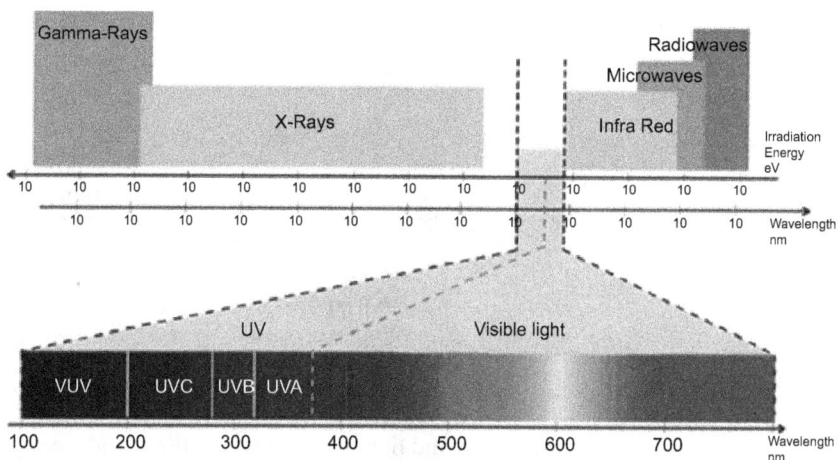

Figure 7.1 Electromagnetic spectrum with different light wavelengths (Koutchma, 2019).

with oxygen atoms can lead to ozone gas formation. Since UV-C is absorbed in a faster rate, it is never observed in nature.

7.2.1 The Genesis of UV light

Nonionizing radiation excites atoms and molecules by causing electrons to flow from a relatively high energy E_2 to a lesser energy E_1, a discrete range of photon energy. According to Planck, each photon possesses an energy E (J) described by Equation 7.1.

$$E = E_2 - E_1 = \frac{hc}{\lambda},$$ (7.1)

where h stands for Planck's constant (6.23×10^{34} J.s), c stands for the speed of light (2.998×10^8 m/s or m s^{-1}), and λ is the wavelength of radiation (m). The difference in energy levels of elements within an atom causes UV light to be emitted. As the energy level of the atom rises, so does the number of electrons, protons, and neutrons within it, and their interaction with external electric fields. These exciting elements emit a distinctive light spectrum. When energy is transmitted to an atom, the kinetic energy of the atom increases, this is known as ionization, and it occurs when atoms or molecules become positively (cations) or negatively charged (free electrons). The ionization energy E_0 of an atom will bind this wavelength (Equation 7.2) at one end depending on the temperature of the cations and free electrons.

$$\lambda_{max} = \frac{hc}{E_0}$$ (7.2)

The energy of a single photon is usually expressed in electron volts (eV) rather than joules (J) while dealing with the particles. One electron volt is equal to 1.60×10^{-19} J. In photochemical terms, the energy of a single photon is expressed in kilocalories per Einstein. In general, one Einstein is defined as the energy in one mole of a photon, which is equal to Avogadro's number, $A = 6.02 \times 10^{23}$. The absorption of one Einstein can incite one mole of the absorbing substance (Equation 7.3).

$$E_E = h\frac{c}{\lambda}A,$$ (7.3)

where E_E denotes Einstein's energy and A denotes Avogadro's number. Nonionizing radiation, as opposed to ionizing radiation, includes UV light. Nonionizing radiation absorption, on the other hand, excites atoms within molecules around or causes them to vibrate but does not completely remove the electrons from atoms.

7.2.2 Gas Discharge

The gas discharge contains a noble gas (argon, neon, krypton, xenon) or a mixture of these gases, and in most cases, a metal vapor emits light at different wavelengths depending on its elemental constituents and the excitation, ionization, and kinetic energy of those elements. When a sufficiently high voltage is applied to a gas, a mixture of excited and nonexcited atoms, cations, and electrons is developed, which is known as gas discharge, which is responsible for UV light emission. As the free electrons collide with an atom, energy is transferred to the atom, and, if sufficient, it is ionized. This atom's ionization causes a rapid increase in the number of electrons and cations and creates a lamp current. At higher temperatures, electron emission begins from the electrode and increases current, reducing the requirement for voltage. Since the voltage is inversely proportional to the applied current, the gas discharges are unsteady and have a negative impedance. As a result, the ballast is connected in series with the gas discharge lamp to provide positive impedance.

7.3 UV LIGHT SOURCES

7.3.1 Continuous UV Source

Conventionally, there are a small number of continuous light UV sources that are available commercially, which include low-pressure mercury (LPM), medium-pressure mercury (MPM), and low pressure, high output (LPHO). The LPM is a monochromatic source, whereas the emission of MPM is polychromatic. LPM and MPM lamps are the most commonly used for UV light treatment of liquid foods, drinks, and beverages, including drinking water processing. The use of pulsed UV light in the production, processing, and handling of food has been approved by the U.S. Food and Drug Administration (US FDA, Code 21CFR179.41). Table 7.1 summarizes the basic characteristics of UV sources.

Mercury emission lamps are made up of mercury and an element and are represented as steam gas lamps with a UV transmission envelope made up of a sealed quartz glass tube at each end (Figure 7.2). The most used fill gas is argon, with an ionization energy of 15.8 eV and the lowest triggered metastable state of 11.6 eV. The argon is used to facilitate the activation of mercury ionization at the beginning of the gas discharge. These sophisticated lamps provide high efficacy

Table 7.1 UV Sources and Their Basic Characteristic (Koutchma et al., 2019)

UV source	Electrical efficiency (%)	UV efficiency (%)	UV irradiance (W/cm^2)	Source surface temperature (°C)	Treatment time scale	Lifetime (month)	Output spectrum
LMP	50	38	0.01	40	s-min	18–24	Mono chromatic 253.7 nm
MPM	15–30	12	600	400–1,000	s	0.5	Poly chromatic 200–600 nm
PUV	15–20	17	30,000	–	ms-min	1 NA	Poly Chromatic 200–400 nm
Excimer	10–35	10–40	–	ambient	s	>6	Mono chromatic tunable 222 nm, 282 nm
UV-LED	60	5–10	0.001	ambient	h	>10	Mono Chromatic 255, 265, 275, and 280 nm)

LOW-PRESSURE MERCURY LAMP - HOT CATHODE TYPE

Tungsten Coil Electrode

Envelope

Electrical Connection

Mercury & Invert Gas Fill

End Seal

LOW-PRESSURE HIGH-OUTPUT MERCURY LAMP - AMALGAM TYPE

Electrical Connection

Tungsten Coil Electrode   Mercury Amalgam

Invert Gas Fill

Envelope

Seal

MEDIUM-PRESSURE MERCURY LAMP

Seal

Electrode - Tungsten Coils on a Tungsten Rod

Mercury & Invert Gas Fill

Envelope

Molybdenum Foil

Electrical Connection

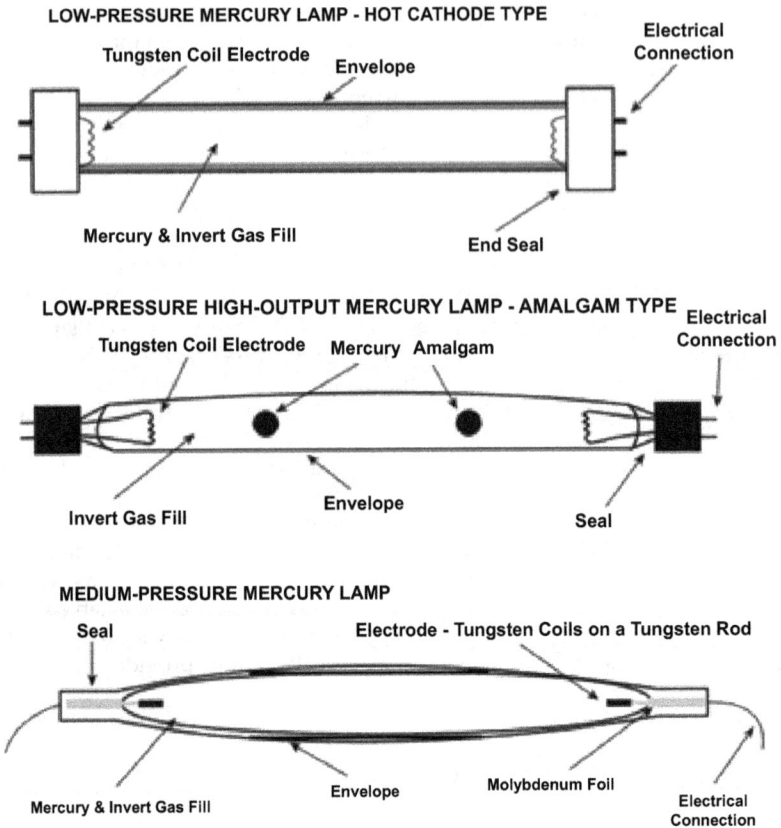

Figure 7.2 Schematic representation of LM, LPHO, MPM lamp (Koutchma, 2019).

and permanence and are compact for numerous applications. LPM lamps have nominal total gas pressures ranging from 102 to 103 Pa (90.01–0.001 mbar) for creating UV light. LPM lamps are typically cylindrical in shape, with diameters ranging from 0.9 to 4 cm and lengths ranging from 10 to 160 cm. Monochromatic light is emitted by low-pressure mercury lamps. About 85 % of the total UV intensity emitted has a wavelength of 253.7 nm, which is more effective in germicidal action. LPM lamps with their well-defined (single 253.7 nm UV-C output) and quantized spectrum are widely available at a low cost. MPM lamps operate at a pressure level of approximately 104 to 106 Pa. MPM lamps operate in the potential gradient range of 5 to 30 W/cm. The emission of MPM lamps is polychromatic with a wider range of UV wavelengths ranging from approximately 250 to nearly 600 nm, culminating from a series of emissions in both the UV and visible ranges. A factor affecting the emission of MPM lamps is the aging of lamp materials, and about 80% of the emission of germicidal wavelengths (above 400 nm) are maintained after 4000 h of operation.

LPM and MPM lamps generally contain mercury, whereas LPHO lamps possess a mercury amalgam. Mercury is also released as very fine particles in the vapor phase, which can easily dissolve in water, as opposed to solid or liquid mercury, which tends to reside. The mercury content of a single UV lamp used for water treatment is typically in the range of 0.005 to 0.4 g (5–400 mg/lamp). LP lamps (5–50 mg/lamp) contain less mercury than LPHO lamps (26–150 mg/lamp) and MP lamps (200–400 mg/lamp). When the UV lamp fails, it is necessary to observe the condition of the mercury (gas, solid, or liquid) to assess potential health risks from exposure to mercury owing to lamp sleeve breakage. As per the U.S. Environmental Protection Agency, short-term mercury exposures above 0.002 mg/L maximum contaminant level (MCL) can potentially cause kidney damage (EPA 1995). Concerns about the effects of mercury discharge into the food plant environment prompted the development and testing of mercury-free special lamps (Koutchma, 2009).

7.3.2 Special UV Lamps

Numerous alternative UV sources like excimer lamps and pulsed UV technologies are used in UV emission lamps that are independent of temperature effects and might be applicable in food treatments. The potential benefits of using microwave lamps over conventional electrode lamps are faster heating and avoidance of the initial degradation process associated with UV lamps. Lamp life is roughly three times that of electrode lamps. Food industries should focus on the benefits of using this technology because it eliminates the need for water and electrodes, as well as the risk of corrosion. This technology has yet to be used in the food industry, and therefore, professionals should consider the aforementioned advantages (Koutchma, 2019).

Excimer is an abbreviated form for "excited dimer". It is essentially a combination of atoms with one excited gas atom. Excimer lamps are created when a rare gas (RG2*) or halogen excimer (X2*) or rare-gas halide exciplex (RDX*) forms. Compared to conventional UV emitters, excimer emitters have properties that enhance and expand UV light use in the food industry. Because of the monochromatic spectrum and the ability to select specific wavelengths, photo processes can be applied with greater precision. It has been reported that an excimer lamp based on a krypton–chlorine (Kr-Cl) gas mixture emits primarily at 222 nm and is essentially as effective as conventional germicidal UV lamps in killing antibiotic-resistant bacteria (254 nm). The light of 222 nm, on the other hand, does not induce typical UV-associated premutagenic DNA lesions in human keratinocytes in a 3D human skin model and appears to be safe for the skin of exposed hairless mice, according to eight cellular and molecular endpoints associated with damaged skin (Buonanno et al., 2017). This suggests that in comparison to conventional UV-C light, far UV-C (222 nm) is safe and useful for germicidal inactivation and does not cause any harm to human health (Bhardwaj et al., 2021). A UV LED is a semiconductor device that emits light through electroluminescence. It has a p-side and an n-side, with a p–n junction as an interface. When an electron–hole interplay induces the electron to fall to a lower energy level, a photon is released, and thus. sunlight with a specific wavelength is emitted. Therefore, as shown in Figure 7.3, LEDs can generate monochromatic light with a narrow wavelength that appears as a specific color to the observer. As the efficiency of UV LEDs improves due to lower production costs and longer life, they will begin to replace the current UV-C light lamp market as an economical, environmentally friendly alternative to avoid wasting energy and improve the safety and shelf lives of produced foods.

7.3.3 PL and Pulse UV Lamp

PL has a wavelength range of approximately 170 to 1000 nm and frequently combines visible UV photons and infrared radiation at energies ranging from

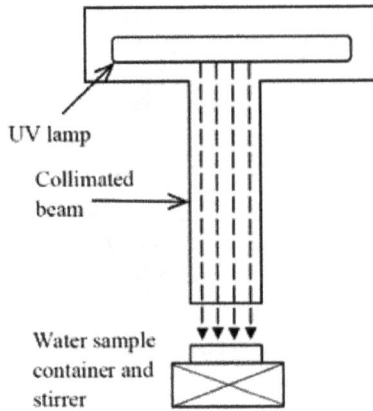

Figure 7.3 Schematic illustration of UV LED lamp (Song et al., 2016).

0.01 to 50 Jcm². PL is delivered in multiple flashes of light per second, allowing high product throughput while using low energy. Xenon lamps are commercial sources of PL that are less harmful to the environment than UV lamps because they do not contain mercury. Pulse UV is created by multiplying power by storing electricity in a capacitor for long durations (fractions of a second) and then releasing it rapidly (thousandths of a second). Following the FDA's approval of "pulsed UV light in the production, processing, and handling of food," numerous scientific studies for PL and PUV in treating foods, food contact surfaces, and the environment have been conducted. This technology is deemed promising, but it has yet to be thoroughly established in the field.

7.4 FACTORS AFFECTING UV TREATMENT
7.4.1 UV Dose

The processing of food by a UV light system is dependent on the delivery of the UV dose, which is influenced by the system's design and the type of food being treated. In liquid food and beverages, UV dose is also affected by UV absorption. Lambert's law (Equation 7.4) expresses the linear relationship between absorbance (A), absorber concentration for electromagnetic radiation (c, mol/L), and extinction coefficient (ε, (L/mol)/cm) or molar absorptivity of the absorbing species. It specifies the units of measurement for light absorbed per unit absorption concentration or optical density, as well as the length of the light path (d, cm).

$$A = \varepsilon \times c \times d \tag{7.4}$$

If a liquid sample contains several light-absorbing species at a given wavelength, the total absorbance is the summation of all absorbers (Equation 7.5)

$$A = \left(\varepsilon_1 \times c_1 \times d_1\right) + \left(\varepsilon_2 \times c_2 \times d_2\right) \tag{7.5}$$

The indices denote the molar absorbance and concentration of the various absorbent species present in the liquid sample.

7.4.2 Wavelength

The variation of microbial UV dose is also affected by the wavelength of UV light. The first-order inactivation constant k_1 has the same wavelength dependence as nucleic acid UV absorption. Strong UV absorption by water below 230 nm limits germicidal action; however, the inactivation constant increases at this wavelength.

7.4.3 Product Parameters

The absorption coefficient (α) is defined as the Naperian absorption coefficient in base e (α_e), and the logarithmic coefficient is called base 10 (α_{10}). This is widely used in computation and is expressed as absorbance divided by path length (m^{-1}) or (cm^{-1}). The absorption coefficient as a function of wavelength is expressed mathematically in Equation 7.6.

$$\alpha_e = 2.303A / dp \tag{7.6}$$

Depth of penetration (d_p) is the depth (cm) at which preliminary flux I_0 decreased at the quartz sleeve through a specific percent of its value, 95% or 99%. The depth of penetration is expressed in Equation 7.7.

$$d_p = \frac{1}{\alpha_e} \tag{7.7}$$

The ratio of transmitted light (I_1) to incident light irradiance (I_0) is termed as the transmittance of a substance (T), expressed in Equation 7.8 (Koutchma et al., 2009).

$$T = \frac{I_1}{I_0} \tag{7.8}$$

Since water is transparent to the wavelengths produced, UV light can easily penetrate it; however, it cannot penetrate milk and other turbid foods such as juices with particles. The opaque foods must be delivered to the UV system as a thin layer, as the absorption coefficient increases, the UV light's penetration capacity reduces. Therefore, it is crucial to understand that enhancing the penetration depth of treated products will benefit the UV light treatment of foods with higher absorption coefficients (Choudhary & Bandla, 2012).

7.5 MECHANISMS AND KINETICS OF UV LIGHT–INDUCED MICROBIAL INACTIVATION

Several mechanisms are involved in microbial inactivation by ultraviolet light, which is reliant on the wavelengths used in the treatment, and can be achieved either directly or indirectly through incident light absorption on cellular DNA (Gayán et al., 2014a).

7.5.1 Mechanism of Microbial Inactivation

7.5.1.1 Continuous UV-C Light

Microorganisms are rendered inactivated by UV light because their nucleic acid is damaged, preventing them from multiplying. The primary mechanism by which UV light kills microorganisms is the development of lesions that interfere with DNA replication. The UV-C wavelength spectra (260–265 nm) are characterized by the highest absorption range by DNA in microbial cells, and it is

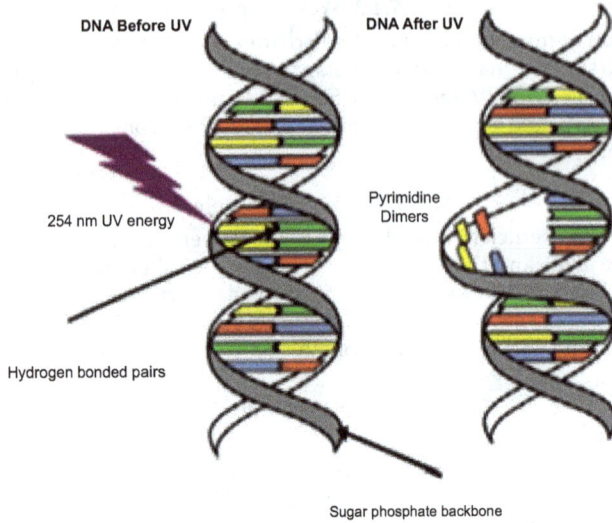

Figure 7.4 DNA structural representation before and after UV light absorption (Koutchma 2019).

the most fatal range for microorganisms (Gayán et al., 2014b). When UV light is applied to DNA within a specific nucleic acid-base, electrons become energized, forming covalent bonds between adjacent bases. UV-C radiation is absorbed by microorganisms, resulting in the formation of DNA photoproducts such as cyclobutene pyrimidine dimers (CPDs) and pyrimidine 6–4 pyrimidone (6–4PP). The structural deformation caused by the formation of these dimers hinders transcription and replication by initiating the formation of new DNA (Figure 7.4), which causes microorganisms to be inactivated (Harm, 1980).

The molecules must absorb the photons emitted and have enough energy to continue the photochemical reaction. LPM lamps are germicidal lamps that transmit UV-C at 254 nm (Figure 7.5), which is very close to the maximum absorption spectrum of DNA of about 260 nm. According to Hamamoto et al. (2007), the germicidal properties of UV light are associated with cellular damage, growth, and DNA lag or damage associated with the production or increase in reactive oxygen species such as superoxide anion radicals (O_2^-), hydroxyl radicals (OH^-), hydrogen peroxide (H_2O_2), and singlet oxygen ($^1O_2$).

7.5.1.2 PL

The photothermal effect is the process of microbial inactivation caused by heat. Hiramoto (1984) proposed that when microorganisms absorb a higher dose of flash discharged by PL, it leads to thermal sterilization. In liquid food processing, the effect of photothermal may be considered minimal because the PL energy can be easily absorbed in the transparent food material resulting in a shorter treatment time and a minimal increase in temperature.

The photophysical effect is the structural impairment caused in microbial cells by the continuous disruption induced by high-energy pulses. Takeshita et al. (2003) observed changes in *Saccharomyces cerevisiae* cells caused by PL greater than a monochromatic UV lamp. Higher energy pulses resulted in cell membrane permeability loss, membrane distortion, and changes in cell shape (Figure 7.6). It

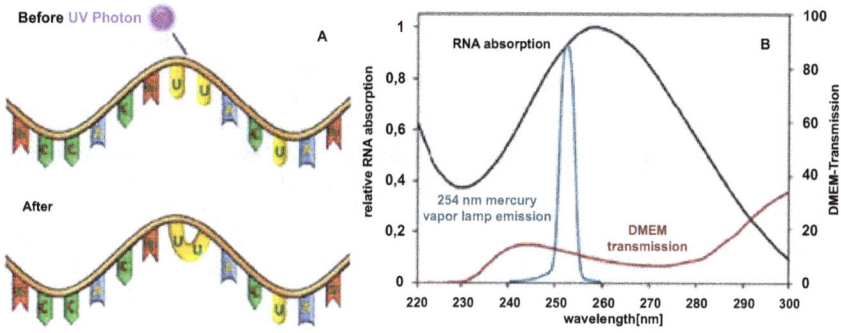

Figure 7.5 Schematic representation of the RNA damage mechanism using UV-C light to form a dimer. RNA relative absorption spectra, an LPM-vapor lamp relative emission spectrum, and transmission of a typical (Eagle) cell culture medium (Heßling et al., 2020).

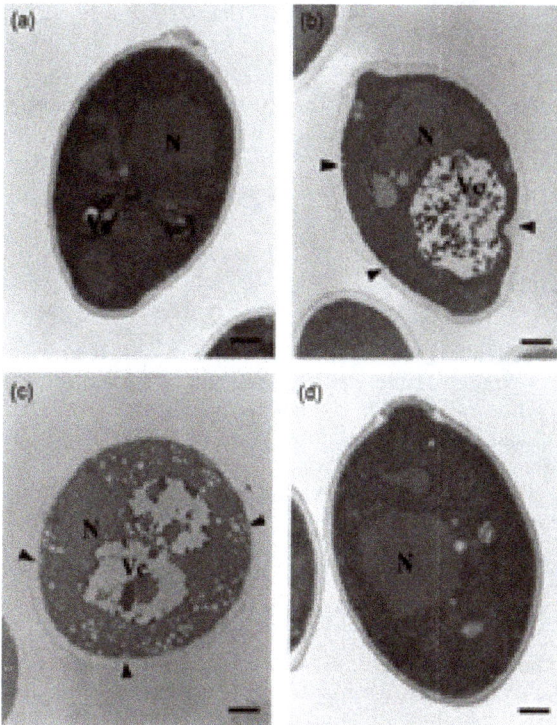

Figure 7.6 *Saccharomyces cerevisiae* transmission electron micrograph. (a) Control, (b) Treatments at 1.4 J/cm² and 2.1 J/cm², (c, d) CW UV light (3 s at 60 mW/cm²/s). The nucleus is represented by the 0.5 N.m bar, and the vacuole is represented by Vc. Damaged/broken membranes are indicated by arrowheads. 40,000 times magnification (Gómez-López et al.2012).

was discovered that PL treatment causes cell leakage, the absence of a cell wall, shrinkage of the cytoplasmic membrane, and the breakdown of internal cellular structure in *Saccharomyces cerevisiae* cells (Krishnamurthy et al., 2010).

7.5.2 Kinetics of Microbial Inactivation

For the study of UV inactivation kinetics, several modeling approaches have been proposed (Collins and Selleck, 1972; Severin et al., 1983). First-order inactivation kinetics is the most basic of the approaches. The inactivation kinetics are assumed to be first order in terms of surviving organism density and light intensity. Therefore, the inactivation rate (organism/cm^3 /sec) is given in Equation 7.9:

$$\frac{dN}{dt} = -k_1 IN,$$

(7.9)

where k_1 is the first-order inactivation rate constant (cm^2/μW-s), I is the intensity at any point (μW/cm^2), and N is the concentration of surviving organism (cfu or pfu/cm^3). The first-order kinetics was also defined as the pseudo-first-order model or mixed second-order model (Koutchma, 2009). By integration, if k_1 and I are constant,

$$\frac{N}{N_0} = exp(-k_1 It),$$

(7.10)

where N_0 is the initial concentration (cfu or pfu/cm^3) and t is the exposure time (s). Although, when the fluence is within certain limits, the first-order model can reasonably predict inactivation. However, the experimental data do not agree with the experiment conducted at low or high fluence (Severin et al., 1984).

Single hit-single target inactivation events are similar to first-order kinetics (Oteiza et al., 2005).

$$\frac{N}{N_0} = e^{-kIt},$$

(7.11)

where k is the inactivation rate coefficient (cm^2/μW.s), I is the UV intensity (μW/cm^2), and t is the exposure time, s. This kinetic equation indicates that the cells get inactivated when subjected to a single harmful event, that is, known as "single hit." UV doses are directly proportional to the number of hits experienced by a single cell.

Another model used for UV inactivation of microorganisms was multi-target kinetics given in Equation 7.12.

$$\frac{N}{N_0} = 1 - \left(1 - e^{-kIt}\right)^n$$

(7.12)

In this model, the particle is assumed to contain a finite number, n, of discrete critical targets, all of which are hit by UV photons, causing cell death and leading to complete microbial inactivation. The probability of attaining a hit on the particle is decreased as the reaction proceeds since the target numbers are infinite. When the value of n is set to 10, single hit–single target kinetics are obtained (Equation 7.11).

Severin et al. (1983) proposed a series event inactivation model designed to tail deviations from the first-order model at low fluence. This model assumes that microbial inactivation occurs systematically, with the rate at each step being of the first order in terms of fluence rate I (Equation 7.13). However, there is a threshold above which all microorganisms are inactivated, while they survive below this level. The threshold level may vary depending on the culturing conditions used before and after irradiation.

$$\frac{dN_i}{dt} = k_{SE}I(N_{i-1} - N_i)$$ (7.13)

The inactivation rate constant, k_{SE}, is denoted by the subscript I in the series event inactivation model. k_{SE} is considered to be constant across levels in this model. The microorganism will become nonviable when n elements of the microorganism, that is, a threshold, are inactivated.

Equation 7.14 determines the concentration of surviving microorganism N if k_{SE} and I are constant.

$$\frac{N}{N_0} = exp(-k_{SE}It)\sum_{i=0}^{n-1}\frac{(k_{SE}It)^i}{i!}$$ (7.14)

where n is a threshold, and if $n = 1$, Equation 7.14 simplifies to the first-order model.

The inactivation curves in UV treatment are typically composed of three phases: a shoulder, a log-linear portion typical of many inactivation curves, and a tail (Figure 7.7). The initial lag of the slope is represented by the shoulder phase. The cell is injured during the "shoulder" phase; however, no death was observed, so the count remains constant until the threshold fluence is attained, at which point the damages are severe enough to cause inactivation. Maximum inactivation occurs at the end portion of the linear curve, where a relatively small number of live microorganisms are present. Tailing is observed as the

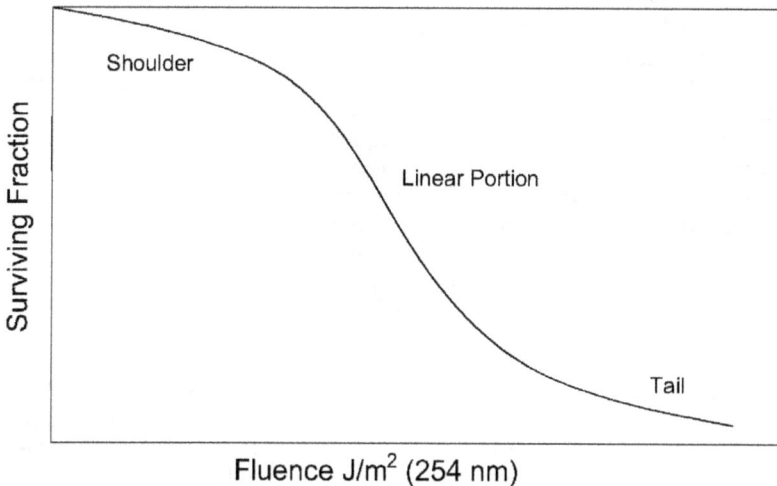

Figure 7.7 A typical bacterial survival curve after UV-C exposure. As a function of UV-C fluence, the fraction of surviving bacterial cells (usually expressed in colony-forming units [CFUs]) is plotted. The three distinct portions of the curve are referred to as the shoulder, the linear portion, and the tail.

suspended material shields the microorganism from UV light, and cell targets cannot be reached. The cell gets aggregated where inactivation of the external cell occurs, thereby protecting the internal cell in illumination. Peleg and Cole (1998) also proposed the Weibull distribution model as a cumulative form of the temporal distribution of lethal events for describing bacterial survival curves from Equation 7.15. As a result, each bacterium dies at a different time. The population exhibits a range of individual resistance to the harmful factors.

$$log \frac{N}{N_0} = kt^{-n}$$ (7.15)

Unluturk et al. (2008) investigated the UV inactivation kinetics of *E. coli* (ATCC 8739) in liquid eggs. Only Chick's disinfection model could be used to predict inactivation parameter *k* from the linear inactivation curve. Sauer and Moraru (2009) reported that all the substrate data fit the Weibull model very well for PL inactivation of *E. coli* O157:H7, and a nonpathogenic *E. coli* strain inoculated in apple juice and apple cider. The two-phase kinetic model was used to inactivate *E. coli* O157:H7 in liquid food, and Ngadi et al. (2003) found that the second slower inactivation phase at higher fluid depths led to low inactivation.

UV doses needed to minimize inhabitants of microbial groups by a single-order scale factor, known as D-values, are commonly used to classify microorganism UV sensitivity. The dose response of microbes to UV light varies with wavelength. A microbe's action spectrum is a plot of its UV sensitivity as a function of wavelength. The wavelength dependence of the first-order inactivation constant k_1 is similar to that of UV absorption in nucleic acids. The inactivation constant has a maximum at or near 260 nm, a minimum at or near 230 to 240 nm, and a zero at or near 300 to 320 nm. The inactivation constant increases below 230 nm; however, the strong absorption of UV light by water at these wavelengths limits germicidal action. (Spikes 1981).

7.6 EFFECTS OF UV LIGHT ON MICROBIAL INACTIVATION

The UV light–based microbial inactivation varies with different microbial species and strain, growth media, and culture stage (Chang et al., 1985; Wright et al., 2000; Koutchma, 2009). Microbial cell size, the density of microorganisms, the location of microorganisms in food, genetic material composition, size, conformation, and microbial ability to regenerate, produce pigments, and irradiation-induced photoproducts are all critical considerations (Tran & Farid, 2004). Other important factors include food type, composition, surface characteristics of the product, and process parameters such as radiation dose, exposure time, and so on. Therefore, understanding these parameters and their impacts on microbial death is critical for determining the appropriate radiation dose required to obtain the required reduction (Guerrero-Beltrán & Barbosa-Cánovas, 2004; Fan et al., 2017).

Owing to the varying penetration capacities of UV light across various physical media, bacteria present in the air have more susceptibility to UV in comparison to that present in liquid (Bintsis et al., 2000). Studies showed that Gram-negative bacteria have a higher susceptibility to UV light than gram-positive bacteria. These variations in UV light susceptibility can be due to innate microbial traits like a thick layer of peptidoglycan on the exterior of the cell walls of gram-positive bacteria making it more resistant to UV light as its penetration becomes difficult (Beauchamp et al., 2012; Kim et al., 2017). After bacteria, yeast is more susceptible to UV light followed by spores of bacteria, fungus, virus, and protozoa (Lopez-Malo et al., 2005; Hijnen et al., 2006).

Yeasts have large cell structures, and their nucleic acid has a lower pyrimidine concentration, owing to a greater likelihood of photons being absorbed by other chemicals before harming DNA, which is associated with a stronger UV resistance than bacteria. Bacterial spores are less susceptible to UV light in comparison to vegetative cells, as they have a thick protein covering, and its dehydrated structure inhibits pyrimidine dimerization (Gayán et al., 2014a). In the case of fungal spores, their color significantly impacts their UV susceptibility since pigmented spores absorb more UV light in comparison to nonpigmented spores, shielding the spore from UV exposure (Gomez-Lopez et al., 2007). In the case of viruses, the inactivation is due to damage in genetic material and/or impact on viral protein. The complex capsid structures of viruses provide a shielding effect against UV and prevent denaturation (Garcia & Barardi, 2019). Furthermore, the state of a cell's physiology, like development stage, growth circumstances, stress prior to treatment, and cell recovery post-treatment, may influence the susceptibility of microorganisms toward UV light (Wassmann et al., 2011). Cells in the logarithmic growth phase show better UV resistance than stationary-phase cells (Lopez-Malo & Palou, 2005; Gayán et al., 2014b). The initial inoculum concentration in food is another crucial factor in determining the efficiency of UV treatment, as a higher initial load diminishes its effectiveness (Fan et al., 2017).

Another important aspect of microorganism recovery circumstances is their ability to restore injury by UV light and endure (Gayán et al., 2014b). Photoreactivation and dark repair are two ways of reversing radiation-induced dimer formation (Salcedo et al., 2007). A key technique called photoreactivation limits the injury to UV-exposed bacterial cells by photolyase enzymes (Yin et al., 2015). The capacity to recuperate enhances the likelihood of microbe survival and shortens the storage life of UV-preserved products. As a result, suitable storage environments for UV-irradiated foods are essential to avert microbial photoreactivation. Guerrero-Beltrán and Barbosa-Cánovas (2004) suggested that limiting exposure to visible light after processing, such as dark packaging to store the product is critical for making food safer and achieving the desired control effects. UV-C light effectively inactivates a wide range of foodborne pathogenic bacteria (e.g., *E. coli*, *Listeria*, *Salmonella*, *Staphylococcus*, *Bacillus*, *Clostridium*, *Yersinia*, *Aeromonas*, *Cladosporium*, *Alicyclobacillus*), spoilage bacteria (e.g., *Lactobacillus*), spoilage yeasts (*Saccharomyces cerevisiae*, *Zygosaccharomyces bailii*); and molds (*Neosartorya fischeri*). Furthermore, the hurdle approach (a combination of UV-C and nonthermal technologies like ultrasound, pulsed electric fields, high hydrostatic pressure, cold plasma, and antimicrobial coatings) has demonstrated excellent synergistic effects against microorganisms.

7.7 EFFECTS OF UV LIGHT ON FOOD QUALITY, NUTRITION, AND SENSORY ATTRIBUTES

7.7.1 Vitamins

Light-sensitive vitamins like retinol (vitamin A), cyanocobalamin (vitamin B_{12}), folic acid (vitamin B_9), riboflavin (vitamin B_2), tocopherols (vitamin E), cholecalciferol (vitamin D), vitamin K, tryptophan, and unsaturated fatty acid residues in solid fats, oils, and phospholipids vary in their photosensitivity (Spikes, 1981). This varying susceptibility of foods to UV light may have a negative impact on food composition, since numerous detrimental light-induced chemical reactions occur because of the generation of free radicals, potentially reducing the number of important nutrients (Lu et al., 2011). Such unwanted reactions deplete vitamin content, degrade proteins, oxidize lipids, destroy antioxidants, alter the texture and color of foods, and produce unpleasant odors (Adhikari and Koutchma,

Table 7.2 Effects of UV Processing on Vitamin Degradation in Fruit and Vegetable Juices (Koutchma, 2019)

Vitamin	Juice	UV dose (kJL-1)	Remaining content (%)
C	Apple[1]	8.1	75
	Apple[2]	1.5	98.3
	Carrot[2]	1.5	100
	Garden vegetable[2]	1.5	100
	Grape (white)[3]	280	1.8
	Orange[4]	48.1	91
	Orange[2]	1.5	83.4
	Orange[4]	4.8	83
	Orange[5]	7.2	82
	Pear[6]	-	79
	Pineapple[7]	7.0	90
A	Carrot[2]	1.5	100
	Garden vegetable[2]	1.5	93.5
	Orange[2]	1.5	89.1
B6	Orange[2]	1.5	83.8
B1	Apple[2]	1.5	100
	Orange[2]	1.5	60
Retinol	Carrot[2]	1.5	100
	Garden vegetable[2]	1.5	100
Carotenoids	Carrot[2]	1.5	100
	Garden vegetable[2]	1.5	93.5
	Orange[2]	1.5	51.4

References: [1]Gayán et al. (2013), [2]Koutchma (2009), [3] Unluturk and Atılgan (2015), [4]Tran and Farid (2004), [5]Torkamani and Niakousari (2011), [6]Falguera et al. (2014), [7]Chia et al. (2012)

2002; Koutchma et al., 2002). Table 7.2 summarizes the effect of UV processing on different vitamin degradation in fruit and vegetable juices.

According to the European Food Safety Authority (EFSA), when milk is treated with UV light, 7-dehydrocholesterol gets converted to vitamin D_3. Aside from this beneficial transition, many vital components can decay at this wavelength, resulting in foul-smelling products. Although the vitamin D level of milk increased due to the UV light treatment, no such effect was observed on concentrations of carotene, vitamin A, and vitamin B_1. Studies in cow and goat milk have reported that vitamin C is more sensitive to UV light, followed by vitamin E, A, and B2, and the amount of loss depends on the initial concentration of vitamins and the treatment dose (Guneser & Yuceer, 2012). In contrast to thermal treatment like high temperature short time (HTST) and ultra-high temperature (UHT), UV exposure did not diminish vitamin B_{12} and riboflavin in bovine milk (Cilliers et al., 2014) and showed less reduction of vitamin A concentration. Several studies reported that vitamin C shows maximum UV absorbance in the germicidal wavelength region (about 260 nm) but not considerably over 300 nm. The absorption coefficient is also influenced by vitamin C concentration, and in general, the higher the vitamin C concentration of a juice, the higher the absorption coefficient (Ye et al., 2007). According to studies, vitamin C–fortified juices require much greater UV irradiation doses for treatment.

7.7.2 Lipids

Studies have reported that UV light speeds up oxidative reactions in oils and fats. UV treatment may significantly impact foods high in unsaturated fatty acids, which are influenced by free radicals and deteriorate, causing a rotting odor and lowering antioxidant effects. A significant level of photodegradation in three essential amino acids, that is, histidine, tryptophan, and phenylalanine, were observed. This photooxidative reaction of fats and unsaturated phospholipids results in the development of foul-smelling compounds (Spikes, 1981).

7.7.3 Proteins

UV light promotes protein degradation by changing solubility, heat sensitivity, mechanical characteristics, and enzyme digestibility by protease. Proteins photodegradation may modify their physical characteristics enough to induce noticeable organoleptic modifications and off-flavors, such as in milk. Studies reported that mild UV treatment does not considerably change the color and lipid oxidation of cheese (Can et al., 2014). However, drastic UV treatments can alter product physicochemical and sensory properties, like color (Can et al., 2014; Keklik et al., 2019), oxidative changes of protein (Fernández et al., 2014), and pH reduction (Keklik et al., 2019). Furthermore, the volatile chemicals like aldehydes and hydrocarbons, resulting from the oxidation of lipids and breakdown of the amino acid were found to be higher in some studies (Fernández et al., 2016).

7.7.4 Carbohydrates

Carbohydrates are not photosensitive in general, but some of their derivatives, such as sugar alcohols or saccharic acid fragments, are sensitive to light. This light-induced depolymerization of polysaccharides may change the properties of foods, for instance, the softening of fruits and vegetables. Foods have distinct colors due to natural pigments and artificial colorants. Exposure to light can modify these food colorants' hues, resulting in food color alterations. Some pigments found in natural foods may help synthesize the photooxidative destruction of various nutrients. Conversely, some dietary ingredients shield other components from these alterations like tocopherols can operate as antioxidants and protect against photodegradation (Spikes, 1981). Most of the research found little to no color difference in most juices and proved that UV treatment preserved the color characteristics of juices far better than thermal treatment.

7.7.5 Phenols

UV treatment reduced considerably the quantity of phenolic components in apple juice; however, it was a smaller reduction than a heat treatment with equivalent efficacy (Tran and Farid, 2004). Enzymatic activity is also affected by UV therapy. Polyphenol oxidase activity in mango nectar lowered to 19%, thus retaining its vibrant natural color for a longer period (Guerrero-Beltrán and Barbosa-Cánovas, 2006). Compared to the equivalent thermal processing, UV treatment showed no or less influence on antioxidant properties or polyphenol oxidase or peroxidase activity.

7.8 MACHINERY/EQUIPMENT

7.8.1 UV Disinfection Systems for Solid Foods and Contact Surfaces

The Claranor system (Avignon, France) disinfects vats and cups in sensitive product processing lines using PL (Figure 7.8a). It is composed of an electronic device that is powered by main currents to generate electric bursts and an optical aperture that guides the beam of light onto the surfaces to be treated. The PL technology provides consistent treatment and 60 strokes per minute high throughput rate. This system is suitable for various cup types and sizes up to a depth of 15 cm, and is also ideal for heat-treated products, fruit-based sweets, dairy products, juices, and cold-filled ready meals.

The first sterilization equipment (stainless steel 304) for infant milk powder cans was also introduced by Claranor (Figure 7.8b). A dry cooling system was designed to meet the demand for a high-quality treatment system for infant milk treatment. Several microbial trials were performed to demonstrate the inactivation efficacy on a wide range of microorganisms (*Cronobacter sakazakii, Bacillus cereus, Candida albicans*). The heat is not released in the environment around the conveyor during the processing but eliminated outside the production room, meeting the food safety standards.

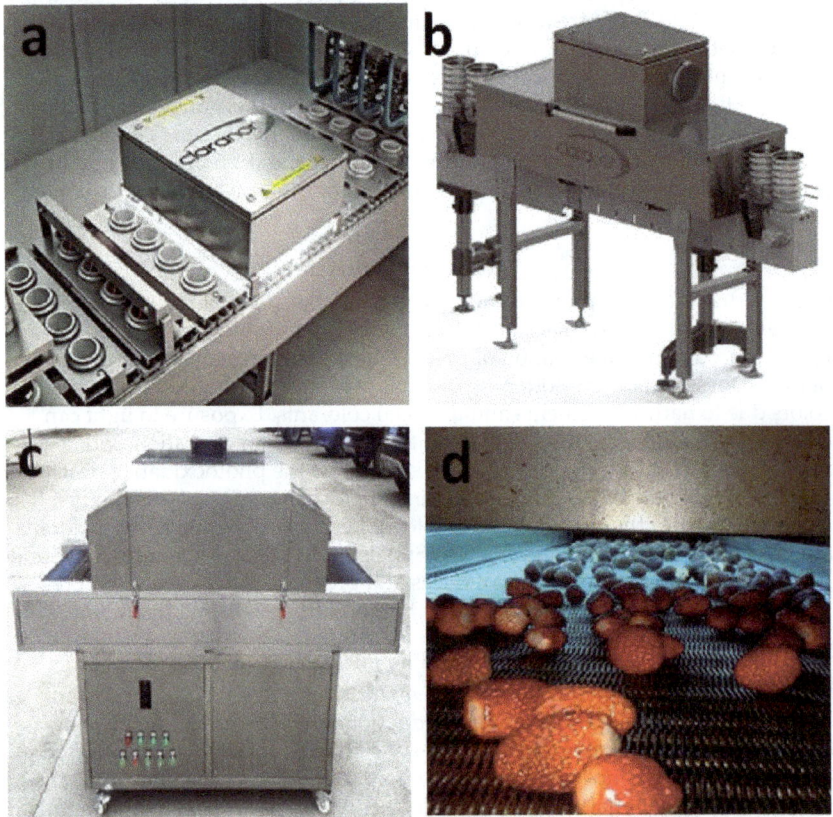

Figure 7.8 UV disinfection systems for solid foods and contact surfaces: (a) PL sterilization for cups, (b) milk powder cans sterilizer, (c) UV food sterilizer, and (d) UV-C treatment of berries.

UV food sterilizer (HT-UV-2000; Figure 7.8c) is mainly designed for rapid sterilization of foods and packaged products that are widely used in hospitals, food-processing units, processing of tea and beverages, and packaging units. This system uses the ultraviolet irradiation sterilization technique to sterilize, treat and kill various microorganisms effectively, and can highly reduce microbial growth in food. The UV sterilizer machine is made up of 304 stainless steel. It mainly consists of a sterilizing device, a protective device, a body bracket, a conveyor belt, and an exhaust fan. It is also effective in reducing chlorides in water by photolysis, making it nonpolluting and environmentally friendly.

The UV-C conveyor system can be used instead of chlorine to reduce pathogens and spoilage microflora on the surface of foods, such as berries, nuts, cheese, dairy products, and others, and prevent odor formation. It is also designed to disinfect the packed food like ready-to-eat cut fruits or packaging during the process. It is available at a comparatively low cost and provides continuous treatment for large-scale food production. It is an effective technique for killing microbial growth in raw meats and poultry. The unique combination of UV-C and ozone gas is also efficient in extending the shelf-life of refrigerated products. The UV-C system is also designed for the treatment of ethnical bread lines such as Pita bread and croissants (Figure 7.8d), as it is a moisture-free system, the growth of molds is prevented, thereby extending the shelf life.

The intense light pulse (ILP) unit is also known as a Tecum-Mobile Decontamination unit (Claranor, Manosque, France). Light pulses of 300 seconds duration and 3 J/cm^2 intensity were used for a voltage of 3000 V. This unit is made up of 23-cm cylindrical xenon flash lamps with 175 mm arc lengths (Flashlamps Verre & Quartz, Bondy, France). The distances between the upper and bottom lights, as well as the left- and right-hand lamps were set at 6 cm and 10 cm, respectively, from the surface of the treated knife. The treatment is followed by one or five flashes and is manually started at a rate of one pulse every 2 seconds. It is a batch lamp system designed for use in laboratories. Rajkovic et al. (2010) reported that using ILP to disinfect the surface of a stainless-steel slicing knife was efficient in reducing *L. monocytogenes* and *E. coli* O157:H7. The findings indicate that this treatment can be beneficial in preventing cross-contamination between the slicing equipment and the final product.

7.8.2 UV-Based Reactors for Liquid Foods and Beverages

UV reactors are widely viable commercially or under development to study UV treatment on liquid foods and beverages. The design of UV reactors includes tubular annular, static, and dynamic mixer systems. In the annular reactors, the liquid is pumped through the gap created by two concentric stationary cylinders containing single or multiple lamps in the annular reactor. An annular type of laminar reactor has been used to treat apple juice and cider (Worobo, 1999), mango nectar, and cheese whey (Guerrero-Beltrén & Barbosa-Cánovas, 2006; Singh & Ghaly, 2007). The reactor's length and gap size vary depending on the type of liquid and flow rate. Tran and Farid (2004) described the UV treatment of orange juice with a vertical single UV lamp thin-film reactor with a thickness of 0.21 to 0.48 mm shown in Figure 7.9a. Keyser et al. (2008) used the Pure UV/Sure Pure reactor to treat apple juice, guava and pineapple juice, mango nectar, and two different orange and tropical juices. A single lamp device is used in this reactor, with a thin film of liquid forming between the surface of the lamp and the surrounding corrugated outer wall. The reactor consists of an inlet and outlet chamber, as well as a corrugated spiral tube connecting the two. Even at the inlet chamber's low flow rate of 3800 L/h, the tangential unit generates turbulence (Re > 7500). The liquid product encounters UV light when flowing through

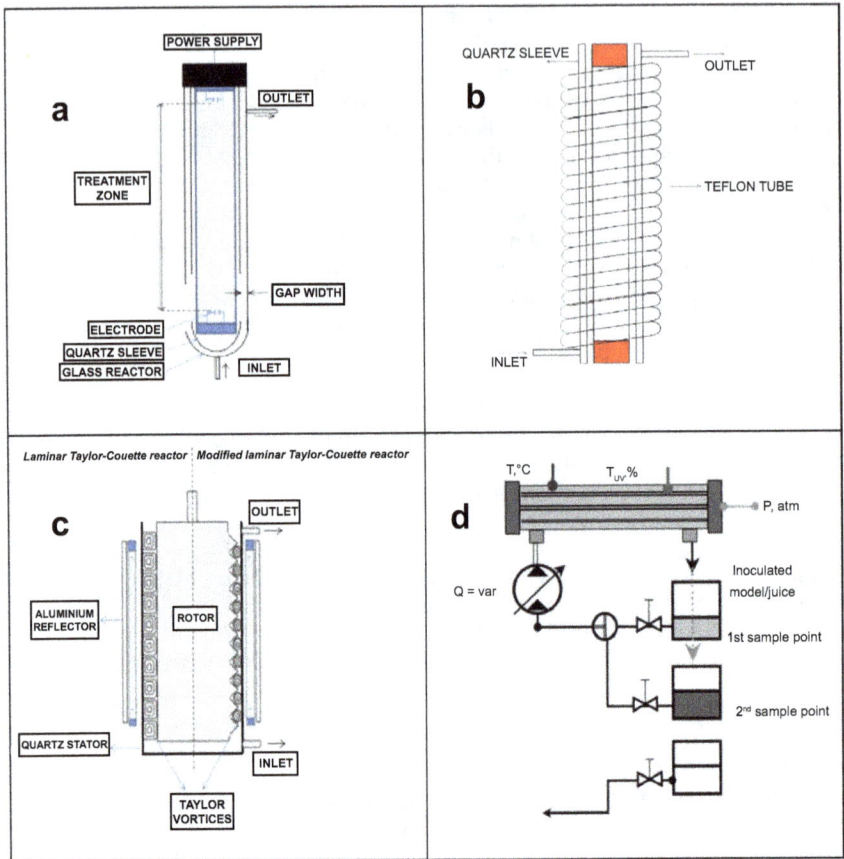

Figure 7.9 Schematics representation of reactors: (a) Thin-film flow concentric tubes reactor (Ye, 2007), (b) coiled tube reactor, (c) Taylor–Couette reactor (Ye et al., 2008), and (d) Aquionics UV reactor (Koutchma et al., 2004).

the sheath and spiral pipe gap as depicted in Figure 7.9b. In the UV treatment of mild fluids, static mixers are used. The best illustration of a static mixer is Latros of Dundee, Scotland. It comprises helical elements that alternate right and left with a 180° rotation, with each element positioned at 90° to the prior element. Flow diversion, flow inversion, circular mixing, and axial variations of the fluid flow are achieved, which constantly evolves the thin film at the internal wall of the pipe, where more microbial cells are exposed to UV light during fluid processing.

Salcor Inc.'s UV reactor, which is an example of a static mixer, is made up of Teflon pipes coiled in a helix with 12 lamps within and 12 outside the helix. The fluid flows through the coiled pipes. Because of the cured flow direction in the coil, a couple of counter-rotating vortices with their axes can be observed the entire length of it. Due to the dean effect, this UV reactor increases the liquid supply by increasing mixing (Koutchma, 2009). Ye et al. (2008) studied the inactivation potency of a Taylor Couette flow reactor, an example of a dynamic mixer shown in Figure 7.9c. Forney et al. (2008) investigated and reviewed the optimal performance of this reactor. This reactor is made up of a stationary outer cylinder and a rotating inner cylinder. This design caused fluid instability, resulting in counter-rotating vortices with the vortex axis operating around the circumference of the cylinders, carrying the fluid with high intensity from the dark outer cylinder to the inner wall region. To compare the efficiency in terms of fluency (mJ/cm^2) or energy (mJ/L), the microbial inhibitory effect was reported as a logarithmic reduction (log_{10}) in the range of target pathogens or a total count.

The Aquionics UV reactor (Slough, England) is a turbulent flow reactor that disinfects juices and beverages with UV light. The liquid is passed through a stainless-steel compartment accompanied by UV emitting low-pressure arc tubes during the processing. The single arc tubes were all assembled on quartz sleeves that were fixed inside the compartment, allowing liquid to pass through all of them (Koutchma et al., 2004). The UV monitor measures the amount of UV light that passes through the arc tube, while the temperature sensor above it measures the temperature inside the chamber. The setup and port samples are depicted in Figure 7.9d.

Two infinity liquid disinfection systems (Atlantic UV, Co., Hauppaunge, N.Y.) were used for UV treatment of fresh mango nectar (Guerrero-Beltrén & Barbosa Cánovas, 2006). It is constructed of stainless steel and is linked in series to form a single UV disinfection system. The annular space between the quartz sleeves and the wall in this system can hold up to 0.25 l (V) of liquid. Inside the disinfection system, a UV lamp (power 25 W, diameter 15 mm, length 556 mm) is installed. The UV lamps are warmed up for at least 30 minutes before circulating the liquid in the system. The disinfection area ($262\ cm^2$) is exposed to UV light at a distance of 46.5 cm. A digital thermometer (Cole-Palmer, Vernon Hills, IL) is used to measure the temperature at the inlet and outlet of the disinfection unit. UV light's intensity ($I = 25\ mW/cm^2$) is measured with a digital radiometer equipped with a silicon photoelectric sensor (Cole Palmer Instrument Co., Vernon Hills, IL). Bio-link chamber (VilberLourmat, France) is a UV-A, UV-C, or UV-B irradiation chamber mainly used to remove contamination by PCR and binding membrane and nucleic acid. According to the publications, the UV-C treatment was given to 20 inoculated berries at different UV doses ($K = 254$ nm) in the Bio-link UV chamber consisting of five lamps of 8 W each. (Maharaj et al., 1999; Nigro et al., 1998). Heraeus Noblelight GmbH (Hanau, Germany) has developed high-quality UV modules, specifically for the food industry. The UV disinfection module is suitable for dairy and beverage filling and sealing machines and conveyor belts, transport containers, work areas, and surfaces of the food. With a working distance of 20 mm and an exposure duration of 210 s, it achieves a disinfection rate of more than 99.9%.

7.8.3 UV Systems for Disinfection of Solid Foods

Pulsed UV light (PUV) treatment of the sliced ham and its effect on the quality was studied by Wambura and Verghese (2011). In this experimental study, the Steri pulse-XL300 Pulse UV-light system (Xenon Corporation, Woburn, MA) was used. The UV lamp pulses three times per second, and each pulse is powered by $1.27\ J/cm^2$ electrical energy located 1.8 cm below the xenon lamp (Figure 7.10). As demonstrated in this study, the temperature of the samples increased with treatment time and distance from the UV lamp. While the xenon flash lamp is an effective UV source for treating the surface of sliced ham, its application to meat may be hampered by undesirable alterations in texture, color, and oxidative stability, all of which have a significant impact on consumer acceptance.

According to the standards of the company and review literature (Krishnamurthy et al., 2004; Sharma & Demirici, 2003), the lamp generates

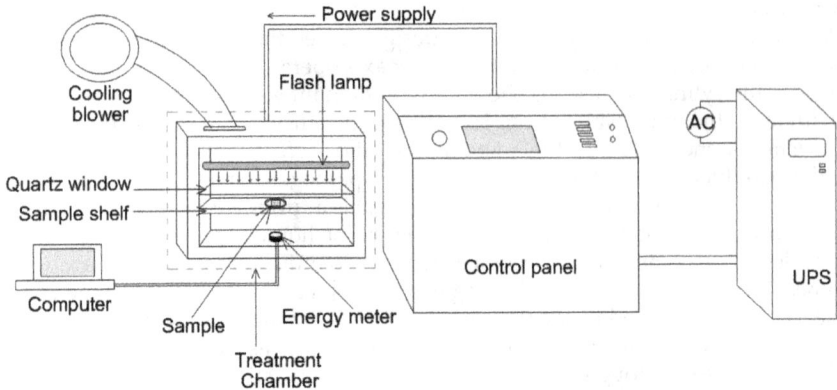

Figure 7.10 Schematic representation of pulsed UV light system.

approximately 0.14 J/cm² per pulse on the strobe surface for an input voltage of 3800 V and three pulses (0.42 J/cm²) per second. The duration of a pulse (pulse width) was 360 μs. No energy loss was observed between pulses, and this energy measure can be compared with the photographic exposure. UV-C radiation equipment was used by Allende and Artés (2003) for the treatment of fresh "Lollo Rosso" lettuce.

It is composed of a stack of 17 stainless-steel reflectors with germicidal emitting lamps that are unfiltered (Sylvania, G30T8, Philips, Netherlands). Above the radiation vessel, the lamp stack is suspended horizontally. Polystyrene trays are placed 60 cm below the lamp, stimulating a processing line. A wooden box encased in aluminum foil and assisted by a metal framework shields the operator from the UV-C lamp, reflectors, and treatment area. A fan was installed on one corner of the wooden box to prevent a temperature rise caused by the illuminated lamp. UV-C treatments were found to inhibit the growth of psychrotrophic bacteria, coliforms, yeast, and molds. Inhibiting microbial growth was more effective with higher UV-C doses than with lower radiation doses. As a result, the dose-dependent effect of UV-C was discovered. After discovering that hydrogen peroxide could oxidize tartaric acid with Fe^{2+} acting as a catalyst, H.J. Fenton proposed the Advanced Oxidation Process (AOP; Giannakis, 2019). The emergence of food safety concerns regarding fresh produce prompted researchers to assess the efficacy of the AOP for surface decontamination. This system consists of a spray-based treatment chamber that sprays peroxide onto fresh produce while being constantly illuminated by UV-C (254 nm). Human pathogens (*E. coli* O157:H7; *Salmonella*) and spoilage bacteria (*Pectobacterium pseudomonas*) were inactivated in fresh produce by AOP. A commercial AOP system from CleanWorks (ON, Canada) for fresh produce treatment. It has been reported that using peroxide or UV alone reduced 1- to 2-log colony-forming units (CFUs), as did AOP at 20°C versus 50°C. (Hadjok et al., 2008). AOP's largest unit was recently installed as a microbial reduction step in a citrus food processor. The 10-m-long unit is made up of hydrogen peroxide spray bars and four modules containing UV-C and ozone-generating lamps decontaminating about 4000 lemons per hour, thereby demonstrating its ability to scale to any size of operation.

7.8.4 UV Systems for Air Disinfection

UVGI (ultraviolet germicidal irradiation) air treatment systems use short-wave UV light energy (UV-C) to inactive bacterial and fungal microorganisms so they cannot replicate further. UV-C LPM lamp devices are widely used in air and surface disinfection. The UV-C system's design is based on the performance data provided by the lamp, ballast, and fixture manufacturers. Upper-air UV-C devices are also used in plant facilities' rooms to irradiate the air in the upper section of the area. The UV-C disinfection treatment can be used in duct systems with the main objective of the preliminary design to uniformly disperse UV energy equally in all directions all across the duct or air-handling unit (AHU) to supply an adequate UV dosage to the airflow through the irradiated region while using the least amount of power. Designed dosage is the primary function of inactivating the target microorganism and disinfecting the airstream at the desired level. The average irradiance of a typical in-duct system ranges from 1000 to 10,000 W/cm^2. In-duct UV-C must be used in combination with appropriate filtration. Filters in the system improve overall air cleaning capabilities by protecting UV lamps from dust and debris deposition, reducing UV output over time.

7.9 APPLICATIONS OF UV LIGHT IN FOOD PRESERVATION

UV has been used to disinfect the air, food contact surfaces, and nonfood contact in the past, and it has lately been employed to enhance the safety of both solid and liquid foods and beverages, and also ingredients. Science and engineering of UV light have recently advanced, making it a feasible choice for commercial use in the food industry to provide chemical and microbiologically safe foods, extend their shelf lives, enhance the functional properties of whole and fresh-cut produce, and provide high-quality products. UV light has been shown to reduce the toxicity of plant-based foods by decreasing toxins like patulin, a mycotoxin present in fresh apple cider (Dong et al., 2010), and controlling the enzymatic browning (Manzocco et al., 2009). Figure 7.11 depicts a schematic design of prospective UV light technology in food applications.

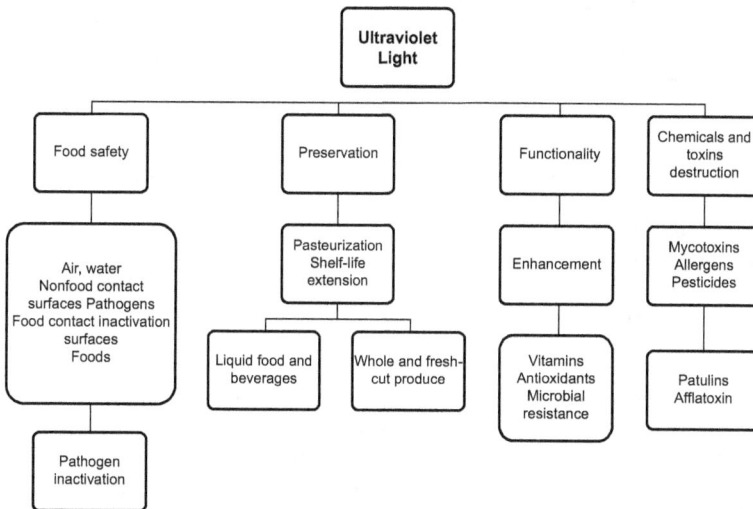

Figure 7.11 Application of UV light technology in areas of food processing (Koutchma, 2019).

181

7.9.1 Surface Disinfection by UV Application

7.9.1.1 Food Contact Surfaces

Millions of microorganisms may reside in food processing equipment and their surfaces. The presence of such microbes impacts the manufacturing of food, and hence the quality and shelf life of the product because of cross-contamination. Therefore, UV light disinfection systems are adopted to improve hygiene in the food industry by eradicating microbial and chemical contamination from such surfaces and consequently enhancing the safety and shelf-life of the finished products. UV light applications comprise surface decontamination of equipment in bakeries, cheese, and meat plants and of conveyor belts in addition to the usual cleaning and sanitization of plant facilities. According to Morey et al. (2010), UV light can destroy up to 99.9% of bacteria on conveyor belts used to transport food products, and UV tunnels are used throughout the production line in the dairy and bakery industries to process food efficiently.

7.9.1.2 Packaging materials

Applying appropriate UV light dosages on food packaging surfaces like boxes, containers, cartons, bottles, caps, clasps, lids, foils, and films may minimize or eradicate the number of harmful or foodborne pathogens. UV can be used to treat both prefilled and post-filled food packaging. The efficiency of UV treatment is better on flat surfaces than on uneven surfaces, because UV light cannot reach every point due to shadowing. The UV light permeability of different packaging materials varies depending on the construction, thicknesses, and distinct qualities. This permeability and transparency are increasingly essential when UV light treatment is done on packaged food (Tarek et al., 2015).

7.9.1.3 Food surfaces

In many circumstances, food surfaces are the principal point of microbial entry. Despite the penetration limitations, research on applying UV light, both continuous and pulsed light treatment, is progressing to minimize microbial contamination on solid food surfaces. In various reports (Koutchma, 2014), the effective reduction in microbial contamination on surfaces of processed foods such as flour, spices, raw whole, and fresh-cut produce, raw animal-origin products, and postharvest treatment of grains has been established.

7.9.2 Meat and Poultry Preservation

UV light is effective against *E. coli*, *Salmonella*, and other non-spore-forming pathogenic microbiota on raw meats, pork skin, and poultry surfaces. Table 7.3 provides the summary of reported works on microbial inactivation of meat and poultry products by UV light technology. Despite the limitation of UV light to penetrate uneven food surfaces like meats, these studies showed that UV light is effective in reducing bacterial contaminations like *L. monocytogenes* and other pathogens of meat and poultry processing units, suggesting its use as a post-lethality treatment for the concerned microorganism. Sommers et al. (2009) have shown the effectiveness of UV light to control *Listeria* contamination on ready to eat (RTE) product surfaces. The UV dosage required to meet specified log reduction standards must first be analyzed to show the efficacy of UV treatment (USDA-FSIS). This information will enable the selection of the proper UV source and the efficient system design based on the application. UV units installed in cold rooms kill the microorganisms on a product's surface and in the air and significantly extend the storage life of chilled carcasses, postponing the beginning of microbial deterioration. UV also decreases shrinkage, preserves fluids, eliminates

Table 7.3 Effects of UV Light Technology on the Microbial Inactivation of Meat and Poultry

Meat & poultry	Target micro-organism	UV treatment parameters	Results	Reference
Pork	B. thermosphacta Y. enterocolitica	253.7 nm Dose: 16.16–19 mJ cm^{-2}	3.0 log$_{10}$ reduction of Y. enterocolitica and B. thermosphacta at 16.16 and 19.30 mJ/cm^2 respectively. No effect on quality parameters	Reichel et al., 2019
Ground beef	Salmonella spp.	254 nm Intensity: 800 µWcm^{-2}	1 log$_{10}$ CFU/g reduction by UV light. 2 log$_{10}$ CFU/g reduction by Bacteriophages + UV light.	Yeh et al., 2018
Goat meat surfaces	E. coli O157:H7	254 nm Intensities: 100 and 200 µW cm^{-2} Treatment time: 2, 4, 6, 8, 10, and 12 min Doses: 0.2–2.4 mJ cm^{-2} Distance: 10 cm	6.66 log$_{10}$ CFU/mL reduction by 1% lemongrass + UV (200 µW cm^{-2}; 2 min) No significant impact was observed on the texture, color, and oxidative stability.	Degala et al., 2018
Chicken thigh Meat (lean surface thigh and skin surface thigh)	E. coli, Salmonella, Campylobacter	Pulsed light 200–1,100 nm Fluence: 1.27 J cm^{-2} Treatment time: 5 and 45 s	1.5 and 2.0 log reductions for E. coli, 1.2 and 2.2 for Campylobacter, and 1.5 and 2.4 for Salmonella, on lean surface thigh for 5 and 45 s respectively. 1.1 and 2.0 log reductions for E. coli, 1.2 and 1.9 for Campylobacter, and 0.9 and 1.8 for Salmonella, on skin surface thigh for 5 and 45 s, respectively.	Cassar et al., 2018

smells in cold rooms, and inhibits surface mold growth. UV promotes metmyo-globin production, and it is important to avoid the browning of red meat.

7.9.3 Bakery Items and Shell Eggs

UV light treatment of 32 kGy has been employed to extend storage life and minimize the post-baking contamination of wrapped partially baked baguettes by 100% (Doulia et al., 2000, EU2016/398). Continuous UV light at 254 nm is efficient in decreasing total aerobic count (Chavez et al., 2002), *Yersinia enterocolitica* (Favier et al., 2001), *S. typhimurium*, and *E. coli* (Coufal et al., 2003) populations on eggshell surfaces. The disinfection of eggshells by pulsed light was studied by Keklik et al. (2010), who observed a logarithmic reduction of 5.3

Table 7.4 Effects of UV Light Technology on the Microbial Inactivation of Shell Eggs

Target microorganism	UV treatment	Log reduction	Reference
Aerobic plate count	Intensity: $7.35 \, mW/cm^2$ Time: 0, 15, 30, and 60 s Intensity: $7.5 \, mW/cm^2$ Time: 0, 12, 36, and 48 s	2–3 $\log_{10}$ CFU/egg at all treatment time 1 to 2 $\log_{10}$ CFU/egg at 45 s	Chavez et al., 2002
Aerobic plate count S. typhimurium E. coli	Intensity: $14 \, mW/cm^2$ Time: 4 min	1.3 log for aerobic plate count 4 log reduction for S. typhimurium 4 to 5 log reduction of E. coli	Coufal et al., 2003
Y. enterocolitica	Intensity: $4{,}573 \, \mu W/cm^2$ Time: >25 min	1.60 log cycles	Favier et al., 2001
Aerobic bacteria	Intensity: $0.62 \, mW/cm^2$ Time: 0, 15, and 30 min	2 log cycle	Kuo et al., 1997
S. enteritidis	Dose: $23.6 \pm 0.1 \, J/cm^2$ Time: 20 s Distance: 9.5 cm below the UV lamp	5.3 log CFU/cm^2	Keklik et al., 2010
Aerobic bacteria E. coli Staphylococcus aureus	UV-C light at 253.7 nm for 4.70 and 18.80 s	4.47 to 3.57 log CFU/egg for uninoculated clean egg shells No significant reduction for very dirty eggs 3.0 log CFU/egg for E. coli 4.0 log CFU/egg for Staphylococcus aureus	De Reu et al., 2006

CFU/cm^2 with no visible damage. De Reu et al. (2006) evaluated the effectiveness of a commercial UV system linked to a conveyor belt in removing bacteria from clean and dirty eggs. They concluded that UV light treatment significantly reduces the bacterial load on eggshells and can minimize contamination of rollers but did not lower the contamination on the inside of the egg. Commercial UV treatment of shelled eggs has yet to be implemented, despite the positive results of UV light–induced microbial inactivation on shell eggs to improve egg safety (Table 7.4).

7.9.4 Postharvest UV Treatment of Grain

After field crops have been harvested, postharvest measures are critical for controlling fungus growth. Mycotoxigenic fungi like Fusarium and Penicillium can generate toxic mycotoxins like zearalenone (ZEN), deoxynivalenol (DON), and ochratoxin A (OTA), in addition to significant yield losses. UV light at 253.7 nm is an emerging nonionizing postharvest strategy for reducing fungal and mycotoxin levels on the surface of grains. Table 7.5 summarizes the effect of UV treatment on microbial inactivation of grains.

Table 7.5 Effects of UV Light Technology on Microbial Inactivation of Grains

Grain	Target microorganism	UV treatment parameters	Results	Reference
Maize	*Fusarium* spp.	UV-C lamps 15 W Treatment time 0, 5, 10, 15, 20, 25 and 30 min Dose: 3.3 to 19.8 J/cm²	42.85 and 52.05% reduction of *Fusarium* species	Paez et al. (2011)
Wheat	Natural occurring bacteria and fungus	Industrial UV-C unit at 254 nm Dose: 195 J cm⁻² UV irradiance: 97 W/m²	90% reduction at 6.3 h for bacteria and 5.6 h for mold. Quality was minimally affected (suggested by germination and amylo-graph tests)	Hidaka and Kotaro Kubota (2006)
Maize and wheat kernels	*Penicillium verrucosum* *Fusarium graminearum* Mycotoxins (zearalenone (ZEN), deoxynivalenol (DON), and ochratoxin A (OTA))	3D UV treatment at 253.7 nm Irradiance: 22.5 mW cm⁻²	On maize *P. verrucosum* and *F. graminearum* were lowered by 79% with 5000 mJ cm⁻² and 60% with 1000 mJ cm⁻² respectively. On maize and wheat, DON (30%; 14%), ZEN (52%; 42%), and OTA (17%; 6%) were reduced.	Popović et al. (2018)

7.9.5 Fresh Produce

UV treatment can be used commercially as surface decontamination of fresh fruits and vegetables, both whole and fresh cut. Its germicidal effect can be more effective than chlorine, ozone, or hydrogen peroxide. Fresh produce is subjected to UV light to fulfill two major objectives. The first and foremost aim is to lower the microbial count on the food surfaces, and the other is to increase host tolerance to bacteria. UV light was tested for surface disinfection of apples, cantaloupe (Beaulieu, 2007; Lamikanra et al., 2005), grapes, kiwifruit, oranges, lemons, nectarines, raspberries, peaches, pears (Lagunas-Solar et al., 2006), and watermelon (Fonseca & Rushing, 2006). According to Fonseca and Rushing (2006), when packed watermelon cubes are exposed to UV light at 4.1 kJ m⁻² a reduction in bacteria populations by more than 1 log was observed without impacting juice leakage, color, or overall appearance.

Various studies have been conducted to justify and prove that UV contributes to the induction of postharvest disease resistance in fruits and vegetables. Table 7.6 summarizes the reported data on impact of UV light processing on microbial inactivation in fresh produce. Obande et al. (2011) investigated the storage life of tomatoes that were exposed to UV-C light at 8 kJ/m² prior to inoculation with *Penicillium digitatum*. The treated fruit samples were firmer

Table 7.6 Effect of UV Light Technology on Microbial Inactivation of Fresh Produce

Fresh produce	Target microorganism	UV treatment	Storage time and temperature	Efficacy, log reduction	Reference
Black peppercorns	Salmonella spp.	5 UV-C lamps Intensity: 616 µW cm^{-2} Time: 22.1 min	–	3.7 log CFU/g when combined with cold plasma treatment at 9.7 kV	Bang et al. (2021)
Fresh-cut strawberries	Mesophilic bacteria, yeasts and molds	3 UV-C lamps Dose: 5.8 kJ m^{-2}	8 days at 0 °C	1.4 log cycle for mesophilic bacteria 1.3 log cycle for yeasts and molds	Avalos-Llano et al. (2020)
Lettuce leaves, apple skin	E. coli L. monocytogenes	Low-pressure mercury lamps and UV-C LEDs 500 mJ cm^{-2}	4°C 25°C	1.19 ± 0.59 and 1.58 ± 0.27 log CFU at 4 and 25°C	Green et al. (2020)
Tomatoes	Penicillium digitatum	Intensity: 1000 W/cm^2 Doses: 3 and 8 kJ/m^2	20 days at 16°C in dark	The diameter of the lesion shows no significant increase after 6 days for 8 kJ/m^2	Obande et al. (2011)
Raspberries, strawberries	E. coli O157:H7, Salmonella. spp.	Fluence per pulse: 1.27 J/cm^2 at 3800 V Treatment time: 5, 10, 30, 45, and 60 s Distance: 3, 8, and 13 cm for raspberries; 5, 8, and 13 cm for strawberries	–	Raspberries: 3.9 and 3.4 log$_{10}$ CFU/g reductions of E. coli O157:H7 and Salmonella at 72 and 59.2 J/ cm^2, respectively. Strawberries: 2.1 and 2.8 log$_{10}$ CFU/g reductions of E. coli O157:H7 and Salmonella at 25.7 and 34.2 J/cm^2, respectively.	Bialka et al. (2008)
Fresh-cut watermelon	Microbial populations	Dose: 1.4, 4.1, 6.9 and 13.7 kJ m^{-2} Time: 1, 3, 5 and 10 min	8 days	More than 1-log reduction at 4.1 kJm^{-2}	Fonseca and Rushing (2006)
Apples, leaf lettuce, Tomatoes	Salmonella spp. E. coli O157:H7	Intensity: 1.5 to 24 mW/cm^2	–	Apple: 3.3 log reduction of E. coli at 24 mW/cm^2 Tomatoes: 2.19 log reduction of Salmonella spp. Green leaf lettuce: Reduction of 2.65 and 2.79 log of Salmonella spp. and E. coli O157:H7 respectively	Yaun et al. (2004)

than the control, and the fungal lesion diameter was significantly less after storage at 20°C for ten days. Besides molds, pathogenic microorganisms, like *Salmonella* spp. and *E. coli* present on the surface of fresh produce, can threaten food safety and human health. Several publications reported that applying UV-C or PL to tomatoes, apples (Yaun et al., 2004), or berries (Bialka et al., 2008) can lower the population of these pathogens. However, because PL light processing was followed by an increase in temperature, microbial inactivation might be the consequence of both light–heat actions. These results showed that UV postharvest processing of a wide range of fresh food might be effective against both pathogenic and nonpathogenic microorganisms. Not only this, but it was also experimentally proved that it delays the ripening of the food to lengthen its life.

7.9.6 Dairy Products

The surface of the dairy products is the primary site for the majority of the chemical, microbial, and enzymatic reactions, resulting in unfavorable changes that may shorten the product's shelf life. UV light application is a more promising technique than chemical preservation for surface decontamination cheese, yogurt, and other dairy products. UV light is adequate for inactivating surface bacteria such as *E. coli, Salmonella typhimurium, Listeria monocytogenes, Pseudomonas* spp., and Enterobacteriaceae to assure safe product and increase storage stability with little impact on the chemical and nutritional contents of the dairy products because of its low penetration. UV treatment on the surface of cheese immediately before packing can be a suitable option for reducing the microbial load on the surface and extending its storage stability with minimum impact on the product. Koca et al. (2018) exposed naturally infected set-type yogurt to UV light at intensity 32.1 W/m^2 for exposure time up to 600 s in a batch-type UV light cabinet. They found that UV treatment is effective for mold inactivation in set-type yogurt, but at higher doses, oxidation takes place. Limited studies have been done on using UV light to decontaminate the surface of various dairy products. Table 7.7 summarizes a few such studies.

7.9.7 Liquid Food Preservation by UV Application

UV light can significantly lower microbial contamination in a diverse variety of liquid foods and drinks. Juices, brines, liquid sugars, process lubricants, and various semitransparent and opaque foods or ingredients fall into this category. UV light was permitted by the U.S. FDA in 2000 as a substitute for heat pasteurization of fresh juice (US FDA, 2000). The FDA hazard analysis critical control points (HACCP) has set a 5-$\log_{10}$ decrease in the target pathogen of concern as a performance requirement for fruit and vegetable juice processing (US FDA, 2000). This UV treatment can preserve a wide range of fruit juices, including orange juices (Tran et al., 2004; Keyser et al., 2008), grapes, cranberries (Guerrero-Beltrán et al., 2009), apple juice (Keyser et al., 2008; Franz et al., 2009), pomegranate juices and pineapple juice. This technology has also shown its efficiency in preserving liquid egg white and increasing the shelf life. Different doses of UV required for lowering the population of various microbial groups, such as bacteria, yeast, fungus, protozoa, and algae, on the basis of their D value as reported by Koutchma (2009), are summarized in Table 7.8. Varying soluble solids content (°Brix), pH levels, and different viscosities and liquid densities of treating liquid food should be considered while deciding the exposure dose and resistance time of UV light.

Table 7.7 Effect of UV Light Technology on the Inactivation of Microorganisms in Dairy Products

Dairy products	Target microorganism	UV treatment	Log reduction/results	Reference
Kashar cheese	Staphylococcus aureus E. coli O157:H7	Pulsed UV light System 3 pulses/s Intensity 44 J cm^{-2} Time: 5, 15, 30, 45, 60 s Lamp-sample Distances: 5, 8, and 13 cm	1.62 and 3.02 log CFU/cm^2 for S. aureus and E. coli O157:H7 respectively.	Keklik et al., 2019
Mozzarella cheese	Enterobacteriaceae Pseudomonas spp.	Lamp: Xenon (200 to 1100 nm) Frequency: 3 Hz Distance: 2.5 cm Time: 1–8 s	1-log reduction at all conditions 6 log CFU/g Pseudomonas spp. (microbiological acceptability limit) was not reached after 2 weeks	Lacivita et al., 2018
Fiordilatte cheese	Pseudomonas spp. Enterobacteriaceae	Four UV-C lamp setup Intensity: 20 W/m^2 Time: 5, 30, 60, 150, 300, 450 and 750 s Distance: 2 cm	1–2-log cycles reduction for Pseudomonas spp. and Enterobacteriaceae Sensory parameters (color, texture, surface appearance) were not influenced	Lacivita et al., 2016
Sliced cheddar cheese	E. coli O157:H7 S. typhimurium L. monocytogenes	Five UV-C lamps Intensity: 3.04 mW/cm^2 Distance: 10 cm Time: 1 min Slices were wrapped in Polypropylene (PP) and Polyethylene (PE)	3.36- and 3.12-log (similar) reductions for E. coli, S. typhimurium, and L. monocytogenes after 1 min exposure PP or PE films in combination with UV-C were effective against foodborne pathogens.	Ha et al., 2016
Gouda, manchego cheese	Listeria spp.	Pulsed light system Intensity: 0.9 to 8.4 J cm^{-2}	3-log-CFU/m^2 reduction in gouda cheese at 0.9 J/cm^2 fluences 1-log-CFU/m^2 reduction in manchego slices at 8.4 J/cm^2	Fernández et al., 2016

Food	Microorganism	Treatment conditions	Results	Reference
White cheddar, processed cheese	Pseudomonas fluorescens, E. coli ATCC 25922, Listeria innocua	Xenon pulsed lamp 3 pulses/s Distance: 5.8 cm Fluence levels: 1.02, 3.07, 6.14, 9.22, and 12.29 J/cm^2	3.37-, 3.74-, and 5.41-log-UFC/g reduction for L. innocua, P. fluorescens, and E. coli, respectively	Proulx et al., 2015
White American cheese	P. roqueforti, L. monocytogenes	Pulsed light 3 pulses/s Distance: 5, 8, and 13 cm Time: 5, 30, and 60 s	Maximum reduction at 5 cm for 40 s Reduction range: 1.32–1.24 log CFU/cm^2	Can et al., 2014
Cottage cheese	Pseudomonas spp.	Pulsed light Intensity: 16 J/cm^2 Pulse duration: 0.5 ms	1.5-log reduction Sensory quality (taste) showed no changes	Dunn et al., 1991
Set-type yoghurt	Molds	Batch UV light cabinet Intensity: 32.1 W/m^2 Time: up to 600 s	No significant changes in color and hardness observed At higher doses, oxidation levels and off-flavor also increased.	Koca et al., 2018

Table 7.8 UV Doses (mJ/cm²) for Inactivation of a Wide Range of Microorganism Groups

S. No	Microorganisms group	D-Value (mJ/cm²)*	References
1	Enteral bacteria	2–8	Koutchma (2009)
2	Cocci and micrococci	1.5–20	
3	Spore formers	4–30	
4	Enteric viruses	5–30	
5	Yeast	2.3–8	
6	Fungi	30–300	
7	Protozoa	60–120	

*The D-value is the dose required for exponential decay of the target microorganism.

7.9.8 Air Disinfection by UV Application

UV light treatment has three essential applications in food production: disinfection of air, air purification from chemical pollutants and allergens, and deodorization of storage and processing areas. A conjugation of UV light and filters has been recommended for providing clean air in food processing units (Koutchma, 2019). The major microbial groups leading to health and safety hazards in the food industry are foodborne and water-borne pathogens. However, some organisms can become airborne and potentially contaminate the ingredients, food, raw or intermediate food, or finished product during the processing and packaging. Microorganisms are killed by UV-C in the air as it passes through the irradiation field and on irradiated surfaces. Viruses and vegetative bacteria are the most vulnerable to UV inactivation, followed by Mycobacteria, bacterial spores, and finally fungus spores.

7.10 CONCLUSION AND FUTURE PROSPECTIVE

UV light processing is a low-cost and eco-friendly technique that has demonstrated great potential in the food industry. It can be an alternative nonthermal treatment for the inactivation of pathogenic, as well as spoilage, microorganisms in both solid and liquid food products, thus extending their shelf lives. Despite the numerous benefits linked with UV treatment, several disadvantages may restrict its usage in commercial food applications. Longtime UV exposure might deteriorate food quality. Optimization of UV dosage along with exposure period is a must for the efficient reduction of microbial contamination in various food products. In addition, the reactivation of microorganisms during storage should be investigated, as bacteria that are not entirely inactivated due to insufficient UV treatment can use photoreactivation to repair their damaged DNA. To determine the rate of UV disinfection, more studies on inactivation kinetics of pathogenic and spoilage microorganisms are needed, considering their interactions with surface materials, like shielding effects and their dependence on topography or structure of the surfaces.

Furthermore, innovative techniques and models for measuring food pathogen inactivation rates or dose-response behavior in extremely absorbent and viscous liquid foods, like juices and beverages must be developed. Identifying surrogate microbes or acceptable pathogen actinometers is required to develop validation procedures for food processing facilities. Due to conflicting information in the literature, the impact of UV light on various quality characteristics such as sensory qualities, vitamins, pigments, and bioactive substances is yet unknown. Moreover, the appropriate labeling can be utilized to increase customer acceptability of UV-irradiated food or ingredients. In combination with other intervention techniques and treatments (such as hurdle technology), UV technology is gaining popularity. Combining food reformulation, active packaging, and postpackaging decontamination technology to produce high-quality, safe, and healthy food items with long shelf lives can be a successful strategy. Thus, UV light is efficient for microbial inactivation and can be commercialized in the food industry as a nonthermal alternative for food preservation.

REFERENCES

Adhikari, C., & Koutchma, T. (2002). Effects of UV radiation on lipids, antioxidants, and vitamins in foods. Abstract of presentation at the AOCS Annual Meeting. In *93rd AOAC Annual Meeting. Montreal.*

Allende, A., & Artés, F. (2003). UV-C radiation as a novel technique for keeping quality of fresh processed 'LolloRosso'lettuce. *Food Research International*, 36(7), 739–746.

Avalos-Llano, K. R., Molina, R. S., & Sgroppo, S. C. (2020). UV-C treatment applied alone or combined with orange juice to improve the bioactive properties, microbiological, and sensory quality of fresh-cut strawberries. *Food and Bioprocess Technology*, 13(9), 1528–1543.

Bang, I. H., In, J., & Min, S. C. (2021). Inactivation of *Salmonella* on black peppercorns using an integrated ultraviolet-C and cold plasma intervention. *Food Control*, 119, 107498.

Beauchamp, S., & Lacroix, M. (2012). Resistance of the genome of *Escherichia coli* and *Listeria monocytogenes* to irradiation evaluated by the induction of cyclobutane pyrimidine dimers and 6–4 photoproducts using gamma and UV-C radiations. *Radiation Physics and Chemistry*, 81, 1193–1197.

Beaulieu, J. C. (2007). Effect of UV irradiation on cut cantaloupe: Terpenoids and esters. *Journal of Food Science*, 72(4), S272–S281.

Bhardwaj, S. K., Singh, H., Deep, A., Khatri, M., Bhaumik, J., Kim, K. H., & Bhardwaj, N. (2021). UVC-based photoinactivation as an efficient tool to control the transmission of coronaviruses. *Science of The Total Environment*, 148548.

Bialka, K. L., & Demirci, A. (2008). Efficacy of pulsed UV-light for the decontamination of *Escherichia coli* O157: H7 and *Salmonella* spp. on raspberries and strawberries. *Journal of Food Science*, 73(5), M201–M207.

Bintsis, T., Litopoulou-Tzanetaki, E., & Robinson, R. K. (2000). Existing and potential applications of ultraviolet light in the food industry – a critical review. *Journal of the Science of Food and Agriculture*, 80(6), 637–645.

Buonanno, M., Ponnaiya, B., Welch, D., Stanislauskas, M., Randers-Pehrson, G., Smilenov, L., . . . & Brenner, D. J. (2017). Germicidal efficacy and mammalian skin safety of 222-nm UV light. *Radiation Research*, 187(4), 493–501.

Can, F. O., Demirci, A., Puri, V. M., &Gourama, H. (2014). Decontamination of hard cheeses by pulsed UV light. *Journal of Food Protection*, 77, 1723–1731.

Cassar, J. R., Mills, E. W., Campbell, J., & Demirci, A. (2018). Pulsed UV light as a microbial reduction intervention for boneless/skinless chicken thigh meat. *Meat and Muscle Biology*, 2(2).

Chang, J. C. H., Ossoff, S. F., Lobe, D. C., Dorfman, M. H., Dumais, C. M., Qualls, R. G., & Johnson, J. D. (1985). UV inactivation of pathogenic and indicator microorganisms. *Applied and Environmental Microbiology*, 49(6), 1361–1365.

Chavez, C., Knape, K. D., Coufal, C. D., & Carey, J. B. (2002). Reduction of eggshell aerobic plate counts by ultraviolet irradiation. *Poultry Science*, 81(8), 1132–1135.

Chia, S. L., Rosnah S., Noranizan M. A., & Wan Ramli W. D. (2012). The effect of storage on the quality attributes of ultraviolet-irradiated and thermally pasteurized pineapple juices. *International Food Research Journal (Malaysia)*, 19(3), 1001–1010.

Choudhary, R., & Bandla, S. (2012). Ultraviolet pasteurization for food industry. *International Journal of Food Science and Nutrition Engineering*, 2(1), 12–15.

Cilliers, F. P., Gouws, P. A., Koutchma, T., Engelbrecht, Y., Adriaanse, C., & Swart, P. (2014). A microbiological, biochemical and sensory characterisation of bovine milk treated by heat and ultraviolet (UV) light for manufacturing Cheddar cheese. *Innovative Food Science & Emerging Technologies*, 23, 94–106.

Collins, H. F., & Selleck, R. E. (1972). *Process kinetics of wastewater chlorination (No. 72)*. University of California, Sanitary Engineering Research Laboratory.

Coufal, C. D., Chavez, C., Knape, K. D., & Carey, J. B. (2003). Evaluation of a method of ultraviolet light sanitation of broiler hatching eggs. *Poultry Science*, 82(5), 754–759.

De Reu, K., Grijspeerdt, K., Herman, L., Heyndrickx, M., Uyttendaele, M., Debevere, J., . . . & Bolder, N. M. (2006). The effect of a commercial UV disinfection system on the bacterial load of shell eggs. *Letters in Applied Microbiology*, 42(2), 144–148.

Degala, H. L., Mahapatra, A. K., Demirci, A., & Kannan, G. (2018). Evaluation of non-thermal hurdle technology for ultraviolet-light to inactivate *Escherichia coli* K12 on goat meat surfaces. *Food Control*, 90, 113–120.

Dong, Q., Manns, D. C., Feng, G., Yue, T., Churey, J. J., & Worobo, R. W. (2010). Reduction of patulin in apple cider by UV radiation. *Journal of Food Protection*, 73(1), 69–74.

Doulia, D., Katsinis, G., & Mougin, B. (2000). Prolongation of the microbial shelf life of wrapped part baked baguettes. *International Journal of Food Properties*, 3(3), 447–457.

Dunn, J. E., Clark, R. W., Asmus, J. F., Pearlman, J. S., Boyer, K., Painchaud, F., & Hofmann, G. A. (1991). U.S. Patent No. 5,034,235. Washington, DC: U.S. Patent and Trademark Office.

Eischeid, A. C., Meyer, J. N., & Linden, K. G. (2009). UV disinfection of adeno-viruses: Molecular indications of DNA damage efficiency. *Applied and Environmental Microbiology*, 75(1), 23–28.

Elmnasser, N., Guillou, S., Leroi, F., Orange, N., Bakhrouf, A., & Federighi, M. (2007). Pulsed-light system as a novel food decontamination technology: A review. *Canadian Journal of Microbiology*, 53(7), 813–821.

EPA, Office of Water. (1995). National primary drinking water regulations contaminant fact sheets inorganic chemicals – Technical version. EPA 811-F-95-002-T, Washington, D.C.

Falguera, V., Garvín, A., Garza, S. et al. (2014). Effect of UV–Vis photochemical processing on pear juices from six different varieties. *Food and Bioprocess Technology*, 7, 84–92.

Fan, X., Huang, R., & Chen, H. (2017). Application of ultraviolet C technology for surface decontamination of fresh produce. *Trends in Food Science & Technology*, 70, 9–19.

Favier, G. I., Escudero, M. E., & de GUZMÁN, A. M. (2001). Effect of chlorine, sodium chloride, trisodium phosphate, and ultraviolet radiation on the reduction of Yersinia enterocolitica and mesophilic aerobic bacteria from eggshell surface. *Journal of Food Protection*, 64(10), 1621–1623.

Fernández, M., Arias, K., & Hierro, E. (2016). Application of pulsed light to sliced cheese: Effect on *Listeria* inactivation, sensory quality and volatile profile. *Food and Bioprocess Technology*, 9(8), 1335–1344.

Fernández, M., Ganan, M., Guerra, C., & Hierro, E. (2014). Protein oxidation in processed cheese slices treated with pulsed light technology. *Food Chemistry, 159,* 388–390

Fonseca, J. M., & Rushing, J. W. (2006). Effect of ultraviolet-C light on quality and microbial population of fresh-cut watermelon. *Postharvest Biology and Technology, 40*(3), 256–261.

Forney, L. J., Ye, Z., & Koutchma, T. (2008). UV disinfection of *E. coli* between concentric cylinders: Effects of the boundary layer and a wavy wall. *Ozone: Science and Engineering, 30*(6), 405–412.

Franz, C. M., Specht, I., Cho, G. S., Graef, V., & Stahl, M. R. (2009). UV-C-inactivation of microorganisms in naturally cloudy apple juice using novel inactivation equipment based on Dean vortex technology. *Food Control, 20*(12), 1103–1107.

Garcia, L. A. T., & Barardi, C. R. M. (2019). Performance of a storage tank coupled with UV light on enteric virus inactivation in drinking water. *Water Supply, 19,* 1103–1109.

Gayán, E., Condón, S., & Álvarez, I. (2014a). Biological aspects in food preservation by ultraviolet light: A review. *Food and Bioprocess Technology, 7,* 1–20

Gayán, E., Condón, S., & Álvarez, I. (2014b). Continuous-flow UV liquid food pasteurization: Engineering aspects. *Food and Bioprocess Technology, 7*(10), 2813–2827.

Gayán, E., Serrano, M. J., Monfort, S. et al. (2013). Pasteurization of apple juice contaminated with *Escherichia coli* by a combined UV–mild temperature treatment. *Food Bioprocess Technol, 6,* 3006–3016.

Giannakis, S. (2019). A review of the concepts, recent advances, and niche applications of the (photo) Fenton process, beyond water/wastewater treatment: Surface functionalization, biomass treatment, combatting cancer and other medical uses. *Applied Catalysis B: Environmental, 248,* 309–319.

Gómez-López, V. M., Devlieghere, F., Bonduelle, V., & Debevere, J. (2005). Factors affecting the inactivation of micro-organisms by intense light pulses. *Journal of Applied Microbiology, 99*(3), 460–470.

Gómez-López, V. M., Koutchma, T., & Linden, K. (2012). Ultraviolet and pulsed light processing of fluid foods. In *Novel Thermal and Non-Thermal Technologies for Fluid Foods* (pp. 185–223). Academic Press.

Gomez-Lopez, V. M., Ragaert, P., Debevere, J., & Devlieghere, F. (2007). Pulsed light for food decontamination: A review. *Trends in Food Science & Technology, 18,* 464–473.

Green, A., Popović, V., Warriner, K., & Koutchma, T. (2020). The efficacy of UVC LEDs and low pressure mercury lamps for the reduction of *Escherichia coli* O157: H7 and *Listeria monocytogenes* on produce. *Innovative Food Science & Emerging Technologies, 64,* 102410.

Guerrero-Beltrán, J. A., & Barbosa-Cánovas, G. V. (2004). Advantages and limitations on processing foods by UV Light. *Food Science and Technology International, 10,* 137–147.

Guerrero-Beltrán, J. O. S. É. Á. N. G. E. L., Welti-Chanes, J. O. R. G. E., & Barbosa-Cánovas, G. V. (2009). Ultraviolet-C light processing of grape, cranberry and grapefruit juices to inactivate *Saccharomyces cerevisiae. Journal of Food Process Engineering, 32*(6), 916–932.

Guerrero-Beltrén, J. A., & Barbosa-Cánovas, G. V. (2006). Inactivation of *Saccharomyces cerevisiae* and polyphenoloxidase in mango nectar treated with UV light. *Journal of Food Protection, 69*(2), 362–368.

Guneser, O. N. U. R., & Yuceer, Y. K. (2012). Effect of ultraviolet light on water- and fat-soluble vitamins in cow and goat milk. *Journal of Dairy Science, 95*(11), 6230–6241.

Ha, J. W., Back, K. H., Kim, Y. H., & Kang, D. H. (2016). Efficacy of UV-C irradiation for inactivation of food-borne pathogens on sliced cheese packaged with different types and thicknesses of plastic films. *Food Microbiology, 57,* 172–177.

Hadjok, C., Mittal, G. S., & Warriner, K. (2008). Inactivation of human pathogens and spoilage bacteria on the surface and internalized within fresh produce by using a combination of ultraviolet light and hydrogen peroxide. *Journal of Applied Microbiology, 104*(4), 1014–1024.

Hamamoto, A., Mori, M., Takahashi, A., Nakano, M., Wakikawa, N., Akutagawa, M.,. . . . & Kinouchi, Y. (2007). New water disinfection system using UVA light-emitting diodes. *Journal of Applied Microbiology, 103*(6), 2291–2298.

Harm, W. (1980). *Biological Effects of Ultraviolet Radiation,* 12, 229. University Press, Cambridge.

Hernández-Hernández, H. M., Moreno-Vilet, L., & Villanueva-Rodríguez, S. J. (2019). Current status of emerging food processing technologies in Latin America: Novel non-thermal processing. *Innovative Food Science & Emerging Technologies, 58,* 102233.

Heßling, M., Hönes, K., Vatter, P., & Lingenfelder, C. (2020). Ultraviolet irradiation doses for coronavirus inactivation – review and analysis of coronavirus photoinactivation studies. *GMS Hygiene and Infection Control, 15.*

Hidaka, Y., & Kubota, K. (2006). Study on the Sterilization of Grain Surface Using UV Radiation – Development and Evaluation of UV Irradiation Equipment –. *Japan Agricultural Research Quarterly: JARQ, 40*(2), 157–161.

Hijnen, W. A. M., Beerendonk, E. F., & Medema, G. J. (2006). Inactivation credit of UV radiation for viruses, bacteria, and protozoan (oo) cysts in water: A review. *Water Research, 40,* 3–22

Hiramoto, T., 1984. Method of sterilization. US Patent 4,464,336

Jermann, C., Koutchma, T., Margas, E., Leadley, C., & Ros-Polski, V. (2015). Mapping trends in novel and emerging food processing technologies around the world. *Innovative Food Science & Emerging Technologies, 31,* 14–27.

Keklik, N. M., Demirci, A., Patterson, P. H., & Puri, V. M. (2010). Pulsed UV light inactivation of *Salmonella Enteritidis* on eggshells and its effects on egg quality. *Journal of Food Protection, 73*(8), 1408–1415.

Keklik, N. M., Elik, A., Salgin, U., Demirci, A., &Koçer, G. (2019). Inactivation of *Staphylococcus aureus* and *Escherichia coli* O157: H7 on fresh kashar cheese with pulsed ultraviolet light. *Food Science and Technology International, 25,* 680–691.

Keyser, M., Müller, I. A., Cilliers, F. P., Nel, W., & Gouws, P. A. (2008). Ultraviolet radiation as a non-thermal treatment for the inactivation of microorganisms in fruit juice. *Innovative Food Science & Emerging Technologies, 9*(3), 348–354.

Kim, D. K., Kim, S. J., & Kang, D. H. (2017). Bactericidal effect of 266 to 279 nm wavelength UVC-LEDs for inactivation of Gram positive and Gram negative foodborne pathogenic bacteria and yeasts. *Food Research International, 97,* 280–287.

Koca, N., Urgu, M., & Saatli, T. E. (2018). Ultraviolet light applications in dairy processing. *Technological Approaches for Novel Applications in Dairy Processing,* 1.

Koutchma, T. (2009). Advances in ultraviolet light technology for non-thermal processing of liquid foods. *Food and Bioprocess Technology, 2*(2), 138–155.

Koutchma, T. (2014). *Food Plant Safety: UV Applications for Food and Non-Food Surfaces.* Elsevier, Academic Press.

Koutchma, T. (2019). *Ultraviolet Light in Food Technology: Principles and Applications.* CRC Press.

Koutchma, T., Englert, R., & Adhikari, C. (2002, June). The effect of browning and suspended particles on the UV treatment of apple juice/cider. In *IFT Annual Meeting, Anaheim, CA.*

Koutchma, T., Forney, L. J., & Moraru, C. L. (Eds.). (2009). *Ultraviolet Light in Food Technology.* CRC.

Koutchma, T., Keller, S., Chirtel, S., & Parisi, B. (2004). Ultraviolet disinfection of juice products in laminar and turbulent flow reactors. *Innovative Food Science & Emerging Technologies, 5*(2), 179–189.

Koutchma, T., & Shmalts, M. (2002). Degradation of vitamin C after alternative treatments of juices. In the Institute *of Food Technologists Annual Meeting, New Orleans. Available at:* http://ift. confex. com/ift/2002/techprogram/ paper_12198.htm. Accessed February (Vol. 25, p. 2015).

Krishnamurthy, K., Demirci, A. L. I., & Irudayaraj, J. (2004). Inactivation of *Staphylococcus aureus* by pulsed UV-light sterilization. *Journal of Food Protection, 67*(5), 1027–1030.

Krishnamurthy, K., Tewari, J. C., Irudayaraj, J., & Demirci, A. (2010). Microscopic and spectroscopic evaluation of inactivation of *Staphylococcus aureus* by pulsed UV light and infrared heating. *Food and Bioprocess Technology, 3*(1), 93–104.

Kuo, F. L., Carey, J. B., & Ricke, S. C. (1997). UV irradiation of shell eggs: Effect on populations of aerobes, molds, and inoculated *Salmonella typhimurium. Journal of Food Protection, 60*(6), 639–643.

Lacivita, V., Conte, A., Lyng, J. G., Arroyo, C., Zambrini, V. A., & Del Nobile, M. A. (2018). High intensity light pulses to reduce microbial load in fresh cheese. *Journal of Dairy Research, 85*(2), 232–237.

Lacivita, V., Conte, A., Manzocco, L., Plazzotta, S., Zambrini, V. A., Del Nobile, M. A., & Nicoli, M. C. (2016). Surface UV-C light treatments to prolong the shelf-life of Fiordilatte cheese. *Innovative Food Science & Emerging Technologies, 36,* 150–155.

Lagunas-Solar, M. C., Pina, C., MacDONALD, J. D., & Bolkan, L. (2006). Development of pulsed UV light processes for surface fungal disinfection of fresh fruits. *Journal of Food Protection, 69*(2), 376–384.

Lamikanra, O., Kueneman, D., Ukuku, D., & Bett-Garber, K. L. (2005). Effect of processing under ultraviolet light on the shelf life of fresh-cut cantaloupe melon. *Journal of Food Science, 70*(9), C534-C539.

Lopez-Malo, A., & Palou, E. (2005). Ultraviolet light and food preservation. *Novel food Processing Technologies,* 405–422

Lu, G., Li, C., & Liu, P. (2011). UV inactivation of milk-related microorganisms with a novel electrodeless lamp apparatus. *European Food Research and Technology, 233*(1), 79–87.

Maharaj, R., Arul, J., & Nadeau, P. (1999). Effect of photochemical treatment in the preservation of fresh tomato (*Lycopersicon esculentum* cv. Capello) by delaying senescence. *Postharvest Biology and Technology, 15*(1), 13–23.

Mala, T., Sadiq, M. B., & Anal, A. K. (2021). Optimization of thermosonication processing of pineapple juice to improve the quality attributes during storage. *Journal of Food Measurement and Characterization,* 1–11.

Manzocco L., Quarta B., & Dri, A. 2009. Polyphenoloxidase inactivation by light exposure in model systems and apple derivatives. *Innovative Food Science and Emerging Technologies, 10,* 506–511.

Morey, A., McKee, S. R., Dickson, J. S. and Singh, M. (2010). Efficacy of ultraviolet light exposure against survival of *Listeria monocytogenes* on conveyor belts. *Foodborne Pathogens and Disease, 7*(6), 737–740.

Ngadi, M., Smith, J. P., & Cayouette, B. (2003). Kinetics of ultraviolet light inactivation of *Escherichia coli* O157: H7 in liquid foods. *Journal of the Science of Food and Agriculture, 83*(15), 1551–1555.

Nigro, F., Ippolito, A., & Lima, G. (1998). Use of UV-C light to reduce *Botrytis* storage rot of table grapes. *Postharvest Biology and Technology, 13*(3), 171–181.

Obande, M. A., Tucker, G. A. and Shama, G. 2011. Effect of preharvest UV-C treatment of tomatoes (*Solanum lycopersicon* Mill.) on ripening and pathogen resistance. *Postharvest Biology and Technology, 62*, 188–192.

Oteiza, J. M., Peltzer, M., Gannuzzi, L., & Zaritzky, N. (2005). Antimicrobial efficacy of UV radiation on *Escherichia coli* O157: H7 (EDL 933) in fruit juices of different absorptivities. *Journal of Food Protection, 68*(1), 49–58.

Páez, C. L. R., Reyes, M. C. P., Aguilar, C. H., Pacheco, F. A. D., Martínez, E. M., Orea, A. C., & Bonilla, J. L. L. (2011). Control of natural mycobiota in maize grains by ultraviolet (UVC) irradiation. *Acta Agrophysica, 18*(2 [193]).

Peleg, M., & Cole, M. B. (1998). Reinterpretation of microbial survival curves. *Critical Reviews in Food Science, 38*(5), 353–380.

Popović, V., Fairbanks, N., Pierscianowski, J., Biancaniello, M., Zhou, T., & Koutchma, T. (2018). Feasibility of 3D UV-C treatment to reduce fungal growth and mycotoxin loads on maize and wheat kernels. *Mycotoxin Research, 34*(3), 211–221.

Priyadarshini, A., Rajauria, G., O'Donnell, C. P., & Tiwari, B. K. (2019). Emerging food processing technologies and factors impacting their industrial adoption. *Critical Reviews in Food Science and Nutrition, 59*(19), 3082–3101.

Proulx, J., Hsu, L. C., Miller, B. M., Sullivan, G., Paradis, K., & Moraru, C. I. (2015). Pulsed-light inactivation of pathogenic and spoilage bacteria on cheese surface. *Journal of Dairy Science, 98*(9), 5890–5898.

Rajkovic, A., Tomasevic, I., Smigic, N., Uyttendaele, M., Radovanovic, R., & Devlieghere, F. (2010). Pulsed UV light as an intervention strategy against *Listeria monocytogenes* and *Escherichia coli* O157: H7 on the surface of a meat slicing knife. *Journal of Food Engineering, 100*(3), 446–451.

Reichel, J., Kehrenberg, C., & Krischek, C. (2019). Inactivation of *Yersinia enterocolitica* and *Brochothrix thermosphacta* on pork by UV-C irradiation. *Meat Science, 158*, 107909.

Salcedo, I., Andrade, J. A., Quiroga, J. M., & Nebot, E. (2007). Photoreactivation and dark repair in UV-treated microorganisms: Effect of temperature. *Applied and Environmental Microbiology, 73*, 1594–1600.

Sauer, A., & Moraru, C. I. (2009). Inactivation of *Escherichia coli* ATCC 25922 and *Escherichia coli* O157: H7 in apple juice and apple cider, using pulsed light treatment. *Journal of Food Protection, 72*(5), 937–944.

Severin, B. F., Suidan, M. T., & Engelbrecht, R. S. (1983). Kinetic modeling of UV disinfection of water. *Water Research, 17*(11), 1669–1678.

Severin, B. F., Suidan, M. T., Rittmann, B. E., & Engelbrecht, R. S. (1984). Inactivation kinetics in a flow-through UV reactor. *Journal (Water Pollution Control Federation)*, 164–169.

Sharma, R. R., & Demirci, A. (2003). Inactivation of *Escherichia coli* O157: H7 on inoculated alfalfa seeds with pulsed ultraviolet light and response surface modeling. *Journal of Food Science, 68*(4), 1448–1453.

Singh, J. P., & Ghaly, A. E. (2007). Effect of flow characteristics on online sterilization of cheese whey in UV reactors. *Applied Biochemistry and Biotechnology, 142*(1), 1–16.

Sommers, C. H., Geveke, D., Pulsfus, S. and Lemmenes, B. 2009. Inactivation of *Listeria innocua* on Frankfurters by ultraviolet light and flash pasteurization. *Journal of Food Science* 74: 3.

Song, K., Mohseni, M., & Taghipour, F. (2016). Application of ultraviolet light-emitting diodes (UV-LEDs) for water disinfection: A review. *Water Research, 94*, 341–349.

Spikes, J. D. (1981). Photodegradation of foods and beverages. *Photochemical and Photobiological Reviews*, 39–85

Takeshita, K., Shibato, J., Sameshima, T., Fukunaga, S., Isobe, S., Arihara, K., & Itoh, M. (2003). Damage of yeast cells induced by pulsed light irradiation. *International Journal of Food Microbiology, 85*(1–2), 151–158.

Tarek, A. R., Rasco, B. A. and Sablani, S. S. 2015. Ultraviolet-C light inactivation kinetics of *E. coli* on Bologna beef packaged in plastic films. *Food and Bioprocess Technology, 8*, 1267–1280.

Torkamani, A. E., & Niakousari, M. (2011). Impact of UV-C light on orange juice quality and shelf life. *International Food Research Journal, 18*(4), 1265–1268.

Tran, M. T. T., & Farid, M. (2004). Ultraviolet treatment of orange juice. *Innovative Food Science & Emerging Technologies, 5*(4), 495–502.

Unluturk, S., & Atılgan, M. R. (2015). Microbial safety and shelf life of UV-C treated freshly squeezed white grape juice. *Journal of Food Science, 80*, 1831–1841.

Unluturk, S., Atılgan, M. R., Baysal, A. H., & Tarı, C. (2008). Use of UV-C radiation as a non-thermal process for liquid egg products (LEP). *Journal of Food Engineering, 85*(4), 561–568.

US FDA. 21CFR179.39 Part 179 – Irradiation In The Production, Processing And Handling Of Food. Subpart B – Radiation and Radiation Sources. Sec. 179.39 Ultraviolet radiation for the processing and treatment of food. Revised April 1, 2018.

US FDA. 2000. 21 CFR Part 179. Irradiation in the Production, Processing and Handling of Food. Fed. Regist. 65, 71056–71058.

Vanga, S. K., Singh, A., & Raghavan, V. (2017). Review of conventional and novel food processing methods on food allergens. *Critical Reviews in Food Science and Nutrition, 57*(10), 2077–2094.

Vollmer, K., Czerny, M., Vásquez-Caicedo, A. L., Iglesias, S. V., Frank, J., Carle, R., & Steingass, C. B. (2021). Non-thermal processing of pineapple (*Ananas comosus* [L.] Merr.) juice using continuous pressure change technology (PCT): HS-SPME-GC – MS profiling, descriptive sensory analysis, and consumer acceptance. *Food Chemistry, 345*, 128786.

Wambura, P., & Verghese, M. (2011). Effect of pulsed ultraviolet light on quality of sliced ham. *LWT-Food Science and Technology, 44*(10), 2173–2179.

Wassmann, M., Moeller, R., Reitz, G., & Rettberg, P. (2011). Growth phase dependent UVC resistance of *Bacillus subtilis*: Data from a short-term evolution experiment. *Archives of Microbiology, 193*, 823–832.

Worobo, R. (1999). Efficacy of the Cider Sure 3500. Ultraviolet light unit in apple cider. *CFSAN Apple cider food safety control workshop*.

Wright, J. R., Sumner, S. S., Hackney, C. R., Pierson, M. D., & Zoecklein, B. W. (2000). Efficacy of ultraviolet light for reducing *Escherichia coli* O157: H7 in unpasteurized apple cider. *Journal of Food Protection, 63*, 563–567

Yaun, B. R., Sumner, S. S., Eifert, J. D., & Marcy, J. E. (2004). Inhibition of pathogens on fresh produce by ultraviolet energy. *International Journal of Food Microbiology, 90*(1), 1–8.

Ye, Z. (2007). *UV Disinfection Between Concentric Cylinders*. PhD thesis, School of Chemical & Biomolecular Engineering, Georgia Institute of Technology, Atlanta.

Ye, Z., Forney, L. J., Koutchma, T., Giorges, A. T., & Pierson, J. A. (2008). Optimum UV disinfection between concentric cylinders. *Industrial & Engineering Chemistry Research, 47*(10), 3444–3452.

Ye, Z., Koutchma, T., Parisi, B., Larkin, J., & Forney, L. J. (2007). Ultraviolet inactivation kinetics of *Escherichia coli* and *Yersinia pseudotuberculosis* in annular reactors. *Journal of Food Science, 72*(5), E271–E278.

Yeh, Y., De Moura, F. H., Van Den Broek, K., & De Mello, A. S. (2018). Effect of ultraviolet light, organic acids, and bacteriophage on *Salmonella* populations in ground beef. *Meat Science, 139*, 44–48.

Yin, F., Zhu, Y., Koutchma, T., & Gong, J. (2015). Inactivation and potential reactivation of pathogenic *Escherichia coli* O157: H7 in bovine milk exposed to three monochromatic ultraviolet UVC lights. *Food Microbiology, 49*, 74–81.

8 Preservation of Food by Gamma-Irradiation Technology

Poulami Sarkar, Dipshikha Tamili, Subhasis Bhattacharya,
and Paramita Bhattacharjee Chakraborti

Corresponding author:
Prof. Paramita Bhattacharjee Chakraborti; Email:
paramita.bhattacharjee@jadavpuruniversity.in

CONTENTS

LIST OF ABBREVIATIONS

AP	Aerobic packaging
Ca-Cas	Calcium caseinate
CAP	Calcium propionate
CAT	Catalase
DM	Demineralized
DPPH	2,2-diphenyl-1-picrylhydrazyl
EVOH – PA	Ethylene vinyl alcohol-polyamide copolymer
FAO	Food and Agricultural Organization
FFA	Free fatty acid
FRAP	Ferric reducing antioxidant power
GMP	Good Manufacturing Practices
GSE	Grape seed extract
IAEA	International Atomic Energy Agency
ISO	International Organization for Standardization
LDPE	Low-density polyethylene

DOI: 10.1201/9781003147978-8

LOX	Lipoxygenase
MAP	Modified atmosphere packaging
MDA	Malonaldehyde
MLP	Moringa leaf powder
MNV-1	Murine Norovirus-1
MUFA	Monounsaturated fatty acid
NADCC	Sodium dichloro-striazinetrione
NE	Nano-emulsion
PAL	Phenylalanine ammonia lyase
PE	Polyethylene
PET	Polyethylene terephthalate
PP	Polypropylene
PUFA	Polyunsaturated fatty acid
RH	Relative humidity
RTE	Ready-to-eat
SDS-PAGE	Sodium dodecyl sulphate – polyacrylamide gel electrophoresis
SOD	Superoxide dismutase
TBA	Thiobarbiturate acid
TBARS	Thiobarbiturate acid reactive substances
TFC	Total flavonoid content
TPC	Total phenolic content
TSS	Total soluble solids
VP	Vacuum packaging
WHO	World Health Organization

8.1 INTRODUCTION

Currently, low-energy consuming food preservation methods such as radiation processing are increasingly preferred owing to escalating global energy crisis. The commonly used conventional food preservation technologies such as fumigation, curing and chemical preservation cause deteriorations in food quality, besides posing hazards to human health and the environment (Van Kooij et al., 1981). To produce safe foods of good quality and reduce postharvest losses, the employment of radiation processing as a food preservation technology is now widely accepted. Compared to the conventional methods of food preservation, especially thermal processing methods, irradiation is energy-conserving and equally effective for extending the shelf lives of agro commodities and food products (Rahman, 2007; da Silva Aquino, 2012).

Radiation processing or irradiation of food involves exposure of food to short energy waves such as gamma rays, electron beam and X-rays to accomplish objectives such as the disinfestation of insects, the eradication of foodborne parasites and pathogens from foods and the enhancement of the shelf lives of food products. Radiations are preferred because they produce desired effects in foods, do not induce radioactivity in them (or in packaging materials of packaged food products) and are available at costs that allow commercial use of the process (da Silva Aquino, 2012). In the case of food processing by ionizing radiation, the energy density per atomic transition is very high, which causes cleavage of molecules by inducing ionization. The use of ionizing radiation on solid foods could thus mimic the same effect as provided by heat pasteurization of liquid foods such as milk and fruit juices, and therefore, this nonthermal processing technology can considerably reduce heat processing or replace completely the use of food additives and fumigants worldwide (Van Kooij et al., 1981).

Codex allows three types of radiation sources for foods: gamma-irradiation from radionuclides such as ^{60}Co and ^{137}Cs, x-rays generated from machines operated at or below an energy level of 5 MeV and high energy electron beam or beta (β) particles, where electrons are generated from machine sources operated at or below an energy level of 10 MeV (Joint FAO/WHO Codex Alimentarius Commission, 1984; WHO, 1988). There are fundamental differences among the three types of irradiations in terms of their energy profiles, mode of production, respective shielding requirements and regulations associated with the usage of these on foods (Pillai and Shayanfar, 2017). Gamma-irradiation technology offers more advantages as a food preservation method over other radiation processing technologies owing to the high penetrability of gamma-rays into food samples, its high efficiency in terms of energy utilization and the low throughput of gamma rays into food samples (Rahman, 2007). This chapter discusses the fundamentals of gamma irradiation of foods, the mechanism of action of gamma radiation in foods, the effects of gamma processing on food products, the essentials of a commercial gamma radiation food processing facility, and commercial uses of gamma irradiation in food processing and preservation. The limitations of gamma-irradiation technology in food processing and its future perspectives are also discussed.

8.2 FUNDAMENTALS AND PRINCIPLES OF GAMMA IRRADIATION

Irradiation is a process by which an object is exposed to radiation. Irradiation generally refers to ionization radiation (Anonymous, 2021a), which can cause ionization and/or excitation of atoms in the incident material but owing to limited energy is incapable of interacting with the nuclei to induce radioactivity in the target (Grandison, 2006). Gamma rays are part of the electromagnetic spectrum, emit from radionuclides in a discontinuous spectrum of radiation intensities and have fixed energies. The two radioisotopes, ^{60}Co and ^{137}Cs, are primarily used commercially as sources of gamma irradiation because of their long half-lives and high energy of gamma rays (IAEA, 2006). ^{60}Co is more commonly used both for industrial and medical purposes than ^{137}Cs. The former is a synthetic radioactive isotope produced artificially in nuclear reactors by neutron bombardment of ^{59}Co ($^{59}_{27}$Co + $^{1}_{0n}$ → $^{60}_{27}$Co) and has a half-life of 5.27 years (da Silva Aquino, 2012; Anonymous, 2021b). A nonradioactive, stable Ni isotope is obtained post-disintegration of ^{60}Co through the emission of a β particle having an energy of 0.313 MeV, which is immediately followed by successive emission of two gamma-ray photons of energies 1.17 MeV and 1.33 MeV to enable the Ni isotope in an excited state to return to its stable state (Anonymous, 2021b; IAEA, 2006). The overall activation and decay reaction of ^{59}Co is as follows:

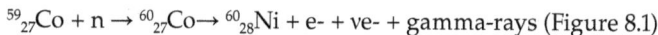

$$^{59}_{27}Co + n \rightarrow {}^{60}_{27}Co \rightarrow {}^{60}_{28}Ni + e- + ve- + \text{gamma-rays (Figure 8.1)}$$

The strength of ^{60}Co decreases with the decay of every ^{60}Co atom following first order reaction kinetics (Dasgupta, 2017), represented by Equation 8.1:

$$I = I_0 e- \lambda^x \qquad (8.1)$$

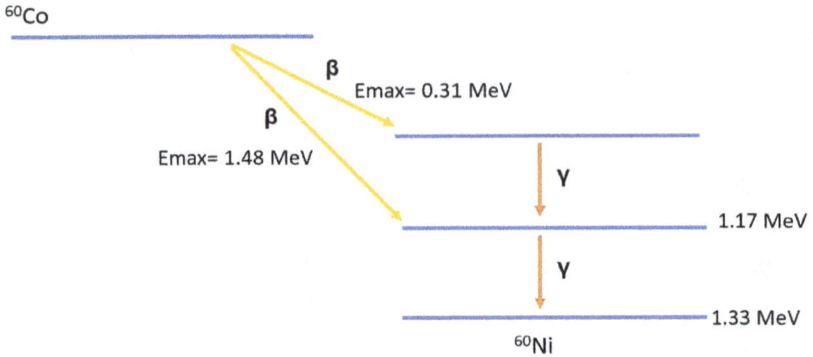

Figure 8.1 Disintegration of cobalt-60.

where I = intensity of incident beam, I_0 = intensity of emergent beam, λ = linear attenuation coefficient (decay rate), and x = thickness of the material. The half-life of a radionuclide (Dasgupta, 2017) follows Equation 8.2:

$$t_{1/2} = 0.693/\lambda \qquad (8.2)$$

where λ = decay rate (yr^{-1}) ($t_{1/2}$ = 5.27 years for ^{60}Co; $t_{1/2}$ = 30.1 years for ^{137}Cs).

8.3 RADIATION UNITS AND DOSES

The concepts related to radiation quantities, their radiation physics, dosimetry, and radiological protection have been reviewed and updated by International Commission on Radiation Units and Measurements (ICRU; IARP, 2016). Table 8.1 summarizes all the units, their definitions and their inter-conversions. Food irradiation normally requires very low doses, namely, 0.1 or 1.0 kGy, to achieve technological purposes. The energy level required for the said low-dose range is equivalent to a heat energy of 0.024°C for 0.1 and 0.24°C for 1.0 kGy. According to the Codex Alimentarius Commission, 10 kGy is the maximum dose for irradiation of agro commodities and food products to render them safe and healthy for human consumption. 10 kGy of ionization energy would be equivalent to 10 J/g of heat energy (Rahman, 2007).

8.4 MECHANISM OF ACTION OF GAMMA RADIATIONS IN FOOD

Ionization radiation–induced effects in a food material are dependent on physical parameters such as dose rate, dose distribution and radiation quality, and physiological parameters of food products such as moisture content, temperature and oxygen concentration, but are primarily associated with biochemical changes in the food (da Silva Aquino, 2012), consequent to ionization events such as ion-pair production, dissociation or electron-releasing generation of hydrogen and excitation of atoms. Irradiation causes significant changes in atomic-cum-molecular structures of biomolecules present in a food. When high-energy ionizing radiation passes through a food, its energy will be absorbed by the food and/or by pests or pathogens present in the food through the process of ionization and/or excitation of electrons. Two theories have been propounded considering the biological effects of gamma radiation on foods: direct effect and indirect effect (Desrosier, 2012).

Table 8.1 Units of Radiation Quantities

Quantity	Definition	Old unit	SI unit	Relationship between units
Radioactivity (A)	The average number of spontaneous nuclear transformation occurring per unit time	Ci	Bq	$1\ Bq = 2.7 \times 10^{-11}$ Ci
Exposure (X)	The quantity of radiation that produces in air 1 coulomb of charge per kg of air	R	$C\ kg^{-1}$	$C\ kg^{-1} = 3876\ R$
Air Kerma (K)	The sum of the initial kinetic energies of all charged particles liberated by radiation in a material of mass 1 kg	–	Gy	1 Gy (Air kerma) = 114 R
Dose (D)	The amount of energy absorbed per unit mass of medium from any kind of ionizing radiation at the point of interest	rad	Gy	1 Gy = 100 rad
Equivalent dose (H_T)	The weighted absorbed dose in any tissue or organ due to radiation	rem	Sievert (Sv)	1 Sv = 100 rem

Reference: IARP, 2016.

8.4.1 Direct Effect

Biologists have proposed the target theory of the direct impact of radiation with the substrate (related to the bioburden of the food product) conferring it to be principally responsible for irradiation effects on food (Desrosier, 2012). This theory was arrived at based on exhaustive investigations of biological changes, namely, cell multiplication, mutation and lethal effects, to name a few. It proposes that if a swiftly moving charged particle hits a molecular complex of a biological material, the function of the material will be irreversibly altered or destroyed (Desrosier, 2012).

8.4.2 Indirect Effect

The indirect effect of irradiation is on the moist tissues of food material caused by the ionization of the water molecules present in the food, transforming them into 'activated water' and resulting in the concomitant formation of highly reactive hydrogen and hydroxyl radicals by free radical mechanism (illustrated in Figure 8.2). H^+ and OH- radicals are chemically very reactive and act as oxidizing or reducing agents and may cleave C–C bonds. Secondary products of irradiation may be of equal importance, since in presence of dissolved oxygen, a hydrogen atom (H) combines with molecular oxygen (O_2) to form very reactive peroxide radicals ($O_2H\ ^\bullet$), which form hydrogen peroxide. The hydroxyl radicals independently can also produce hydrogen peroxide (Desrosier, 2012). These radicals contribute substantially to the biological effects of ionizing irradiation.

In food products, both these effects can occur simultaneously affecting insects, pathogens, and microbes (such as bacteria, yeast and molds). Since living tissues have more than 75% water, a majority of the energy is utilized in forming activated water, and only a relatively small amount of energy directly affects the biomolecules in the food material (IARP, 2016). Thus, the indirect effect would profoundly affect the biochemical composition of foods and significantly alter the food quality.

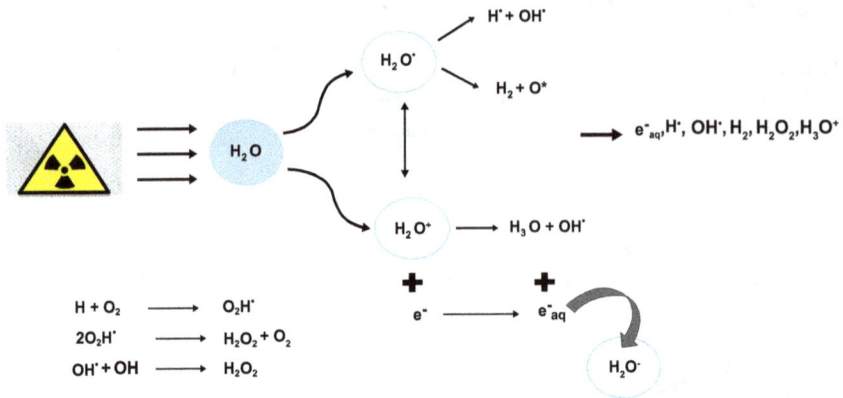

Figure 8.2 Effect of gamma irradiation on water molecules.

Table 8.2 Typical Energy Values Used for Various Food-Processing Technologies

Food processing	Energy value (kJ/kg)
Radpasteurization (2.5 kGy)	21
Radpasteurization (30 kGy)	157
Heat appertization (heat sterilization)	918
Freezing of chicken in a blast freezer from 4.4 to 23.3°C	7552
Cold storage of the above product at −25°C for 24.5 days	5149
Refrigerated storage of the above product at 0°C for 5.5 days	318
Refrigerated storage of the above product at 0°C for 10.5 days	396
Cooking at 93°C, post-thawing of whole chicken	2558

Reference: Chakraborty, 2010

8.4.3 Energy Utilized in Gamma Irradiation

The energy utilized for gamma irradiation of foods is low vis-à-vis conventional thermal (canning, retorting) and nonthermal (freezing and cold storage, supercritical carbon dioxide processing, pulsed electric field processing, high-pressure processing, oscillating magnetic field technology) food-processing technologies. Table 8.2 lists energy consumption in various food processing operations (Chakraborty, 2010). There are two types of industrial gamma processing of foods. In radpasteurization (pasteurization by radiation), 2.5 kJ/kg of energy is absorbed by the food products, but the product temperature rises less than 1°C. Radappertization (sterilization by radiation), on the other hand, corresponds to absorbing much greater energy of about 30 kJ/kg. Considering the commercial operation of gamma-irradiation plants (Section 8.6), irradiation efficiency is typically considered to be 35%; that is, the food product absorbs 35% of the radiation energy emitted by the source. The food irradiation processing (Section 8.5) time generally utilizes 68% of plant running time, resulting in an overall plant efficiency of 24%. Thus, radiation energy of 10 kJ/kg is used for radpasteurization of foods at a 2.5 kGy dose of gamma radiation. The energy for conveyors, air blowers and mechanical systems contributes to an additional energy consumption of 11 kJ/kg. Thus, 2.5-kGy radpasteurization would entail an energy consumption of 21 kJ/kg, while the same for 30-kGy radappertization would be around 160 kJ/kg (Chakraborty, 2010).

8.4.4 Effects of Gamma Irradiation on Food Materials

The effects of gamma irradiation on food materials are as follows (Willemoti, 1996; Rahman, 2007; Erkmen and Bozoglu, 2016; IARP, 2016):

- Nucleic acids are the principal targets of the free radicals generated during gamma irradiation. DNA lesions cause irreparable DNA damage in cells and non-lethal DNA mutations that occur could be passed on by subsequent cell divisions.

- Ionization radiation also cleaves chromosomes, which leads to a rearrangement of chromosome structure and number and abnormality in chromosome replication, broadly known as chromosomal aberrations. Alterations in the DNA, that is, gene mutation, may also occur.

- Membrane lipids are the second target of gamma radiation after DNA. Alterations in membrane lipids, particularly in polyunsaturated fatty acids (PUFA)–rich lipids have deleterious consequences on various membrane functions, chiefly permeability.

- In the case of proteins, principally the activities of membrane enzymes are affected.

- Cell walls of bacteria and fungi may be destroyed, leading to their cell injuries.

- All the previously mentioned changes not only affect cell division but can also cause cell death (apoptotic in microbes) and/or cell cycle arrest (in vegetables and fruits).

- Radiation can kill or reduce total microbial counts of bacteria and yeasts, delay mold growth, sterilize insect eggs or kill parasites, insects and larvae in food products.

8.4.5 Advantages of Gamma-Irradiation Technology in Food Processing

The following are distinct advantages of gamma irradiation in food processing and preservation (Bose, 2000; Chakraborty, 2010; Bhattacharjee and Singhal, 2009; Desrosier, 2012):

- Exposure of foods to gamma rays produces effects similar to those by pasteurization, cooking and other forms of thermal processing with fewer detrimental effects on nutritional value and sensory attributes such as flavor, texture and appearance of food.

- Gamma-irradiation technology has proved to be equally efficient in eliminating foodborne disease-causing pathogens such as *Salmonella sp.*, *Campylobacter sp.*, *E. coli* and viruses, as thermal processing methods are.

- Owing to the high degree of homogeneous penetration of gamma rays into food materials, completely packaged food products can be subjected to radiation treatment.

- Gamma radiation does not cause a rise in temperature of food products during processing and is therefore commonly referred to as 'cold sterilization' (of late, this term is less used since sterilization and radiation processing are fundamentally different). Thus, gamma irradiation can safely be applied to agricultural commodities since there would be minimal deleterious changes in the sensory attributes of foods vis-à-vis those caused by thermal processing. Gamma irradiation can better preserve fresh-like character of agro commodities and food products.

- Radiation processing of foods is not influenced by environmental factors such as the humidity of the surrounding atmosphere and temperature.

- Gamma radiation cannot induce radioactivity in foods, and therefore, gamma-irradiated foods are free from any harmful or toxic radioactive residues, unlike foods treated with chemical fumigants. Gamma-ray-processed products are therefore ready for immediate consumption or use post-irradiation.

- The radiation processing facilities are completely safe for both workers and the surrounding environment.

- Gamma irradiation could be a successful alternative to hazardous pesticides, fumigants and preservatives.

Gamma radiation processing of foods embraces the basic principles of green chemistry and thus can be categorized as a 'green technology of food preservation'. The main applications of irradiation are extension of shelf lives of agro and food products, improvements in their nutritional quality and disinfestation of microbial-cum-insect populations in the same. Different doses of gamma-irradiation are required to achieve different purposes. The doses permitted vary according to the type of food products, and the radiation effects desired in the same. In general, low doses of irradiation (up to 1 kGy) are designed to control the proliferation of insects in grains, inhibit sprouting in white potatoes, kill or prevent the maturation of *Trichinella* sp. in pork, prevent decay and control and/or kill insects in fruits and vegetables (Rahman, 2007). Medium doses in the range of 1–10 kGy are designed to control *Salmonella*, *Shigella*, *Campylobacter* and *Yersinia* (causal organisms of food-borne illnesses and spoilage) counts in meat, poultry and fish and to delay mold growth on strawberries and other fruits. Generally, high doses (greater than 10 kGy) are designed to kill microorganisms and insects in spices (DAE, 2014) and to commercially sterilize foods with the aim to completely destroy all microorganisms of public health concern. Although doses above 10 kGy are commercially used to achieve sterilization to the same extent (DAE, 2014) as thermal sterilization (such as canning), these doses have not been approved by the FDA for use in the United States or elsewhere, with the exception of some special hospital diets for immune-deficient patients in the United States (Narvaiz, 2015).

Table 8.3 summarizes research findings on gamma irradiation of food products highlighting the physicochemical and biochemical parameters investigated in food products and the effects of radiation on the said parameters. From the review of the literature (published from 2016 to present), the following findings are summarized: doses as low as 0.25 kGy are needed to reduce fungal populations in raw agro commodities and as high as 40 kGy for complete sterilization of RTE food products; significant alterations in proximate, physicochemical and sensory properties occur in the dose ranges of 0.12–0.2 kGy in roots and tubers, 0.04–7 kGy in fruits and vegetables, 2.5–13 kGy in cereals and grains, 0.5–6 kGy in nuts, 0.5–10 kGy in pulses, 0.5–15 kGy in spices and herbs, 0.5–20 kGy in oils, 0.2–2 kGy in mushrooms, 0.75–3 kGy in fish and fish products and 1–40 kGy in raw to processed meat products. Considerable work has also been conducted by various researchers on gamma-radiation-induced alterations in bioactive compounds in foods that have shown promising effects, the most significant of these being the enhancement in the biosynthesis of bioactives in their respective produce when irradiated in the dose range of 0.04–10 kGy. Other major benefits of radiation processing of foods include the phytosanitization of fruits and vegetables, a reduction in cooking time, an improvement of malting properties of grains and the rehydration characteristics of pulses and dried vegetables (Mostafavi et al., 2012).

Table 8.4 summarizes the effects of combination treatments of gamma radiation with other food-processing technologies, such as sample treatments with

Table 8.3 Studies on Gamma-Irradiation Treatment of Food Products

Gamma-irradiation-induced alterations in food microbiology

S. No	Food item	Irradiation dose range	Parameters investigated	Effects	Reference
1	Mushroom (*Volvariella volvacea*)	0.2–1 kGy	Microbial load during storage	Decrease in microbial populations at 0.8 kGy compared to the control samples when stored at 16 °C, 55–60% RH	Hou et al. (2018)
2	Ponkan mandarin (*Citrus reticulata* Blanco) fruit	0.25–1 kGy	Fungal (*Penicillium digitatum*) growth during storage	Marked inhibition of *P. digitatum* with increase in dose, causing compromised growth of mycelium, intracellular content leakage because of breakdown of membrane integrity; inhibition of enzyme activities of *P. digitatum* at 0.5 kGy while kept at room temperature (8–11 °C)	Zhao et al. (2020)
3	Quince fruit (*Cydonia oblonga* Mill)	0.3–2.1 kGy	Inhibition of fungal (*Penicillium expansium*) decay during storage	1.8 and 2.1 kGy irradiation doses lowered yeast and mold (*P. expansium*) counts below their detection levels up to 15 days under ambient storage conditions (15 ± 2°C and 85% RH)	Hussain et al. (2019)
4	Pistachios (*Pistacia vera* L.)	0.5–5 kGy	Inactivation of *Escherichia coli* O157:H7, *Salmonella typhimurium*, and *Listeria monocytogenes*	Decrease in the three pathogen populations (below their respective detection limits) in 5 kGy-irradiated pistachios without disagreeable color (Lab) changes; *L. monocytogenes* exerts more gamma-irradiation resistance compared to *E. coli* O157:H7 and *S. typhimurium*	Song et al. (2018)
5	Kimchi seasoning mixture	1–10 kGy	Microbiological safety against *Escherichia coli* O157:H7 and *Salmonella typhimurium*	Decrease in microbial population proportional to radiation dose; decrease in growth rate of aerobic bacteria by 3.29 log CFU/g at 10 kGy; elimination of lactic acid bacteria and foodborne pathogens when irradiated above 5 kGy	Jeong et al. (2020)

(*Continued*)

Table 8.3 (Continued)

S. No	Food item	Irradiation dose range	Parameters investigated	Effects	Reference
6	Beef	2–6 kGy	Microbial changes	Significant reduction in total viable count, total coliform count, total yeast, and mold counts with increasing doses from 2 to 6 kGy	Haque et al. (2017)
7	Parsley leaves (Petroselinum crispum [Mill.] Fuss var. neapolitanum)	0.7–2.7 kGy	Microbial (bacteria, yeast, and mold) load	Decrease in microbial load proportional to irradiation dose; count reduction by 5 log CFU g^{-1} when irradiated at 2.7 kGy (destruction of molds better than yeast)	Cătunescu et al. (2019)
8	Isabel Precoce (Vitis labrusca), Bordô (Vitis labrusca) and BRS Violeta (hybrid of Vitis labrusca) grape juice	1–2 kGy	Microbial (coliforms 35 °C and 45 °C, enterobacteria, filamentous fungi, and yeast) population during storage	Reduction of microflora in 2 kGy gamma-irradiated grape juice blends stored at room temperature (21 ± 3 °C) with concomitant preservation of nutritional and sensory characteristics	Mesquita et al. (2020)
9	Tilapia (Oreochromis sp.) fillets	1 and 3 kGy	Microbial load during storage	Complete elimination of coliform bacteria at 3 kGy and 1 kGy; longest acceptability (70 days) of 3 kGy-irradiated fillets (according to the criterion for acceptable microbial counts when stored in ice)	Surendra et al. (2018)
10	Pomegranate (Punica granatum L.) arils cv. 'Malas Saveh'	1–5 kGy	Microbial (bacteria, molds, and yeast) population	Significant reduction in the population of bacteria, fungi, and yeasts at all said doses of gamma-irradiation; the lowest microbial population at 5 kGy	Ashtari et al. (2019)
11	Date fruits (Phoenix dactylifera L.)	1–5 kGy	Microbial load during storage	Maximum reduction of microbial population in 5kGy-irradiated samples when stored at 25 ± 2 °C for 4 months	Zarbakhsh and Rastegar (2019)
12	Dairy-like product (Kemi block)	1–10 kGy	Microbiological properties	Decrease in microbial load with increasing dose of gamma-irradiation with significant reduction at 3, 5, and 20 kGy doses	Odueke et al. (2018)

No.	Food	Dose	Parameter studied	Findings	Reference
13	Salt-reduced hot dog wieners	1.5–4.5 kGy	Microbiological (lactic acid bacteria, anaerobic psychrotrophic bacteria, thermotolerant coliforms, coagulase-positive *Staphylococci*, sulfite-reducing *Clostridia*, *Salmonella*) safety	Reduction in microbial development and absence of negative effect of lipid oxidation at 1.5 kGy; greater reduction with higher doses of 3.0 and 4.5kGy	Rodrigues et al. (2021)
14	Brown rice-based weaning food	2–10 kGy	Microbiological properties during storage	Reduction in the microbial count with increasing irradiation dose when kept under ambient conditions (32 ± 2 °C) for 12 months	Jan et al. (2020)
15	Peanut (*Arachis hypogaea* L) kernels	2.5–10 kGy	Microbial population (bacteria, yeast, mold, E. coli, and *Salmonella*) during storage	Complete elimination of total plate counts as well as yeast and mold count at 7.5 kGy; complete elimination of E. coli at 5.0 kGy	Gojiya et al. (2020)
16	Sesame (*Sesamum indicum* L.) seeds	3–9 kGy	Microbial properties (bacteria and fungi) during storage	Complete elimination of fungi at 33 kGy; complete elimination of bacteria at 9 kGy even after 12 months of storage at room temperature (18–25 °C, 50–70% RH)	Al-Bachir (2016)
17	*Gwamegi* (half-dried Pacific herring or Pacific saury) and semi-dried squid	3–10 kGy	Inactivation of MNV-1 (murine norovirus-1)	Dose of ≥7 kGy effectively reduced MNV-1 titers by >90% in half-dried seafood products without any concomitant changes in color (Lab) and sensory properties when stored at 10 °C	Kang et al. (2016_
18	Yellow noodle with addition of rice bran	4–8 kGy	Microbiological quality during storage	Irradiated noodles were microbiologically safe to consume even after day 7 when stored at 8 °C	Rahim et al. (2021)
19	Fresh pasta	5–13 kGy	Microbial population (total aerobic mesophilic, psychrotrophic bacteria, molds and yeasts) during storage	No growth at 13 kGy when stored at 7, 15, and 25 °C and at 10 kGy when stored under refrigerated conditions at 7 °C for87 days	Cassares et al. (2020)
20	Soybean and soybean oil	10–30 kGy	Fungal load	Complete elimination of fungi at 10-, 20-, 30-kGy doses of irradiation when incubated at 28 °C for 60 days	Zhang et al. (2018)
21	Ready-to-eat (RTE) stir-fried chicken dices with hot chili	10–40 kGy	Microbial (total aerobic bacteria, *Bacillus* spp. yeasts, molds, coagulase positive *Staphylococci*, *Escherichia coli*, *Salmonella*) population during storage	Complete sterilization at 10 and 20 kGy doses when stored at 25 °C for one year	Chen et al. (2016)

(Continued)

Table 8.3 (Continued)

Gamma-irradiation-induced alterations in proximate, physicochemical and sensory properties

S. No	Food item	Irradiation dose range	Parameters investigated	Effects	Reference
			Roots, tubers and related products		
1	Onion (*Allium cepa* L.)	0.01–0.2 kGy	Physiological, physicochemical, and biochemical analyses during shelf-life	Inhibition of loss in weight, rotting, and sprouting; retention of desirable total soluble solids (TSS), ascorbic acid, and pyruvic acid contents; acceptable texture and no major changes in color (L, a, b and ΔE) up to 84 days of storage at ambient conditions (24 °C–34 °C, 40–75% RH) post-curing of harvested bulbs and gamma-irradiation at 0.12 kGy	Sharma et al. (2020)
2	Onion (*Allium cepa* L.) bulbs	0.03–0.15 kGy	Dry matter and protein contents; peroxidase (POD) and phenylalanine ammonia-lyase (PAL) activities; physiological characteristics; expression of PAL gene	Increase in POD activity; decrease in protein content and PAL activity; dry matter content was found to be genotype-dependent when stored at 10–15 °C, 70% RH for 4 months; expression of PAL gene did not change with irradiation	Miladilari et al. (2019)
3	Potato (*Solanum tuberosum*), variety *Kufri Jyoti*	0.1 and 0.2 kGy	Physical parameters (weight, specific gravity, texture, and color in terms of ΔL, Δa, Δb) and sprout inhibition during storage	Prevention of sprouting and no change in physical parameters at 0.1-kGy irradiation after 30 days of harvest when stored at 15 °C for five months; unfavorable changes (skin darkening, black spot, shriveling) observed at 0.2 kGy	Sarkar and Mahato (2020)
4	Green plantains (*Musa paradisiaca*)	0.04–1 kGy	Sensory, physicochemical, biochemical parameters and extension of shelf-life	Shelf-life of 0.04 kGy gamma-irradiated green plantains were extended by 22 days with a lead of 8 days when stored at 23 ± 2 °C, 70% R.H	Sarkar et al. (2021)

Fruits, vegetables and related products

No.	Product	Dose	Quality parameters	Findings	Reference
5	Guava (*Psidium guajava* L.)	0.1–0.3 kGy	Shelf-life and post-harvest quality	Increase in post-harvest shelf-life (93.8%); no adverse effect in fruit quality (firmness, % titratable acidity against NaOH, TSS, and vitamin C contents) and sensory quality attributes (appearance, taste, texture, and flavor) at 0.2 kGy when stored under ambient conditions (25 ± 2 °C, 85 ± 4% RH)	Sau et al. (2018)
6	Ponkan mandarin (*Citrus reticulata* Blanco) fruit	0.25–1 kGy	Physiological changes; antioxidant and enzyme activities	Decrease in rate of fruit decay, nutrient and weight losses; reduction in malondialdehyde production; improvements in activities of superoxide dismutase (SOD), catalase (CAT), and POD at 0.5 kGy when stored at room temperature (8–11 °C) for 26 days	Zhao et al. (2020)
7	Quince fruit (*Cydonia oblonga* M)	0.3–2.1 kGy	Physicochemical parameters (firmness, color in terms of Lab, titratable acidity [% malic acid]; total sugars, ascorbic acid, chlorophyll, total carotenoids, total phenolic content (TPC) and water-soluble pectin contents; polyphenol oxidase activity; % weight loss of fully ripe fruits; % decay)	Positive correlation between radiation and retention of firmness; inverse correlation between irradiation and water-soluble pectin contents; shelf-life extension up to 40 days at 1.8- and 2.1-kGy doses when stored under ambient (temperature 15 ± 2 °C, 85% RH) conditions	Hussain et al. (2019)
8	Cucumber (*Cucumis sativus* L.)	0.5 and 0.7 kGy	Physicochemical properties (shelf life, weight loss, pH, moisture, and vitamin C contents)	Shelf-life extension by 4 days at 0.5 kGy and 0.7 kGy when stored at 29 °C, 74 % RH; loss of weight but retention of freshness; no change in vitamin C content and pH values	Lwin (2020)
9	Oranges (*Citrus sinensis* L.)	0.5–2 kGy	Physicochemical (firmness, titratable acidity (% citric acid), TSS, vitamin C, and TPC; radical-scavenging activity) and volatile characteristics	Major volatile compounds identified were aldehydes and acetic esters; 0.5–1.0-kGy preserved sensorial and nutritional qualities of oranges during storage at 4 °C	Jo et al. (2019)

(Continued)

211

Table 8.3 (Continued)

S. No	Food item	Irradiation dose range	Parameters investigated	Effects	Reference
10	Jujube (*Ziziphus jujuba* var *vulgaris*) fruit	0.5–5 kGy	Physicochemical properties	No changes in TSS; increase in total acidity; decrease in L* value and significant increase in a* and b*values; at≤ 2.5 kGy irradiation is efficient for preservation when stored at 4 °C, 40% RH for one month	Najafabadi et al. (2017)
11	Isabel Precoce (*Vitis labrusca*), Bordô (*Vitis labrusca*) and BRS Violeta (hybrid of *Vitis labrusca*) grape juice	1–2 kGy	Physical, physicochemical (pH, titratable acidity in terms of % tartaric acid, color in terms of L*, C*, and h*, antioxidant activity, TSS, and vitamin C contents), and sensory properties	Highest antioxidant and vitamin C contents until 90 days at 2 kGy; unchanged sensory and nutritional qualities at 2 kGy when stored at room temperature (21 °C ±3 °C) for 120 days	Mesquita et al. (2020)
12	Tomato (*Solanum ycopersicon* L.)	1–4 kGy	Physicochemical properties (pH, total titratable acidity (% citric acid), TSS and moisture contents, physiological weight loss)	Significantly low weight loss; total titratable acidity was 0.23%–0.51%, TSS was 3.5%–5.0% and moisture content was 94.43%–96.53% during storage; better preservation at 1 kGy when stored at 10 ± 1 °C instead of 28 ± 1 °C	Gyimah et al. (2020)
13	Watercress (*Nasturtium officinale* R. Br.)	1–5 kGy	Visual, physicochemical (pH, proximate compositions, TSS, free sugar, organic acids, fatty acids, and tocopherol contents) and antioxidant activity	No change in color (CIE L*, a*, b*); high levels of PUFA and reducing sugar contents at 2 kGy; better antioxidant activity and high levels of sucrose, monounsaturated fatty acid (MUFA), tocopherol, and total flavonoid content (TFC) at 5 kGy when stored in the dark at4 ± 1 °C for 7 days	Pinela et al. (2016)
14	Pomegranate (*Punica granatum* L.) arils cv. 'Malas Saveh'	1–5 kGy	TPC, ascorbic acid, total anthocyanin and hydrogen peroxide contents, titratable acidity in terms of % citric acid, polyphenol oxidase activity	Lower TSS, TPC, anthocyanin and ascorbic acid contents, titratable acidity (% citric acid), antioxidant capacity, the activity of polyphenol oxidase; higher hydrogen peroxide content at >1-kGy dose of irradiation; 1-kGy increase shelf life when stored at 4 °C, >80% RH	Ashtari et al. (2019)

15	Date fruits (*Phoenix dactylifera* L.)	1–5 kGy	Physicochemical properties (TSS and moisture contents, titratable acidity in terms of % malic acid, pH) phytochemical properties (TPC and TFC, total antioxidant and enzyme activities), and sensory properties	Enhancement in shelf life and retention of total antioxidant content, sensory attributes, pH value, and POD activity at 5 kGy when stored at 25 ± 2 °C for 4 months	Zarbakhsh and Rastegar (2019)
16	Grape seeds (*Vitis vinifera* L.)	1–7 kGy	Proximate analysis of seed, peroxide, and free fatty acid (FFA) values, fatty acid profile, TPC, sterol contents, and antioxidant activity of seed oil	Decrease in C18:1 and C18:2 fatty acids, β-sitosterol and TPC and antioxidant capacity; increase in FFA and peroxide values with increase in irradiation dose from 1–7 kGy; unchanged chemical composition at 5 kGy	Apaydin et al. (2017)

Cereals and grains and their food products

17	Soybean	0.5–5 kGy	Protein quality attributes and parameters affecting protein modification	Reduction in turbidity, surface hydrophobicity, protein oxidation and increase in protein solubility, sulfhydryl contents, free phenolics, DPPH (2,2-diphenyl-1-picrylhydrazyl) and FRAP (Ferric reducing antioxidant power) activities at 2.0 kGy; decrease in lipoxygenase (LOX) and bound phenolics with increasing doses	Kumari et al. (2020)
18	Whole wheat (*Triticum aestivum* L.) flours	2.5 and 5 kGy	Thermal, rheological, and antioxidant properties	Decrease in thermal properties and dynamic moduli with increasing dose from 0–5 kGy and increase in antioxidant properties at 5 kGy	Bhat et al. (2020)
19	Brown rice–based weaning food with mung bean, apple pulp powder, and walnut powder	2–10 kGy	Physicochemical, functional, antinutritional and sensory attributes	Decrease in sensory characteristics, browning index, rehydration ratio, swelling power, phytate, and oxalate contents; increase in solubility index with increasing irradiation dose and storage time; improvements in solubility, antioxidant activity, TPC, iron, calcium, and β-carotene contents with increasing irradiation dose but decreased during storage (12 months) at 32 ± 2 °C in the said dose range	Jan et al. (2020)

(Continued)

Table 8.3 (Continued)

S. No	Food item	Irradiation dose range	Parameters investigated	Effects	Reference
20	Quinoa (*Chenopodium quinoa*) flour	3 and 6 kGy	Phytochemicals (saponin, TPC, and TFC) and antioxidant potency	Increase in TPC and TFC; maintained antioxidant content and activity at 3 kGy	Abdelaleem and Elbassiony (2021)
21	Yellow noodle with addition of rice bran	4–8 kGy	Physicochemical parameters (moisture content, water activity, pH, color (L*, a*, b*), texture profile, microstructure of noodles)	Significant decrease in pH, moisture content, hardness, lightness, and redness; significant increase in breaking length of noodles; minimum 6 kGy required for extension of shelf-life up to 7 days at 8 °C	Rahim et al. (2021)
22	Fresh pasta	5–13 kGy	Physical, chemical, and sensory characteristics	Unchanged cooking loss and significant reduction in water absorption after irradiation; no changes in sensory and cooking quality; good shelf stability at room temperature (25 °C) for 90 days at 13 kGy; similar effects with combination treatment of irradiation-cum-refrigeration at 10 kGy	Cassares et al. (2020)
Nuts					
23	Hazelnut kernel	0.5–1.5 kGy	Sensory properties and chemical characteristics (total FFA, total fat, and protein contents)	Significant increase in total fat value; no change in FFA, crude protein, crude cellulose, and moisture contents and water activity; optimum enhancement of shelf-life and retention of nutrition with highest vitamin E at 0.5 kGy when stored at 20 ± 0.5 °C, 55–60% RH for 18 months	Koç Güler et al. (2017)
24	Pistachio (*Pistacia vera* L.) nuts	1–6 kGy	Antioxidant activity; TPC, pigments and soluble protein contents; protein binding and fatty acid profiles; hardness; aromatic and sensorial features	Significant increase in TPC, antioxidant activity, and sensory attributes up to 2 kGy; significant reduction at doses of 4 and 6 kGy; reduction in soluble protein, chlorophylls, carotenoids, and unsaturated fatty acids contents; brightness parameters of color (L*, a*, b*) by irradiation; increase in number and concentrations of volatile components and saturated fatty acids; ≤ 2 kGy well preserved physicochemical, structural and sensory properties	Alinezhad et al. (2020)

Pulses

No	Food item	Dose	Properties studied	Findings	References
25	Chickpea (*Cicer arietinum*) flour	0.5–10 kGy	Proximate, physicochemical, pasting, thermal, and functional properties (water and oil absorption capacities)	No significant change in proximate composition; significant decrease in pasting properties (peak, final, set-back, and trough viscosities), pasting temperature, pH, texture profile, and enthalpy of gelatinization with increasing dose; increase in swelling, syneresis, solubility, oil, and water absorption capacities and gelatinization temperature with increasing dosage in the said range	Bashir and Aggarwal (2016)

Spices and herbs

No	Food item	Dose	Properties studied	Findings	References
26	Sesame (*Sesamum indicum* L.) seeds	0.5–2 kGy	Protein characteristics and functional (foaming capacity, foam stability, emulsion capacity, stability) properties	Significant increase in globulin and albumin fractions of proteins at high doses (>1.0 kGy); increase in *in vitro* protein digestibility; maximum protein solubility at 1.0 kGy; no changes in SDS-PAGE electrophoretic patterns of proteins and in foaming stability; significant enhancements in emulsification capacity and emulsion stability at 1.0- and 1.5–2.0-kGy doses respectively; maximum increase in foaming capacity at 1.0 kGy; enhancement in protein and functional properties	Hassan et al. (2018)
27	Parsley leaves (*Petroselinum crispum* (Mill.) Fuss var. neapolitanum)	0.7–2.7 kGy	Sensory and biochemical properties (total chlorophyll, chlorophyll a and b, and water contents)	High consumer preference for 0.7–1.4 kGy-irradiated samples when stored at 4 °C; shelf-life extension up to 30 days but disagreeable odor at higher doses (2.7 kGy)	Cătunescu et al. (2019)
28	Garden thyme (*Thymus vulgaris* L.) and peppermint (*Menta ×piperita* L.)	2–10 kGy	Cytotoxicity (in tumor and non-tumor cells) and TPC	TPC not affected; no changes in cytotoxic properties of peppermint on tumor cell lines (CF-7, NCI-H460, HeLa, and HepG2) up to 10 kGy; increase in cytotoxicity of thyme in tumor cell lines at 10-kGy; optimized dose for microbial decontamination and disinfestation without modification in phenolic compositions and bioactive properties was 10 kGy	Pereira et al. (2016)

(Continued)

Table 8.3 (Continued)

S. No	Food item	Irradiation dose range	Parameters investigated	Effects	Reference
29	Sesame (*Sesamum indicum* L.) seeds	3–9 kGy	Chemical and sensorial properties	No effects on moisture, ash, and fat contents and sensory properties; slight effects in protein and sugar contents; decrease in % total acidity in terms of lactic acid and a significant increase in total volatile basic nitrogen when stored for 12 months at room temperature 18–25 °C, 50–70% RH at the said dose range	Al-Bachir (2016)
Oils					
30	Dried thyme (*Thymus vulgaris* L.)	5–15 kGy	Chemical composition of essential oils; total antioxidants, TFC and TPC; antioxidant activity; thiobarbituric acid (TBA) value	Increase in TPC, TFC and total antioxidants contents; moderate changes in essential oils at 5 and 10 kGy; thymol found to be sensitive at 15 kGy; decrease in DPPH radical-scavenging activity; increase in TBA content at 10 and 15 kGy	Al-Kuraieef and Alshawi (2020)
31	Virgin coconut oil	4.2 kGy	Removal of rancid-acid odor	Elimination of rancid-acid odor from virgin coconut oil stored at 23 ± 1 °C for 28 days with no alteration in its antioxidant efficacy	Ghosh et al. (2016)
32	Soybean (*Glycine max* L.) oil	0.5 and 1 kGy	Lipase and LOX activities; oil quality (MDA content, fatty acid profile); carotenoid and tocopherol contents; SOD, peroxide, and CAT activities; total antioxidant and ROS contents; radical scavenging activity	Inhibition of rancidity development and enhancement in levels of tocopherols during storage at 40 °C at low dose of irradiation; absence of significant alteration in fatty acid composition; deactivation of LOX enzymes and lipase; enhancement of stability of oil during storage at 0.5 kGy due to lower contents of primary and secondary oxidative products	Bansal et al. (2020)
33	Mesocarp oil of Macauba or macaw palm [*Acrocomia aculeata* (Jacq.) Lodd. Ex. Mart]	4 and 8 kGy	Physicochemical and biochemical properties (mesocarp water and oil contents, water activity, lipase and POD activities, FFA and total carotenoids contents, peroxide value, oxidative stability, molar absorptivity)	Maintenance of oil acidity at acceptable levels; improvement of oxidative stability when stored at 23 ± 1 °C; higher FFA content and oxidative stability index at 4 kGy during storage	Tilahun et al. (2020)

34	Soybean and soybean oil	10–30 kGy	Aflatoxin B_1 content, proximate composition and quality (acid value, iodine value, peroxide value, fatty acid composition, tocopherols profile, and oxidation stability)	Reduction in aflatoxin B_1 at ≥10 kGy; no significant changes in quality attributes (acid value, peroxide value, iodine value), tocopherol(s) contents and oxidation stability of oils at ≤20 kGy when stored for 30 days; no significant change in the proximate and fatty acid composition of oil but negatively affect the quality parameters (peroxide value, acid value, iodine value), the content of tocopherols and oxidation stability of oils at 30 kGy	Zhang et al. (2018)

Mushrooms

35	Mushroom (*Volvariella volvacea*)	0.2–1 kGy	Sensory (odor, firmness, visual assessment of surface color, veil opening, rotting decay) and physicochemical properties (respiratory rate, weight loss, browning) and selected enzyme (malonaldehyde [MDA], SOD, CAT) activities	Best commercial quality at 0.8 kGy when stored at 16 ± 0.5 °C, 55% RH for 7 days; decrease in postharvest softening, browning, weight loss, respiration rate, and MDA content; higher CAT and SOD activities	Hou et al. (2018)
36	Wild mushroom (*Boletus pinophilus* *Boletus pinophilus* Pilát & Dermek and *Clitocybe subconnexa* Murrill)	2 kGy	Chemical and antioxidant characteristics	Significant changes in chemical (specially sugars and fatty acids contents) and antioxidant parameters particularly in *B. pinophilus*; no changes in overall characteristics; increase in antioxidant activity	Fernandes et al. (2016)

Fish and fish products

37	Mullet fish (*Mugil cephalus*) fillets	0.75–3 kGy	Fatty acids profile	Increase in total saturated and MUFA with increasing irradiation dose in the said dose range; significant difference in PUFA contents (linoleic, arachidonic, and eicosapentaenoic acids) between irradiated and non-irradiated samples when stored at -20°C	Abd El-Ghafour et al. (2018)

(Continued)

Table 8.3 (Continued)

S. No	Food item	Irradiation dose range	Parameters investigated	Effects	Reference
38	Tilapia (*Oreochromis* sp.) fillets	1 and 3 kGy	Fatty acids profile and chemical parameters (lipid content, peroxide, thiobarbiturate acid reactive substances (TBARS) and FFA values, fatty acid composition)	Safest irradiation dose was 1 kGy when stored in ice (0 °C); shelf-life extension up to 70 days at 3 kGy and up to 56 days at 1 kGy; significant reduction in saturated fatty acids at 1 kGy; significant increase in MUFA and PUFA at 1 kGy and signification reduction at 3 kGy; significant reduction in palmitic and linoleic and docosahexaenoic acids contents with storage	Surendra et al. (2018)
			Meat and meat products		
39	Salt-reduced hot dog wieners	1.5–4.5 kGy	Physical, chemical stability and sensory acceptance	Significant reduction in red color (a*); increase in lipid oxidation at the initial stage of storage at higher doses of the said range; no change in sensory properties; no negative effects of lipid oxidation at 1.5 kGy	Rodrigues et al. (2021)
40	Goat meat	1–6 kGy	Identification of lipid molecular species and quantification of significant lipids	Twelve subclasses of 174 lipids were identified; significant increase in total triacylglycerol, phosphatidylcholine, phosphatidylethanolamine, lysophosphatidylethanolamine, ceramide, lysophosphatidylcholine, sphingosine bases, and DHA-enriched phosphatidylcholine; reduction in total diacylglycerol, phosphatidylserine, phosphatidylglycerol, phosphatidylinositol and sphingomyelin	Jia et al. (2021)
41	Beef	2–6 kGy	Proximate, sensory, physicochemical and biochemical characteristics	Increase in color (visual) values, dry matter, cooking loss, TBA, peroxide and FFA values; decrease in ash content and pH with increasing dose and storage time; best results in terms of overall acceptability and shelf-life extension at 6 kGy when stored for 60 days at −20 °C	Haque et al. (2017)

S. No	Food item	Irradiation dose range	Parameters investigated	Effects	Reference
42	RTE stir fry chicken dices with hot chili	10–40 kGy	Sensory quality and protein content	No changes in protein content and sensory parameters; most shelf-stable at 10 and 20 kGy when stored for one year at 25 °C	Chen et al. (2016)
Others					
43	Kimchi seasoning mixture	1–10 kGy	Physicochemical properties	Significant delay in changes in pH, acidity, reducing sugar content, and headspace gas concentrations; no significant change in color (L*, a*, b*) values 10 kGy when stored at 4 °C	Jeong et al. (2020)
44	Guar gum and Locust Bean gum	2.5–5 kGy	Proximate, physicochemical, structural, functional (capacity and stability of emulsion, stability and capacity of foam, viscosity, water absorption capacity, swelling power, solubility index), and rheological properties	Reduction in final viscosity at neutral and acidic pH; significant decrease in viscosity and gel formation, swelling and thickening abilities, pH and bright appearance; increase in oil and water absorption, solubility, emulsion capacity, and flow properties in the said dose range	Hamdani et al. (2018)
45	Wild barnyard grass (*Echinochloa crusgalli* L.) seed flour	2.5 and 5 kGy	Proximate, physicochemical, functional and antioxidant attributes	Modifications in physicochemical, pasting, functional and antioxidant properties; decrease in swelling index, amylose content, peak and setback viscosities, syneresis, foaming capacity and foaming stability; increase in water and oil absorption capacities, emulsion capacity and antioxidant properties at the said doses	Wani et al. (2021)
Gamma-irradiation-treatment-induced alteration in bioactive compounds					
S. No	Food item	Irradiation dose range	Parameters investigated	Effects	Reference
1	Green plantains (*Musa paradisiuca*)	0.04–1 kGy	Serotonin and melatonin contents	Enhancement of serotonin content in 1 kGy-irradiated green plantains on day 10 when stored at 23 ± 2 °C, 70% R.H; enhancement of melatonin content in 0.6-kGy irradiated plantains on day 18 when stored at 23 ± 2 °C, 70% R.H.	Sarkar et al. (2021)

(Continued)

Table 8.3 (Continued)

S. No	Food item	Irradiation dose range	Parameters investigated	Effects	Reference
2	Chickpeas, kidney beans and green lentils	0.25–1 kGy	Total carotenoids, oligosaccharides raffinose and stachyose, vitamins thiamine (B1) and riboflavin (B2) contents	No significant effects on riboflavin, thiamine, raffinose, and stachyose and on the sensory qualities and alterations observed on total carotene content by irradiation; tolerable losses in the nutrients studied in chickpeas, kidney beans, and green lentils at 1 kGy	Aylangan et al. (2017)
3	Fenugreek (*Trigonella Foenum-graceum* L.) and spinach (*Spinacia oleracea* L.) leaves	0.25–1.5 kGy	Bioactive compounds and antioxidant activities	Enhancement in individual and total bioactive components; positive correlation between gamma-irradiation dose and TPC; increase in antioxidant content and activity when stored under refrigerated conditions (3 ± 1 °C, RH 80%) for 4 days	Hussain et al. (2016)
4	Jujube (*Ziziphus jujuba* var *vulgaris*) fruit	0.5–5 kGy	Bioactive compounds such as TPC, organic acids, anthocyanins, and vitamins (water-soluble) contents	Increase in total monomeric anthocyanin and TPC up to 2.5 kGy of irradiation; no effect on malic, citric and succinic acids; decrease in ascorbic acid, vitamins C and B1 and cyanidin-3,5-diglucosidecontents at 5 kGy when stored at 4 °C, 40% RH for one month	Najafabadi et al. (2017)
5	Persian pistachio nuts	1–4 kGy	TPC, anthocyanin and antioxidant properties	Significant increase in antioxidant activity and TPC at 1–2 kGy; decrease in TPC at 4 kGy; significant increase in anthocyanin at 1–4 kGy	Akbari et al. (2018)
6	Rye grains	3 and 10 kGy	TPC, TFC and tannins contents, potential antioxidant capacity, quantitative and qualitative profile of polyphenols	Increase in phenolic acids especially sinapic, ferulic, *p*-coumaric, and vanillic depending on the irradiation dose and variety; unchanged or reduced antiradical and antioxidant activities when stored at ambient laboratory temperature in the dark for 2 weeks	Gumul and Berski (2021)
7	Bulgarian teas (mixed herbal teas)	5 kGy	TPC and tannin contents and antioxidant activity	No change in radical scavenging activity; increase in antiradical activity, tannin, and TPC	Janiak et al. (2017)

Table 8.4 Studies on Gamma-Irradiation Treatment in Combination with Other Treatments

Combination treatment–induced alterations in microbiological properties

S. No	Food item	Combination treatment	Irradiation dose	Purpose of treatment	Effects	Reference
1	Hummus	Gamma-irradiation followed by refrigeration (4 ±0.5 °C)	0.1–2.5 kGy	Microbial load during storage (total mesophilic aerobic, lactic acid bacteria, mold and yeast counts)	Significant decrease in total mesophilic aerobic, lactic acid bacteria, mold and yeast counts in a dose-dependent manner; microbial count at ≥1.5 kGy combined with refrigeration (4 ±0.5 °C) treated samples complied with national and international microbial criteria of RTE foods even after 35 days	Al-Qadiri et al., 2020
2	Guava (*Psidium guajava* L.)	Gamma-irradiation followed by application of 4% $CaCl_2$ and 2 dm^3 kg^{-1} lemongrass oil	0.2–0.6 kGy	Inhibition of fungal growth (*Alternaria alternata, Alternaria solani, Aspergillus niger, Botrytis cinerea, Fusarium solani* and *Rhizopus tolonifera*) during cold storage	*A. alternata, A. solani, A. niger, B. cinerea, F. solani* and *R. tolonifera* were identified in decayed fruit; irradiation alone was beneficial for preservation at temperature of 8 ± 1°C, 90% RH for 20 days, compared to combination treatment	Hassanein et al., 2018
3	Satsuma mandarin	Treatment with sodium dichloro-striazinetrione (NaDCC) solution of various concentration prior to gamma-irradiation	0.2 and 0.4 kGy	Antifungal potency against *Penicillium digitatum*	Complete prevention of mycelial growth, spore germination and germ tube elongation of *P. digitatum*; inhibition of green mold decay by combined treatment when 10 ppm NaDCC pre-treated samples were subjected to 0.4 kGy irradiation	Jeong et al., 2016

(Continued)

Table 8.4 (Continued)

S. No	Food item	Combination treatment	Irradiation dose	Purpose of treatment	Effects	Reference
4	Carrot	Edible coating (nano-emulsion formulated by essential oils, citrus extract and cranberry juice encapsulation into calcium caseinate [Ca-Cas]–based matrix) followed by gamma-irradiation	0.5 kGy	Microbial population (total mesophilic flora, yeast and mold counts)	Highest reduction in total mesophilic flora, yeast and mold count in 0.5 kGy irradiated nano-emulsion incorporated film-coated carrots during storage at 4 °C	Ben-Fadhel et al., 2021
5	Strawberry (*Camarosa* cv.)	Gamma irradiation followed by refrigeration	0.5 and 1 kGy	Presence of fungal decay during storage	Reduction of fungal decay by 31% and 43% at 0.5 kGy and 1.0 kGy respectively when stored at 4 ± 1 °C, 90% RH for 20 days; decrease in fungal decay with increasing dose	Panou et al. (2020)
6	Kiwi fruits	Transparent film packaging followed by gamma-irradiation and refrigerated storage	0.5–2 kGy	Microbial proliferation during refrigerated storage	Reduction in yeast (by 5.7 log) and mold counts and by 6.0 log in the bacterial count at 2 kGy after 16 days of storage when stored at 3 ± 1 °C, 80% RH	Hussain et al. (2021)
7	*Tuber aestivum* truffels	Gamma-irradiation followed by modified atmosphere packaging (MAP) in polypropylene (PP) trays	0.5–2.5 kGy	Microbiological quality	Reduction in microbial counts by more than 3 log cfu/g when irradiated; slowdown in microbial growth with increasing dose of irradiation; slight growth of yeast and molds at the end of storage	Tejedor-Calvo et al. (2020)
8	Pomegranate fruit (*Punica granatum* L.)	Aerobic packaging (AP) with PP film followed by gamma-irradiation and refrigerated storage	0.5–3 kGy	Microbial population during refrigerated storage	Significant reduction in the growth of bacteria and fungi at ≥ 0.75 kGy when stored at 4°C, 80–85% RH	Pourebrahimi et al. (2020)

No.	Food	Treatment	Dose	Parameter studied	Observation	Reference
9	Bottom mushroom (*Agaricus bisporus* L.)	Gamma-irradiation followed by packaging in Ag- nanoparticles polyethylene (PE) films	1 and 2 kGy	Microbial population (*Escherichia coli* ATCC 25922 and *Staphylococcus aureus* ATCC 29523)	PE films with Ag nanoparticles could significantly reduce microbial load in samples when stored at 4 °C for 21 days	Ghasemi-Varnamkhasti et al. (2018)
10	Chicken meat	Pretreated with 3 % turmeric powder, AP or vacuum packaging (VP) and then subjected to gamma-irradiation	1 and 2 kGy	Microbiological parameters (total aerobic bacteria and coliforms) during storage	Decrease in total bacteria and coliform counts with increasing dose; no contamination in turmeric powder coated-cum-2 kGy irradiated samples for both AP and VP treated samples	Arshad et al., 2019
11	Pork	Gamma-irradiation followed by antimicrobial treatment (acetic acid, chitosan, carvacrol, tea polyphenols, chitosan-and-tea polyphenols, chitosan-and-carvacrol) and VP using food-grade PE	1 and 2 kGy	Antimicrobial activity against *Salmonella typhimurium* ATCC 14028 and *Staphylococcus aureus* ATCC 25923	Inactivation of populations of *S. aureus* and *S. typhimurium* by combination treatment with chitosan/ chitosan-and-tea polyphenols/ chitosan-and-carvacrol-cum gamma-irradiation at 2 kGy; no regrowth in VP treated meat samples during storage for 14 days at 4 °C	Hu et al., 2021
12	Millet (*Sorghum bicolor, Pennisetum glaucum, and Setaria italica*)	Packaged in plastic bag followed by heat treatment and gamma-irradiation	1 and 2.5 kGy	Microbiological (total fungal load) characteristics during storage	Significant reduction in fungal growth in combination treatment; improved storability at 2.5 kGy (over 90 days)	Huang et al., 2021

(Continued)

Table 8.4 (Continued)

S. No	Food item	Combination treatment	Irradiation dose	Purpose of treatment	Effects	Reference
13	Silver carp (*Hypophtha-lmichthys molitrix*) fish fillets	0.5 % rosemary (*Rosmarinus officinalis*) essential oil and Ca-Cas coating followed by low density PE packaging and gamma-irradiation	1–5 kGy	Microbiological counts (Lactic acid bacteria, Enterobacteriaceae, *Staphylococcus aureus*, *Bacillus cereus*, *Vibrio* spp. *Salmonella* spp.) during cold storage	Significant reduction in total bacteria, psychrophilic bacteria and lactic acid bacteria counts by combination treatment; reduction in Enterobacteriaceae, *S. aureus* and *B. cereus* counts, elimination of *Vibrio* sp. And *Salmonella* spp. In combination treatment with 1 kGy irradiation; complete elimination of all bacteria when coated and irradiated at 3 or 5 kGy and kept at 4 ± 1°C for 27 days	Abdeldaiem et al., 2018
14	Chicken meat balls	Moringa leaf powder (MLP) treatment followed by AP or VP and irradiation	1.5 and 3 kGy	Microbial population (total aerobic bacteria and coliform counts)	Reduction in total aerobic bacteria and coliforms counts at a higher dose of irradiation and MLP treated samples when stored in refrigerator at 4°C up to 14 days	Nisar et al., 2020
15	Chicken meat	Chitosan coating (2%) containing 0.1% grape seed extract (GSE), packaging in sterile PE bags and then subjected to gamma-irradiation	2.5 kGy	Microbial (aerobic mesophilic and psychrotrophic bacteria) counts during storage	Significant reduction in bacterial growth by combination treatment of gamma-irradiation and chitosan coating containing GSE	Hassanzadeh et al., 2017
16	Rainbow trout (*Oncorhynchus mykiss*)	TiO$_2$ (1%) nanocomposite packaging followed by gamma-irradiation	3 and 5 kGy	Microbiological population	Significant reduction in the microbial count in TiO$_2$ nanocomposite packaged and 3 kGy-irradiated samples during cold storage (4 °C) for 20 days	Hashemabad et al., 2018

17	Peanuts (*Arachis hypogaea* L.)	Gamma-irradiation followed by microwave heating (360 or 480 or 600 W) and vice-versa	5–9 kGy	Microbiological count (total bacterial, aerobic spore, yeast and mold counts)	Complete elimination of microbes after exclusive irradiation or combination (7 kGy and 360–600 W); Reduction in theutagenicity of *S. typhimurium* (85% in strain TA98 and 75–80% in strain TA100)	Patil et al., 2019
18	Turmeric (*Curcuma longa* L.) powder	MAP (VP or N$_2$) or AP in multilayered film (polyethylene terephthalate [PET]/ethylene vinyl alcohol-polyamide copolymer [EVOH – PA]/low-density polyethylene [LDPE]) packaged samples followed by gamma irradiation	5–15 kGy	Microbial load (total aerobic bacteria, mesophiles, spore-forming bacteria, total yeast/mold and coliform counts)	Significant reduction in microbial population at the lowest dose of 5 kGy; no detectable population at higher doses	Esmaeili et al., 2018
19	Thyme (*Thymus vulgaris*), rosemary (*Rosmarinus officinalis* L.), black pepper (*Piper nigrum* L.), and cumin (*Cuminum cyminum* L.)	MAP or AP followed by gamma irradiation	6 and 14 kGy	Antimicrobial properties against *Bacillus cereus*, *Escherichia coli*, *Aspergillus niger* and *Staphylococcus aureus*	Enhancement in antimicrobial activity against *B. cereus* and *S. aureus* by thyme essential oil in combination treatments of irradiation-cum-MAP and irradiation-cum-AP; antimicrobial activity against all microorganisms by thyme extract; increase in antimicrobial properties of cumin essential oil against all microorganisms	Kirkin and Gunes (2018)

(*Continued*)

Table 8.4 (Continued)

S. No	Food item	Combination treatment	Irradiation dose	Purpose of treatment	Effects	Reference
				Fruits, vegetables and related products		
1	Fried potato chips	Gamma irradiation and blanching prior to frying	0.05–0.15 kGy	Reduction in acrylamide content	Reduction in acrylamide formation by 20–54 % with only gamma-irradiation; reduction by 67–78% after employing combination treatment (irradiation at 0.15 kGy)	Abboudi et al. (2016)
2	Green chilies (*Capsicum annuum* L.)	MAP using anti-fog (RD45) filmfollowed by gamma-irradiation	0.1–0.5 kGy	Extension of shelf life during storage at low temperature	Insignificant changes in weight, firmness, total chlorophyll, carotenoids and ascorbic acid contents and in antioxidant activity after combination treatment; extension of shelf life up to 42 days at 0.25 kGy when stored at 8 °C ± 1 °C	Chitravathi et al. (2020)
3	Guava (*Psidium guajava* L.)	Gamma-irradiation followed by application of 4% $CaCl_2$ and 2 dm^3 kg^{-1} lemongrass oil	0.2–0.6 kGy	Extension of shelf life; physicochemical changes in weight loss, TSS, fruit firmness, vitamin C contents, % decay, % titratable acidity (citric acid)	Reduction in % water loss; prevents losses in TSS and vitamin C contents and total acidity (% citric acid) at 0.4 kGy; increase in fruit softness at a higher dose (0.6 kGy); preservation of fruit quality in combination treatment with 0.4 kGy when kept at 8 ± 1°C and 90% RH for 20 days	Hassanein et al. (2018)

#	Food	Treatment	Dose	Parameters	Observations	Reference
4	Plantain (*Musa* sp AAB)	Packaging in whole sealable PE or perforated sealable PE or jute sac followed by gamma-irradiation	0.2–1 kGy	Ethylene production and rate of respiration	More pronounced effect on respiratory rate than on ethylene production by irradiation; least amounts of ethylene and CO_2 production at 0.8 kGy when packaged in whole PE bags and stored at 27°C ± 3 °C, 75% RH	Ofosu et al. (2020)
5	Tamarillo (*Solanum betaceum Cav.*)	Gamma-irradiation followed by application of an edible coating (Sta-Fresh 2505) and vice-versa	0.25–3 kGy	Physicochemical properties (pH, respiration rate, TSS, firmness, pulp appearance, weight loss, sensory properties such as off-flavor, aroma and hardness)	Maintenance of postharvest properties with reduced weight loss; better firmness and appearance; lower respiration rate at combination treatment with 0.5-kGy irradiation; extension of shelf life up to 2 weeks than single treated fruit and up to four weeks w.r.t. control	Abad et al. (2017)
6	Carrot	Edible coating (nano-emulsion [NE] made of cranberry juice, citrus extract and essential oils incorporated into Ca-Cas-based encapsulation matrix) followed by gamma-irradiation	0.5 kGy	Extension of shelf-life (texture, weight loss and color [ΔE*] changes)	Extension of shelf life beyond 7 days by combination treatment (Ca-Cas + NE + irradiation at 0.5 kGy) during storage at 4 °C	Ben-Fadhel et al. (2021)
7	Strawberry (*Fragaria x ananassa*) (*Camarosa* cv.)	Gamma-irradiation followed by refrigerated storage	0.5 and 1 kGy	Sensory properties (appearance, texture, taste, odor), physico-chemical attributes (L*, a*, b* color parameters, weight loss, firmness, TSS, pH, and % titratable acidity in terms of citric acid)	Prolongation of shelf-life by 3 days at 0.5 kGy (total shelf life of 10 days); acceleration in degradation (increase in weight loss and decrease in firmness) at 1 kGy; best suitable dose was 0.5 kGy when stored under refrigeration (4 ± 1 °C, 90% RH) for 20 days	Panou et al. (2020)

(Continued)

227

Table 8.4 (Continued)

Combination treatment–induced alterations in proximate, physicochemical and sensory properties

S. No	Food item	Combination treatment	Irradiation dose	Purpose of treatment	Effects	Reference
8	Kiwi fruits	Transparent film packaging followed by gamma-irradiation and refrigerated storage	0.5–2 kGy	Storage quality during refrigerated storage	Significant inhibition of polyphenol oxidase activity, browning of surface and preservation of green color (L, a, b) and other quality parameters (texture; odor; TPC, ascorbic acid and chlorophyll contents) at 2 kGy up to 16 days under refrigerated conditions (3 ± 1°C, RH 80%)	Hussain et al. (2021)
9	Pomegranate fruit (*Punicagranatum* L.)	AP using PP film followed by gamma irradiation and refrigerated storage	0.5–3 kGy	Physicochemical (weight loss, pH, titratable acidity, TSS, TPC, total anthocyanin content) and sensory characteristics during refrigerated storage	Total anthocyanin, TPC and color (L*, a*, b*) qualities affected by irradiation and storage in a dose- and time-dependent manner; unacceptable sensorial qualities at 0–3 kGy after day 14; extension of shelf-life by 14 days at 1–2 kGy when stored in refrigerator (4°C, 80–85% RH)	Pourebrahimi et al. (2020)
10	Yellow corn (*Zea mays*) kernels	PP/PET/ aluminium foil trilamellate packaging followed by gamma-irradiation	1–10 kGy	Extension of shelf-life (sensory, physiological parameters)	Best shelf-stable (120 days) at 3 kGy with a lead of 30 days when kept at ambient conditions (23 ± 2°C, 80% R.H)	Pal and Bhattacharjee (2020)
11	Peach (*Prunus persica*l)	Gamma-irradiation followed by box packaging	2.5 and 5 kGy	Storage quality (proximate analyses) and shelf-life extension	Extension of shelf life up to 17 days when stored at 25 ± 2 °C; no adverse effects on irradiation; storage for more than 2 weeks showed loss of firmness	Khan et al. (2018)

Cereals and grains

12	Millets (*Sorghum bicolor, Pennisetum glaucum,* and *Setaria italica*)	Packaged in plastic bag followed by Heat treatment and gamma-irradiation	1 and 2.5 kGy	Pasting and rheological characteristics during storage	Reduction in pasting (pasting and peak temperature and peak viscosity etc.) and rheological (storage moduli, loss moduli, yield point, etc.) properties by combination of heat and gamma irradiation during 90 days' storage at ambient laboratory temperature; slight change of color (L, a, b, ΔE); increase in protein and carbohydrate contents during storage	Huang et al. (2021)
13	Chiffon cake	Pre-treatment of batter with calcium propionate (CAP) (1.5 g or 3 g/kg dough prior to baking followed by gamma irradiation)	2–10 kGy	Extension of shelf life	Reduction of crumb yellowness, specific volume, chewiness, hardness, springiness; increase in crust yellowness and peroxide values at 2–10 kGy; significant decrease in acceptability score above 4 kGy; no significant difference in scores between 2 kGy and 4 kGy; extension of shelf life to 90 days at 4 kGy, with or without pre-treatment with 1.5 g/kg CAP when stored at 25 ± 5 °C	Sirisoontaralak et al. (2017)
14	Hummus	Gamma-irradiation followed by refrigeration (4 ±0.5 °C)	0.1–2.5 kGy	Extension of shelf-life (% titratable acidity [citric acid], pH and water activity)	Extension of shelf-life up to 35 days at ≥ 1.5 kGy	Al-Qadiri et al. (2020)

(Continued)

229

Table 8.4 (Continued)

S. No	Food item	Combination treatment	Irradiation dose	Purpose of treatment	Effects	Reference
				Nuts		
15	Peanuts (*Arachis hypogaea* L.)	Gamma-irradiation followed by microwave heating (360 or 480 or 600 W) and vice-versa	5–9 kGy	Physicochemical (moisture content, hardness, color [L, a, b, ΔE], peroxide value and FFA), aflatoxin B₁ (AFB₁) content and sensory properties	Retention of physicochemical properties post combined treatments; sensory analysis (appearance, color, texture, taste, after-taste and overall acceptability) after treatment exerts satisfactory scores; maximum reduction in AFB1 at 7 kGy and 600 W	Patil et al. (2019)
				Spices and herbs		
16	Garlic ('PG-18' variety)	Blanching followed by packaging [PP or perforated PP or VP], gamma-irradiation and refrigeration	0.5–2.5 kGy	Extension of shelf-life by physicochemical analysis	Significant changes in TSS, weight, firmness, ascorbic acid, allicin content and color (L, a, b, C, H, ΔE and browning index) with irradiation dose and storage days; significantly high reduction in weight in VP treated samples and enhancement with storage days; significant enhancement in ascorbic acid with higher irradiation dose; increase in allicin content with increasing dose; highest overall acceptability score at 2.5 kGy in all types of packaging but slight decrease in quality attributes; best quality retention for 77 days in refrigerated storage (4 ± 1 °C; 65 ± 5% RH) at 1.5 KGy after packaging in PP packages of 150 gauge	Sharma et al. (2021)

	Food	Treatment	Dose	Parameters studied	Results	Reference
17	Turmeric (*Curcuma longa* L.) powder	MAP (VP or N$_2$) or AP in multilayered film (PET / EVOH-PA/LDPE) packaged samples followed by gamma irradiation	5–15 kGy	Antioxidant activities (DPPH•, FRAP, and ABTS•+) and color parameters (L*, a*, b*, ΔE)	Highest antioxidant properties in AP-treated samples (irradiated up to 10 kGy); maximum discoloration at 10-kGy AP-treated samples	Esmaeili et al. (2018)
18	Thyme (*Thymus vulgaris* L.), rosemary (*Rosmarinus officinalis* L.), black pepper (*Piper nigrum* L.), and cumin (*Cuminum cyminum* L.)	MAP or AP followed by gamma-irradiation	6 and 14 kGy	Essential oil yield, TPC and antioxidant properties	Decrease in TPC except in thyme; increase in DPPH activity in thyme extract and in cumin essential oil (AP treated) but decrease of the same in cumin extract; increase in FRAP values in 6 and 14 kGy-MAP treated rosemary essential oil and extract; also increase of the same in 14 kGy-AP-treated rosemary extract and 6- and 14-kGy AP-treated black pepper extract; increase in β-carotene–linoleic acid bleaching-based antioxidant activity in essential oils of thyme (AP-treated) and black pepper (MAP-treated), but decrease in cumin extract (MAP-treated)	Kirkin and Gunes (2018)
Fungus						
19	*Tuber aestivum* truffels	Gamma-irradiation followed by MAP in PP trays	0.5–2.5 kGy	Quality, shelf life and sensory evaluation (firmness, gleba aspect, aroma, flavor and general acceptability)	Reduction in postharvest sensory losses after irradiation; extension of shelf life up to 42 days with a combination of MAP and irradiation (2.5 kGy) when stored at 4 °C; no remarkable changes in total carbohydrates, protein and total sterol concentrations post-irradiation; significant increase in TPC at 1.0–2.5 kGy	Tejedor-Calvo et al. (2020)

(Continued)

231

Table 8.4 (Continued)

S. No	Food item	Combination treatment	Irradiation dose	Purpose of treatment	Effects	Reference
20	Bottom mushroom (*Agaricus bisporus* L.)	Gamma-irradiation followed by packaging in Ag- nanoparticle-incorporated PE films	1 and 2 kGy	Physicochemical properties	Lowest reduction in pH and L* when 2 kGy irradiated samples were packaged in Ag-nanoparticles PE films and stored at 4 °C; negligible changes in weight loss, b* and browning index in the same condition	Ghasemi-Varnamkhasti et al. (2018)
				Fish		
21	Rainbow trout (*Oncorhynchus mykiss*)	TiO$_2$ (1%) nanocomposite packaging followed by gamma-irradiation	3 and 5 kGy	Enhancement of shelf-life (total volatile base-nitrogen, pH, peroxide and TBA values; oxidation of protein through evaluating carbonyl index)	Longer retention of chemical and sensory characteristics and extension of shelf life for 20 days when TiO$_2$ (1%) nanocomposite pretreated samples were subjected to irradiation at 3 kGy and kept under 4 °C	Hashemabad et al. (2018)
				Meat and meat products		
22	Chicken meat	Pretreated with 3 % turmeric powder, AP or VP and then subjected to gamma irradiation	1 and 2 kGy	Physicochemical, antioxidant parameters and sensory attributes	Higher TPC and antioxidant potency (DPPH) in 2 kGy meat pre-treated with turmeric powder on day 0 of storage; higher antioxidant potency in VP- vis-à-vis AP-treated samples; higher peroxide and TBARS values in 2-kGy AP-treated samples stored for 14 days stored at 4 °C	Arshad et al. (2019)

| 23 | Chicken meatballs | MLP treatment followed by AP or VP and irradiation | 1.5–3 kGy | Improvement of shelf-life | Preservation of sensory parameters, functional and nutritional at lower doses; stabilization and protection of fats and amino acids, and other functional and physicochemical components exert by phytochemicals present in MLP; best shelf stability (14 days) at 1.5 kGy with pretreated MLP-cum-VP-treated samples stored at 4 °C | Nisar et al. (2020) |
| 24 | Chicken meat | Chitosan coating (2%) containing 0.1% grape seed extract (GSE), packaging in sterile PE bags and then subjected to gamma irradiation | 2.5 kGy | Extension of shelf-life with chemical (lipid oxidation, pH, water activity) properties | Extension of shelf life up to 14 days in refrigerated conditions by combination treatment; improvement in sensorial properties by combination treatment; irradiation-induced lipid oxidation in GSE treated samples; increase in pH was prevented by chitosan coating containing GSE | Hassanzadeh et al. (2017) |

(Continued)

Table 8.4 (Continued)

S. No	Food item	Combination treatment	Irradiation dose	Purpose of treatment	Effects	Reference
1	*Tuber aestivum* truffels	Gamma-irradiation followed by MAP in PP trays	0.5–2.5 kGy	Contents of sterols and TPC	Decrease in sterol content particularly ergosterol, brassicasterol and stigmasterol at 2.5 kGy; levels of TPC doubled at 1.5 and 2 kGy during storage at 4 °C	Tejedor-Calvo et al. (2020)
2	Turmeric (*Curcuma longa* L.) powder	MAP (VP or N_2) or AP in high-barrier multilayered film (PET/EVOH-PA/LDPE)–packaged samples followed by gamma irradiation	5–15 kGy	Contents of TPC and curcuminoid contents (curcumin, demethoxycurcumin, and bisdemethoxycurcumin)	Combination of AP and irradiation had a synergistic effect (enhanced activity) on the yields of curcuminoids; highest contents of curcuminoids and TPC in 15-kGy AP-treated samples when stored at 4 °C	Esmaeili et al. (2018)
3	Yellow corn (*Zea mays*) kernels	Packaging in trilamellate (3-ply PP/PET/Al) film followed by gamma irradiation	1–10 kGy	Enhancement of lutein	Enhancement in lutein content with minimum aflatoxin content in 3-ply PP/PET/Al-laminate packaged and 3-kGy gamma-irradiated corn kernels when stored at 23 ± 2 °C, 80% R.H.	Pal and Bhattacharjee (2020)

antimicrobial agents, namely, chemicals or essential oils; refrigeration; packaging in flexible and nanocomposite materials; and application of MAP/AP/VP. While coating foods with antimicrobial solutions/nano-emulsions/essential oils, packaging in flexible trilamellates or nanocomposites and heat treatment (including microwave heating) are applied either pre– or post–gamma irradiation of food products, refrigerated storage is conducted only post-irradiation processing for extending the shelf lives and preserving nutritional-cum-sensory quality attributes of foods.

8.5 ESSENTIALS AND CATEGORIES OF COMMERCIAL GAMMA-RADIATION FOOD-PROCESSING FACILITY

8.5.1 Essentials of Gamma-Radiation Plants

In a commercial/industrial gamma-irradiation facility, the irradiation room in which food products are treated with gamma radiation is the primary zone that is protected with a radiation shield. The facility also houses a shielded storage room (dry or wet) for radiation source (^{60}Co), the source *per se*, a product transport mechanism, a source hoist mechanism and a radiation processing cell. The accessory components of a gamma-irradiation commercial facility are control console (room), product containers (totes), control and safety interlock systems, dedicated spaces for loading and unloading of products and accessory service equipment. The facility should also have quality control and analytical laboratories for analyses of irradiated food products and may have a microbiology and a material testing laboratory (IAEA, 2006).

Irradiation room: Food irradiation is conducted inside a gamma-irradiation chamber shielded (biological shielding) by 1.5–1.8-m high-density cement concrete walls (DAE, 2014). This ensures sufficient attenuation of radiation emanating from the source (IAEA, 2006) prevents leakage of radiation outside the chamber to the working area and, hence, poses no occupational/health hazards to radiation workers (operators) and maintains the radiation level inside the room close to that of the surrounding.

Shielded source storage: The source used for gamma irradiation is placed inside a shielded storage room, which is usually located beneath the irradiation room. The shielded storage room can be either dry (shielded by solid wall) or wet (shielded by water). Both dry and wet storage of source in shielded storage room permits free access to personnel for working inside the gamma-irradiation room for regular plant maintenance and related activities (IAEA, 2006). Water shielding is generally used in wet shielded room since water is transparent, is readily available, allows fast heat transfer and, most important, because, water shielding does not allow leakage of radiation from the storage room (IAEA, 2006). In commercial food irradiation plants, ^{60}Co alone is used as the gamma-radiation source globally and operated preferably in wet storage mode (also refer to Section 8.6).

The radiation source when not in operating condition is stored in a 6- to 7-m-deep stainless steel–lined water pool for wet storage. The water pool is rendered corrosion-free and water seepage from it is blocked by an inner lining of stainless steel or ceramic (DAE, 2014). The accessory materials in wet storage room such as a guide system, source containers and racks are also manufactured with stainless steel to avoid galvanic corrosion. In the wet storage room, it must be ensured that water loss in the pool surface in 24 h should be within 0.1% of the total volume of water in the pool; if exceeded, provision must be provided to restore water automatically to the normal level. The water to be used for the wet storage of source must meet the following specifications: it must be demineralized (DM); free from turbidity, dissolved solids and microbial load chiefly fungus; and should have electrical conductivity exceeding 10 μ mho/cm (DAE, 2014). These requirements must be maintained by an online cleanup system. The pool is provided with a radiation-resistant physical barrier to ensure safety of work personnel. The continuous removal of decay of the stored radiation source must be ensured by the use of forced ventilation or water circulation in the irradiation room. Radiation monitors are also installed in the water treatment system (DAE, 2014).

Gamma-irradiation source: ^{60}Co source, in the form of ISU code W-91, doubly encapsulated in stainless steel tubes of AISI (L) grade of equivalent metallurgical properties as approved by the Atomic Energy Regulatory Board is used in wet-source storage-type plants. The encapsulation allows efficient dissipation of the decay heat to the external surface during storage and use. Between the two radioisotopes used for generating gamma rays, ^{60}Co is more commonly used for both industrial and medical purposes (Section 8.2), while the source of ^{137}Cs is restricted to small self-contained, dry-storage irradiators and are used primarily for insect sterilization and more commonly for blood irradiation (IAEA, 2006). ^{60}Co used commercially as a radiation source is manufactured as unit sources commonly referred to as pencils (IAEA, 2006). These pencils are manufactured from slugs (Figure 8.3), which are small cylinders or pellets constructed from 99.9% pure Co-sintered powder welded into Zircaloy capsules and are subsequently allowed to reside in a nuclear power reactor for about 18–24 months (duration varies depending on the neutron flux) for production of ^{60}Co. In the commercial irradiation facility, these unit sources are loaded in a source rack that can accommodate 100 kCi ^{60}Co (maximum source loading). The source rack (Figure 8.4) is provided with numerous source positions such that existing sources can be stacked in the rack for a period of 15 years with provision for additional sources. A commercial food irradiation facility is permitted to have a source of capacity 1000 kCi. The throughput of a food item in a radiation facility using ^{60}Co or other power sources can be related to the dose requirements of a process and source strength by the following equation (DAE, 2014):

$$\text{throughput (kg/h)} = [(\text{power in kW} \times 3600)/\text{dose in kGy}]' \text{ efficiency of irradiator}, \tag{8.3}$$

where E is the efficiency of the radiation energy utilized in the process expressed in fraction; MCi or 37.5 PBq of ^{60}Co = 15.0 kW of power (DAE, 2014).

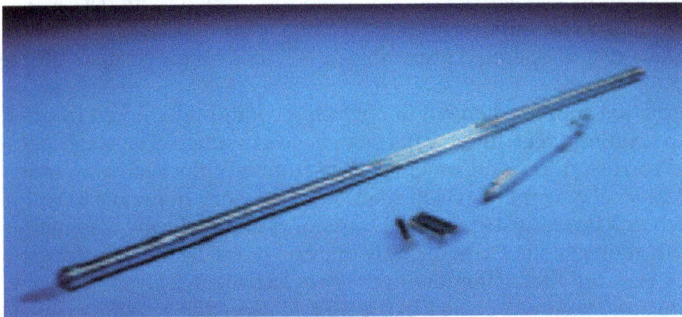

Figure 8.3 Slugs and pencil of cobalt-60, which are the building blocks of a radiation source rack (IAEA, 2006).

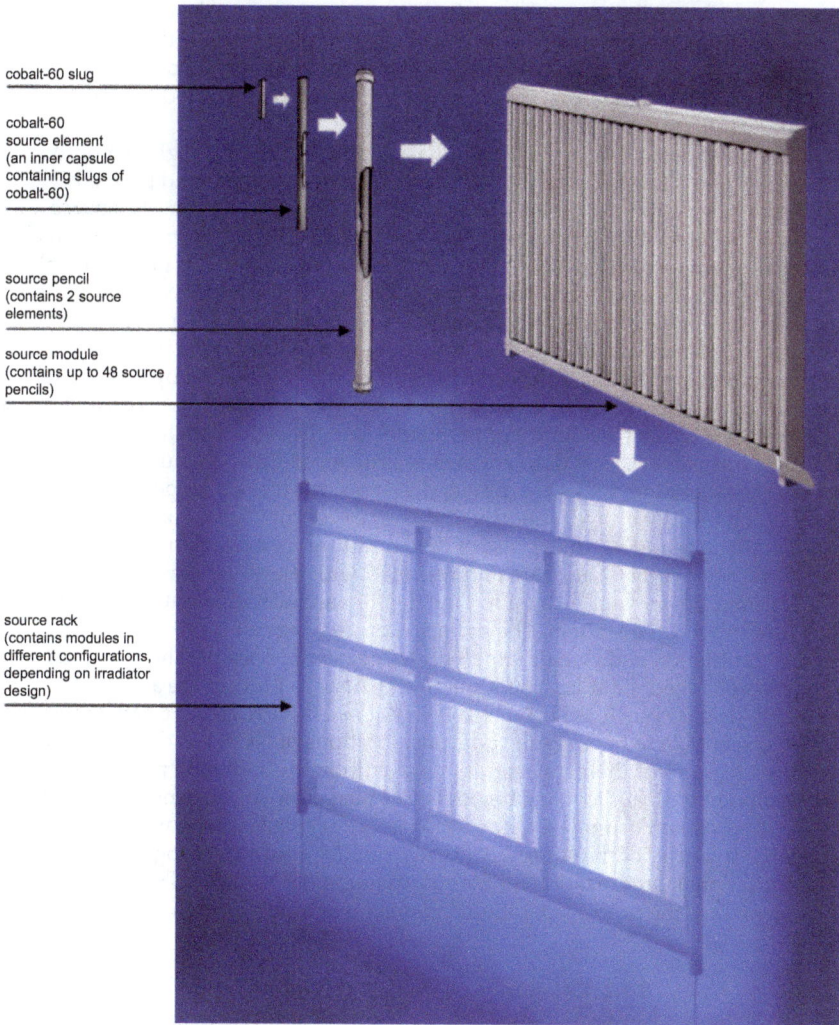

cobalt-60 slug

cobalt-60 source element (an inner capsule containing slugs of cobalt-60)

source pencil (contains 2 source elements)

source module (contains up to 48 source pencils)

source rack (contains modules in different configurations, depending on irradiator design)

Figure 8.4 Schematic diagram representing a typical cobalt source rack built from slugs, pencils and modules (IAEA, 2006).

Food product transport mechanism: According to the irradiator design, transport mechanisms of products vary. For a continuous irradiator, food products packed in crates or boxes are placed on automated conveyer belts and mechanically positioned around the source rack and propelled through the concrete wall labyrinth into the sample chamber located in the gamma-irradiation chamber, where the crates are exposed to the radioactive source (DAE, 2014). These boxes are mechanically turned for perfect alignment (or attenuation) with the radiation source to allow uniform irradiation of food products. For stationary irradiation, the radiation source is propelled toward the food products arranged inside the irradiation room. In the case of a malfunction of the product conveying system, the radiation processing is terminated, and the source is lowered back to its storage position. All components of the transport system are designed to achieve a perfect fail-safe operation. The inadvertent entry of persons at the product entry or exit ports is prohibited during radiation processing. Additionally, facilities are so designed that radiation operation can be thwarted or terminated inside the radiation cell if desired (DAE, 2014).

Source hoist mechanism: For food irradiation, the source is brought out of the DM water pool after necessary activation of all safety devices and perfectly aligned to the product to be irradiated by an automatic conveyor. The source is raised to the operating position by hydraulic/pneumatic power and lowered to its safe storage position after usage (i.e., plant shutdown) with gravitational force. The radiation source is placed inside the irradiation chamber only for the duration the food products undergo gamma-irradiation treatment. The determination of the absorbed dose depends on the residence time of the carrier (or tote box) in its irradiation position, which, in turn, is monitored by dose meters positioned at several positions in the tote box (DAE, 2014).

Radiation processing cell: Entry of personnel into the radiation chamber is through a personal access door and safety is ensured through multiple interlocks. The safety interlocks are provided to interlock the source position electrically, mechanically or hydraulically to thwart or terminate the source irradiation process when required. In the case of power failure for more than 10 s, irradiation is automatically terminated and resumed with diesel generators in the event of persistent power loss. A radiation incident on an individual must undergo a minimum of three scatters. The radiation processing cell is provided with proper ventilation and is maintained under negative pressure. The air inside the cell is changed for at least six times every hour to prevent the accumulation of toxic gases such as O_3, NOx at levels above their respective safe levels. In zones susceptible to have significant concentrations of NOx, gas detectors are installed to monitor their concentrations. Additionally, radiation processing cells are equipped with audiovisual devices-cum-heat-cum-smoke sensors. The fire exhausting system is rendered free from chemical substances that can adversely affect the integrity of the source. Adequate quality assurance programs such as International Organization for Standardization (ISO) 9000 or ISO 14000 are established for the proper operation and design of all irradiation plants (DAE, 2014).

8.5.2 Categories of Food-Irradiation Plants

Based on design, irradiation plants can be classified into two categories, that is, a machine-driven plant and natural source–driven plant. The machine-driven plants generate a radiation field of X-rays while the natural source–driven plant generates gamma radiation using ^{60}Co and ^{137}Cs and are employed for food irradiation, and the former sources are preferred for food irradiation due to ease of manufacture. Furthermore, based on radiation procedure, food irradiation plants could be of two categories: a product-overlapping plant and a source-overlapping plant. In the former, the length of the product box is greater than that of the source pencil, this radiation procedure makes the plant more efficient because all the radiation energy generated from the source is used to irradiate the products. In source-overlapping plants, the source pencil length is greater than the product box length; hence, the entire room is filled with electromagnetic waves, ensuring a good dose uniformity ratio (IAEA, 2006). Furthermore, based on the universal categorization, food irradiation plants are of the following categories

Category I self-contained dry storage irradiator: In this category, the radiation source remains stationary, and the food product to be irradiated moves into the irradiation cell or chamber for a predetermined period depending on the dose of gamma irradiation to be used. This type of irradiator is used for research purpose since the irradiator is of relatively small throughput and needs small doses (IAEA, 2010).

Category II panoramic dry source storage irradiator: Panoramic irradiators are more suitable where the source consists of several ^{60}Co pencils or cylinders, which can be moved into a large irradiation room (IAEA, 2006). The sealed source is fully shielded since it is enclosed within a dry container (constructed of lead or other appropriate high atomic number material) when not in use and, when in use, is exposed within an area that is kept inaccessible during operation by an entry control system. This procedure allows products in product boxes to either remain stationary or move toward the source. The source is a stationery pencil and is surrounded by product boxes (IAEA, 2010).

Category III self-contained wet source storage irradiator: In this category also, the source is stationary and resides in the underwater shield. Product boxes placed inside temperature-cum-water-resistant chambers move towards the source (IAEA, 2010).

Category IV panoramic wet source storage irradiator: The sealed source, when not in use, is submerged within a DM water pool and thereby completely shielded. For use, the source is brought out of the water pool, and similar to category II, it radiates a fixed volume of space, which is rendered completely inaccessible to operators by an entry control system, implying that it is a controlled human access irradiation facility (Figure 8.5). It is a regulatory requirement that a food irradiation facility must provide separate storage areas for both nonirradiated and irradiated food products without any possibility of them being mixed as shown by a separating fence in Figure 8.5 (IAEA, 2010).

The commercial food irradiation facilities operating in India are listed in Table 8.5. The primary design features (type of category, storage, design capacity) of the irradiation plants, current source activity, operational status and examples of food products commercially irradiated have been presented in the table. Internationally, there are about 61 facilities in different countries that actively irradiate agricultural and food products including spices, animal feeds, apiarian products, horticulture products and food packaging materials (such as plastic bags, glass and metal containers, and cheesecloth, to name a few). The primary purpose is to perform decontamination, microbial load reduction, pest control, shelf-life extension, sprout inhibition (IAEA, 2004). A type IV irradiation plant category is the most used for food irradiation commercially.

8.6 COMMERCIAL APPLICATIONS OF FOOD IRRADIATION

The first irradiated food that was commercially available and successfully introduced in the market was irradiated frozen ground beef in Minnesota in May 2000. The Minnesota Department of Health was compelled to play a proactive role in encouraging retailers to market irradiated ground beef consequent of a series of massive product recalls of nonirradiated beef and beef products contaminated by *E. coli* O157:H7 and subsequent disease outbreak. Nationwide food service providers and renowned meat companies such as Minnesota-based Schwan's, Inc.; Omaha Steaks of Nebraska; and Wegmans of Rochester, New York have since then been successfully marketing irradiated ground beef (Eustice, 2017). Meat retailers in the United States strongly believe that food irradiation is a value-addition process that offers consumers additional food safety. The interest of these leading companies in food irradiation catalyzed breaking the barrier of concern for retailers in several other countries (Eustice, 2017).

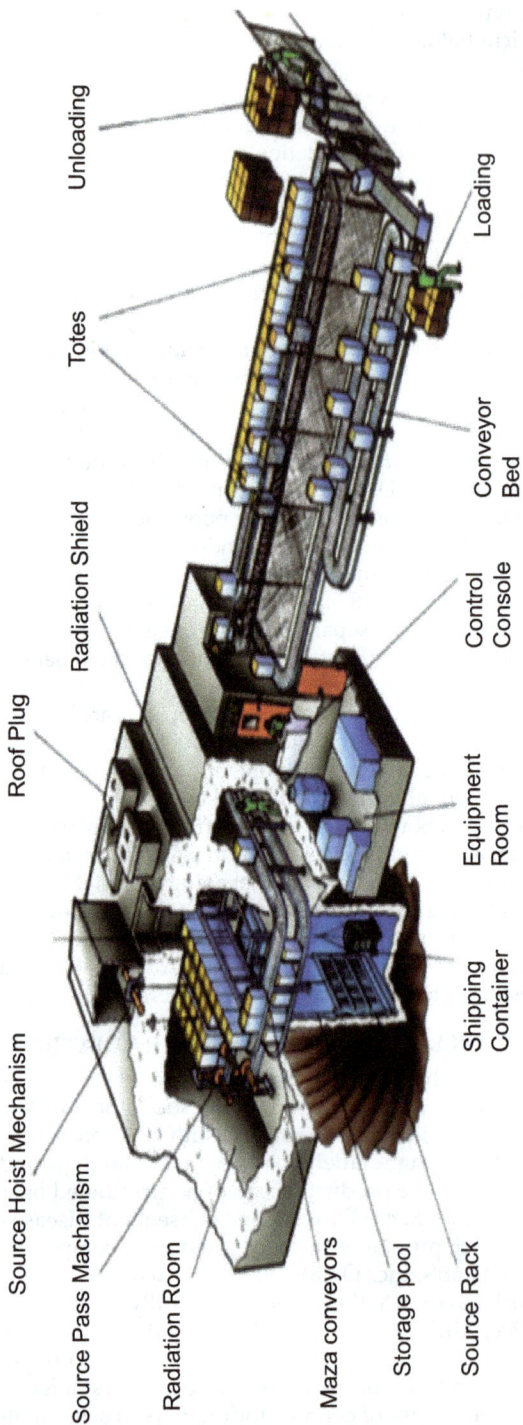

Figure 8.5 Schematic diagram of a Category IV panoramic, wet-storage gamma-irradiation facility (IAEA, 2006).

Table 8.5 Irradiation Facilities at National Level

List of gamma-irradiators installed and under operation in India with their present status of source activity (as of 16.08.2021)

S. No	Name of the radiation facility	Design features	Year of commission-ing	Purpose of usage	Operating agency/body	Present activity and operational status
01	Isomed, BARC, Mumbai, Maharashtra	• Category II • Dry storage • Design capacity: 1 MCi	1974	Sterilization of medical and health care products	Government of India	• Shut down for up-gradation and design validation
02	Shri Ram Applied Radiation Center, New Delhi	• Category IV • Wet storage • Design capacity: 800 kCi	Started in 1984 for R&D; Commercial application started in 1990	Sterilization of medical and health care products	Private trustee	• Commercial service provider • Present activity: 745 kCi
03	Radiation Processing Plant, Vashi, Maharashtra, Board of Radiation and Isotope Technology	• Category IV • Wet storage • Design capacity: 1 MCi	2000	Disinfestations and sterilization of spices, pet feed and allied products	Government of India (GoI)	• Commercial Service provider within the range of 6 to 15 kGy of dose • Present activity: 640 kCi
04	Krushak Plant, Lasalgaon, Nashik, Maharashtra	• Category IV • Wet storage • Source overlapping type low dose plant • Design capacity: 400 kCi	2002	Low dose application for food and agro products, sprout inhibition of bulbs and tubes, phytosanitization of fruits for export	BRIT, GoI with private operation and maintenance (O & M) partner	• Commercial service provider • Maximum dose use up to 10 kGy • Present activity: 360 kCi; • Registered with National Plant Protection Organization (NPPO) for National Standards for Phytosanitary Measures (NSPM)-21

(Continued)

Table 8.5 (Continued)

S. No	Name of the radiation facility	Design features	Year of commissioning	Purpose of usage	Operating agency/ body	Present activity and operational status
05	Organic Green Food Ltd. Kolkata, West Bengal	• Category IV • Wet storage • Product overlapping type • Design capacity: 1 MCi. • Hanging carrier type design • First private plant of India	2004	Sterilization of medical and health care products; disinfestations of laboratory equipments at medium dose	Private operator	• Commercial service provider • Present activity: 470 kCi
06	A.V.Processor Pvt Ltd, Sterico, MIDC Ambernath, Mumbai, Maharashtra	• Category IV • Wet storage • Product overlapping type • Design capacity 1.2 MCi • Hanging carrier type design	2005	Sterilization of medical and health care products; disinfestations of laboratory equipment at medium dose	Private operator	• Commercial service provider • Present activity: 1150 kCi
07	Universal Medicap Ltd, Dashrath, Vadodara, Gujarat	• Category IV • Wet storage • Product overlapping type • Design capacity: 1 MCi • Hanging carrier type design.	2005	Spices, psyllium husk, turmeric powder for domestic market; dehydrated onion flakes and powder for export; rubber stoppers and Al vials for pharmaceutical use	Private operator	• Commercial service provider • Present activity: 640 kCi
08	Gamma Agro and Medical Pvt. Ltd., Hyderabad, Telengana	• Category IV • Wet storage • Product overlapping type • Design capacity: 1 MCi • Hanging carrier type design	2007	Sterilization of medical and healthcare products; pet feed; chili powder and herbal products	Private operator	• Commercial service provider • Present activity: 900 kCi

No.	Name/Location	Features	Year	Products	Operator	Notes
09	Microtrol Sterilisation Services Pvt. Ltd., Bangalore, Karnataka	• Category IV • Wet storage • Product overlapping batch type plant • Tote Box design • Design capacity: 1.4 MCi	2006	Sterilization of medical and healthcare products and food products in medium dose range	Private operator	• Commercial service provider • Present activity: 1330 kCi
10	Agrosurg Irradiators, Vasai, Thane, Maharashtra	• Category IV • Wet storage • Product overlapping–type plant • Overhanging carrier-type design • Design capacity: 1 MCi	2007	Sterilization of medical and healthcare products, dehydrated onion powder and flakes; pet feed; raw materials of herbal products	Private operator	• Commercial service provider • Present activity: 950 kCi
11	Jhunson Chemical Pvt Ltd, Bhiwadi, Rajasthan	• First multipurpose plant of India • Category IV • Wet storage • Product overlapping type • Design capacity: 1 MCi • Multipurpose plant with double-tier roller conveyor with aluminum tote box • Split type source frame	2009	Spices, raw materials of herbal products, dry fruits and vegetables; pet feed; medical and surgical goods	Private operator	• Commercial service provider • Present activity: 210 kCi
12	Innova Agri Bio Park Ltd., Malur, Dist. Kolar, Karnataka	N/A	2011	N/A	Private operator	• Commercial service provider • Maximum dose uses up to 10 kGy • Present activity – Not known • Registered with NPPO for NSPM-21

(Continued)

Table 8.5 (Continued)

S. No	Name of the radiation facility	Design features	Year of commission-ing	Purpose of usage	Operating agency/body	Present activity and operational status
13	Hindustan Agro Co-Operative Ltd., Rahuri, Ahmednagar, Maharashtra	N/A	2012	Designed for agro products	Co-operative sector	• Present status not available
14	Nipro India Corporation Pvt Ltd, Satara, Maharashtra	• Design capacity: 5 MCi	2012	Designed for captive use for medical and health care products	Private operator	• Captive use • Present activity: 1100 kCi
15	Impartial Agro Tech Pvt Ltd., Unnao, Lucknow, Uttar Pradesh	• Category IV • Wet storage • Product-overlapping type • Design capacity: 1 MCi • Multipurpose plant with double-tier roller conveyor with aluminum tote box • Split-type source frame	2014	Pet feed	Private operator	• Commercial service provider • Present activity: Less than 100 kCi • Plant not in operation
16	Gujarat Agro Industries Corporation Ltd., Bavla – Bagodara Highway Village: Ramnagar, Gujarat	• Category IV • Wet storage • Product-overlapping type • Design capacity: 1 MCi • Multipurpose plant with double-tier roller conveyor with aluminum tote box • Split-type source frame	2014	Multipurpose plant for food and agro products including spices, onion powder and flakes, fruits and vegetables; pet feed	Government of Gujarat with private O & M partner	• Commercial service Provider • Present activity: 610 kCi • Registered with NPPO for NSPM-21

#	Facility	Specifications	Year	Products	Operator	Status
17	Maharashtra State Agricultural Board, Vashi, Navi Mumbai, Maharashtra	• Category IV • Wet storage • Source-overlapping type • Design capacity: 1 MCi • Designed with moving shield	2015	Spices, onion powder and flakes, fruits and vegetables; pet feed	Government of Maharashtra	• Commercial service provider • Present activity: 304 kCi • Registered with NPPO for NSPM-21
18	Aligned Industries Dharuhera, Rewari, Haryana	• Category IV • Wet storage • Product overlapping type. • Design capacity: 1 MCi • Multipurpose plant with double-tier roller conveyor with aluminum tote box • Split-type source frame	2015	Sterilization of surgical items; containers for pharmaceutical industries; spices; raw materials for herbal products; pet feed	Private operator	• Commercial service provider • Present activity: 510 kCi
19	Electromagnetic Industries, Vadodara, Gujarat	• Category IV • Wet storage • Product-overlapping type • Design capacity: 3MCi • Multipurpose plant with double-tier roller conveyor with aluminum tote box • Split-type source frame	2018	Dehydrated onion flakes and powder, psyllium husk, chili and turmeric powder; pet feed	Private operator	• Commercial service provider • Present activity: 760 kCi
20	Pinnacle Therapeutics Private Limited, Ahmedabad, Gujarat	• Category IV • Wet storage • Product-overlapping type • Design capacity: 3 MCi • Multipurpose plant	2018	Medical and surgical products; pet feed	Private operator	• Commercial service provider • Present activity: 220 kCi

(Continued)

Table 8.5 (Continued)

S. No	Name of the radiation facility	Design features	Year of commission-ing	Purpose of usage	Operating agency/ body	Present activity and operational status
21	Jamnadas Industries, Betma. Indore, Madhya Pradesh	• Category IV • Wet storage • Product-overlapping type • Design capacity: 3 MCi • Multipurpose plant with double-tier roller conveyor • Close loop with automated loading/unloading station • Aluminum tote box • Split-type source frame • First irradiator installed integrated with 5000 Mt multipurpose cold storage for agro product	2020	Spices, soya chunk, pulses, corn grits, dehydrated onion flakes and powder; pet feed	Private operator	• Commercial service provider • Present activity: 410 kCi
22	Andhrapradesh Medtech Zone, Govt of Andhra Pradesh, Vishakapatnam	• Category IV • Wet storage • Product overlapping type • Design capacity: 3 MCi • Double-tier roller conveyor • Close loop with automated loading/unloading station • Aluminum tote box • Split type source frame	2020	Multipurpose plant	Joint venture of Government of Andhra Pradesh and GoI	• Present activity: 300 kCi • Commercial activity yet to start

No.	Facility	Technical features	Year	Plant type	Operator	Activity
23	AV GammaTech LLP, Ambernath, Mumbai	• Category IV • Wet storage • Product-overlapping type • Design capacity: 5 MCi • Double tier roller conveyor • Close loop with automated loading/unloading station • Aluminum tote box • Split-type source frame	2021	Multipurpose plant	Private operator	• Present activity: 210 kCi • Commercial activity yet to start
24	Microtrol Sterilisation Services Pvt. Ltd., Bawal, Haryana	• Category IV • Wet storage • Product-overlapping-batch type plant • Tote box design • Design capacity: 1.5 MCi	2021	Multipurpose plant	Private operator	• Commercial service provider • Present activity: 700 kCi
25	Akshar Gamma Steriles LLP, Anand Nagar, MIDC Ambernath, Maharashtra	• Category IV • Product overlapping • Design capacity 3 MCi with double-tier roller conveyor • Close loop with automated loading/unloading station • Aluminum tote box • Split-type source frame	2020	Multipurpose plant	Private operator	• Commercial service provider • Present activity: 240 kCi
26	Hi Media Laboratories Ltd., Anandnagar MIDC Ambernath, Maharashtra	• Category IV • Wet storage • Product overlapping • Design capacity: 3 MCi 2021	Plant for captive use	Private operator	• Present activity: 300 kCi	

(Continued)

247

Table 8.5 (Continued)

S. No	Name of the radiation facility	Design features	Year of commission-ing	Purpose of usage	Operating agency/ body	Present activity and operational status
27	Solas Industries, Koshi Kalan, Mathura, Uttar Pradesh	• Design capacity: 3MCi • Double-tier roller conveyor • Close loop with automated loading/unloading station • Aluminum tote box • Split-type source frame 2021	Multipurpose plant	Private operator	• Present activity: 300 kCi • Commercial activity yet to start	
28	Gammatech Pvt Ltd., Kayathar, Tirunelveli Tamil Nadu	• Design capacity: 3 MCi • Double-tier roller conveyor • Close loop with automated loading/unloading station • Aluminum tote box • Split-type source frame 2020	Multipurpose plant	Private operator	Source loading yet to be completed	

Source: Data from Mr. Subhasis Bhattacharya, Designer and Consultant of gamma radiation processing plants in India and abroad; Director, Danver Hydromatics Pvt. Ltd., Kolkata, West Bengal, India.

In the first decade of 2000, irradiated products from mostly Hawaii, Mexico and Asia captured the US supermarket produce section. In 2004–05, Australian irradiated produce items (mostly mangos) occupied a core niche in New Zealand markets, which by 2011 catapulted to more than 1000 metric tons annually. In 2008, Mexico marketed large volumes of irradiated produce, mostly guavas to the United States (Maherani et al., 2016; Eustice, 2017). In the present day, almost 22 countries including the United Kingdom, China, India, Japan, France, Finland, Germany and the Republic of Korea are operating about 515 radiation plants based on Russian technology. Moreover, the Rosatom State Atomic Energy Corporation of Russia have planned to extend their technology to the United Arab Emirates, the Republic of Mauritius and Malaysia (Eustice, 2017) to boost use of gamma-radiation processing of foods in these countries.

8.7 PACKAGING AND LABELING OF IRRADIATED PRODUCE

8.7.1 Packaging

Irradiated foods in flexible packaging are of concern, since it would favor migration of synthetic chemicals (such as monomers, additives, etc.) from synthetic plastic packaging directly into foods. For this reason, the FDA (Morehouse and Komolprasert, 2004) mandates that packaging materials of foods that simultaneously undergo irradiation along with foods must comply with regulations based on appropriate testing. Thin plastic films are objectionable since they allow microorganisms to enter through microscopic flaws or openings caused by damage during sealing and rough handling. Laminated foils and composites of plastics are more appropriate for irradiation (Chmielewski, 2006). Materials approved for use for food irradiation are listed in 21 CR 179.45 which specifies that glassine paper, coated cellophane, wax-coated paperboard, kraft paper, nylon 11, multilayer PET, PVDC-VC copolymer, PS and polyolefin films can be used up to 10 kGy; ethylene-vinyl acetate copolymers can be used up to 30 kGy; vegetable parchments; and multilayer PE, nylon 6, PET and PVC-VA copolymer films up to 60 kGy (Bose, 2000). At a dosage of more than 20 kGy, the physical properties of plastic films such as PE, Mylar (biaxially oriented polyethylene terephthalate), vinyl and polystyrene change with minor consequences to the products packaged within. At doses more than 30 kGy, Saran, Pliofilm and cellophanes become brittle, and thus, rigid containers such as tin-coated cans and aluminum containers are more preferred for packaged food irradiation at these doses. Off-odors are produced in plastic containers (such as those made of PE) upon irradiation at sterilizing doses owing to formation of short fragments of polymer, which migrate to foods and pose disastrous health consequences to humans. However, nylon is an exception that permits very low odor formation during sterilization irradiation and is therefore the most preferred for the irradiation of fish, meat and poultry products (Desrosier, 2012; Keay, 1968, Chmielewski, 2006).

8.7.2 Labeling

All irradiated foods must be labeled with the 'Radura' symbol, a stylized flower, in accordance with Food Safety Standards (Food Product Packaging and Labeling) Regulations, 2011. According to the FDA the phrase 'treated with radiation' or 'treated by irradiation' must be displayed along with the 'Radura' symbol (Morehouse, 2002). The logo is green in color and resembles a plant in circle, top half of which is dashed. The word *Radura* is derived from the word *radurization*. The initial letters were derived from the word *radiation* and the last letters from *durus*, which is the Latin word for hard-cum-lasting. The Radura logo symbolizes the following: the central dot is the radiation source, the two

PROCESSED BY IRRADIATION METHOD
DATE OF IRRADIATION..
LICENCE NO.................. ....
PURPOSE OF IRRADIATION..

....
....

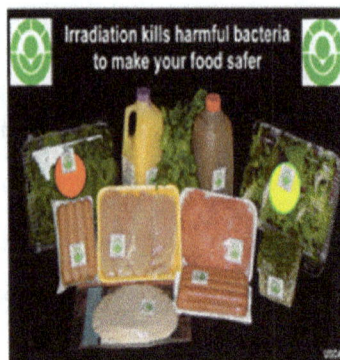

Figure 8.6 Labeling of packaged irradiated foods (Anonymous, 2021c).

circle segments ('leaves') are the biological shield to protect the workers and the environment, the outer ring is the transport system, the lower half is the shielded loading area and the upper broken half represents the rays hitting the target goods on the transport system (Ehlermann, 2009). The label on the packaged food product (Figure 8.6) must bear this logo along with the information: processed by irradiation, the date of irradiation, the license number and the purpose of irradiation. Additionally, the label must also declare that irradiated commodities such as potato and onion that are consumed for their nutritional values must not be used for cultivation purposes (Chakraborty, 2010).

8.8 LIMITATIONS OF FOOD IRRADIATION

As a method of food preservation, food irradiation has a few limitations. Irradiation has a high capital cost and requires a critical minimum capacity and product volume for economic operation (Urbain, 1982; Vas, 1979), although it entails low operating costs and low energy consumption (Loaharanu, 1995). The success of the treatment depends on the commodities to be irradiated, including the types of cultivars/variants, their physiological status, the degree of maturity (i.e., stage of senescence), pre- and postharvest treatments, the dose of radiation, the conditions of storage of the product post-irradiation (temperature and relative humidity) and the susceptibility of the agro produce/food products to microbial infestation and of the microorganisms to radiation damage (Willemoti et al., 1996). Most important, radiation processing is a needs-based technology and cannot be ubiquitously applied to all agro commodities or food products. It cannot render a spoiled food restore its pre-irradiated organoleptic value or destroy pesticide residues and toxins present therein.

The amenability of a particular food commodity to radiation processing must be tested in a laboratory prior to subjecting the same to gamma irradiation. Only those foods for which specific benefits are achievable at appropriate radiation doses and those duly permitted to be irradiated under the Atomic Energy (Radiation Processing of Food and Allied Products) Rules, 2012 can be processed by radiation (DAE, 2014). There are also safety issues concerning food irradiation such as residual radioactivity of the irradiated products, including the presence of free radicals and radiolytic products therein; the occurrence of polyploidy in—and carcinogenic and mutagenic properties, nutritional quality, toxicity and microbiological status of—the irradiated produce; and safety of the plant operators who handle radiation processing (Moy, 1993).

Another serious limitation of irradiation processing of food products is that unless food packages are labeled prior to irradiation, there is difficulty in identifying an irradiated food product and distinguish it from its nonirradiated counterpart. Irradiation is known to break chemical bonds forming free radicals that combine with other molecules to form radiolytic products, such as benzene (from potassium benzoate) in irradiated meat (Zhu et al., 2004), and MDA, formaldehyde and acetaldehyde in carbohydrate-rich (sugars) foods (EFSA, 2011). The compounds produced by radiolysis of fats such as long-chain hydrocarbons and 2-alkyl-cyclobutanones have also been identified by gas chromatography-mass spectrometry (GC-MS) analysis of irradiated foods by several researchers, which can unambiguously serve as biomarkers of irradiation (Bhattacharjee et al., 2003; Kim et al., 2004; Bose, 2000, Nawar et al., 1996). However, rigorous extraction-cum-downstream processing is required to obtain these compounds from fat-rich matrices, where co-extracted fat renders biomarker(s) identification challenging. Solvent-less supercritical CO_2 extraction is warranted to selectively extract the biomarkers to forgo cumbersome chromatographic separation of the extracts obtained by conventional extraction procedures (Bhattacharjee et al., 2003). These analytical techniques are yet to be adopted in place in irradiation plants in our country and worldwide.

Other concerns of gamma irradiation include irradiation of spices conducted commercially at high doses are detrimental to their phenolic contents, which are responsible for their antioxidative potency; there is a decline in fecundity in fruit flies when fed with irradiated meat (chicken); there are chances of production of aflatoxins on irradiated grains owing to elimination of competing microorganisms by irradiation and the possibility of recontamination of irradiated foods if they encounter unclean surfaces or raw foods during inappropriate storage, handling and processing. Thus, the complete safety of irradiated products for human consumption is still questionable (Vas, 1979).

However, the foremost challenge in the production and consumption of irradiated foods still remains to be poor knowledge and a lack of awareness among consumers about gamma-irradiation technology and the safety of irradiated foods. A lack of harmonization among countries that have approved irradiation of foods impedes intercountry trade of irradiated foods, which is a crucial challenge (Loaharanu and Ahmed, 1991) affecting process economics. Griffith (1992) identified few major factors that impede commercialization of irradiated foods, such as a lack of commitment of the food industry for use of this processing technology; the development of a less complicated, reliable analytical procedure for distinguishing irradiated foods from their nonirradiated counterparts; the conservative attitude of consumers for the consumption of irradiated food products; and the nonexistence of harmonization of legislation needed to determine the future of food irradiation. Moy (1993) also identified four causes for slow commercialization of this technology, such as the time-consuming approval process for food irradiation, the prevalence of global antinuclear activism, insufficient consumer education and industry's hesitation in acceptance of this technology. Loaharanu (1995) studied the results of consumer attitude surveys conducted in different countries and concluded that in advanced countries, consumers who are aware of radiation processing demand precise information about safety, benefits, and limitations of food irradiation, while a large section of the population lack knowledge about the radiation treatment of foods and their suitability for consumption. Hashim et al. (1996) reviewed consumer attitudes about irradiated poultry and recommended educating consumers about safety of irradiated foods through seminars/television programs; the distribution of pamphlets, brochures and the like; and scientific investigations, including the implementation of in-store sampling and surveys of irradiated foods, such as cooked irradiated poultry (Rahman, 2007). A survey conducted in Santiago, Chile, in 2011 with 497 people assessed their knowledge and acceptance level of irradiated foods and concluded that about 46% of consumers expressed their belief that irradiated food is synonymous to radioactive food and 91% of consumers agreed to consume irradiated food only after being completely assured of its safety (Junqueira-Goncalves et al., 2011).

Thus, consumer awareness and acceptance of irradiated foods as safe for consumption would be the chief criteria that would decide future developments in the field of food irradiation considering its huge potential in extending the shelf lives of agro commodities postharvest and in delivering healthy foods safe for human health and for the environment. It is necessary to disseminate *en masse* the positive outcomes of scientific toxicological studies and animal feeding trials, which have shown no evidence of toxicity on consumption of irradiated food products (Rahman, 2007), to escalate consumer acceptance globally. Implementing a harmonized international regulation among all countries can possibly overcome the limitations regarding marketing and consumption of irradiated foods and would allow hassle-free international trade of irradiated food products.

8.9 CONCLUSION AND FUTURE PERSPECTIVES

In 1980, the Expert Committee on the Wholesomeness of Irradiated Foods- a joint body of the WHO, FAO and IAEA concluded that the "irradiation of any commodity up to an overall average dose of 10 kGy presents no toxicological hazards and introduces no special nutritional or microbiological problems". Subsequent reviews conducted by Canada, Denmark, France, Sweden, the United Kingdom and the United States and by the Scientific Committee for Food of the European Economic Commission reaffirmed the above finding. In 1968, a WHO/FAO/IAEA Task Force Committee created to assess the use of irradiation in ensuring hygienic quality of foods concluded that

at present, and in the foreseeable future, no known technology can guarantee the production of raw foods of animal origin, particularly poultry and pork, without the presence of certain pathogenic organisms and parasites, such as *Salmonella*, *Campylobactor*, *Toxoplasma* and *Trichinella*, where such foods are important in the epidemiology of foodborne diseases, decontamination/disinfection by irradiation must be seriously considered.

Loaharanu and Murrell in1994 opined that a ban on the usage of ethylene oxide for foods by the European Union could boost the irradiation processing of spices and vegetable seasonings. Fumigants such as ethylene oxide are reportedly carcinogenic, and methyl bromide could be harmful to the ozone layer (Loaharanu, 1992). In May 1992, the WHO reviewed the safety and nutritional concerns of irradiated foods by food safety specialists. The participants unanimously agreed to the following:

Irradiated food produced under established good manufacturing practices (GMP) is to be considered safe and nutritionally adequate because the process:

- will not introduce changes in the composition of the food which, from a toxicological point of view, would impose an adverse effect on human health;

- will not introduce changes in the microflora of the food which would increase the microbiological risk to consumers;

■ will not introduce nutrient losses in the composition of the food which, from a nutritional point of view, would impose an adverse effect on the nutritional status of individuals or populations.

According to IAEA, 40 types of food and agricultural products are currently allowed for irradiation in about 40 countries, whereas in the United States, around 60 commercial irradiation facilities are in operation (Sommers, 2004). More rigorous research is needed in the following domains of food irradiation for wider acceptance of gamma-irradiation technology as a safe and effective method of food processing and preservation: the identification and quantification of radiolytic products as biomarkers of irradiated foods as a function of dose and storage time; a comparison of the toxicological properties of volatile components of irradiated and nonirradiated foods; a comparison of the losses of nutritional values of foods consequent to irradiation vis-à-vis those processed by the other technologies; the effects of combinations of irradiation and other processing technologies on the nutritional value of food products; chemical, nutritional and toxicological studies of the radiolytic products of lipids, with special emphasis on peroxide and epoxide formation and *cis-trans* isomerization; the collection and scrutiny of data from long-established cohort studies on animals for possible *in vivo* changes that may have resulted from consumption of irradiated laboratory feed; and development of novel radiation-stable rigid and flexible food packaging composites/laminates. These endeavors would certainly enhance the knowledge of the consequences of food processing by irradiation and would thus facilitate an expansion of the use of this green technology in food preservation.

Acknowledgment: Authors are grateful to Ms. Sahida Sultana for assisting us with the bibliography of the chapter.

REFERENCES

Abad, J., S. Valencia-Chamorro, A. Castro, and C. Vasco. 2017. Studying the effect of combining two nonconventional treatments, gamma irradiation and the application of an edible coating, on the postharvest quality of tamarillo (*Solanum betaceum Cav.*) fruits. *Food Control* 72:319–323.

Abboudi, M., M. Al-Bachir, Y. Koudsi, and H. Jouhara. 2016. Combined effects of gamma irradiation and blanching process on acrylamide content in fried potato strips. *International Journal of Food Properties* 19:1447–1454.

Abd El-Ghafour, S., A. H. Zakar, and A. S. Mohamad. 2018. Changes of fatty acid profile of mullet fish (*Mugil cephalus*) fillets as influenced by gamma irradiation. *Egyptian Journal of Aquatic Research* 44:241–244.

Abdelaleem, M. A., and K. R. A. Elbassiony. 2021. Evaluation of phytochemicals and antioxidant activity of gamma irradiated quinoa (*Chenopodium quinoa*). *Brazilian Journal of Biology* 81:806–813.

Abdeldaiem, M. H., H. G. Mohammad, and M. F. Ramadan. 2018. Improving the quality of silver carp fish fillets by gamma irradiation and coatings containing rosemary oil. *Journal of Aquatic Food Product Technology* 27:568–579.

Akbari, M., M. Farajpour, M. Aalifar, and M. S. Hosseini. 2018. Gamma irradiation affects the total phenol, anthocyanin and antioxidant properties in three different persian pistachio nuts. *Natural Product Research* 32:322–326.

Al-Bachir, M. 2016. Some microbial, chemical and sensorial properties of gamma irradiated sesame (*Sesamum indicum L.*) seeds. *Food Chemistry* 197:191–197.

Alinezhad, M., M. Hojjati, H. Barzegar, S. Shahbazi, and H. Askari. 2020. Effect of gamma irradiation on the physicochemical properties of pistachio (*Pistacia vera L.*) nuts. *Journal of Food Measurement and Characterization* 15:199–209.

Al-Kuraieef, A. N., and A. H. Alshawi. 2020. The effect of gamma irradiation on the essential oils and antioxidants in dried thyme. *International Journal of Food Studies* 9:203–212.

Al-Qadiri, H., A. Amr, M. A. Al-Holy, and M. Shahein. 2020. Effect of gamma irradiation against microbial spoilage of hummus preserved under refrigerated storage. *Food Science and Technology International* 0:1–10.

Anonymous. 2021a. Irradiation. Retrieved from https://en.wikipedia.org/wiki/Irradiation (accessed on October 25, 2021).

Anonymous. 2021b. Cobalt-60. Retrieved from https://en.wikipedia.org/wiki/Cobalt-60 (accessed on October 25, 2021).

Anonymous. 2021c. Retrieved from www.google.com/search?q=labelling+of+irradiated+foods+by+pfa&rlz=1C1VDKB_enIN929IN929&sxsrf=AOaemvKyWEZJz41OG4tZYzkHFQ9_4WCn1w:1635496439067&source=lnms&tbm=isch&sa=X&ved=2ahUsKEwi-iKCCm-_zAhX-xjgGHcsNBy0Q_AUoAXoECAEQAw&biw=1280&bih=520&dpr=1.5 (accessed on October 29, 2021).

Apaydin, D., A. S. Demirci, and U. Gecgel. 2017. Effect of gamma irradiation on biochemical properties of grape seeds. *Journal of the American Oil Chemists Society* 94:57–67.

Arshad, M. S., Z. Amjad, M. Yasin, F. Saeed, A. Imran, M. Sohaib, F. M. Anjum, and S. Hussain. 2019. Quality and stability evaluation of chicken meat treated with gamma irradiation and turmeric powder. *International Journal of Food Properties* 22:154–172.

Ashtari, M., O. Khademi, M. Soufbaf, H. Afsharmanesh, and M. A. Askari Sarcheshmeh. 2019. Effect of gamma irradiation on antioxidants, microbiological properties and shelf life of pomegranate arils cv. 'Malas Saveh'. *Scientia Horticulturae* 244:365–371.

Aylangan, A., E. Ic, and B. Ozyardimci. 2017. Investigation of gamma irradiation and storage period effects on the nutritional and sensory quality of chickpeas, kidney beans and green lentils. *Food Control* 80:428–434.

Bansal, N., P. Dahiya, G. R. Prashat, D. Goswami, S. Kumari, S. Pushkar, A. Kumar, B. Singh, A. Sachdev, T. Vinutha, and S. Praveen. 2020. Effects of gamma irradiation on soybean oil stability by enhancing tocopherol content in soybean. *Journal of Radioanalytical and Nuclear Chemistry* 326:1617–1629.

Bashir, K., and M. Aggarwal. 2016. Effects of gamma irradiation on the physicochemical, thermal and functional properties of chickpea flour. *LWT – Food Science and Technology* 69:614–622.

Ben-Fadhel, Y., M. C. Cingolani, L. Li, G. Chazot, S. Salmieri, C. Horak, and M. Lacroix. 2021. Effect of γ-irradiation and the use of combined treatments with edible bioactive coating on carrot preservation. *Food Packaging and Shelf Life* 28:100635.

Bhat, N. A., I. A. Wani, A. M. Hamdani, and F. A. Masoodi. 2020. Effect of gamma-irradiation on the thermal, rheological and antioxidant properties of three wheat cultivars grown in temperate Indian climate. *Radiation Physics and Chemistry* 176:108953.

Bhattacharjee, P., and R. S. Singhal. 2009. Effect of irradiation on food texture and rheology. *Novel Food Processing*, ed. J. Ahmed, H. S. Ramaswamy, S. Kasapis and J. I. Boye, 103–125. Boca Raton: CRC Press.

Bhattacharjee, P., R. Singhal, P. Iyer, and A. Gholap. 2003. Hydrocarbons as marker compounds for irradiated cashew nuts. *Food Chemistry* 80:151–157.

Bose, M. 2000. Nashik to house country's first onion irradiation centre. *Indian Express Newspapers (Bombay) Ltd.*

Cassares, M., N. L. Sakotani, L. Kunigk, P. A. S. Vasquez, and C. Jurkiewicz. 2020. Effect of gamma irradiation on shelf life extension of fresh pasta. *Radiation Physics and Chemistry* 174:108940.

Cătunescu, G. M., M. Muntean, O. Marian, A. P. David, and A. M. Rotar. 2019. Comparative effect of gamma irradiation, drying and freezing on sensory, and hygienic quality of parsley leaves. *LWT – Food Science and Technology* 115:108448.

Chakraborty, P. 2010. Radiation processing of food – safety & quality. *Indian Food Industry* 29: 23–32.

Chen, Q., M. Cao, H. Chen, P. Gao, Y. Fu, M. Liu, Y. Wang, and M. Huang. 2016. Effects of gamma irradiation on microbial safety and quality of stir fry chicken dices with hot chili during storage. *Radiation Physics and Chemistry* 127:122–126.

Chitravathi, K., O. P. Chauhan, and J. Kizhakkedath. 2020. Shelf life extension of green chillies (*Capsicum annuum* L.) using passive modified atmosphere packaging and gamma irradiation. *Journal of Food Processing and Preservation* 44:1–8.

Chmielewski, A. G. 2006. *Packaging for Food Irradiation*, 1–26. Warsaw, Poland: Institute of Nuclear Chemistry and Technology.

da Silva Aquino, K.A. 2012. Sterilization by gamma irradiation. *Gamma Radiation*, ed. Prof. F. Adrovic, 953–978. Rijeka, Croatia: In Tech.

DAE (Department of Atomic Energy), Board of Radiation & Isotope Technology. 2014. *Radiation Processing of Food and Medical Products-Technical Document*, 1–89. Mumbai: Board of Radiation and Isotope Technology.

Dasgupta, C. R. 2017. *A Handbook of Degree Physics: Volume II*. Kolkata, India: Book syndicate Pvt. Ltd.

Desrosier, N. W. and D. K. Tressler. 2012. *Fundamentals of Food Freezing*. Westport, CT: AVI Publishing Company, INC.

EFSA Panel on Food Contact Materials, Enzymes, Flavourings and Processing Aids (CEF). 2011. Scientific Opinion on the chemical safety of irradiation of food. *EFSA Journal* 9:4:1930.

Ehlermann, D. A. E. 2009. The Radura terminology and food irradiation. *Food Control* 20:526–528.

Erkmen, O., and T. F. Bozoglu. 2016. Food preservation by irradiation. *Food Microbiology: Principles into Practice*, 106–126. Chichester, UK: John Wiley and Sons, Ltd.

Esmaeili, S., M. Barzegar, M. A. Sahari, and S. Berengi-Ardestani. 2018. Effect of gamma irradiation under various atmospheres of packaging on the microbial and physicochemical properties of turmeric powder. *Radiation Physics and Chemistry* 148:60–67.

Eustice, R. F. 2017. Global status and commercial applications of food irradiation. *In: Food Chemistry, Function and Analysis*, ed. I. C. F. R. Ferreira, A. L. Antonio and S. C. Verde, 397–424. The Royal Society of Chemistry.

Fernandes, Â., J. C. M. Barreira, A. L. Antonio, M. B. P. P. Oliveira, A. Martins, and I. C. F. R. Ferreira. 2016. Extended use of gamma irradiation in wild mushrooms conservation: Validation of 2 kGy dose to preserve their chemical characteristics. *LWT – Food Science and Technology* 67:99–105.

Ghasemi-Varnamkhasti, M., A. Mohammad-Razdari, S. H. Yoosefian, and Z. Izadi. 2018. Effects of the combination of gamma irradiation and Ag nanoparticles polyethylene films on the quality of fresh bottom mushroom (*Agaricus bisporus* L.). *Journal of Food Processing and Preservation* 42:1–8.

Ghosh, P. K., S. Chatterjee, P. Bhattacharjee, and N. Bhattacharyya. 2016. Removal of rancid-acid odor of expeller-pressed virgin coconut oil by gamma irradiation: Evaluation by sensory and electronic nose technology. *Food and Bioprocess Technology* 9:1724–1734.

Gojiya, D. K., S. P. Cholera, and A. M. Joshi. 2020. Influence of gamma irradiation on microbial load of peanut (*Arachis hypogaea* L) kernels. *International Journal of Current Microbiology and Applied Sciences* 9:589–602.

Grandison, A. S. 2006. *Food Processing Handbook*. Ed. J. G. Brennan. Weinheim: Wiley-VCH Verlag GmbH and Co. KGaA.

Griffith, G. 1992. Irradiated foods: New technology, old debate. *Trends in Food Science and Technology* 3:251.

Gumul, D., and W. Berski. 2021. The polyphenol profile and antioxidant potential of irradiated rye grains. *International Journal of Food Science* 2021:1–7.

Gyimah, L. A., H. M. Amoatey, R. Boatin, V. Appiah, and B. T. Odai. 2020. The impact of gamma irradiation and storage on the physicochemical properties of tomato fruits in Ghana. *Food Quality and Safety* 4:151–157.

Hamdani, A. M., I. A. Wani, and N. A. Bhat. 2018. Effect of gamma irradiation on the physicochemical and structural properties of plant seed gums. *International Journal of Biological Macromolecules* 106:507–515.

Haque, A., A. Hashem, M. M. Hossain, F. J. Rima, and A. A. Hossain. 2017. Effect of gamma irradiation on shelf life and quality of beef. *Journal of Meat Science and Technology* 5:20–28.

Hashemabad, Z. N., B. Shabanpour, H. Azizi, S. M. Ojagh, and A. Alishahi. 2018. Effects of TiO$_2$ nanocomposite packaging and gamma irradiation on the shelf-life of rainbow trout stored at (+4°C). *Turkish Journal of Fisheries and Aquatic Sciences* 18:1387–1397.

Hashim, I. B., A. V. Resurreccion, and K. H. McWatters. 1996. Consumer acceptance of irradiated poultry. *Poultry Science* 74:1287–1294.

Hassan, A. B., N. S. Mahmoud, K. Elmamoun, O. Q. Adiamo, and I. A. Mohamed Ahmed. 2018. Effects of gamma irradiation on the protein characteristics and functional properties of sesame (*Sesamum indicum* L.) seeds. *Radiation Physics and Chemistry* 144:85–91.

Hassanein, R. A., E. A. Salem, and A. A. Zahran. 2018. Efficacy of coupling gamma irradiation with calcium chloride and lemongrass oil in maintaining guava fruit quality and inhibiting fungal growth during cold storage. *Folia Horticulturae* 30:67–78.

Hassanzadeh, P., H. Tajik, S. M. R. Rohani, M. Moradi, M. Hashemi, and J. Aliakbarlu. 2017. Effect of functional chitosan coating and gamma irradiation on the shelf-life of chicken meat during refrigerated storage. *Radiation Physics and Chemistry* 141:103–109.

Hou, L., J. Lin, L. Ma, S. Qu, H. Li, and N. Jiang. 2018. Effect of ^{60}Co gamma irradiation on postharvest quality and selected enzyme activities of *Volvariella volvacea*. *Scientia Horticulturae* 235:382–390.

Hu, Z., Y. Xiao, B. Wang, T. Z. Jin, W. Lyu, and D. Ren. 2021. Combined treatments of low dose irradiation with antimicrobials for inactivation of foodborne pathogens on fresh pork. *Food Control* 125:1–7.

Huang, H. H., P. K. Dikkala, K. Sridhar, H. T. Yang, J. T. Lee, and F. J. Tsai. 2021. Effect of heat and γ-irradiation on fungal load, pasting, and rheological characteristics of three whole and dehulled millets during storage. *Journal of Food Processing and Preservation* 45:1–12.

Hussain, P. R., P. Suradkar, S. Javaid, H. Akram, and S. Parvez. 2016. Influence of postharvest gamma irradiation treatment on the content of bioactive compounds and antioxidant activity of fenugreek (*Trigonella foenum-graceum* L.) and spinach (*Spinacia oleracea* L.) leaves. *Innovative Food Science and Emerging Technologies* 33:268–281.

Hussain, P. R., S. A. Rather, P. P. Suradkar, and A. Omeera. 2021. Gamma irradiation treatment of minimally processed kiwi fruit to maintain physicochemical quality and prevent microbial proliferation during refrigerated storage. *Journal of Food Processing and Preservation* 45:1–10.

Hussain, P. R., S. A. Rather, P. P. Suradkar, and O. Ayob. 2019. Gamma irradiation treatment of quince fruit (*Cydonia oblonga* Mill): Effect on post-harvest retention of storage quality and inhibition of fungal decay. *Journal of Radiation Research and Applied Sciences* 12:118–131.

IAEA (International Atomic Energy Agency). 2004. Directory of gamma radiation processing facilities in Member States. *Industrial Applications and Chemistry Section*. Vienna, Austria: IAEA.

IAEA (International Atomic Energy Agency). 2006. Gamma irradiators for radiation processing (INIS-XA–862). *Industrial Applications and Chemistry Section.* Vienna, Austria: IAEA.

IAEA (International Atomic Energy Agency). 2010. IAEA safety standards for protecting people and the environment: Radiation safety of gamma, electron and x ray irradiation facilities; Specific safety guide: No.-SSG-8. *Safety Fundamentals, Safety Requirements and Safety Guides:* IAEA, Vienna, Austria.

IARP (Indian Association for Radiation Protection). 2016. *RSO Certification Course for Gamma Irradiation Chamber*, 3.1–4.4. Mumbai: Indian Association for Radiation Protection.

Jan, A., M. Sood, K. Younis, and R. U. Islam. 2020. Brown rice based weaning food treated with gamma irradiation evaluated during storage. *Radiation Physics and Chemistry* 177:109158.

Janiak, M. A., A. Slavova-Kazakova, M. Karamać, V. Kancheva, A. Terzieva, M. Ivanova, T. Tsrunchev, and R. Amarowicz. 2017. Effects of gamma-irradiation on the antioxidant potential of traditional Bulgarian teas. *Natural Product Communications* 12:181–184.

Jeong, R. D., E. H. Chu, G. W. Lee, C. Cho, and H. J. Park. 2016. Inhibitory effect of gamma irradiation and its application for control of postharvest green mold decay of Satsuma mandarins. *International Journal of Food Microbiology* 234:1–8.

Jeong, S. G., J. E. Yang, J. H. Park, S. H. Ko, I. S. Choi, H. M. Kim, H. H. Chun, M. J. Kwon, and H. W. Park. 2020. Gamma irradiation improves the microbiological safety and shelf-life of kimchi seasoning mixture. *LWT – Food Science and Technology* 134:110144.

Jia, W., Q. Shi, and L. Shi. 2021. Effect of irradiation treatment on the lipid composition and nutritional quality of goat meat. *Food Chemistry* 351:129295.

Jo, Y., K. Ameer, N. Chung, and J. H. Kwon. 2019. Influence of gamma irradiation on volatile flavour profiles and physicochemical attributes of navel oranges (*Citrus sinensis* L.) under post-irradiation storage. *International Food Research Journal* 26:1035–1043.

Joint FAO/WHO Codex Alimentarius Commission. 1984. Codex general standard for irradiated foods and recommended international code of practice for the operation of radiation facilities used for the treatment of foods. *Codex Alimentarius* 15: 5–75, FAO/WHO, Rome.

Junqueira-Goncalves, M. P., M. J. Galotto, X. Valenzuela, C. M. Dinten, P. Aguirre, and J. Miltz. 2011. Perception and view of consumers on food irradiation and the Radura symbol. *Radiation Physics and Chemistry* 80:119–122.

Kang, S., S. Y. Park, and S. D. Ha. 2016. Application of gamma irradiation for the reduction of norovirus in traditional Korean half-dried seafood products during storage. *LWT – Food Science and Technology* 65:739–745.

Keay, J. N. 1968. The effect of doses of gamma radiation up to 16 Mrad on plastic packaging materials for fish. *International Journal of Food Science and Technology* 3:123–129.

Khan, Q. U., I. Mohammadzai, Z. Shah, I. Ullah, T. N. Khattak, H. Noreen, and W. Hassan. 2018. Effect of gamma irradiation on nutrients and shelf life of peach (*Prunus persical*) stored at ambient temperature. *The Open Conference Proceedings Journal* 09:8–15.

Kim, J. H., H. Y. Seo, and K. S. Kim. 2004. Analysis of radiolytic products of lipid in irradiated dried squids (*Todarodes pacificus*). *Journal of Food Protection* 67:1731–1735.

Kirkin, C., and G. Gunes. 2018. Modified atmosphere packaging and gamma-irradiation of some herbs and spices: Effects on antioxidant and antimicrobial properties. *Journal of Food Processing and Preservation* 42:1–11.

Koç Güler, S., S. Z. Bostan, and A. H. Çon. 2017. Effects of gamma irradiation on chemical and sensory characteristics of natural hazelnut kernels. *Postharvest Biology and Technology* 123:12–21.

Kumari, S., O. P. Gupta, C. B. Mishra, V. Thimmegowda, V. Krishnan, B. Singh, A. Sachdev, and A. Dahuja. 2020. Gamma irradiation, an effective strategy to control the oxidative damage of soy proteins during storage and processing. *Radiation Physics and Chemistry* 177:109134.

Loaharanu, P. and D. Murrell. 1994. A role for irradiation in the control of food-borne parasites. *Trends in Food Science and Technology* 5:190–195.

Loaharanu, P. and M. Ahmed. 1991. Advantages and disadvantages of the use of irradiation for food preservation. *Journal of Agricultural and Environmental Ethics* 4:14–30.

Loaharanu, P. 1992. The status and prospects of food irradiation. *The IFTEC Symposium S-14 on Food Irradiation: Recent Developments and Future Prospects.* Hague, The Netherlands: IFTEC.

Loaharanu, P. 1995. Food irradiation: Current status and future prospects. *New Methods of Food Preservation*, ed. G. W. Gould, 90–111. Boston, MA: Springer.

Lwin, T. T. 2020. A study on effects of gamma irradiation on fresh cucumber fruit (*Cucumis sativus L* .). *Journal of the Myanmar Academy of Arts and Science* XVIII: 147–153.

Maherani, B., F. Hossain, P. Criado, Y. Ben-Fadhel, S. Salmieri, and M. Lacroix. 2016. World market development and consumer acceptance of irradiation technology. *Foods* 5:79.

Mesquita, T. C., M. C. E. V. Schiassi, A. M. T. Lago, Í. Careli-Gondim, L. M. Silva, N. de Azevedo Lira, E. E. N. Carvalho, and L. C. de Oliveira Lima. 2020. Grape juice blends treated with gamma irradiation evaluated during storage. *Radiation Physics and Chemistry* 168:108570.

Miladilari, S., M. Ahmadi, A. Kashi, A. Mousavi, and Y. Mostofi. 2019. Physiological responses of gamma-irradiated onion bulbs during storage. *Journal of Agricultural Sciences (Tarim Bilimleri Dergisi)* 26:442–451.

Morehouse, K. M. 2002. Food irradiation – US regulatory considerations. *Radiation Physics and Chemistry* 63:281–284.

Morehouse, K. M., and V. Komolprasert. 2004. Irradiation of food and packaging: An overview. *ACS Symposium Series* 875:1–11.

Mostafavi, H. A., S. M. Mirmajlessi, and H. Fathollahi. 2012. The potential of food irradiation: Benefits and limitations. *Trends in Vital Food and Control Engineering, ed.* Prof. A. A. Eissa, 43–68. Rijeka, Croatia: InTech.

Moy, J. H. 1993. Food irradiation-lessons and prospects for world food preservation and trade. *Development of Food Science and Technology in South East Asia, ed.* O.B. Liang, A. Buchanan, and D. Fardiaz, 86. Bogor: IPB Press.

Najafabadi, N. S., M. A. Sahari, M. Barzegar, and Z. H. Esfahani. 2017. Effect of gamma irradiation on some physicochemical properties and bioactive compounds of jujube (*Ziziphus jujuba var vulgaris*) fruit. *Radiation Physics and Chemistry* 130:62–68.

Narvaiz, P. 2015. Irradiated food for special diets. *Stewart Postharvest Review* 3.3:1–7.

Nawar, W. W., Z. Zhu, H. Wan, E. DeGroote, Y. Chen, and T. Aciukewicz. 1996. Progress in the detection of irradiated foods by measurement of lipid-derived volatiles. *Detection Methods for Irradiated Foods: Current Status.* Ed. R. Gray and J. Pearce. Belfast, Northern Ireland, United Kingdom: Royal Society of Chemistry.

Nisar, M. F., M. S. Arshad, M. Yasin, M. K. Khan, M. Afzaal, S. Sattar, and H. A. R. Suleria. 2020. Evaluation of gamma irradiation and moringa leaf powder on quality characteristics of meat balls under different packaging materials. *Journal of Food Processing and Preservation* 44:1–11.

Odueke, O. B., S. A. Chadd, R. N. Baines, K. W. Farag, and J. Jansson. 2018. Effects of gamma irradiation on the shelf-life of a dairy-like product. *Radiation Physics and Chemistry* 143:63–71.

Ofosu, D. O., F. Appiah, and B. K. B. Banful. 2020. Interactive effect of different fruit packaging materials and doses of gamma irradiation on postharvest behaviour of Apantu (*Musa sp* AAB) fruits from the green stage to the onset of ripening. *International Journal of Postharvest Technology and Innovation* 7:109–117.

Pal, S., and P. Bhattacharjee. 2020. Gamma irradiation of trilamellate packaged yellow corn (*Zea mays*) kernels enhances lutein content and shelf life. *Journal of Food Processing and Preservation* 44:1–14.

Panou, A. A., I. K. Karabagias, and K. A. Riganakos. 2020. Effect of gamma-irradiation on sensory characteristics, physicochemical parameters, and shelf life of strawberries stored under refrigeration. *International Journal of Fruit Science* 20:191–206.

Patil, H., N. G. Shah, S. N. Hajare, S. Gautam, and G. Kumar. 2019. Combination of microwave and gamma irradiation for reduction of aflatoxin B_1 and microbiological contamination in peanuts (*Arachis hypogaea* L.). *World Mycotoxin Journal* 12:269–280.

Pereira, E., A. I. Pimenta, R. C. Calhelha, A. L. Antonio, S. C. Verde, L. Barros, C. Santos-Buelga, and I. C. F. R. Ferreira. 2016. Effects of gamma irradiation on cytotoxicity and phenolic compounds of *Thymus vulgaris* L. and *Mentha x piperita* L. *LWT – Food Science and Technology* 71:370–377.

Pillai, S. D., and S. Shayanfar. 2017. Electron beam technology and other irradiation technology applications in the food industry. *Topics in Current Chemistry* 375:249–268.

Pinela, J., J. C. M. Barreira, L. Barros, S. C. Verde, A. L. Antonio, A. M. Carvalho, M. B. P. P. Oliveira, and I. C. F. R. Ferreira. 2016. Suitability of gamma irradiation for preserving fresh-cut watercress quality during cold storage. *Food Chemistry* 206:50–58.

Pourebrahimi, Y., M. Fazel, and H. R. Akhavan. 2020. Effect of gamma irradiation on the quality characteristics and shelf life of pomegranate arils during refrigerated storage. *Journal of Agricultural Science and Technology* 22:723–735.

Rahim, F. N. A., W. Z. Wan-Ibadullah, A. Zainuri, M. Dzomir, R. Shukri, N. Hanani, Z. Abedin, N. A. Mustapha, K. Mahmud, A. Rashid, and I. F. M. Rashedi. 2021. Effect of gamma irradiation on physicochemical properties and microbiological quality of yellow noodle with addition of rice bran. *Journal of Advanced Research in Applied Sciences and Engineering Technology* 22:26–45.

Rahman, M. S. 2007. *Handbook of Food Preservation*. Boca Raton: CRC Press/Taylor and Francis.

Rodrigues, I., A. Baldini, M. Pires, J. C. Barros, R. Fregonesi, C. G. de Lima, and M. A. Trindade. 2021. Gamma ray irradiation: A new strategy to increase the shelf life of salt-reduced hot dog wieners. *LWT – Food Science and Technology* 135:110265.

Sarkar, P. and S. K. Mahato. 2020. Effect of gamma irradiation on sprout inhibition and physical properties of *Kufri jyoti* variety of potato. *International Journal of Current Microbiology and Applied Sciences* 9:1066–1079.

Sarkar, P., D. Tamili, and P. Bhattacharjee. 2021. Low dose gamma-irradiation enhances shelf-life and contents of serotonin and melatonin in green plantains (*Musa paradisiaca*): A study involving antioxidant synergy. *Journal of Food Processing and Preservation* 00:e15934.

Sau, S., P. Datta, T. Sarkar, and S. Sarkar. 2018. Impact of low doses of gamma irradiation on off-season guava at ambient storage condition. *International Journal of Current Microbiology and Applied Sciences* 7:295–307.

Sharma, P., S. R. Sharma, R. K. Dhall, and T. C. Mittal. 2020. Effect of γ-radiation on post-harvest storage life and quality of onion bulb under ambient condition. *Journal of Food Science and Technology* 57:2534–2544.

Sharma, P., S. R. Sharma, R. K. Dhall, T. C. Mittal, and S. Kumar. 2021. Variation in quality and acceptability of minimally processed garlic in response to γ-irradiation and packaging during refrigerated storage. *Radiation Physics and Chemistry* 180:109193.

Sirisoontaralak, P., P. Suthirak, K. Papaka, and P. Vongsawasdi. 2017. Development of shelf stable chiffon cake using gamma irradiation. *LWT – Food Science and Technology* 75:78–84.

Sommers, C. H. 2004. Food irradiation is already here. *Food Technology* 58:22.

Song, W. J., Y. H. Kim, and D. H. Kang. 2018. Effect of gamma irradiation on inactivation of *Escherichia coli* O157:H7, *Salmonella typhimurium* and *Listeria monocytogenes* on pistachios. *Letters in Applied Microbiology* 68:96–102.

Surendra, I. H. W., E. M. R. K. B. Edirisinghe, and R. M. N. P. Rathnayake. 2018. Effect of low-dose gamma irradiation on the quality of tilapia fish muscle with storage at 0°C. *American Journal of Food Science and Technology* 6:237–246.

Tejedor-Calvo, E., D. Morales, S. García-Barreda, S. Sánchez, M. E. Venturini, D. Blanco, C. Soler-Rivas, and P. Marco. 2020. Effects of gamma irradiation on the shelf-life and bioactive compounds of *Tuber aestivum* truffles packaged in passive modified atmosphere. *International Journal of Food Microbiology* 332:108774.

Tilahun, W. W., J. A. Saraiva Grossi, and S. P. Favaro. 2020. Mesocarp oil quality of macauba palm fruit improved by gamma irradiation in storage. *Radiation Physics and Chemistry* 168:108575.

Urbain, W. M. 1982. *Food Irradiation*. New York: Academic Press.

Van Kooij, J., H. Leveling, and J. Schubert. 1981. Food preservation by irradiation. *IAEA Bulletin* 23: 33–36.

Vas, K. 1979. Food preservation. *Developments in Food Science* 2, ed. H. Chiba, M. Fujimaki, K. Iwai, H. Mitsuda, and Y. Morita, 205. Tokyo, Japan: Kodansha Ltd.

Wani, I. A., N. Sultan, N. Qadir, and T. Ahmad. 2021. Physico-chemical, functional and antioxidant properties of wild barnyard grass (*Echinochloa crusgalli L.*) seed flour as affected by gamma-irradiation. *Radiation Physics and Chemistry* 183:109409.

WHO (World Health Organization). 1988. *Food Irradiation: A Technique for Preserving and Improving the Safety of Food*. Geneva, Switzerland: World Health Organization (WHO).

Willemoti, C., M. Marcotte, and L. Deschenes. 1996. Ionizing radiation for preservation of fruits. *Processing Fruits: Science and Technology*, ed. L.P. Somogyi, H.S. Ramaswamy, and Y.H. Hui, 221. Lancaster, PA: Technomic Publishing Company.

Zarbakhsh, S., and S. Rastegar. 2019. Influence of postharvest gamma irradiation on the antioxidant system, microbial and shelf life quality of three cultivars of date fruits (*Phoenix dactylifera L.*). *Scientia Horticulturae* 247:275–286.

Zhang, Z. S., Q. F. Xie, and L. M. Che. 2018. Effects of gamma irradiation on aflatoxin B_1 levels in soybean and on the properties of soybean and soybean oil. *Applied Radiation and Isotopes* 139:224–230.

Zhao, B., S. Hu, D. Wang, H. Chen, and M. Huang. 2020. Inhibitory effect of gamma irradiation on *Penicillium digitatum* and its application in the preservation of Ponkan fruit. *Scientia Horticulturae* 272:109598.

Zhu, M. J., A. Mendonca, E. J. Lee, and D. U. Ahn. 2004. Influence of irradiation and storage on the quality of ready-to-eat turkey breast rolls. *Poultry Science* 83:1462–1466.

9 Food Preservation by Ohmic Heating

Abhijit Chatterjee and Uttara Mahapatra

Corresponding author:
Dr. Abhijit Chatterjee, Assistant Professor,
Email: abhijitchatterjee1729@gmail.com

CONTENTS

9.1 INTRODUCTION

Ohmic heating (OH) involves the flow of electric current through the liquid food material that leads to the generation of thermal energy through the entire volume due to the joule effect of the electrical resistance of the food material (Makroo et al. 2020). In the early twentieth century, OH was applied for commercial pasteurization of milk (Goullieux and Pain 2014), but due to a lack of technical know-how and suitable electrode materials, its application was limited in 1980. Then, OH gained popularity to implement the high-temperature, short-time (HTST) principle in pasteurization and sterilization of liquid suspension with particulate matter.

In conventional heating (CH) of food material, a temperature gradient is generated either by using an electrical heating element in a surface or indirectly by using a hot medium (e.g., steam) in a heat exchanger so that flow of thermal energy can happen from higher to lower temperature by conduction and convection. In the preparation of canned food by conventional heating, a significant portion of food is overprocessed (with consequent quality deterioration) due to nonuniform heating. Uniform heating is even more challenging to achieve

DOI: 10.1201/9781003147978-9

for viscous liquid and biphasic foods, that is, liquids with particulates. Unlike conventional heating, OH does not depend on heat-transfer coefficients of the food material; faster and uniform heating is possible in OH assuming similar conductivities of both phases (Müller et al. 2020). The literature reports a 90–95% savings on heating time and an 82–97% savings on energy expense in OH compared to conventional heating (Sakr and Liu 2014). OH is even more energy-efficient (90% conversion to heat) than microwave heating (70% conversion to heat), another example of a volumetric heating system.

An OH system comprises a power supply, heating chamber, electrodes, current sensor, thermocouples, and data acquisition system (Figure 9.1). As dynamic cold spots may be generated depending on local variation in electric conductivities, and for the dielectric properties of the food material, a real-time, accurate, in situ measurement of temperature is essential to evaluate the nonhomogeneous heating. This is served by infrared, insulated, or optic fiber sensors. A complete unit of food processing through OH should contain an efficient pumping device and aseptic packaging machinery along with the OH system (Figure 9.1). There are three generic configurations available for the OH system (Goullieux and Pain 2014):

a. **Batch:** A batch ohmic heater is a thermally and electrically insulated chamber having a static fluid medium (feed) to be heated. Electrodes may be parallel plates or cylinders. Due to a simple configuration and for ease of operation, it is mostly used in research laboratories to find out the optimum composition of feed and the influence of different process parameters on the quality of end products. Its industrial application is limited except for the production of breadcrumbs in Japan.

b. **Transverse:** In this configuration, food products flow parallel to the electrode plates but perpendicularly to the electric field. The contact area between food material and electrodes is large, the spacing between plates is narrow, the voltage between electrodes is low (<96 V), and the current is high. This geometry provides a constant electric field but may lead to overheating, boiling, and electrode erosion. This configuration is most suitable for foods without suspended solids such as milk.

c. **Collinear:** In this configuration, electrodes are widely spaced, and food material flows from one electrode to another parallel to the electric field, maintaining a constant current density. In this process, the area of contact between food products and the electrode surface is small and the voltage is high (up to 4500 V). It may be proved that for the same amount of electrical heat generation, OH under a constant electric field strength will lead to a higher increase in temperature compared to the one under constant current density.

Different conductive materials such as titanium, platinized titanium, stainless steel, aluminum, and graphite have been reported in the design of OH (Sakr and Liu 2014). A major side effect of ohmic heating is electrochemical reactions occurring at the electrode/solution interface. As a result, corrosion of electrodes may occur due to the dissolution of metallic electrodes, especially at low frequency, that is, 50–60 Hz, which is readily available from power lines. As food products remain in direct contact with the electrode, the metal ions may be released into food material due to a typical anodic half-reaction for the metallic electrode (M (solid) $\Leftrightarrow$ M n + (aqueous) + n e, where n = 1, 2, 3, . . .) and may act as a toxic contaminant (Assiry et al. 2003). Also, water electrolysis from food material may happen at low-frequency alternating current, leading to the release of hydrogen and oxygen gases at the electrode/solution interface. The molecular oxygen produced during electrolysis can oxidize all oxidizable food components, particularly lipids and vitamins like ascorbic acid (vitamin C; Assiry et al. 2003). However, a higher frequency of 100 kHz^{-1} MHz may be used to overcome both issues (Kubo et al. 2020). In general, the use of stainless steel electrodes at a high-frequency alternating current is satisfactory (Sakr and Liu 2014).

The electrical conductivity (σ) of the food sample is a crucial parameter as it is directly related to the electrical heat generation. This is measured as a ratio of electric current density and electric field strength and may be calculated from Equation 9.1, where I = alternative current passing through the material (A), L = distance between electrodes (m), A = cross-sectional area of the material in the heating cell (m^2), and V = voltage across the material (V). The SI unit of electrical conductivity is Siemens per meter (S m^{-1}). The electrical power supply to the ohmic heater is calculated from Equation 9.2, where P = electric power generated in the heater (W). However, to emphasize the volumetric nature of the process, this expression can be modified to get Equation 9.3, where, E = electric field (V m^{-1}) and V_m = volume of the material ($V_m = A * L$) (m^3) so that the G, heating power per unit volume of the material, can be defined as Equation 9.4 (Goullieux and Pain 2014).

The approximate thermal power required for a heating process is given by Equation 9.5, where $\dot{m}$ = mass flow rate of the product (kg s^{-1}), C_p = heat capacity (J kg^{-1} °C^{-1}), T_i = initial temperature (°C), and T_f = final temperature (°C). The ratio of thermal energy taken up by the food material (Equation 9.5) and the electrical energy supplied to the system (Equation 9.2) dictates the energy efficiency of the system.

$$\sigma = \frac{I}{A}\frac{L}{V} \tag{9.1}$$

$$P = VI = I^2R \tag{9.2}$$

$$P = \sigma E^2 V_m \tag{9.3}$$

$$G = \sigma E^2 \tag{9.4}$$

$$P_{thermal} = \dot{m}C_P\left(T_f - T_i\right) \tag{9.5}$$

The temperature distribution in fluid during OH is based on an energy balance as follows:

$$\rho C_p v_z \frac{\partial T_f}{\partial z} = \nabla.\left(k_f \nabla T_f\right) - n_p A_p h_{fp}\left(T_f - T_{ps}\right) + \dot{u}f, \tag{9.6}$$

where ρ is the density (kg m^{-3}); Cp, the specific heat (J kg^{-1} k^{-1}); T, the temperature (°C); z, the distance (m); v_z, the fluid velocity (m s^{-1}); f, the fluid; p, the particle; ps, the particle surface; k, the thermal conductivity (W m^{-1} k^{-1}); np, the number of particles; A, the surface area of particles (m^2); h_{fp}, the fluid to particle heat transfer coefficient (W m^{-2} k); and u_f, the internal energy generation rate of the fluid. The temperature distribution in a particle during ohmic heating can be predicted with the conduction heat-transfer equation with internal energy generation:

$$\dot{\nabla}\left(k\nabla T\right) + \dot{u} = \rho C_p \frac{\partial T}{\partial t} \tag{9.7}$$

The internal energy generation is

$$\nabla.(\sigma\nabla V)=0. \tag{9.8}$$

Numerous models have been developed based on the numerical solution of these equations with appropriate boundary conditions and assumptions and from dimensional groupings.

9.2 MECHANISM OF INACTIVATION

There are several thermal and nonthermal mechanisms of microbial inactivation, and a few recent investigations have been given in Table 9.1.

9.2.1 Thermal Mechanism

The major effect of OH on food products is thermal in nature. Regardless of the method of heating, the kinetics of microbial and enzyme inactivation during thermal processing of food is described by parameters such as decimal reduction times (D, min), inactivation rate constant (k, s^{-1}), z-values (z °C), and activation energy (E_a, kJ mol^{-1}; Makroo et al. 2020). When a suspension of microbial cells (vegetative or spore) is heated at a constant lethal temperature, the decrease in the number of living cells with time typically follows a logarithmic curve, which can be explained by assuming a first-order model for thermal destruction of microorganisms (Equation 9.9). This equation follows the definition of decimal reduction time, that is, the heating time required to reduce the original microbial population or enzyme activity by a factor of 10 at a specific lethal temperature. As the lethal temperature is increased, the rate of thermal inactivation is increased with a consequent decrease in the D value. The effect of temperature on the D value is also logarithmic and is expressed by defining z (Equation 9.10): the increase in temperature required for a 10-fold increase in the rate constant (k) (i.e., to reduce D value by a factor of 10). This is in accordance with Arrhenius's definition of activation energy (E_a; Equation 9.11). For OH, the z-value may be expressed in terms of voltage gradient (V_g, V cm^{-1}) as defined by Equation 9.12 (Brochier et al. 2016). However, depending on the heating method, the values of z and D may be different, which is discussed in subsequent sections (Saxena et al. 2017).

$$\log\frac{N_0}{N}=\frac{kt}{2.303}=\frac{t}{D} \tag{9.9}$$

$$z=\frac{T_2-T_1}{\log D_1-\log D_2} \tag{9.10}$$

$$k_T=\alpha exp\left(-\frac{E_a}{RT}\right) \tag{9.11}$$

$$z_v=\frac{V_{g_2}-V_{g_1}}{\log D_1-\log D_2} \tag{9.12}$$

Here N_0 is the initial population, N is the number of surviving cells at time t, and α is the pre-exponential factor.

The basic objective of the thermal processing of food is to provide safe and high-quality food to consumers. The thermal process is designed to ensure commercial sterilization. This implies that the coldest location of the food products

Table 9.1 Effect of OH Parameters on Inactivation of Key Microorganisms and Enzymes Found in Various Food Products

Voltage	Frequency	Temperature	Target microorganisms/enzymes	Findings	Reference
26.7 Vcm^{-1}	60, 200, 500, and 1000 Hz		*Escherichia coli* O157:H7, *Salmonella typhimurium*, *Listeria monocytogenes*, MS-2 phage in buffered peptone water and tomato juice	• Higher fat content negatively affects performance of OH • Use of thermal infrared camera to monitor temperature distribution in nonhomogeneous fat-containing samples • The protective effect of milk fat on *Escherichia coli* O157:H7 and *Listeria monocytogenes*	Kim et al. (2017)
	1000 Hz	100°C	*Alicyclobacillus acidoterrestris* (sporulated) in Apple juice	• OH treatment showed better preservation of organoleptic qualities (°Brix, color, and pH) of commercial apple juice and lesser treatment time compared to CH	Kim et al. (2017)
5, 10, and 20 Vcm^{-1}	20, 40, and 60 kHz	101°C	*Bacillus subtilis* (sporulated) in sodium chloride solution	• Inactivation of *Bacillus subtilis* spores increased with electric field and frequencies.	Murashita et al. (2017)
		57.5, 60, 62.5 and 65°C	*Staphylococcus aureus* *Escherichia coli* (vegetative) in infant formula	• Higher efficiency of OH over CH. • Evidence of electroporation. • Most efficient treatment at 65°C.	Rodrigues et al. (2018)
9.43, 10.93 and 12.14 Vcm^{-1}			*Escherichia coli* O157:H7 *Salmonella Typhimurium*, *Listeria monocytogenes* (vegetative) in buffered peptone water and tomato juice	• 5 log reductions using sequential continuous-type ohmic heating	Kim et al. (2018)
5, 7.5 and 10 V cm^{-1}			*Escherichia coli* O157:H7 (vegetative) in broth	• The temperature of *Escherichia coli* broth increased linearly with the increasing of time in OH. • Inactivation history of the broth using OH differs from CH.	Shao et al. (2019)

(Continued)

Table 9.1 (Continued)

Voltage	Frequency	Temperature	Target microorganisms/enzymes	Findings	Reference
0–120 Vcm⁻¹		40°C and 50°C	*Escherichia coli* K12 (vegetative) in apple juice	• The combinatorial process (OH and shear stress) at 50°C resulted in highest lethal rate with better retention of nutritional quality of apple juice.	Mok et al. (2019)
		75°C	*Listeria monocytogenes* (vegetative) in sausage	• Both OH and CH affect color and texture of sausage. • OH didn't alter chemical composition, pH, lipid oxidation, cooking loss, or water holding capacity of sausage • Consumers were unable to detect any sensory differences between the methods.	Inmanee et al. (2019)
150, 200, and 250 V		99.4°C	*Escherichia coli,* *Staphylococcus aureus,* *Salmonella enterica,* *Salmonella Typhimurium,* *Shigella dysenteriae,* *Shigella flexneri* PTCC (vegetative) in blended citrus juice	• Inactivation depends on applied voltage (increasing voltage resulted reduced processing time) and type of pathogen. • Maximum inactivation rate was observed for *Staphylococcus aureus*, followed by *Escherichia coli, Salmonella enterica* subsp. *enterica* serovar Paratyphi A, *S. Typhi, S. dysenteriae* and *S. flexneri*.	Hashemi and Roohi (2019)
0–300 V		21–86°C	*Leuconostoc mesenteroides,* (vegetative) *Lactobacillus acidophilus,* (vegetative) *Lactobacillus plantanarum,* (vegetative) *Sacharomyces cerevisae,* (vegetative) *Byssochlamys fulva,* (sporulated) in sour orange juice	• Out of four OH variables studied (temperature, current, voltage, electrical conductivity), first two were found to affect the inactivation of spoilage microorganisms most.	Hashemi et al. (2019b)
100 and 200V		27–75°C	*Escherichia coli,* *Salmonella typhimurium,* *Salmonella enteritidis,* *Staphylococcus aureus* (vegetative) in cantaloupe juice	• Comparison among CH, OH, and microwave heating • Shorter processing time for OH than CH • Higher loss of vitamin C in OH, higher degradation of β–carotene and phenolic compounds in CH	Hashemi et al. (2019a)

Sample	Voltage	Frequency	Temperature	Observations	Reference
Bacillus subtilis (sporulated)		50 Hz		• Both heat and electric fields were necessary for enhanced spore inactivation. Thus, it is hypothesized that either the heat treatment makes the spore susceptible to the electric field, or vice versa.	Schottroff et al. (2019)
Polyphenoloxidase in watermelon juice		50 Hz	90°C	• Faster enzyme inactivation in OH • Lycopene content remained unaltered in both CH and OH. • Better retention of color in OH than CH.	Makroo et al. (2017)
–	25 and 50	50 Hz	−20 °C to 65 °C	• Comparison between OH and sous-vide treatment on scallops • Less processing time in OH by increasing frequency and temperature with higher retention of sensory properties and reduced denaturation of actin (muscle protein)	Llave et al. (2018a)
Peroxidase in pumpkin cubes	24–36	60 Hz		• Blanching time reduction (4 min in CH vs. 2 min in OH) for 90% inactivation of peroxidase activity • Similar color retention in OH and CH.	Gomes et al. (2018)
Different enzymes in whey-raspberry flavored beverage	25 V	60–1000 Hz		• OH treated samples (10, 100, and 1000 Hz at 25 V; 45, 60, and 80 V at 60 Hz) presented higher antioxidant, angiotensin-converting enzyme, antidiabetic values, and more diversified volatile compounds.	Ferreira et al. (2019)
Polyphenol oxidase in viscous carboxymethyl cellulose solution		60 Hz		• OH affects the polyphenol oxidase activity in temperatures higher than 50°C (reduction to 20 % at 50°C, 12 min; complete inactivation after 2 min at 80°C), • CH requires temperature above 60°C (reduction to 90% at 60°C, 8 min and complete inactivation after 2 min at 80°C). • Higher rate of enzyme inactivation by OH than CH.	Leite et al. (2019)

(Continued)

Table 9.1 (Continued)

Voltage	Frequency	Temperature	Target microorganisms/enzymes	Findings	Reference
50–100 V	60 Hz	100°C	Total bacterial count in fermented red pepper paste	• OH of a viscous food in a continuous system with 2-log reduction. • OH system was available for 57 cycles during 1 cycle of CH, and it could process 104.4 kilotonnes of product.	Cho et al. (2017)
4.6	60 Hz	57.5–65.0°C	*Escherichia coli* and *Staphylococcus aureus* in infant formula	• Suggested evidence of electroporation. • Considering both microbial species, the best thermal treatment for this infant formula was obtained by OH at 65°C.	Rodrigues et al. (2018)
5–10 V cm^{-1}	50 Hz	72°C	*Escherichia coli* in buffer	• Comparison among CH, high voltage short time OH, and low voltage long time OH. • Evidence of nonthermal mechanism in OH. • Quantitative evaluation of proteomic profiles using capillary-HPLC-MS/MS technology	Tian et al. (2018c)
3.9 and 20.5 V cm^{-1}	10–10^5 Hz	75°C	Peroxidase (1.1.1.1) (sugarcane)	• No evidence on nonthermal mechanism on inactivation of the enzyme (voltage gradient up to 20.5 V cm^{-1}, with different frequencies and waveforms). • No significant effect on stability of phenolic compounds by variation of electric field, frequency, voltage gradient but by variation in waveforms. • Major color changes at 10 Hz/25 V and 60 Hz/130 V	Brochier et al. (2018)

(e.g., at the geometric center of a can heated by conduction, the central axis of a tube in continuous processing in a tubular heat exchanger) should achieve the desired level of lethality calculated for inactivation of the reference microorganism, which may be the most heat resistant foodborne pathogen or spoilage microorganism likely to grow under normal storage and handling conditions of the food products. The integrated lethality of a thermal process (F) is calculated as follows:

$$F = \int 10^{\frac{\left(T - T_{ref}\right)}{z}} dt,$$
(9.13)

where process temperature, T, is a function of time t; T_{ref} is the reference temperature. The F value signifies the time duration for an equivalent constant temperature process at T_{ref} having an identical reduction in the microbial population.

Although the microbial/enzymatic inactivation and consequent food safety are increased at a higher temperature, the nutritional and organoleptic quality of the product may deteriorate. So optimizing the process is required, which means finding the processing conditions that will provide the required shelf life with minimum loss of nutrients and retention of sensory characteristics (appearance, color, flavor, texture). As sterilization requires higher activation energy (345 kJ mole⁻¹ for killing spores of *Clostridium botulinum*) than that of nutrient degradation (131 kJ mole⁻¹ for Maillard reaction), the z-value of product degradation reactions is usually larger than that of thermal death of microorganisms. It follows that for an equal amount of sterilization, that is, same inactivation efficiency, processing at a higher temperature for a shorter time results in less thermal damage to quality, which conforms to the theoretical background of the HTST concept (Eisner 2021). For an effective HTST process, the product needs to be heated and cooled rapidly, which may be accomplished using a plate heat exchanger in conventional heating. However, HTST processing of food products

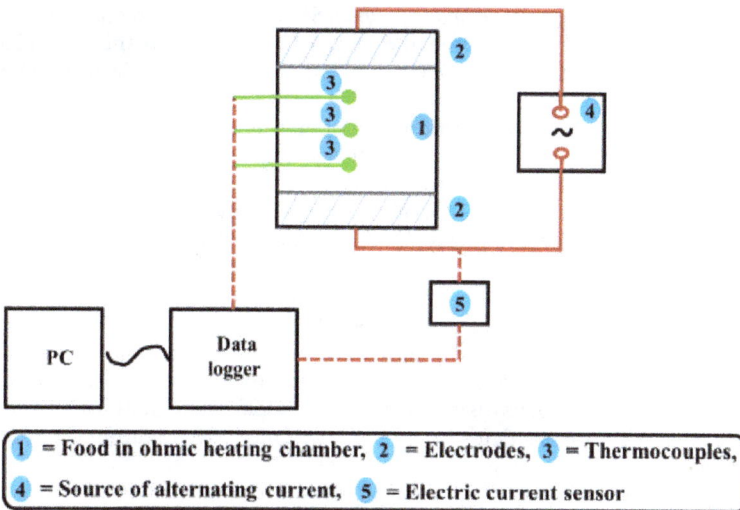

1 = Food in ohmic heating chamber, **2** = Electrodes, **3** = Thermocouples, **4** = Source of alternating current, **5** = Electric current sensor

Figure 9.1 Schematic of an OH system.

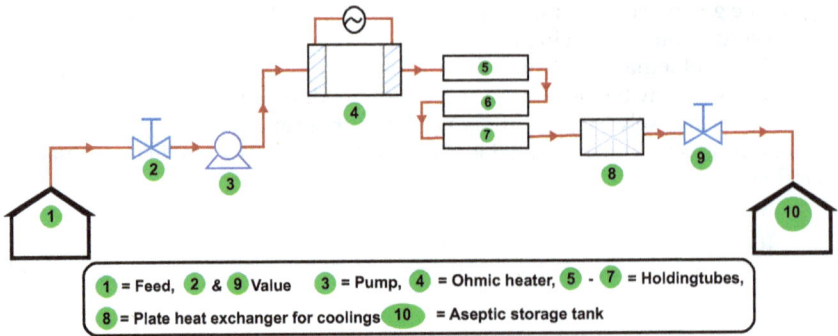

Figure 9.2 Schematic of a continuous thermal processing unit with OH facility.

containing large particulate matter by conventional heating is challenging. As heat transfer within the solid food particle is controlled by conduction and biological materials are usually poorly conductive, a sharp gradient may exist between the surface and center of the particle even for a high convective heat-transfer coefficient of the liquid matrix. When such food products reside in the holding section of a continuous sterilization unit (Figure 9.2) to achieve thermal equilibrium, a considerable portion of the food outside the coldest point will be overprocessed, and thermal damage to the product quality will be more extensive due to higher processing temperature. The main difference between the thermal effect produced by conventional heating and OH is that OH, being a volumetric method, eliminates the temperature gradient within the broth, assuming the conductivity of the solid and liquid phases are comparable. A recent investigation applied the computational fluid dynamics (CFD) method to compare the three-dimensional (3D) temperature distribution of ohmic heating (150, 200, and 250 V; 120 s; 99.4°C) and conventional heating (90 °C; 15 min) for the inactivation of *Escherichia coli*, *Staphylococcus aureus*, *Salmonella enterica* subsp. *enterica* serovar Paratyphi A, *Salmonella typhi*, *Shigella dysenteriae*, and *Shigella flexneri* in a mixture of sweet lemon and orange juice (Hashemi and Roohi 2019). The 3D simulation revealed that OH utilized natural convective heat transfer through homogeneously distributed vortices created due to the buoyancy effect. This led to more uniform distribution of temperature compared to conventional heating, where a stratified temperature pattern was found due to gradual 'conductive' transfer of heat from the wall to the center of the container without any convective effect. It was also found that due to the volumetric generation of heat, the temperature elevation (26–99.4°C) was more rapid in OH compared to conventional heating.

9.2.2 Nonthermal Mechanism

In addition to the thermal effects, applying an electric current through food products may induce the inactivation of microorganisms or enzymes through nonthermal mechanisms (Müller et al. 2020; Kubo et al. 2020; Tian et al. 2018b). Numerous articles have been published investigating the role of nonthermal effects during the OH process. Most of these articles compare conventional heating and OH by treating the food sample under similar conditions followed by measurement of residual microbial or enzymatic activity. To claim 'intrinsic' nonthermal effect, thermal and electrical effects need to be decoupled. So the following points are to be considered during the design of such experiments:

a. For a fair comparison between conventional heating and OH, the time–temperature profile for both processes must be identical (Schottroff et al. 2020; Kubo et al. 2020; Tian et al. 2018b). This is not easy to achieve as the rate of conventional heating decreases over time due to a decrease in the temperature gradient, whereas temperature rises almost linearly in the case of OH under constant power. The electric field strength or power input to the system must be varied to match the temperature profile with conventional heating. If time–temperature profiles for conventional heating and OH are significantly different, inactivation kinetics need to be mathematically modeled for each process. The fitted models are to be analyzed to determine if there is any significant difference between thermal death curves for the same reference temperature (Kubo et al. 2020). Inactivation data during heating and cooling times in the CH process to be incorporated in calculation.

b. Temperature distribution within the food sample is fundamentally different for conventional heating and OH. The most published article relied on the measurement of temperature at the geometric center of the sample only for both conventional heating and OH neglecting the temperature inhomogeneities within the treatment chamber. As the final microbial or enzymatic activity of the food material is measured as an average over the whole sample and temperature distribution within the sample is not monitored during the investigation, comparable treatment conditions for both methods can only be obtained if the uniformity of temperature is ensured (Kubo et al. 2020). The uniformity of temperature in the food material can be improved either by using a small quantity of sample (negligible temperature gradient) or by using liquid food with stirring (Tajchakavit and Ramaswamy 1995).

9.2.2.1 Nonthermal Mechanism of Microbial Inactivation

The most frequently cited nonthermal mechanism for microbial inactivation is electroporation (Müller et al. 2020). Microbial cells contain electrically charged species such as cations of light metals (Na^+, K^+), negatively charged proteins, and other charged molecules. An external electric field applied on the OH system exerts a transmembrane potential in the lipid bilayer of the cell membrane so that charged intracellular species will be attracted toward oppositely charged electrodes. When the applied electric field exceeds a threshold value, intracellular material will be exuded by making pores in the cellular structure, leading to lysis of the cell membrane followed by the death of the microbial cells. The intensity of electroporation is closely related to operating parameters such as frequency and electric field strength, and the lethal effect of electroporation is intensified by the effect of high temperature (Park and Kang 2013). In general, thermal effects (mostly involving protein denaturation) and nonthermal effects (electroporation) can be differentiated easily at a lower temperature only. At higher temperatures, both conventional heating and OH tend to produce similar inactivation results. Rodrigues et al. (2019) evaluated the inactivation kinetics of *Staphylococcus aureus* and *Escherichia coli* in infant formula under conventional heating and OH, ensuring identical thermal histories. Compared to conventional heating, lower D values of *Staphylococcus aureus* were reported for OH at all temperatures investigated (57.5–65°C), whereas significantly lower D value of *Escherichia coli* was obtained at 57.5°C only. Another study hypothesized that sublethal injuries of *Escherichia coli* were caused by the formation of toxic hydrogen peroxide within the cell, as a consequence of the alternating current used in OH (Tian et al. 2018a). Murashita et al. (2017) investigated the nonthermal effects of OH by comparing the kinetic model of *Bacillus subtilis* spores between conventional heating and OH. The inactivation effect was greater in OH at all electric fields studied (5, 10, 20 Vcm^{-1}) than conventional heating, and increases in the electric field or frequency resulted in higher inactivation.

Somavat et al. (2012) evaluated nonthermal effects by comparing the inactivation kinetics of *Geobacillus stearothermophilus* spores (ATCC 7953) between conventional heating (at temperatures 121, 125, 130°C) and OH (at frequency 60 Hz and 10 kHz). A specially designed ohmic test chamber containing a cell holder and glass capillary cells (placed parallel to the electric field) was used to ensure identical temperature profiles for both conventional heating and OH treatments. At 130°C, that is, the highest temperature of the investigation, all *D* values were statistically equal. At all other temperatures, lower *D* values were obtained in OH compared to conventional heating. The nonthermal effect was most pronounced (highest difference of *D* values between conventional heating and OH) at 121°C, that is, the lowest temperature studied. It was hypothesized that dipicolinic acid (DPA) and Ca-DPA complex, a major component of spore and responsible for protective coating, was released from the spore due to a synergistic effect of electric field and high temperature. A similar nonthermal effect was also reported in the inactivation of *Bacillus coagulans* spores within tomato juice. In a recent study, Schottroff et al. (2019) investigated inactivation mechanisms of bacterial endospores by electric fields using *Bacillus subtilis* spores and mutants lacking components involved in resistance: small–acid soluble proteins (SASPs) protecting DNA (PS578); the coat covering the spore (PS3328); and the spore-germination enzyme SleB (FB122(+). Similar to previous studies by Somavat et al. (2012) the study differentiated thermal and electric field effects by using a small volume of samples encased in capillary glass tubes immersed in an ohmically heated liquid. However, instead of having a thermal holding time at the reduced electric field, which may diminish the effectiveness of OH, sample temperature was constantly increased up to the predefined maximum value at a constant electric field of 60 Vcm^{-1} at 10 kHz frequency followed by immediate cooling of the samples. Compared to conventional heating, the study confirmed up to 2.4 log10 of additional inactivation of *Bacillus subtilis* endospores by OH. It was revealed that the mechanism of additional inactivation could not be explained by electroporation, induction of germination, or increased resistive heating of the spore. It was concluded that the action of the electric field might involve the spore core only as it was not targeted to the spore coat and the SleB.

9.2.2.2 Nonthermal Mechanism of Enzymes Inactivation

Although extensive literature support is available focusing on the nonthermal effect of OH on microorganisms, only a limited number of studies are available to explore the nonthermal effect of OH on food proteins and enzymes (Kubo et al. 2020; Samaranayake and Sastry 2016a, 2016b). To investigate the difference in mechanism of enzyme inactivation between OH and conventional heating, the following food materials were utilized: (a) fresh apple juice; (b) a mash of two batches of cloudberry jams having a different fraction of pulp, water, and sugar; (c) raw bovine milk diluted by ultra-high temperature (UHT) milk; and (d) three batches of vegetables having carrot, broccoli, and potato in different proportions. Three different enzymes, for example, pectin methyl esterase, alkaline phosphatase, and peroxidase, were assayed to evaluate the effect of thermal treatment (OH or conventional heating) at different incubation temperatures ranging from 54–78°C. Although experimental data for thermal inactivation of enzymes by OH or conventional heating followed the same kinetic model implying the same mechanism of inactivation, parameter values were different to accommodate the higher rate of inactivation of enzymes observed in OH. For alkaline phosphatase (ALP), Arrhenius activation energy was higher in OH inactivation, whereas in the case of pectin methylesterase (PME), the OH method did not affect this value but increased the pre-exponential value. The OH process showed an entropic effect in peroxidase inactivation. As the activation enthalpy values were the same for both OH and conventional heating, it was concluded that the tertiary structure of the enzyme was not modified by an electric field. However, the influence of the electric field was not explained by modeling enzyme activity using molecular dynamics.

Under a given electric field, the electrophoretic motion of an enzyme molecule in an aqueous medium is controlled by its net charge and dipole moment. The translational and rotational motions of the molecule are determined by the net charge and net dipole moment of the molecule, respectively. Samaranayake and Sastry 2016a studied the non-thermal effect of OH on PME and polygalacturonase (PG) activities in tomato homogenate at a low field strength (0.4 Vcm^{-1}) at 65°C. It was suggested that pure enzyme solutions in a buffer medium would behave differently than those in a real food matrix, as most of the enzyme activity in the suspension was observed in the solid phase and not in the liquid phase of the suspension. Significant (up to 26%) deactivation of PME was reported at a low-frequency window of 1–60 Hz, whereas inconclusive data was obtained for PG. The temperature of 65°C was selected as PME showed maximum activity at this temperature. So statistically significant inactivation at this temperature proved the existence of a nonthermal effect in OH. Later, a similar investigation was conducted to find the effect of increasing the electric field (5, 8, 10.5 Vcm^{-1}) at 60 Hz for PME only (Samaranayake and Sastry 2016b). Due to electrophoretic motion, PME appears to 'feel' a slightly higher 'electrical temperature' than the actual bulk temperature during OH. Consequently, there was a significant increase in the enzyme activity near its peak activity temperature (at 70°C) compared to conventional heating. When processing temperature was increased beyond this threshold value, inactivation of the enzyme was observed, and the rate of inactivation was higher than the conventional heating due to its higher apparent elevation of temperature within the OH treatment chamber due to molecular motions induced by the electric field. This effect of enhanced activation and deactivation was also increased with increasing field strength. All these articles focus on the postprocessing residual activity of the enzyme concerned. Samaranayake and Sastry (2018) studied the nonthermal effect of OH on the *in situ* alpha-amylase activity at various frequencies (1 Hz to 1 MHz). The electric field (1 Vcm^{-1}) was kept low to minimize the OH effect. Compared to the control experiment without an electric field at the same temperature, up to a 41% increase in enzymatic activity was reported at low frequency (1–60 Hz). At higher frequency (>100 Hz), however, enzymatic activity remained unaffected or mildly inhibited by the electric field. A molecular dynamics simulation results showed that active translational motion occurred in the frequency window 1–60 Hz, leading to increased mobility of the enzyme with a consequent increase in the frequency of molecular collisions. Due to rapid reversal of the direction of the electric field, translational motion becomes restricted at the higher frequency; that is, the amplitude of oscillatory motion was too small to induce an effect on enzyme activity.

None of these studies predicted conformational changes within the enzyme molecule. Rodrigues et al. (2019) studied the nonthermal effect of OH on functional properties of the β-lactoglobulin (β–lg) protein, the major fraction in whey protein ingredients. A pure fraction of the protein in nonaggregating conditions (low concentration and ionic strength) was thermally treated with and without the presence of the electric field for 10 min at various temperatures (50, 60, 70, 80, 90°C) and pH 3–7. During OH, the frequency was adjusted to 20 kHz, the electric field was kept at 80 Vcm⁻¹ during heating and 20 Vcm⁻¹ during holding. The characterization of protein was done by using techniques such as circular dichroism, intrinsic and extrinsic fluorescence, and free thiol groups reactivity. Significant changes in secondary structural elements, tryptophan (Trp) fluorescence, hydrophobic site accessibility, sulphohydryl (SH) reactivity, and disulfide arrangement were observed for all treatments except those conducted at the lowest temperature. As discussed before, a high-frequency electric field for a short time of exposure is expected to reduce the effectiveness of the electroporation, which explains the absence of conformational changes in low-temperature treatments. The result confirmed that under experimental conditions used, that is, moderate electric field strength, high frequency, and short time of exposure, the electrical effect was not sufficient to interrupt protein folding. Once the unfolding of protein was induced through thermal action, the electric field prompted structural changes in the unfolded protein. The nonthermal effect was dependent on the pH and was more pronounced at neutral pH, which accelerated thermal action. The authors hypothesized that the mechanism of nonthermal effects was related to the vibrational effects induced by molecular motion at a high-frequency electric field.

9.3 FACTORS AFFECTING OH

9.3.1 Extrinsic Factors

9.3.1.1 Current

According to Joule's law of heating, the amount of heat generated during OH is proportional to the square of the electric current. Studies on the inactivation of *Saccharomyces cerevisiae* in phosphate buffer solution at pH 7.1 using direct current have found that *D*-value was decreased from 1547 to 140 min with an increase in the current from 0.1 to 1 A (Tian et al. 2018b). However, instead of controlling electric current directly, the OH system typically controls the electric current indirectly by fixing electric field strength.

9.3.1.2 Temperature and Time

A major mechanism of inactivation using OH is through the thermal route. It is obvious from the definition of *D*- and *z*-values that longer heating times and higher temperatures should increase the inactivation of microorganisms. A comparison between conventional heating and OH under a similar heating curve was discussed in Section 9.1. An increase in processing temperature also increases the conductivity of the food material, which indirectly increases heat generation in OH, which is discussed later in this section.

9.3.1.3 Frequency

Lee et al. (2015) studied the effect of frequency (60 Hz–20 kHz) and waveform (sine, square, sawtooth) on the inactivation of two microorganisms (*Escherichia coli* O157:H7, nontyphoidal serovars of *Salmonella enterica*, such as serovar *Typhimurium*) in salsa by OH at a constant electric field of 12.5 Vcm⁻¹. The shape of the waveform did not influence inactivation. Electrode corrosion was prevented at a frequency higher than 1 kHz. The electrical conductivity was increased at a higher frequency (with a consequent increase in heating rate) but reached a plateau at 500 Hz, after which the effect of change in frequency was not significant. The heating time required to reduce microbial population 1 log CFU g^{-1} was also decreased with increased frequency. Higher conductivity was reported at a higher frequency in OH of beef (Llave et al. 2018b). Ambiguous results have often been reported in studies focusing on the effect of frequency on OH inactivation of microbial cells. The inactivation of spores of *Geobacillus stearothermophilus* in tomato juice was studied in high (60 kHz) and low (10 kHz) frequencies (Somavat et al. 2012), and at each frequency, three different temperature values were selected: 121, 125, and 130°C. At the lowest temperature, the D-value was higher at a higher frequency, and with an increase in temperature, the difference in D-values between the high and low frequencies was reduced. At the highest temperature, no significant difference was found between high and low frequency OH. However, in another work concerning the same food material and spores, the D-value was found to be greater at a lower frequency at 95 and 100°C (Somavat et al. 2013) due to difference in the intracellular content of the spore.

9.3.1.4 Electric Field

Heat generation in OH is related to the electric field strength by Equation 9.4, which showed that increasing electric field strength would increase heat generation leading to a reduction in processing time. A number of research studies supported this hypothesis, for example, in doubling the electric field from 10 to 20 Vcm^{-1}, approximately 75% reduction in processing time for 5D reduction of the microbial load was noted (Sagong et al. 2011). A recent study comparing parameters of conventional heating and OH for a dairy product showed that the electric field was increased from 1.82 to 9.1 Vcm^{-1}, the volumetric rate of heat generation was increased from 0.003 to 0.014 Wm^{-3} with a consequent reduction in processing time from 94.28 to 5.25 min, whereas conventional heating for the same product required 9.2 min (Kuriya et al. 2020). The energy consumption was decreased at increasing electric fields due to higher heating rates (Kuriya et al. 2020). The heating rate for conventional heating was 9.9 °Cmin^{-1}, which was increased to 17.1 °Cmin^{-1} for OH at the highest electric field. Another study found the presence of a significant number of viable microbial cells after 60 s of OH at 30 Vcm^{-1}, whereas no microbial cells were able to resist 30 s of OH when field strength was doubled (Park et al. 2017). Another study by Shao et al. (2019) showed slower recovery of injured cells at higher electric fields. Studies on three different microbial species *Escherichia coli* O157:H7, *Salmonella typhimurium*, and *Listeria monocytogenes* in tomato juice and orange juice showed that when treatment time was kept constant, higher efficacy of inactivation by OH was found at higher field strength within the range of 10–40 Vcm^{-1} at 60 Hz frequency (Lee et al. 2012; Sagong et al. 2011). Similar results were also found for the inactivation of *Bacillus subtilis* spores within 5–20 Vcm^{-1} electric field at 20 kHz (Murashita et al. 2017).

9.3.2 Intrinsic Factors

9.3.2.1 pH

In general, when all other conditions are the same, low pH increases the inactivation rate by OH. This has been observed in a number of studies involving microorganisms such as *Salmonella typhimurium*, *Listeria monocytogenes,* and *Escherichia coli* O157:H7 in orange juice (Kim and Kang 2015a; Lee et al. 2015). However, at high temperatures, the effect of low pH may be masked by the thermal effect found in a study focusing on the inactivation of *Bacillus licheniformis* spores in carrot juice.

9.3.2.2 Species and Growth Stage

Similar to pH, the z-value of a microbial species obtained through OH processing is related to its thermal resistance. A recent study by Schottroff et al. (2020) on the inactivation of six different microorganisms showed that *Listeria innocua* was more resistant to OH than *Staphylococcus carnosus*. Another study showed cells of *Salmonella typhimurium* were most resistant in the stationary phase, followed by the log phase and lag phase during OH. Gavahian and Tiwari (2020) applied OH for the growth of *Lactobacillus, Streptococcus,* and *Saccharomyces* in a fermenter, and found that the lag phase of these microorganisms was reduced compared to CH.

9.3.2.3 Composition of Food

The conductivity of the food material plays an instrumental role in determining the success of the OH. The electrical conductivity of a food depends on its composition, and may be classified as having good conductivity ($\sigma > 0.05$ Sm^{-1}, e.g., condiments, eggs, yogurt, milk desserts, fruit juices, wine, gelatin, hydrocolloids), low conductivity ($0.005 < \sigma < 0.05$ Sm^{-1}, e.g., margarine, marmalade, powders) and poor conductivity ($\sigma < 0.005$ Sm^{-1}, e.g., frozen foods, foam, fat, syrup, liquor) (Goullieux and Pain 2014). The more conductive it is, the less electric field it needs to achieve a given heating rate. In general, the conductivity of a liquid (e.g., fruit juice) is increased linearly with increasing temperature, as expressed in Equation 9.14, and is influenced by the presence of particulate matter, nonconducting components, electrolyte, and soluble solid content (Goullieux and Pain 2014).

$$\sigma = \sigma_{in}\left[1 + m\left(T - T_{in}\right)\right], \tag{9.14}$$

where σ_{in} is the electrical conductivity at the starting temperature T_{in} and m a proportionality constant °C^{-1}.

9.4 PRESENCE OF PARTICULATE MATTER

In conventional heating, when a suspension containing food particles in water such as a piece of potato ($3 \times 4 \times 0.75$ cm^3) was heated, the heating of liquid was faster than that of solid (Samprovalaki et al. 2007). OH, being a volumetric heating technology, can overcome this issue and allows uniform heating rate provided electrical conductivity of solid phase is comparable to that of liquid. The electrical heat generation in a solid–liquid mixture of food has been extensively studied in the early 1990s, which can be summarized as follows. The ratio of electrical heat generation in any phase with the control, that is, liquid without the particulates, was expressed in terms of the following factors: (a) the local solid:liquid conductivity ratio and (b) the fraction of local solids content in the sample (Samprovalaki et al. 2007). A change in the solid:liquid ratio will change the rate of heating in each phase. If the electrical conductivity of the particulate matter is lower than the liquid in a suspension, the solid phase may be heated faster or slower than the liquid depending on the concentration of the particulate matter. When the concentration of the solid is low, the liquid phase is heated faster due to bypassing the current from solid to liquid; the fraction of the current passing through the solid will be increased with the increasing solid concentration leading to faster heating of solid after a threshold value. When particulate is heated faster than the liquid, a temperature gradient would exist in the liquid surrounding solid particles and within the particulate matter itself. For example, a finite element analysis reported a sharp temperature gradient, that is, 120°C in the center versus 80°C at the surface of a spherical particle (Samprovalaki et al. 2007). This differential heating effect would be pronounced in the liquid phase for a highly viscous solution. For OH of foods having high solid content, the electric field may be distorted around the solid–liquid interface, and the magnitude of distortion will depend on the following factors: (a) electrical conductivities of fluid and the particles; (b) particle size, shape, concentration, and specific heat; (c) viscosity of the fluid; and (d) orientation relative to the electrodes and to other particles (Samprovalaki et al. 2007). This distortion of the electric field leads to differential heating rates and affects the processing time of the food.

9.5 PRESENCE OF NONCONDUCTING COMPONENTS

The fat content of food acts as electrical insulation and is protective against microbial inactivation through OH. For example, in a study comparing OH for conventional (5% fat) and wagyu beef (18% fat), lower electrical conductivity was reported for the latter one (Llave et al. 2018b). In another study involving minced beef samples differing in fat content (2–15%), the lowest electrical conductivity was found for the sample having the highest fat content (Bozkurt and Icier 2010). Milk samples with increasing fat content (0–10%) were studied to compare microbial inactivation by conventional heating and OH. It was found that the inactivation rate was not influenced by the presence of fat content in conventional heating as heating occurs through conduction and convection. For OH, however, the heating rate and consequent inactivation rate were decreased with increasing fat content and the extent of inhibition was dependent on the microbial species, as inactivation of *Escherichia coli* was more affected than *Salmonella typhimurium* (Kim and Kang 2015b).

9.6 PRESENCE OF ELECTROLYTES

The addition of electrolytes such as common salt increases the heating rate by OH due to an increase in the conductivity of the materials such as shredded cabbage, daikon radish, and meatballs. However, the filling pattern can influence this effect; if the electrolyte particles are preferentially located on the current path between the electrodes of an ohmic heater, this portion of the food may be heated more slowly than others due to the low voltage drop across this path. Similarly, nonuniform salt distribution in food products may lead to local hot and cold spots (Tian et al. 2018b).

9.7 SOLUBLE SOLID CONTENT

Park et al. (2017) applied voltage gradients (30–60 Vcm^{-1}) to apple juices differing in soluble solid content (18–72 °Brix). At any given voltage gradient, the rate of heating was increased with an increase in soluble sugar concentration up to 36 °Brix, after which the trend was reversed. The slowest rate of heating was observed for 72 °Brix.

9.8 EFFECT ON FUNCTIONAL AND SENSORY PROPERTIES OF FOOD

Most studies on OH claimed better retention of natural color, pH, sensory quality and bioactive compounds (e.g., ascorbic acid, phenolic compounds, anthocyanins, carotenoids, chlorophyll and betalains in fruits and vegetables) due to the HTST effect compared to conventional heating (Shao et al. 2021). Recent studies comparing quality attributes of carrot juice (Negri Rodríguez et al. 2021) and pomelo juice (Doan et al. 2021) pasteurized by OH and conventional heating showed an improved retention of bioactive compounds in OH processing. Pasteurization of an orange juice with suspended alginate particle using OH showed only a 4.2% decrease in vitamin C content (43.5 mg/100 g) compared to a 13.6% decrease in conventional heating (39.2 mg/100 g) with respect to the untreated control (45.4 mg/100 g; Wattanayon et al. 2021). Another study reported lower drip loss in OH of beef, presumably due to low myofibrillar protein denaturation due to quicker heating (Llave et al. 2018b; Tian et al. 2016).

A recent investigation evaluated qualities of blueberry-flavored dairy desserts such as physicochemical characteristics (fatty acids profile, volatile compounds, rheological assay, color parameters) and bioactive compounds (total phenolics, anthocyanins, antioxidant activity, inhibition of α-glycosidase, α-amylase, angiotensin-converting enzymes [ACEs]) processed by OH (using the electric field at 1.82–9.1 V cm^{-1}) and conventional heating (Kuriya et al. 2020). Similar to conventional heating, bioactive compounds with antioxidant properties such as anthocyanin, ascorbic acids, phenolic compounds were degraded in OH, and degradation was increased with an increase in the electric field, especially above 7.3 Vcm^{-1}. However, OH processing increased inhibition of α-glycosidase and ACE. A higher electrical field resulted in a negative effect on these functional characteristics as well. Several other studies (pasteurization of goat milk, whey acerola–flavored drink), including this one, found no significant difference in the fatty acid profile between OH and conventional heating processing. Depending on the electric field used, the OH caused both the formation and degradation of many volatile compounds and induced formation of Maillard reaction products such as furfural, 5-hydroxymethylfurfural, furanone, and pyranone.

9.9 NOVEL APPLICATIONS OF OH

As the OH process offers uniform and rapid heating along with retention of heat-labile nutrients, most studies focused on the feasibility of replacing conventional heating by OH in pasteurization or commercial sterilization. However, it was suggested that electroporation might facilitate the mass transfer as well. The diffusion coefficient of beet dye from beet cubes increased with an increase in electric field and decreased frequency (Kulshrestha and Sastry 2003). This prompted research on the applicability of OH in various other unit operations in food processing requiring both heat and mass transfer such as blanching, evaporation, dehydration, fermentation, and extraction. A predictive CFD model was developed to explain transport phenomena during OH-assisted dehydration (Turgut et al. 2021). By combining OH with convectional drying, the drying time of potato slices was shortened by 20–60% (Turgut et al. 2021). Compared to conventional heating, simultaneous extraction of pectin and essential oil from citrus fruits using OH was found to produce a higher yield with a higher concentration of limonene (Tunç and Odabaş 2021). Similarly, the extraction of C-phycocyanin, phenolic, and other intracellular compounds from *Spirulina* using OH at a higher temperature (44°C) for a shorter time (30 min) has reported a higher yield than conventional heating (Ferreira-Santos et al. 2020). Evaporation under vacuum has already been established as a standardized process for moisture removal, and several research studies have been conducted to evaluate the feasibility of using OH in vacuum evaporation of fruit juices and tomato paste to reduce the processing time and energy consumption (Alkanan et al. 2021). Similarly, the processing time of black garlic, a product of Maillard reaction under controlled temperature (70°C) and humidity (94%), was reduced from 30 days to 12 days by adding a pretreatment by OH (Ríos-Ríos et al. 2021). It was concluded that the nonthermal effect of electric current induced hydrolysis of fructans to fructose leading to an accelerated Maillard reaction. OH has also been applied in lab-scale lactic acid and alcoholic fermentation to reduce the lag phase of

Lactobacillus, Streptococcus, and *Saccharomyces* species and to obtain a higher yield of the desired fermentation products (Gavahian and Tiwari 2020).

9.10 CONCLUSION

A huge volume of research work has been done in the last few decades regarding the utilization of OH in thermal food preservation. Most of these articles assessed the efficacy of the process by comparing OH with conventional heating in terms of reduction in processing time, quality of the food products obtained (retention of nutrients, texture, color, flavors of food products), and mechanism of lethality. Almost all these articles conducted experiments to study the effect of process parameters (variation of the electric field, frequency, etc.). Most of these articles investigated different kinetic modeling to fit experimental data for the inactivation of target microorganisms or enzymes whereas a few articles employed numerical modeling using CFD or other similar tools to quantify the spatial and temporal distribution of temperature, electric field, the relative velocity of phases involved in the process for design and optimization of the OH process. However, only a handful of companies currently manufacture industrial OH systems (e.g., APV Baker Ltd. [UK], C-Tech Innovation [UK], Agro process [IAI Group, Canada], Yanagiya Machinery Co. Ltd. [Japan], Kasag [Switzerland], Alfa Laval [Sweden], Raztek [USA], Emmepiemme SRL [Italy]) to cater a limited number of commercial OH plants operated throughout the world for pasteurization and sterilization of pumpable foods such as dried fruits and vegetables, tomato paste, various dairy products. To bridge the gap between lab and industry, the future direction of the research should be aimed at developing models that can correlate OH design parameters with the design goals, that is, treatment homogeneity and desired quality attributes of the final product. In addition, further technical knowledge is required to commercialize the application of OH in other areas of food processing such as drying, extraction, fermentation, evaporation, and gelatinization.

REFERENCES

Alkanan, Z.T., A.R.S. Al-Hilphy, A.B. Altemimi, R. Mandal, and A. Pratap-Singh. 2021. Comparison of quality characteristics of tomato paste produced under ohmic-vacuum combination heating and conventional heating. *Applied Food Research*: 100014. www.sciencedirect.com/science/article/pii/S2772502221000147.

Assiry, A., S.K. Sastry, and C. Samaranayake. 2003. Degradation kinetics of ascorbic acid during ohmic heating with stainless steel Electrodes. *Journal of Applied Electrochemistry* 33, no. 2: 187–196. https://doi.org/10.1023/A:1024076721332.

Bozkurt, H., and F. Icier. 2010. Electrical conductivity changes of minced beef – fat blends during ohmic cooking. *Journal of Food Engineering* 96, no. 1: 86–92. www.sciencedirect.com/science/article/pii/S0260877409003471.

Brochier, B., G.D. Mercali, and L.D.F. Marczak. 2016. Influence of moderate electric field on inactivation kinetics of peroxidase and polyphenol oxidase and on phenolic compounds of sugarcane juice treated by ohmic heating. *LWT* 74: 396–403. www.sciencedirect.com/science/article/pii/S002364381630487X.

Brochier, B., G.D. Mercali, and L.D.F. Marczak. 2018. Effect of ohmic heating parameters on peroxidase inactivation, phenolic compounds degradation and color changes of sugarcane juice. *Food and Bioproducts Processing* 111: 62–71. www.sciencedirect.com/science/article/pii/S0960308518305017.

Cho, W.-I., E.-J. Kim, H.-J. Hwang, Y.-H. Cha, H.S. Cheon, J.-B. Choi, and M.-S. Chung. 2017. Continuous ohmic heating system for the pasteurization of fermented red pepper paste. *Innovative Food Science & Emerging Technologies* 42: 190–196. www.sciencedirect.com/science/article/pii/S1466856416307020.

Doan, N.K., Q.D. Lai, T.K.P. Le, and N.T. Le. 2021. Influences of ac frequency and electric field strength on changes in bioactive compounds in ohmic heating of pomelo juice. *Innovative Food Science & Emerging Technologies* 72: 102754. www.sciencedirect.com/science/article/pii/S1466856421001557.

Eisner, M.D. 2021. Direct and indirect heating of milk – a technological perspective beyond time – temperature profiles. *International Dairy Journal* 122: 105145. www.sciencedirect.com/science/article/pii/S0958694621001734.

Ferreira, M.V.S., L.P. Cappato, R. Silva, R.S. Rocha, J.T. Guimarães, C.F. Balthazar, E.A. Esmerino, et al. 2019. Ohmic heating for processing of whey-raspberry flavored beverage. *Food Chemistry* 297: 125018. www.sciencedirect.com/science/article/pii/S0308814619311203.

Ferreira-Santos, P., R. Nunes, F. De Biasio, G. Spigno, D. Gorgoglione, J.A. Teixeira, and C.M.R. Rocha. 2020. Influence of thermal and electrical effects of ohmic heating on c-phycocyanin properties and biocompounds recovery from *Spirulina platensis*. *LWT* 128: 109491. www.sciencedirect.com/science/article/pii/S0023643820304801.

Gavahian, M., and B.K. Tiwari. 2020. Moderate electric fields and ohmic heating as promising fermentation tools. *Innovative Food Science and Emerging Technologies* 64: 102422. https://doi.org/10.1016/j.ifset.2020.102422.

Gomes, C.F., J.R. Sarkis, and L.D.F. Marczak. 2018. Ohmic blanching of tetsukabuto pumpkin: Effects on peroxidase inactivation kinetics and color changes. *Journal of Food Engineering* 233: 74–80. www.sciencedirect.com/science/article/pii/S0260877418301493.

Goullieux, A., and J.-P. Pain. 2014. *Chapter 22 – Ohmic Heating. Emerging Technologies for Food Processing.* Second Edi. Elsevier Ltd. http://dx.doi.org/10.1016/B978-0-12-411479-1.00022-X.

Hashemi, S.M.B., A. Gholamhosseinpour, and M. Niakousari. 2019a. Application of microwave and ohmic heating for pasteurization of cantaloupe juice: Microbial inactivation and chemical properties. *Journal of the Science of Food and Agriculture* 99, no. 9: 4276–4286. https://onlinelibrary.wiley.com/doi/abs/10.1002/jsfa.9660.

Hashemi, S.M.B., M.R. Mahmoudi, R. Roohi, I. Torres, and J.A. Saraiva. 2019b. Statistical modeling of the inactivation of spoilage microorganisms during ohmic heating of sour orange juice. *LWT* 111: 821–828. www.sciencedirect.com/science/article/pii/S0023643819303895.

Hashemi, S.M.B., and R. Roohi. 2019. Ohmic heating of blended citrus juice: Numerical modeling of process and bacterial inactivation kinetics. *Innovative Food Science & Emerging Technologies* 52: 313–324. www.sciencedirect.com/science/article/pii/S1466856418314887.

Inmanee, P., P. Kamonpatana, and T. Pirak. 2019. Ohmic heating effects on *Listeria monocytogenes* inactivation, and chemical, physical, and sensory characteristic alterations for vacuum packaged sausage during post pasteurization. *LWT* 108: 183–189. www.sciencedirect.com/science/article/pii/S0023643819302087.

Kim, Sang-Soon, W. Choi, and D.-H. Kang. 2017. Application of low frequency pulsed ohmic heating for inactivation of food-borne pathogens and MS-2 phage in buffered peptone water and tomato juice. *Food Microbiology* 63: 22–27. www.sciencedirect.com/science/article/pii/S074000201530191X.

Kim, Sang-Soon, and D.-H. Kang. 2015a. Effect of milk fat content on the performance of ohmic heating for inactivation of *Escherichia coli* O157:H7, *Salmonella enterica* serovar *typhimurium* and *Listeria monocytogenes*. *Journal of Applied Microbiology* 119, no. 2 (August): 475–486.

Kim, Sang-Soon, and D.-H. Kang. 2015b. Comparison of pH effects on ohmic heating and conventional heating for inactivation of *Escherichia coli* O157:H7, *Salmonella enterica* serovar *typhimurium* and *Listeria monocytogenes* in orange juice. *LWT – Food Science and Technology* 64, no. 2: 860–866. www.sciencedirect.com/science/article/pii/S0023643815300050.

Kim, Sang-Soon, and D.-H. Kang. 2017. Synergistic effect of carvacrol and ohmic heating for inactivation of *E. Coli* O157:H7, *S. Typhimurium*, *L. Monocytogenes*, and MS-2 bacteriophage in salsa. *Food Control* 73: 300–305. www.sciencedirect.com/science/article/pii/S0956713516304492.

Kim, Sang-Soon, and D.-H. Kang. 2017. Combination treatment of ohmic heating with various essential oil components for inactivation of food-borne pathogens in buffered peptone water and salsa. *Food Control* 80: 29–36. www.sciencedirect.com/science/article/pii/S095671351730186X.

Kim, Sang-Soon, S.-H. Park, and D.-H. Kang. 2018. Application of continuous-type pulsed ohmic heating system for inactivation of food-borne pathogens in buffered peptone water and tomato juice. *LWT* 93: 316–322. www.sciencedirect.com/science/article/pii/S0023643818302470.

Kubo, M.T., É.S. Siguemoto, E.S. Funcia, P.E. Augusto, S. Curet, L. Boillereaux, S.K. Sastry, and J.A. Gut. 2020. Non-thermal effects of microwave and ohmic processing on microbial and enzyme inactivation: A critical review. *Current Opinion in Food Science* 35: 36–48.

Kulshrestha, S., and S. Sastry. 2003. Frequency and voltage effects on enhanced diffusion during moderate electric field (MEF) treatment. *Innovative Food Science & Emerging Technologies* 4, no. 2: 189–194. www.sciencedirect.com/science/article/pii/S1466856403000031.

Kuriya, S.P., R. Silva, R.S. Rocha, J.T. Guimarães, C.F. Balthazar, R.P.S. Pires, E.R. Tavares Filho, et al. 2020. Impact assessment of different electric fields on the quality parameters of blueberry flavored dairy desserts processed by ohmic heating. *Food Research International* 134, no. January.

Lee, J.-Y., S.-S. Kim, and D.-H. Kang. 2015. Effect of pH for Inactivation of *Escherichia coli* O157:H7, *Salmonella typhimurium* and *Listeria monocytogenes* in orange juice by ohmic heating. *LWT – Food Science and Technology* 62, no. 1, Part 1: 83–88. www.sciencedirect.com/science/article/pii/S0023643815000365.

Lee, S.-Y., H.-G. Sagong, S. Ryu, and D.-H. Kang. 2012. Effect *of Continuous Ohmic Heating to Inactiv*ate *Escherichia coli* O157:H7, *Salmonella typhimurium* and *Listeria monocytogenes* in orange juice and tomato juice. *Journal of Applied Microbiology* 112, no. 4 (April): 723–731.

Leite, T.S., C.P. Samaranayake, S.K. Sastry, and M. Cristianini. 2019. Polyphenol oxidase inactivation in viscous fluids by ohmic heating and conventional thermal processing. *Journal of Food Process Engineering* 42, no. 5: e13133. https://doi.org/10.1111/jfpe.13133.

Llave, Y., K. Morinaga, M. Fukuoka, and N. Sakai. 2018a. characterization of ohmic heating and sous-vide treatment of scallops: Analysis of electrical conductivity and the effect of thermal protein denaturation on quality attribute changes. *Innovative Food Science & Emerging Technologies* 50: 112–123. www.sciencedirect.com/science/article/pii/S1466856418302613.

Llave, Y., T. Udo, M. Fukuoka, and N. Sakai. 2018b. Ohmic heating of beef at 20 kHz and analysis of electrical conductivity at low and high frequencies.

Journal of Food Engineering 228: 91–101. www.sciencedirect.com/science/article/pii/S0260877418300748.

Makroo, H.A., N.K. Rastogi, and B. Srivastava. 2020. Ohmic heating assisted inactivation of enzymes and microorganisms in foods: A review. *Trends in Food Science and Technology* 97, no. January: 451–465. https://doi.org/10.1016/j.tifs.2020.01.015.

Makroo, H.A., J. Saxena, N.K. Rastogi, and B. Srivastava. 2017. Ohmic heating assisted polyphenol oxidase inactivation of watermelon juice: Effects of the treatment on pH, lycopene, total phenolic content, and color of the juice. *Journal of Food Processing and Preservation* 41, no. 6: e13271. https://ifst.onlinelibrary.wiley.com/doi/abs/10.1111/jfpp.13271.

Mok, J.H., T. Pyatkovskyy, A. Yousef, and S.K. Sastry. 2019. Combined effect of shear stress and moderate electric field on the inactivation of *Escherichia coli* K12 in apple juice. *Journal of Food Engineering* 262: 121–130. www.sciencedirect.com/science/article/pii/S0260877419302134.

Müller, W.A., L.D. Ferreira Marczak, and J.R. Sarkis. 2020. Microbial inactivation by ohmic heating: Literature review and influence of different process variables. *Trends in Food Science and Technology* 99, no. May 2019: 650–659. https://doi.org/10.1016/j.tifs.2020.03.021.

Murashita, S., S. Kawamura, and S. Koseki. 2017. Effects of ohmic heating, including electric field intensity and frequency, on thermal inactivation of *Bacillus subtilis* spores. *Journal of Food Protection* 80, no. 1: 164–168.

Negri Rodríguez, L.M., R. Arias, T. Soteras, A. Sancho, N. Pesquero, L. Rossetti, H. Tacca, N. Aimaretti, M.L. Rojas Cervantes, and N. Szerman. 2021. Comparison of the quality attributes of carrot juice pasteurized by ohmic heating and conventional heat treatment. *LWT* 145: 111255. www.sciencedirect.com/science/article/pii/S0023643821004084.

Park, I.-K., and D.-H. Kang. 2013. Effect of Electropermeabilization by ohmic heating for inactivation of *Escherichia coli* O157:H7, *Salmonella enterica* serovar *typhimurium*, and *Listeria monocytogenes* in buffered peptone water and apple juice. *Applied and Environmental Microbiology* 79, no. 23: 7122–7129.

Park, I.K., J.W. Ha, and D.H. Kang. 2017. Investigation of optimum ohmic heating conditions for inactivation of *Escherichia Coli* O157:H7, *Salmonella Enterica* serovar *typhimurium*, and *Listeria monocytogenes* in apple juice. *BMC Microbiology* 17, no. 1: 1–8.

Ríos-Ríos, K.L., M. Gaytán-Martínez, D.M. Rivera-Pastrana, E. Morales-Sánchez, M. Villamiel, A. Montilla, E.M. Mercado-Silva, and M.E. Vázquez-Barrios. 2021. Ohmic eating pretreatment accelerates black garlic processing. *LWT* 151: 112218. www.sciencedirect.com/science/article/pii/S0023643821013712.

Rodrigues, R.M., A.A. Vicente, S.B. Petersen, and R.N. Pereira. 2019. Electric Field Effects on β-lactoglobulin thermal unfolding as a function of pH – impact on protein functionality. *Innovative Food Science and Emerging Technologies* 52: 1–7. https://doi.org/10.1016/j.ifset.2018.11.010.

Sagong, H.-G., S.-H. Park, Y.-J. Choi, S. Ryu, and D.-H. Kang. 2011. Inactivation of *Escherichia coli* O157:H7, *Salmonella typhimurium*, and *Listeria monocytogenes* in orange and tomato juice using ohmic heating. *Journal of Food Protection* 74, no. 6 (June 1): 899–904. https://doi.org/10.4315/0362-028X.JFP-10-552.

Sakr, M., and S. Liu. 2014. A comprehensive review on applications of ohmic heating (OH). *Renewable and Sustainable Energy Reviews* 39: 262–269. http://dx.doi.org/10.1016/j.rser.2014.07.061.

Samaranayake, C.P., and S.K. Sastry. 2016a. Effect of moderate electric fields on inactivation kinetics of pectin methylesterase in tomatoes: The roles of electric field strength and temperature. *Journal of Food Engineering* 186: 17–26. www.sciencedirect.com/science/article/pii/S0260877416301212.

Samaranayake, C.P., and S.K. Sastry. 2016b. Effects of controlled-frequency moderate electric fields on pectin methylesterase and polygalacturonase activities in tomato homogenate. *Food Chemistry* 199: 265–272. www.sciencedirect.com/science/article/pii/S0308814615302880.

Samaranayake, C.P., and S.K. Sastry. 2018. In-situ activity of α-Amylase in the presence of controlled-frequency moderate electric fields. *LWT – Food Science and Technology* 90, no. October 2017: 448–454. https://doi.org/10.1016/j.lwt.2017.12.053.

Samprovalaki, K., S. Bakalis, and P.J. Fryer. 2007. Ohmic heating: Models and measurements. *Heat Transfer in Food Processing* 13: 159–186.

Saxena, J., H. Ahmad Makroo, and B. Srivastava. 2017. Effect of ohmic heating on polyphenol oxidase (PPO) inactivation and color change in sugarcane juice. *Journal of Food Process Engineering* 40, no. 3 (June 1): e12485. https://doi.org/10.1111/jfpe.12485.

Schottroff, F., D. Biebl, M. Gruber, N. Burghardt, J. Schelling, M. Gratz, C. Schoenher, and H. Jaeger. 2020. Inactivation of vegetative microorganisms by ohmic heating in the kilohertz range – evaluation of experimental set-ups and non-thermal effects. *Innovative Food Science & Emerging Technologies* 63: 102372. www.sciencedirect.com/science/article/pii/S1466856420303180.

Schottroff, F., T. Pyatkovskyy, K. Reineke, P. Setlow, S.K. Sastry, and H. Jaeger. 2019. Mechanisms of enhanced bacterial endospore inactivation during sterilization by ohmic heating. *Bioelectrochemistry* 130: 107338. https://doi.org/10.1016/j.bioelechem.2019.107338.

Shao, L., X. Tian, Q. Yu, L. Xu, X. Li, and R. Dai. 2019. Inactivation and recovery kinetics of *Escherichia Coli* O157:H7 treated with ohmic heating in broth. *LWT* 110, no. 17: 1–7. https://doi.org/10.1016/j.lwt.2019.04.062.

Shao, L., Y. Zhao, B. Zou, X. Li, and R. Dai. 2021. Ohmic heating in fruit and vegetable processing: Quality characteristics, enzyme inactivation, challenges and prospective. *Trends in Food Science & Technology* 118: 601–616. www.sciencedirect.com/science/article/pii/S0924224421005689.

Somavat, R., H.M.H. Mohamed, Y.K. Chung, A.E. Yousef, and S.K. Sastry. 2012. Accelerated inactivation of *Geobacillus Stearothermophilus* spores by ohmic heating. *Journal of Food Engineering* 108, no. 1: 69–76. http://dx.doi.org/10.1016/j.jfoodeng.2011.07.028.

Somavat, R., H.M.H. Mohamed, and S.K. Sastry. 2013. Inactivation kinetics of *Bacillus Coagulans* spores under ohmic and conventional heating. *LWT – Food Science and Technology* 54, no. 1: 194–198. http://dx.doi.org/10.1016/j.lwt.2013.04.004.

Tajchakavit, S., and H.S. Ramaswamy. 1995. Continuous-flow microwave heating of orange juice: Evidence of nonthermal effects. *Journal of Microwave Power and Electromagnetic Energy* 30, no. 3 (January 1): 141–148. https://doi.org/10.1080/08327823.1995.11688270.

Tian, X., W. Wu, Q. Yu, M. Hou, F. Jia, X. Li, and R. Dai. 2016. Quality and proteome changes of Beef *M. Longissimus Dorsi* cooked using a water bath and ohmic heating process. *Innovative Food Science and Emerging Technologies* 34: 259–266.

Tian, X., Q. Yu, L. Shao, X. Li, and R. Dai. 2018a. Sublethal injury and recovery of *Escherichia Coli* O157:H7 after ohmic heating. *Food Control* 94: 85–92. https://doi.org/10.1016/j.foodcont.2018.06.028.

Tian, X., Q. Yu, W. Wu, and R. Dai. 2018b. Inactivation of microorganisms in foods by ohmic heating: A review. *Journal of Food Protection* 81, no. 7: 1093–1107.

Tian, X., Q. Yu, W. Wu, X. Li, and R. Dai. 2018c. Comparative proteomic analysis of *Escherichia Coli* O157:H7 following ohmic and water bath heating by capillary-HPLC-MS/MS. *International Journal of Food Microbiology* 285: 42–49. www.sciencedirect.com/science/article/pii/S0168160518303106.

Tian, X., Q. Yu, D. Yao, L. Shao, Z. Liang, F. Jia, X. Li, T. Hui, and R. Dai. 2018d. New insights into the response of metabolome of *Escherichia Coli* O157:H7 to ohmic heating. *Frontiers in Microbiology* 9, no. December: 1–12.

Tunç, M.T., and H.I. Odabaş. 2021. Single-step recovery of pectin and essential oil from lemon waste by ohmic heating assisted extraction/hydrodistillation: A multi-response optimization study. *Innovative Food Science & Emerging Technologies* 74: 102850. www.sciencedirect.com/science/article/pii/S1466856421002514.

Turgut, S.S., E. Küçüköner, A.H. Feyissa, and E. Karacabey. 2021. A novel drying system – simultaneous use of ohmic heating with convectional air drying: System design and detailed examination using CFD. *Innovative Food Science & Emerging Technologies* 72: 102727. www.sciencedirect.com/science/article/pii/S1466856421001284.

Wattanayon, W., P. Udompijitkul, and P. Kamonpatana. 2021. Ohmic heating of a solid-liquid food mixture in an electrically conductive package. *Journal of Food Engineering* 289: 110180. www.sciencedirect.com/science/article/pii/S0260877420302776.

10 Food Preservation by Microwave Heating

Debasmita Dutta and Debjani Dutta

Corresponding author:
Dr. Debjani Dutta, Associate Professor, Email: debjani.dutta@bt.nitdgp.ac.in
First Author: *Debasmita Dutta, Email: mails.ddutta@gmail.com*

CONTENTS

10.1 INTRODUCTION

Food products are required to be preserved for commercialization, and food preservation is the art of maintaining the nutritional quality of food and making it consumable for a certain period. The preservation method primarily depends on the type of food and processing technology. Numerous food preservation treatments exist in the food-processing industry like drying, pasteurization, freezing, by maintaining the pH, blanching, gas removal, texture modification, and others (Amit et al., 2017). Thermal treatments have several adverse effects on the heat-labile bioactive compounds present in the food (Al-Juhaimi et al., 2018). Improper heating parameters like high energy and prolonged heating may produce low-quality food products and a significant amount of waste. These problems generally occur during conventional thermal treatment. To overcome these flaws, "green technology" has been developed, such as pulsed electric field, supercritical fluid extraction, ultrasonication, and microwave processing (Uzel, 2018). These technologies have low processing times, high heating rates, and are environment-friendly, as no water is required for their operation. In addition, they cause lesser nutritional loss and maintained sensory properties compared to conventional heat treatment. Replacing traditional processing and preservation techniques enhances production efficiency, decrease water and fossil energy usage, and reduce the production of hazardous substances. In conventional heat treatment, heat transfers from the source (heating device) to the object (heating item), whereas in microwave heating, the dissipation of heat occurs inside the irradiated object.

Microwave energy was mainly used for communication purposes at the time of World War II. In 1949, Percy Spencer discovered the utilization of microwave energy in the food processing industry (Fito et al., 2004). Microwave ovens have either monomode or multimode cavities. The application of monomode in the food

DOI: 10.1201/9781003147978-10

industry is limited as it excites only one mode of resonance that is only applicable for low-volume food products. In contrast, multimode can heat a large volume of food products (Rana & Rana, 2014). Only dipoles can absorb the microwave and transform it into heat. Microwave heating depends on the dielectric properties of the objects (Sutar & Prasad, 2008), so the energy distribution is not uniform at the time of heating and is considered a limiting factor. Apart from this disadvantage, microwave treatment has several benefits like a noncontact treatment, low thermal gradients, and rapid energy absorption. This chapter aims to deal with the preservation ability of microwave energy combined with or without other treatments and its application in food processing and preservation. More attention has been given to ongoing research on microwave-assisted preservation and heating kinetics to understand the importance of microwave usage in food sectors.

10.2 PRINCIPLE AND MECHANISM OF MICROWAVE HEATING

The electromagnetic radio waves range a frequency band from 300 MHz to 300 GHz (Orsat et al., 2017). The most commonly used frequency in the domestic and industrial microwave oven is 2450 MHz, with a 3.8-cm penetration depth. For commercial purposes, 917 MHz frequency with 8- to 22-cm penetration ability is used only (Decareau, 1985). Microwave heating among food materials occurs due to their ability to transform microwave energy into thermal energy through two mechanisms: dipolar movement and ionic polarization. The magnetron is the device that attains the transformation inside an oven. Microwave heating is highly influential on the molecules possessing a higher dipole moment (water) present in the food materials and causing dielectric heating. In an alternating-current electric field, the bipolar molecules try to realign in the direction of the electric field by rotating 2.45×10^9 times/s. The friction among bipolar molecules due to the rapid movement results in heat generation throughout the food materials (Sutar & Prasad, 2008; Orsat et al., 2017). In an oscillating electric field, the fast movement of ionic molecules to align themselves causes the polarization of ions (Von Hippel, 1954; Decareau & Peterson, 1986).

Moreover, microwave heating can be influenced by the state of the constituents, like free ions having higher microwave absorptivities than bound ones. Figure 10.1 demonstrates the heating mechanism of a microwave oven. Linn

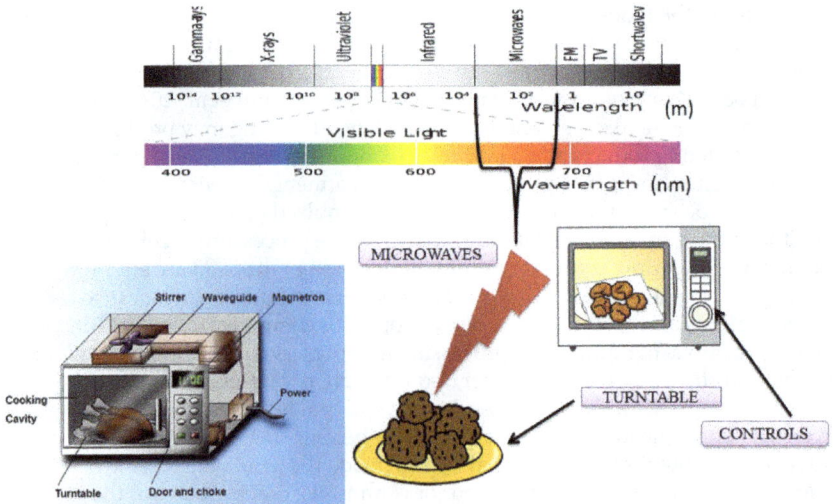

Figure 10.1 The heating mechanism of a microwave oven.

and Moller (2003) postulated an equation for transforming microwave energy to thermal energy.

$$P = 2\pi E^2 f\varepsilon''\varepsilon_0 V,$$

where P is the power applied (W), E is the electric field strength (V/m), ε_0 is the permittivity of free space (8.854188×10^{-12} F/m), f is the frequency Hz, V is the volume of the material m^3, and ε'' is the dielectric loss factor.

10.3 REACTION KINETICS OF MICROWAVE-ASSISTED FOOD PRESERVATION

Understanding kinetic modeling of heat treatment parameters is beneficial in the food industry, as it reveals the primary reaction mechanism. Two reactions occur at the time of microwave-assisted food processing—physical and chemical; both reactions make changes, which could be explained by kinetic modeling (Boekel et al., 2001). There are various models to understand kinetics (Vagenas & Marinos-kouris, 1991). Good knowledge about reaction kinetics helps fortify food products or prevent nutrient degradation. Emerging demand for the application of microwaves in food preservation increases research efforts on the optimization of the process parameters. In the first step, the basic kinetic principles of microwave-assisted treatment are determined, and then the generation of the kinetic data is elucidated. In the 19th century, performing the kinetics study of microwave treatment was difficult due to irregular heat distribution, improper chemical reactions, and difficulty in temperature monitoring (Jahngen et al., 1990; Constable et al., 1992). A study was done based on reaction kinetics under microwave irradiation; the power level was kept constant at 750 W, and a magnetic stirring device, a pressure gauge, and a thermometer were fitted out to ensure the uniform distribution of heat. The acid-catalyzed isomerization transformed carvone into carvacrol, and a Diels–Alder reaction occurred within anthracene and diethyl maleate. The rate of the reaction remained the same for both conventional and microwave heating treatments. The activation parameters for isomerization were E* = 89KJ/mol, lnA = 21.5, and E* = 83KJ/mol, lnA = 19.7 for traditional and microwave treatment, respectively, whereas, in the Diels–Alder reaction, the parameters were E* = 94KJ/mol, lnA = 15.6, and E* = 88KJ/mol, lnA = 13.9 for conventional and microwave treatment, respectively (Raner et al., 1993). In another study, the drying kinetics and physicochemical properties of Pomelo were analyzed. The result showed the combined microwave convective drying decreased the drying time compared to a microwave drying treatment. The optimized condition for the drying of the Pomelo slices was 90W power at 75°C temperature, where the maximum retention of antioxidant and phenolic content occurred (Yildiz & Izli, 2019). In this study, the rehydration kinetics of dried basil were studied. Here Peleg's model, the Weibull equation, the first-order kinetics model, and the exponential association equation were applied. The result exhibited that the kinetic rate constants (k_1) and (k_2) belong to Peleg's model and the scale parameter (β) belonging to the Weibull model were inversely proportional to the microwave output power used at the time of drying. The kinetic constant (k_1) and characteristic constant (k_2) belong to Peleg's model, and scale parameters (β) of the Weibull model were directly proportional to the amount of sample used. All four models were showing a good relationship with experimental data. Still, the Weibull model had a high coefficient value to determine the R^2 value, so it was chosen as the best-fitted model to identify the rehydration kinetics (Demirhan & ÖZbek, 2010). In this experiment, dehydration of Asiatic Himalayan black carrot was performed. Drying kinetics and moisture diffusivity were checked using different mathematical models to identify the most appropriate model for the drying process with experimental data. As the Page model had the highest coefficient value (R^2) and the lowest root mean square value, so it was selected as the best-fitted model (Haq et al., 2018). The microwave-assisted degradation kinetics of water-soluble vitamins was calculated. It followed first-order reaction kinetics. The result showed that vitamin C was the most degraded vitamin, followed by thiamin, riboflavin, and niacin. However, this degradation was much lower than conventional heating (Okmen & Bayindirli, 1999). In order to optimize the drying parameters of corn to obtain a good quality product, a kinetic study was performed at various power levels, moisture content, and exposure time. The drying rate equation,

$$\text{Drying rate} = (M_t + dt - M_t)/dt,$$

where drying rate is the amount of water removed per unit time per unit dry matter (g water/g dry matter/seconds), M_t is the moisture content at a specific time (g water/g dry matter); t is the drying time (seconds); and $M_t + dt$ is the moisture content at t + dt (g water/g dry matter) showed an increased drying rate, which was proportional to stress crack index and inversely proportional to germination, true density, and bulk density (Choudhary, 2013). Table 10.1 illustrates the kinetics study performed by various microwave-assisted treatments.

10.4 DIFFERENT TYPES OF MICROWAVE-ASSISTED FOOD PRESERVATION TECHNIQUES

Food preservation is a process that involves the elimination of microorganisms, the inhibition of enzymatic reactions, and the lowering the moisture content of food. Preservation improves the shelf lives of food products, makes the food available in the offseason, and reduces transportation costs. The food-processing industry practices numerous preservation techniques like canning, drying, and others of food to decrease the water activity (a_w). But most of these methods showed a deleterious effect on food properties (color, flavor, texture, nutrient) and on the environment. The microwave preservation technique is an emerging treatment that consumes less energy, saves time, and has a lower carbon footprint than traditional treatment (Mukhopadhyay et al., 2017). The volumetric heating process lowers the heating time and inactivates microorganisms without rendering the desired nutrients in the food (Decareau, 1985; Meredith, 1998). There are four microwave-assisted food preservation techniques (Figure 10.2).

10.4.1 Drying

Worldwide, more than 20% of foods are dried for preservation purposes (Orsat et al., 2006). In this process, water is removed to prevent microbial growth and inactivates some of the degrading enzymes (polyphenol oxidase, or PPO, and peroxidase, or POD) present in agriculture produce (Grabowski et al., 2003). The water activity (a_w) threshold for microbial inactivation is 0.86 for bacteria, 0.7 for yeasts, and 0.6 for molds. The a_w also influences enzymatic reactions. The a_w of 0.4 in dried products provides good stability (Schiffmann, 2001). Several drying techniques exist, among which convective drying is the most popular. But convective drying treatment significantly changes organoleptic properties of food (Sufer et al., 2017). Microwave drying is a good alternative as it prevents the degradation of the bioactive compounds present in food and saves treatment time and energy (Sorour et al., 2014; Demiray et al., 2017). During microwave

Table 10.1 Kinetic Study of Different Food Preservation Techniques

Best-fitted model	Equation	Food sample	Purpose of the kinetics study	References
Midilli model	$M = a.exp\left(-kt^n\right) + bt$	Ginger	Effect of pulse microwave drying on ginger	Nema et al. (2013)
Empirical model	$X_s = X_{s0}\exp\left(-k_M t\right)$	Apple, kiwi, and pear	Effect of different parameters on vacuum drying	Kiranoudis et al. (1997)
First-order reaction	$A = A_0\exp\left(-k_t\right)$	Vegetables (spinach, bell peppers, and carrots)	To know the reaction rate of peroxidase inactivation	Ramesh et al. (2002)
Page model Modified Page model	$MR = \exp\left(-kt^n\right)$ $MR = \exp\left[\left(-kt\right)^n\right]$	Kumquats slices	To understand the drying characteristics of kumquats	Ozcan-Sinir et al. (2019)
Weibull model	$M = M_e + \left(M_0 - M_e\right)\exp\left[-\left(\frac{t}{\beta}\right)^{\alpha}\right]$	Basil	Effect of drying conditions and rehydration temperature on rehydration kinetics of microwave–dried basil	Demirhan & Özbek (2010)
First-order reaction kinetics	$-\dfrac{d(C)}{dt} = k \cdot (c)$	Water-soluble vitamins	Effect of microwave heat treatment on the degradation reaction of water-soluble vitamins.	Okmen & Bayindirli (1999)
Exponential model Page model	$MR = \exp\left(-kt\right)$ $MR = \exp\left(-kt^n\right)$	Mushrooms	Effect of drying Parameters (microwave power, system pressure, and product thickness) on drying kinetics by microwave vacuum drying	Giri & Prasad (2007)
Page model	$MR = \exp\left(-kt^2 1\right)$	Asiatic Himalayan black carrot	Effect of microwave treatment on dehydration properties of Asiatic Himalayan black carrot	Haq et al. (2018)

(Continued)

Table 10.1 (Continued)

Best-fitted model	Equation	Purpose of the kinetics study	Food sample	References
First-order fractional conversion model / First-order kinetics	$\dfrac{A}{A_{untreated}} = \dfrac{A_{rf}}{A_{untreated}} + \dfrac{A_{untreated} - A_{rf}}{A_{untreated}} \cdot e^{-k_T \cdot t}$ $\dfrac{C}{C_{treated}} = \exp(-K_T \cdot t)$	Effect of blanching on the inactivation of ascorbic acid oxidation & degradation of L-ascorbic acid	Frozen mangoes	Xanthakis et al. (2018)
Gompertz model	$C(F_{80}) = C_{max}\left(\dfrac{C_O}{C_{max}}\right)^{\exp(-k \cdot F_{80})}$	Effect of different processing conditions on antioxidant activity and phenolic content	Cloudy apple juice	Siguemoto et al. (2019)
Pseudo-first-order kinetic	$T = T_b - (T_b - T_0)\exp(-mt)$	Effect of microwave heating on the physical properties of green beans	Green beans	Ruiz-Ojeda & Peñas (2013)
Alibas	$MR = a.\exp(-kt^n + bt) + g$	Effect of different microwave power on drying rate	Vegetable amaranth	Mujaffar et al. (2016)
Page's model	$MR = \exp(-kt^n)$	Determine the moisture ratio as a result of drying time	Turkey okra	Dadali et al. (2007)
Page's model	$MR = \exp(-kt^n)$	Drying kinetics of parsley leaves	Parsley leaves	Soysal (2004)
Weibull Aghbashlo et al	$MR = \exp\left(-\left(\dfrac{t}{b}\right)^a\right)$ $MR = \exp\left(-\dfrac{k_1 t}{1 + k_2 t}\right)$	Different drying power on drying kinetics	Cherry tomatoes	Kipcak et al. (2020)
Weibull	$MR = \exp\left(-\left(\dfrac{t}{b}\right)^a\right)$	Effects of process parameters on drying kinetics	Mussels	Kipcak et al. (2017)
Page's model Linear model	$\dfrac{(X_w^t - X_w^e)}{(X_w^0 - X_w^e)} = \exp(-k^* t^n)$ $X_w^t = X_w^0 - a^* t$	Microwave drying parameter and combined with hot air-drying parameters	Apricot	Igual et al. (2012)
Midilli model	$M = a.\exp(-kt^n) + bt$	Effect of microwave power on drying kinetics	Pepper	Darvishi et al. (2014)

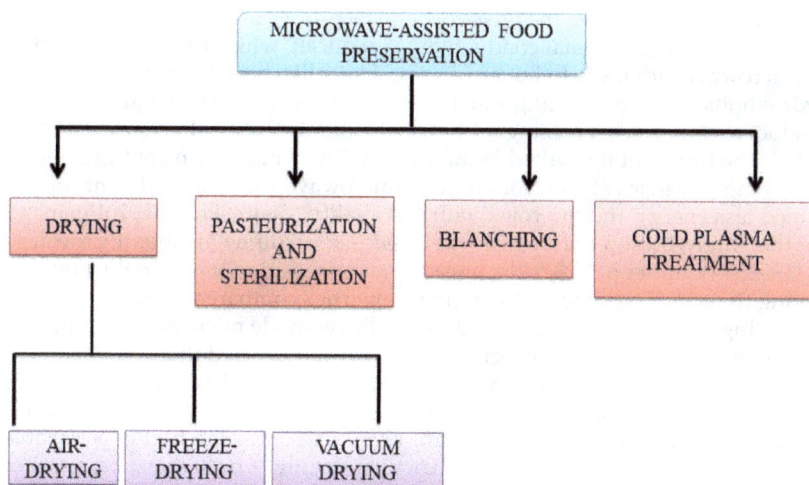

Figure 10.2 Classification of microwave-assisted food preservation.

drying, electromagnetic energy vaporizes the water present in food leading to an internal pressure that drives the water to the surface of the food material and lowers the drying time five times lesser than traditional drying. It also prevents food shrinkage and hardening of the end product (Onwude et al., 2016; Morais et al., 2018). In convective drying, a long drying period and the low thermal conductivity of food limit the heat transfer to the inner section of food (Demirhan & ÖZbek, 2010). There are three drying applications that the food sector have been practicing: air-drying, vacuum drying, and freeze-drying.

Microwave hot air drying was proved to be a rapid drying technique with high healthy compound retention (Ren & Chen, 1998). Garlic cloves were microwave air-dried at a temperature ranging from 50–70°C, with 1.0–2.0 m/s air velocities using continuous microwave power at 40 W. It exhibited an 80–90% reduction in moisture content, and the dried garlic had high quality (Sharma & Prasad, 2001). Microwave drying of kumquat, a citrus fruit, at 375 W showed the highest antioxidant activity, moisture diffusivity, and rehydration capacity compared to the conventional treatment. It also reduced the drying time; however, discoloration of the fruit was observed (Ozcan-Sinir et al., 2019). Granny Smith apple puree was mildly treated with microwave heating at 652 W for 35 s, which resulted in the reduction of microbial population and lower vitamin C degradation. But the low temperature was unable to inactivate undesirable enzymes responsible for enzymatic browning (POP). This was avoided by the addition of natural antibrowning agents (Picouet et al., 2009). Microwave air-drying of cumin seeds at 730 W for 10 min helped retain the characteristic flavor compounds of cumin (Behera et al., 2004). Microwave air-drying was applied for the preservation of green tea, and the retention of the primary components like vitamin C, tea polyphenols, and chlorophyll was much higher than oven drying (Huang et al., 2007). Microwave-assisted air-drying played a significant role in the inactivation of E. coli at 900-W microwave power for 90 s, and yeast growth at 450-W microwave power for 60 s in apple juice (Pragalyaashree et al., 2019).

Vacuum drying is one of the most efficient drying techniques for oxidation-prone bioactive compounds. The air is absent during the treatment, preventing oxidation and improving the physicochemical properties of products. But in vacuum drying, thermal conductivity is reduced, which can be enhanced by microwave radiation. In hot air-drying, fruits like berries formed a waxy hydrophobic layer, preventing water transport from the interior part of the product to its surface. Freezing or thawing could overcome this hazard and reduce the time, but it resulted in inferior quality of berries. In contrast, hot air drying of berries at 80°C followed by microwave vacuum treatment consumed less energy and improved nutrient quality (Zielinska et al., 2015). In another study, button mushroom was dried in a vacuum chamber microwave at 115–285 W under 6.5–23.5 kPa pressure and showed a 70–90% reduction in drying time and improved rehydration properties compared to traditional air-drying treatment (Giri & Prasad, 2007). Pulse-mode microwave vacuum drying was subjected to cranberries, and the result showed that when the cranberries were kept at a 250-W power level, under 5.33 kPa pressure with less power-on time and higher off time were ideal conditions for the drying treatment (Yongsawatdigul & Gunasekaran, 1996). Raspberry is a highly perishable fruit, and treating it with microwave vacuum drying, a puffed raspberry snack was produced. In this experiment, 2.68 kW power was applied for 81 s, and it retained a significant amount of anthocyanin, a bioactive compound present in this fruit. The vacuum condition played a vital role in removing oxygen, so oxidation was prevented (Yanqiu et al., 2013). Garlic slices were subjected to microwave-vacuum drying at various power levels, and the result showed increased microwave power directly proportional to breaking stress, water absorption capacity, and drying rate. The slicing could increase water absorption capacity and drying time, but the drying time was still 10 times lesser than conventional drying (Figiel, 2009).

Microwave is known as the fourth-generation drying technology due to the high thermal efficiency (Vega-Mercado & Marcela, 2001). Microwaves can heat a substance volume under vacuum conditions and can be used for freeze-drying (Zhang et al., 2006). In microwave freeze-drying, drying is performed when the product is in the frozen state, and the dried product has a porous structure that leads to a suitable rehydration property (Ambros et al., 2018). The main problem that has been observed in microwave freeze-drying is the uneven distribution of the microwave field in the drying cavity (Vriezinga & Sanchez-Pedreno, 2002). To overcome this problem, various mathematical models have been developed, and numerous studies have been done to understand the parameters of this technique. With advanced technology, the microwave freeze-drying method can produce superior quality dried food products at a low cost. In an experiment, cabbage was dried using the microwave freeze-drying technique. Based on all the loopholes that are there, a customized microwave freeze-drying equipment was built. Two magnetrons were applied, which worked alternatively. Drying time was decreased to almost half of the time required for the conventional freeze-drying treatment. The drying rate was directly proportional to the microwave power and inversely proportional to the cavity pressure and material thickness in the sublimation phase. In microwave freeze-drying treatment, the effect of microwave radiation and less sublimation phase time showed a low microbial growth with enhanced nutritional properties (Duan et al., 2007). In another freeze-drying treatment, onion slices were freeze-dried at −20°C for 2 hours and subjected to microwave vacuum drying at 300-mbar pressure with various microwave power applications (120–1200W). For this experiment, the vacuum drier was designed, and this experiment saved >96% processing time. Energy consumption and cost were also reduced with improved product quality (Abbasi & Azari, 2009). Some bananas were kept at

−30°C for 5 hours in an ultra-low temperature freezer for microwave freeze-drying treatment. Bananas contained approximately 7 g/100gm reducing sugar and 3 g/g. db moisture content and showed the highest organoleptic properties when exposed to 2-W/g microwave power (Jiang et al., 2010). The instant veg soup was microwave freeze-dried at −40°C to −45°C temperature under 100-Pa vacuum pressure. For uniform heat distribution, three magnetrons were fixed at 60° angles. Material thickness also played an important role which was directly proportional to drying time. The optimized condition was 450–675-W microwave power and 15–20-mm material thickness to obtain a high-quality product (Wang et al., 2009).

10.4.2 Pasteurization and Sterilization

Pasteurization is a traditional method for preserving foods by killing unwanted microorganisms and inhibiting undesirable enzymes. Microwave heating can be considered an optimized thermal treatment; microwave heating can solve the problem that occurs by slow thermal diffusion of traditional approaches. To enhance the shelf life of citrus juice, microwave pasteurization was performed at the commercial level (Schlegel, 1992). Different studies were performed to understand the lethality of microwave irradiation, and the efficiency of pasteurization in the home microwave oven (Harlfinger, 1992; Thompson & Thompson, 1990). It exhibited similar advantages as the conventional pasteurization process. Various studies showed that this treatment diminished the degradation of physicochemical properties and reduced exposure to high energy (Harlfinger, 1992). In a study, a domestic microwave was used to pasteurize apple juice at various power levels (270–900 W) to inhibit *Escherichia coli*. The result showed a reduced population within the range of 2–4 logs. Between 720–900-W microwave power levels, the D-values were 0.42 min at 900 W (70.3°C) and 0.48 min at 720 W (76.2°C). This proved that microwave power level was solely responsible for the inactivation of the *E. coli* and not the heat (Celis et al., 2002). Tomato puree was pasteurized at 900-W power, followed by rapid cooling. The yeast and mold counts were acceptable, and the *E. coli* population decreased by 5 log cycles as recommended by the U.S. Federal Drug Administration (FDA). Soluble solids and sensory properties remained constant though a slight color change was observed after 31 days of storage (Pérez-Tejeda et al., 2016). The pasteurization of apple cider by applying power ranges between 900–2000 W reduced *E. coli* 25922 by 5 log cycles (Gentry & Roberts, 2005). A similar result was observed in pasteurized mango puree without changing acidity and color (Rodriguez-Salinas et al., 2011). The pasteurization of guava nectar retained 94% vitamin C with unaltered color, and no microbial growth was observed (Salazar-Gonzalez et al., 2011). Undesirable enzymes present in food often used as an indicator by which the adequacy of pasteurization can be understood. PPO and POD, present in green coconut water, initiates oxidation when exposed to air (Andrey et al., 2008). Microwave pasteurization was performed to prevent microbial growth and deterioration of nutritional quality and sensory properties. The result showed inactivation of these two enzymes was more rapid than traditional treatment. Pasteurization of nonacidic juice blends was done at 3 W/g power and offered the maximum reduction of the PPO and pectin methyl esterase (PME) activities. The energy cost was cheaper than the conventional method, and complete inhibition of fungus, *E. coli*, and other bacteria was also observed (Math et al., 2014). Red palm oil rich in carotenoids are heat-labile due to the presence of conjugated double bonds. Sterilization of this oil by microwave irradiation required less temperature (<60°C) and more time (>20 min), which led to the deactivation of the enzyme lipase. This microwave irradiation technique prevented the degradation of carotenoids (4000 ppm) than standard methods (Sarah, 2018). The pasteurization of sour cherry juice was done using the microwave-ultrasonic system for the complete inactivation of *E. coli*. However, it degraded a significant amount of vitamin C content in the juice. The optimized treatment condition was approximately 352-W microwave output power at 49.94°C temperature and about 475-W ultrasonic energy for 6 min (Samani, Khoshtaghaza, Minaee, et al., 2015).

10.4.3 Blanching

Blanching is another conventional preservation pretreatment done before performing canning or freezing. This pretreatment inhibits undesirable enzymes or enzymatic browning (PPO and others), decreases microbial load, and removed trapped air and metabolic gases from the food. These air and gases are replaced by water which enhances the formation of uniform crystal growth during freezing. Parameters of microwave blanching like treatment time, temperature gradients, penetration depth, and sample properties on the heating cavity (electromagnetic field) should be carefully controlled. During microwave blanching, the leaching losses are low leading to an improved nutritional quality of foods. POD enzyme oxidizes polyphenols into dihydroxy-phenylalanine (DOPA) orthodiphenol from which orthoquinone is produced. Polymerization of orthoquinone finally produces the brown pigment named melanin. This pretreatment has been practiced by the food industry since the late 1920s (Lee, 1958; Downing, 1996; Günes & Bayindirh, 1993). To inactivate these enzymes, potato slices were pretreated with acid saline solution (lactic acid and sodium or calcium chloride solution) and subjected to microwave blanching. Calcium chloride treated lactic acid solution showed more inhibitory effect against PPO (Severini et al., 2004). A comparative experiment was performed between microwave and conventional blanching by using pretreated potato slices and cubes. Sliced potatoes were microwaved at 850 W power for 20–180 s, and they retained the desirable color more than the traditional treatment. Simultaneously, potato cubes were microwave blanched by immersing them in the water for uniform heat distribution. Rapid starch gelatinization was overcome by adding calcium chloride and sodium chloride aqueous solution during microwave-assisted blanching (Severini et al., 2001). Microwave blanching of green beans at 490-W power for 2 min or 700-W power for 1 min led to decreased PPO activity about 88% and enhanced ascorbic acid retention up to 70% (Brewer et al., 2003). Microwave blanching of vegetables (carrot, bell peppers, spinach) prevented the loss of water-soluble components (Ramesh et al., 2002). Mushrooms were blanched before canning for 4 min at 40°C, which enhanced the yield about 3.41% than conventional treatment (Baldwin et al., 1986). In another study, avocado was blanched at high temperatures for a short time using a microwave. The result showed that at pH 5.5, heating for 30 sec with 1% avocado leaves decreased the flavor loss and increased some odorants (Guzmán-Gerónimo et al., 2008). In our lab (Chakraborty & Dutta, 2012), microwave blanching of papaya puree was performed by applying different power levels (250 W, 300 W, 350 W, 450 W, 500 W, and 600 W) for 80 s. Almost 96% of PPO inactivation was observed at 600 W, and 70% inactivation was observed at 250-W power (Figure 10.3). Based on the k_d value (deactivation rate constant), obtained by following first-order kinetics, known as single-step two-stage theory, the deactivation of PPO was assumed. Structural analysis was also performed using SEM, and the results were compared with the thermally treated puree (Figures 10.4 and 10.5). At 600 W, a significant amount of cell damage was observed. The low decimal log reduction value (*D*-value) obtained in the

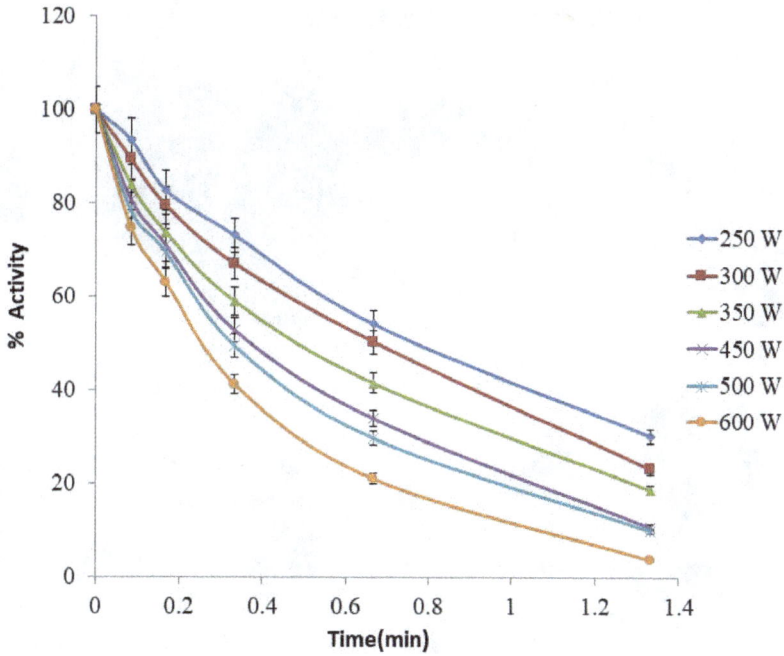

Figure 10.3 Microwave-assisted PPO deactivation (Chakraborty & Dutta, 2012).

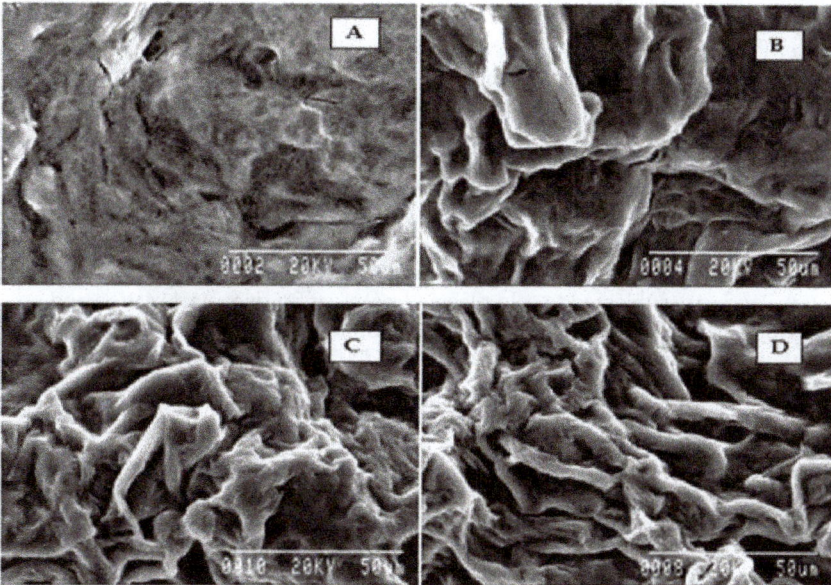

Figure 10.4 SEM analyses images for microwave-treated puree for 80 s at (A) control, (B) 250 W (C) 450 W, and (D) 600 W (Chakraborty & Dutta, 2012).

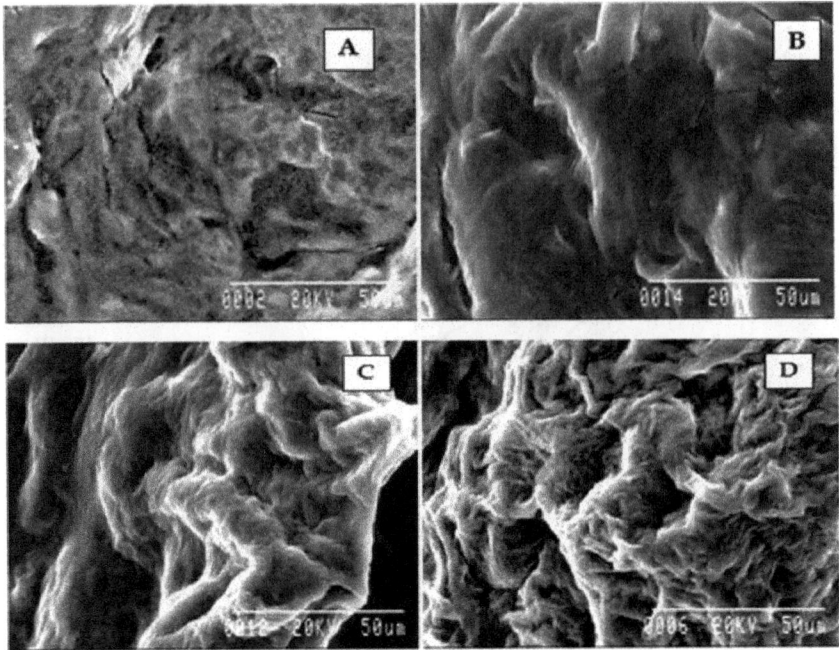

Figure 10.5 SEM analyses images for conventionally treated puree for 1280 s at (a) control, (b) 50°C, (c) 70°C, and (d) 90°C (Chakraborty & Dutta, 2012).

microwave condition compared to the conventional treatment demonstrated better PPO deactivation. Still, more work is required to fill the knowledge gap in industrial blanching technology. Various parameters like appropriate applicator, heating time, and temperature need to be explored to obtain an optimized process parameter.

10.4.4 Cold Plasma Treatment

Cold plasma treatment is an emerging technology that grabs the attention of researchers, and in this approach, gas is excited by microwave radiation and radiofrequency without altering the temperature and produces plasma (Niemira, 2012). Foods are subjected to cold plasma treatment for decontamination and enzyme inactivation. In general, microwave radiation, plasma jet, and corona discharge form the cold plasma (Toshifuji et al., 2003). Electromagnetic waves consist of frequencies more than hundreds of MHz and produce microwave cold plasma (Scholtz et al., 2015). This microwave cold plasma has more energetic electrons than other sources (Iza et al., 2007; Kwon et al., 2012). Figure 10.6 illustrates the mechanisms of enzyme inactivation and microbial inhibition by microwave cold plasma treatment. Microwave cold plasma treatment of potato slices at 900 W for 40 min showed the maximum inactivation of the enzyme (PPO) without altering nutrient properties (Kang et al., 2019). In another study, the cold plasma treatment of onion powder using helium gas was done at low and high microwave densities (170 mW.m^{-2} and 250 mW.m^{-2}), and the result showed that high microwave density at 400 W for 40 min resulted in the inhibition of *B. cereus* and *E. coli* 0157H7 up to 60 days of storage at 4°C. Onion contains some bioactive compounds

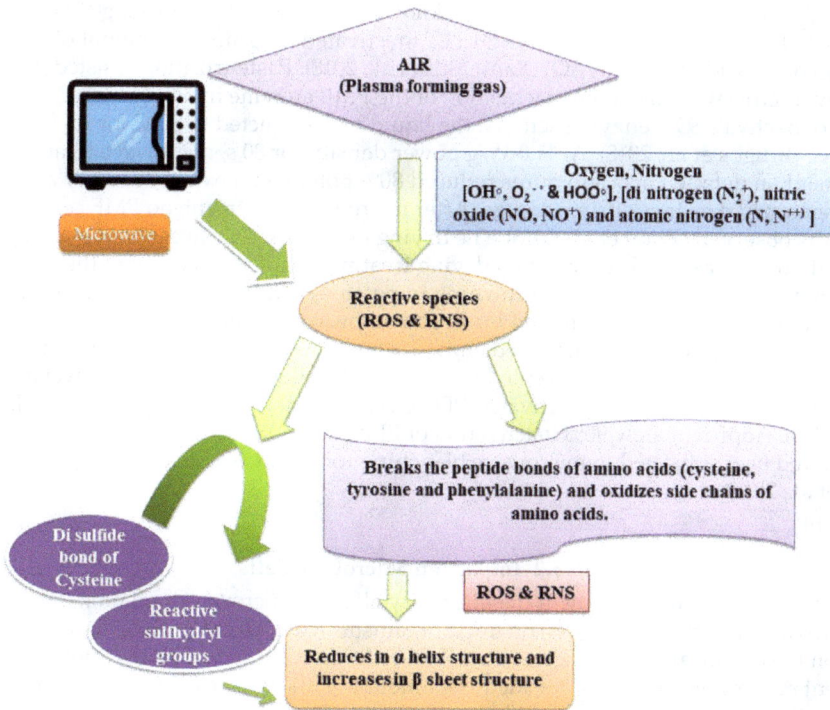

Figure 10.6 Mechanism for inactivation of microbial growth enzymes by using Microwave-cold plasma treatment.

(flavonoids-quercetin) that also played a role in the inhibition of *E. coli* (Kim et al., 2017). In a similar study, microwave-assisted plasma treatment showed an excellent inhibition activity against molds in different spices (black pepper seeds, juniper, allspice berries), and no molds were found after irradiating the sample for 45 and 60 s, although bacteria survived this treatment as they were more resistant (Wiktor et al., 2020). Red pepper flakes were treated with microwave-combined cold plasma technique. Low microwave power density ($0.17 W/m^2$) reduced 0.7 and 1.4 log spores/cm^2 of *B. cereus* and *A. flavus*, respectively, when kept for 150 days at 25°C, whereas high microwave power density ($0.25 W/m^2$) reduced 1.5 log spores/cm^2 of *B. cereus* and *A. flavus*. High and low microwave power density did not alter the physicochemical properties and antioxidant activity and prevented capsaicin and dihydrocapsaicin degradation (Kim et al., 2019). A similar study was also performed on mandarin to understand the effect on the inhibition of *Penicillium italicum*. Nitrogen cold plasma treatment was used by applying 0.7-kPa pressure, 900-W power for 10 min, and it showed the most effective protection against *P. italicum* (Won et al., 2017).

10.5 IMPACT OF MICROWAVE HEATING

10.5.1 Impact on Enzymes

Frozen mangos were subjected to microwave heating under 120 W with 14.39 W/cm^3 power density for 5 min, known as HTST treatment, and exposed under

100-W power 11.99 W/cm^3 for 12 min, known as low-temperature, long-time (LTLT) treatment. Both HTST and LTLT inactivated a significant amount of ascorbic acid oxidase (AAO) (Xanthakis et al., 2018). Pasteurization assisted by microwave caused the inactivation of the PME enzyme in orange juice. To inactivate 99% enzyme activity, the liquid was subjected to 80°C for 20 sec (Brugos et al., 2018). At 11.0-W/g power density for 80 s, microwave treatment on defatted avocado puree reduced 80% polyphenol oxidase activity and remained constant during storage. After microwave treatment, no PME activity was observed (Zhou et al., 2016). The drying of purple flesh sweet potato using microwave-assisted spouted bed drying treatment rapidly inactivated the enzymes responsible for the browning in potatoes (Liu et al., 2015). An osmotic steam blanching treatment and infrared microwave hot air-drying technique together significantly inactivated enzymes (pectinesterase and POD) present in the lemon slices and prevented browning (Deepika et al., 2018). Microwave heating at 89°C for 40 s inactivated PPO enzyme in acai berry pulp (Costa et al., 2020). Approximately 90% inactivation of PPO and POD enzymes was observed in red beet subjected to microwave blanching for 5 min at 200-W power (Latorre et al., 2012).

10.5.2 Impact on Microbial Cells

Various experiments explain the inhibitory effect of microwave radiation (non-thermal and thermal effects) on microorganisms (Gedikli et al., 2008). Based on the chemical composition of microbial cells, they are treated with different microwave radiation. An adequate amount of microwave energy absorbed by the intracellular component led to structural deformation, instability, and dysfunction of microbial cells. These further make the cell wall and membrane permeable and inhibit microbial growth (Vergani et al., 2011). Table 10.2 summarizes the effectiveness of microwave radiation on the inhibition of microorganisms present in various food items.

10.5.3 Impact on Physicochemical Properties

Microwave blanching on mangoes slightly reduced the total soluble solids (sugars) and vitamin C content, and these amounts were still comparatively higher than the conventional blanching treatment (Xanthakis et al., 2018). In another study, microwave heating increased phenolic content by 29.41% in defatted avocado, and no chlorophyll degradation was observed (Zhou et al., 2016). Microwave blanching retained 59% of anthocyanin in purple flesh sweet potato, but mild color degradation was also observed (Liu et al., 2015). The rate of nutrient and color degradation was directly proportional to drying temperature and power. An optimum condition of 1.5 W/g power and 25°C air temperature obtained bitter melon with higher nutritional quality (Nguyen et al., 2019). Microwave blanching of red beetroot at 100–200 W caused shrinkage and mild color degradation and altered elastic properties. Shrinkage can be prevented by immersing beetroots in the water and subjecting them to microwave treatment at <935 W (Latorre et al., 2012). Similar results were observed in green beverages and tomato puree caused by microwave pasteurization (González-Monroy et al., 2018; Pérez-Tejeda et al., 2016). Pasteurized rice (ready to eat) had a shelf life of 42 days when kept at 7°C temperature of storage. No significant difference was observed in the sensory property, but it became slightly soft due to the freezing treatment (Montero et al., 2020).

Table 10.2 Microbial Inhibition Using Different Microwave Treatments

Microwave treatment	Food	Microorganisms	Microbial reduction	Treatment condition	References
Heat treatment	Paprika power	Aerobic mesophilic bacteria	4.8 log units	650 W power for 60 sec followed by oven drying at 98°C for 20 min	Eliasson et al. (2015)
Heat treatment	Black pepper	Total mesophilic aerobic bacteria	4.11 log units	663 W power for 12.5 min	Jeevitha et al. (2016)
		Total yeast and mold count	4.02 log units		
		Total coliform, *Salmonella*, *Shigella*, and *E. coli*	Not detectable		
Heat treatment	Powder infant formula milk	*Cronobacter sakazakii*	5 $\log_{10}$ cycles	800-W and 900-W power level	Pina-Pérez et al. (2014)
Microwave heating	peanuts	Total bacterial count, aerobic spore count, yeast, and mold count	Complete elimination	360 W to 600 W power with 7-kGy γ-irradiation	Patil et al. (2019)
Microwave heating	Rough rice kernels	Aerobic bacteria	4.56 log CFU	915-MHZ power	Smith et al. (2018)
		A. flavus	2.93 log CFU		
Microwave heating	Granny Smith apple puree	*E. coli*	1.01 log CFU/g	652-W power for 35 s	Picouet et al. (2009)
		Total aerobic mesophilic count	1.16 log CFU/g		
		L. innocua	Not detectable		
Microwave-powered cold plasma treatment	Cabbage and lettuce	*Salmonella typhimurium*	1.5 log CFU/g	900-W power for 10 min with nitrogen	Lee et al. (2015)
	Cabbage	*L. monocytogenes*	0.3 to 2.1 log CFU/g	400–900-W power, 667 Pa pressure for 1 to 10 min with helium–oxygen gas mixture	
	Lettuce	*L. monocytogenes*	1.8 log CFU/g	400-W power for 10 min with helium–oxygen gas mixture	
	Figs	*E. coli*	0.5 to 1.3 log CFU/g	900-W power for 10 min with nitrogen	
		L. monocytogenes	1.0–1.6 log CFU/g	400-W power for 10 min with helium–oxygen gas mixture	

(Continued)

Table 10.2 (Continued)

Microwave treatment	Food	Microorganisms	Microbial reduction	Treatment condition	References
Sterilization	Apple juice	E. coli	Inactivated	900-W power for 90 s	Pragalyaashree et al. (2019)
Cold plasma treatment	Red pepper flakes	B. cereus	0.7 log spores/cm² and 1.4 log spores/cm²	900-W power for 20 min (low microwave power density and high microwave power density)	Kim et al. (2019)
		Aspergillus flavus	1.5 log spores/cm² and 1.5 log spores/cm²	900-W power for 20 min (low microwave power density and high microwave power density)	
Cold plasma treatment	Onion powder	B. cereus A. brasiliensis E. coli	2.1 log spores/cm² 1.6 log spores/cm² 1.9 CFU/cm²	400-W power for 40 min. High microwave cold plasma treatment	Kim et al. (2017)
Pasteurization	Tomato puree	E. coli	5 log cycles	950-W power	Pérez-Tejeda et al. (2016)
Pasteurization	Apple juice	E. coli	2–4 logs population	720–900-W power for 60–90 s	Celis et al. (2002)
Pasteurization	Green bean	Listeria innocua	9 log CFU/g	70C for 2 min	Inanoglu et al. (2021)
Microwave treatment	Egg white foam – chocolate pie lemon pie	Salmonella typhimurium	After 42 h storage – 2.9 × 10⁹ cells/g 0.0 cells/g	< 2 min	Baldwin et al. (1968)
Microwave heating	Orange juice	Total bacterial count	0	140 s	Lin et al. (1971)
Pasteurization	Milk	Bacterial count	0 per mL	192°F for 12 s	Hamid et al. (1969)
Microwave heating	Beef loaf Green beans	Aerobic plate counts	≤30 CFU/g 10¹–10⁴ CFU/g 10⁵–10⁶ CFU/g	110 s 50, 80, and 110 s 20 s	Dahl et al. (1980)

Process	Food	Microorganism	Result	Conditions	Reference
Microwave heating	Raw poultry	Salmonella enteritidis	6.4 log cycles / 5 log cycles / Not detectable	800-W power 10–140 s / High power for 95 s / Medium power for 140 s / High power for 110 s	Pucciarelli et al. (2005)
Microwave treatment	Soup	Salmonella typhimurium / E. coli	Reduction observed on the top region	915 MHz	Culkin et al. (1975)
Microwave heating	Meat tissue	Pseudomonas putrefaciens	Two decimal reduction	63°C	Crespo et al., 1977
		Streptococcus faecalis	Two decimal reduction	74°C	
		Lactobacillus plantarum	Resistant to microwave heating	Resistant	
Microwave radiation	Fish meal (cod and salmon)	E. coli / Salmonella enteritidis	Complete inactivation	$140 \text{ kJ} \times \text{g}^{-1}$ power	Bauza-Kaszewska et al. (2014)
Pasteurization	Orange juice	S. cerevisiae	Complete destruction	350-W power, 35°C temperature, and 778-W ultrasonic power for 11 min.	Samani, Khoshtaghaza, Lorigooini, et al. (2015)
Pasteurization	Sour cherry juice	E. coli	Zero count	352 W at 49.94°C temperature and 475-W ultrasonic power for 6 min	Samani, Khoshtaghaza, Minaee, et al. (2015)

10.6 APPLICATIONS OF MICROWAVE HEATING IN FOOD PRESERVATION

Microwave treatment has proved better than other conventional heat treatments of food. Many companies have worked on this technology and provided advanced microwave processing equipment. Some of these companies are EnWave Corporation (Canada), Stalam (Vicenza, Italy), Advanced Microwave Technologies (AMT, Edinburgh, UK), and Defreeze Corporation (USA). The volumetric heat distribution of microwave coagulates the meat protein uniformly and reduces nitrosamine formation leading to enhancement of the stability of the meat fat in sliced bacon (Edgar, 1986; Ohlsson & Bengtsson, 2000). Industrial microwave bacon cookers by Defreeze Corporation (USA) are also available in the market. Processing of heat-labile vitamins was done by microwave treatment to prevent degradation (Villanueva et al., 2000). There are many commercial microwave dryers available in the market, like Ferrite's industrial microwave system (USA), Microdry Incorporated (USA), Sairem (France), and many more. The world's most efficient microwave drying system is the MIP9 dryer from Ferrite's industrial microwave systems. This dryer has rapid and uniform heat distribution technology (Ferrite, 2015). EnWav also designed two radiant energy vacuum (REV) technologies for microwave drying of food samples; nutraREV and quantaREV. The equipment of nutraREV is commercially used to dry several food products like vegetables, herbs, dairy products, meat, and others. This equipment can provide up to 150-kW power and ≤37°C temperature for drying purposes. For the production of vacuum-dried cheese snacks, 100-kW power was applied. The quantaREV is used to dry a large volume of food products. For microwave freeze-drying, EnWave designed a pilot-scale platform called freezeREV (Enwave, 2015). Advanced microwave technologies (UK, Edinburgh) produced microwave pasteurizers for liquid materials at the pilot and industrial levels. Stalam (Italy) has microwave pasteurization equipment for fresh semolina pasta and packaged bread loaves. MicvacAB (Sweden) and Tops foods nv (Belgium) provide equipment for the pasteurization of ready-to-eat meals. AMT (UK) units are used for the microwave pasteurization of tomato juice. This unit consists of six magnetrons for the production of microwave radiation. The pasteurization treatment was standardized by raising the temperature of 85°C for 81.8 sec with 100 L/h flow capacity followed by rapid cooling (Stalam, 2015; Stratakos et al., 2016).

10.7 CONCLUSION AND PERSPECTIVES

Several food industries have widely accepted eco-friendly microwave technology with a low carbon footprint. This chapter discussed various microwave-assisted food preservation techniques and kinetics modeling of microwave heat treatment parameters. Microwave treatments in pasteurization, blanching, drying, cold plasma prevent food wastage, produce food products with better nutritional quality, and have higher decontamination ability than the others. The synergistic effect of microwave and other treatments like ultrasound, air drying may also provide a better-quality food product. Despite being widely used by various food processing industries, some parts of microwave energy remain unexplored. On that account, further investigation needs to be carried out to develop new and advanced technologies.

REFERENCES

Abbasi, S., & Azari, S. (2009). Original article Novel microwave – freeze drying of onion slices, 974–979. https://doi.org/10.1111/j.1365-2621.2008.01774.x

Al-Juhaimi, F., Ghafoor, K., Özcan, M. M., Jahurul, M. H. A., Babiker, E. E., Jinap, S., . . . & Zaidul, I. S. M. (2018). Effect of various food processing and handling methods on preservation of natural antioxidants in fruits and vegetables. *Journal of Food Science and Technology*, 55(10), 3872–3880.

Ambros, S., Mayer, R., Schumann, B., & Kulozik, U. (2018). *US. Innovative Food Science and Emerging Technologies*. Elsevier Ltd. https://doi.org/10.1016/j.ifset.2018.05.020

Amit, S. K., Uddin, M. M., Rahman, R., Islam, S. M. R., & Khan, M. S. (2017). A review on mechanisms and commercial aspects of food preservation and processing. *Agriculture and Food Security*, 6(1), 1–22. https://doi.org/10.1186/s40066-017-0130-8

Andrey, J., Gut, W., Vitoriano, P., Oliveira, D., & Cecilia, C. (2008). Inactivation kinetics of polyphenol oxidase and peroxidase in green coconut water by microwave processing, 88, 169–176. https://doi.org/10.1016/j.jfoodeng.2008.02.003

Baldwin, D. R., Anantheswaran, R. C., Sastry, S. K., & Beelman, R. B. (1986). Effect of microwave blanching on the yield and quality of canned mushrooms. *Journal of Food Science*, 51(4), 965–966. https://doi.org/10.1111/j.1365-2621.1986.tb11209.x

Baldwin, R. E., Cloninger, M., & Fields, M. L. (1968). Growth and destruction of Salmonella typhimurium in egg white foam products cooked by microwaves. *Applied Microbiology*, 16(12), 1929–1934. https://doi.org/10.1128/aem.16.12.1929-1934.1968

Bauza-Kaszewska, J., Skowron, K., Paluszak, Z., Dobrzański, Z., &ʻrutek, M. (2014). Effect of microwave radiation on microorganisms in fish meals. *Annals of Animal Science*, 14(3), 623–636. https://doi.org/10.2478/aoas-2014-0020

Behera, S., Nagarajan, S., & Jagan Mohan Rao, L. (2004). Microwave heating and conventional roasting of cumin seeds (Cuminum cyminum L.) and effect on chemical composition of volatiles. *Food Chemistry*, 87(1), 25–29. https://doi.org/10.1016/j.foodchem.2003.10.012

Boekel, M. A. J. S. V., & Tijskens, L. M. M. (2001). *Kinetics Modelling*. Chapter Three. Wageningen University.

Brewer, M. S., & Begum, S. (2003). Effect of microwave power level and time on ascorbic acid content, peroxidase activity and color of selected vegetables. Journal of Food Processing and Preservation, 27, 411–426

Brugos, A. F. O., Gut, J. A. W., & Tadini, C. C. (2018). Inactivation kinetics of pectin methyl esterase in the microwave-assisted pasteurization of orange juice. *LWT*, 97, 603–609. https://doi.org/10.1016/j.lwt.2018.07.042

Celis, E., Bruijn, J. De, Vidal, L. V, & Can, J. A. (2002). Pasteurisation of apple juice by using microwaves. *LWT*, 392, 389–392. https://doi.org/10.1006/fstl.2001.0865

Chakraborty, A., & Dutta, D. (2012). Polyphenol oxidase from *Carica papaya*: Isolation, characterization and deactivation study, M.Tech. thesis. NIT, Durgapur.

Choudhary, R. (2013). Microwave drying kinetics and quality characteristics of corn. *International Journal of Agricultural and Biological Engineering*, 6(1), 90–99. https://doi.org/10.3965/j.ijabe.20130601.009

Constable, D., Raner, K., Somlo, P., & Straw, C. J. (1992). A new microwave reactor suitable for organic synthesis and kinetics studies. *Microwave Power and Electromagnetic Energy*, 27, 195.

Costa, H. C. de B., Siguemoto, É. S., Cavalcante, T. A. B. B., de Oliveira Silva, D., Vieira, L. G. M., & Gut, J. A. W. (2021). Effect of microwave-assisted processing on polyphenol oxidase and peroxidase inactivation kinetics of açai-berry (Euterpe oleracea) pulp. *Food Chemistry*, 341(October 2020), 128287. https://doi.org/10.1016/j.foodchem.2020.128287

Crespo, F. L., Ockerman, H. W., & Irvin, K. M. (1977). Effect of Conventional and Microwave Heating on *Pseudomonas putrefaciens*, *Streptococcus faecalis* and *Lactobacillus plantarum* in Meat Tissue1. *Journal of Food Protection*, 40(9), 588–591. https://doi.org/10.4315/0362-028x-40.9.588

Culkin, K. A., & Fung, D. Y. C. (1975). Destruction of *Escherichia coli* and *Salmonella typhimurium* in microwave cooked soups. *Journal of Milk Food Technology*, 38(1), 8–15. https://doi.org/10.4315/0022-2747-38.1.8

Dadali, G., Apar, D. K., & Özbek, B. (2007). Microwave drying kinetics of okra. *Drying Technology*, 25(5), 917–924. https://doi.org/10.1080/07373930701372254

Dahl, C. A., Matthews, M. E., & Marth, E. H. (1980). Cook/chill foodservice system with a microwave oven: Aerobic plate counts from beef loaf, potatoes and frozen green beans. *Journal of Microwave Power*, 15(2), 94–105. https://doi.org/10.1080/16070658.1980.11689192

Darvishi, H., Asl, A. R., Asghari, A., Azadbakht, M., Najafi, G., & Khodaei, J. (2014). Study of the drying kinetics of pepper. *Journal of the Saudi Society of Agricultural Sciences*, 13(2), 130–138. https://doi.org/10.1016/j.jssas.2013.03.002

Decareau, R. V. (1985). *Microwaves in the Food Processing Industry*, Academic Press.

Decareau, R. V., & Peterson, R. A. (1986). *Microwave Processing and Engineering*. Ellis Horwood Ltd. & VCH Publishers.

Deepika, S., & Sutar, P. P. (2018). Combining osmotic – steam blanching with infrared – microwave – hot air drying: Production of dried lemon (citrus limon l.) slices and enzyme inactivation. *Drying Technology*, 36(14), 1719–1737. https://doi.org/10.1080/07373937.2017.1422744

Demiray, E., Seker, A., & Tulek, Y. (2017). Drying kinetics of onion (*Allium cepa L.*) slices with convective and microwave drying. *Heat Mass Transfer*, 53, 1817–1827. https://doi.org/10.1007/s00231-016-1943-x

Demirhan, E., & ÖZbek, B. (2010). Rehydration kinetics of microwave-dried basil. *Journal of Food Processing and Preservation*, 34(4), 664–680.

Downing, D. L. (1996).Canning Operations. In *A Complete Course in Canning and Related Processes: Book 1. Fundamental Information on Canning*, 13th Ed. (pp. 269–272). CTI Publications, Inc.

Duan, X., Zhang, M., & Mujumdar, A. S. (2007). Studies on the microwave freeze drying technique and sterilization characteristics of cabbage. *Drying Technology*, 25(10), 1725–1731. https://doi.org/10.1080/07373930701591044

Edgar, R. (1986). The economics of microwave processing in the food industry. *Food Technology*, 106112.

Eliasson, L., Isaksson, S., Lövenklev, M., & Ahrné, L. (2015). A comparative study of infrared and microwave heating for microbial decontamination of paprika powder. *Frontiers in Microbiology*, 6, 1071

Enwave.(2015). Enwave Corporation. www.enwave.net/nutrarev.php. Accessed in October 2015.

Ferrite. (2015). Ferrite Incorporated. www.ferriteinc.com/products_drying.html. Accessed in October 2015.

Figiel, A. (2009). Drying kinetics and quality of vacuum-microwave dehydrated garlic cloves and slices. *Journal of Food Engineering*, 94(1), 98–104. https://doi.org/10.1016/j.jfoodeng.2009.03.007

Fito, P., Chiralt, A., & Martín, M. E. (2004). Current state of microwave applications to food processing. *Novel Food Processing Technologies*, 525–537.

Gedikli, S., Tabak, Ö., Tomsuk, Ö., & Çabuk, A. (2008). Effect of microwaves on some gram negative and gram positive bacteria. *Journal of Applied Biological Sciences*, 2(1), 67–71.

Gentry, T. S., & Roberts, J. S. (2005). Design and evaluation of a continuous flow microwave pasteurization system for apple cider. *LWT*, 38, 227238.

Giri, S. K., & Prasad, S. (2007). Drying kinetics and rehydration characteristics of microwave-vacuum and convective hot-air dried mushrooms. *Journal of Food Engineering*, 78(2), 512–521. https://doi.org/10.1016/j.jfoodeng.2005.10.021

González-Monroy, A. D., Rodríguez-Hernández, G., Ozuna, C., & Sosa-Morales, M. E. (2018). Microwave-assisted pasteurization of beverages (tamarind and green) and their quality during refrigerated storage. *Innovative Food Science and Emerging Technologies*, 49(August), 51–57. https://doi.org/10.1016/j.ifset.2018.07.016

Grabowski, S., Marcotte, M., & Ramaswamy, H. S. (2003). Drying of fruits, vegetables, and spices. In A. Chakraverty, A. S. Mujumdar, G. S. V. Raghavan, & H. S. Rawaswamy (Eds.), *Handbook of Postharvest Technology: Cereals, Fruits, Vegetables, Tea, and Spices* (pp. 653–695). Marcel Dekker.

Günes, B., & Bayindirh, A. (1993). Peroxidase and lipoxygenase inactivation during blanching of green beans, green peas, and carrots. *Lebensm-Wiss und-Technol, 26*, 406–410.

Guzmán-Gerónimo, R. I., López, M. G., & Dorantes-Alvarez, L. (2008). Microwave processing of avocado: Volatile flavor profiling and olfactometry. *Innovative Food Science and Emerging Technologies*, 9(4), 501–506. https://doi.org/10.1016/j.ifset.2008.05.003

Hamid, M. A. K., Boulanger, R. J., Tong, S. C., Gallop, R. A., & Pereira, R. R. (1969). Microwave Pasteurization of Raw Milk. *Journal of Microwave Power*, 4(4), 272–275. https://doi.org/10.1080/00222739.1969.11688733

Haq, R. ul, Kumar, P., & Prasad, K. (2018). Effect of microwave treatment on dehydration kinetics and moisture diffusivity of Asiatic Himalayan black carrot. *Journal of the Saudi Society of Agricultural Sciences*, 17(4), 463–470. https://doi.org/10.1016/j.jssas.2016.11.004

Harlfinger, L. (1992). *Microwave Sterilization*. Food Technology (USA)

Huang, Y., Sheng, J., Yang, F., & Hu, Q. H. (2007). Effect of enzyme inactivation by microwave and oven heating on preservation quality of green tea. *Journal of Food Engineering*, 78(2), 687–692. https://doi.org/10.1016/j.jfoodeng.2005.11.007

Igual, M., García-Martínez, E., Martín-Esparza, M. E., & Martínez-Navarrete, N. (2012). Effect of processing on the drying kinetics and functional value of dried apricot. *Food Research International*, 47(2), 284–290. https://doi.org/10.1016/j.foodres.2011.07.019

Inanoglu, S., Barbosa-Cánovas, G. V., Patel, J., Zhu, M. J., Sablani, S. S., Liu, F., . . . & Tang, J. (2021). Impact of high-pressure and microwave-assisted thermal pasteurization on inactivation of *Listeria innocua* and quality attributes of green beans. *Journal of Food Engineering, 288*, 110162.

Iza, F., Lee, J. K., & Kong, M. G. (2007). Electron kinetics in radio-frequency atmospheric- pressure microplasmas. *Physical Reviews Letters, 99*, 075004.

Jahngen, E. G., Lentz, R. R., Pesheck, P. S., & Sackett, P. H. (1990). Hydrolysis of adenosine triphosphate by conventional or microwave heating. *The Journal of Organic Chemistry*, 55(10), 3406–3409.

Jeevitha, G. C., Sowbhagya, H. B., & Hebbar, H. U. (2016). Application of microwaves for microbial load reduction in black pepper (Piper nigrum L.). *Journal of the Science of Food and Agriculture*, 96(12), 4243–4249. https://doi.org/10.1002/jsfa.7630

Jiang, H., Zhang, M., & Mujumdar, A. S. (2010). Microwave freeze-drying characteristics of banana crisps. *Drying Technology*, 28(12), 1377–1384. https://doi.org/10.1080/07373937.2010.482702

Kang, J. H., Roh, S. H., & Min, S. C. (2019). Inactivation of potato polyphenol oxidase using microwave cold plasma treatment. *Journal of Food Science, 84*(5), 1122–1128. https://doi.org/10.1111/1750-3841.14601

Kim, J. E., Oh, Y. J., Song, A. Y., & Min, S. C. (2019). Preservation of red pepper flakes using microwave-combined cold plasma treatment. *Journal of the Science of Food and Agriculture, 99*(4), 1577–1585. https://doi.org/10.1002/jsfa.9336

Kim, J. E., Oh, Y. J., Won, M. Y., Lee, K. S., & Min, S. C. (2017). Microbial decontamination of onion powder using microwave-powered cold plasma treatments. *Food Microbiology, 62*, 112–123. https://doi.org/10.1016/j.fm.2016.10.006

Kipcak, A. S. (2017). Microwave drying kinetics of mussels (*Mytilus edulis*). *Research on Chemical Intermediates, 43*(3), 1429–1445. https://doi.org/10.1007/s11164-016-2707-4

Kipcak, A. S., & Doymaz, İ. (2020). Microwave and infrared drying kinetics and energy consumption of cherry tomatoes. *Chemical Industry and Chemical Engineering Quarterly, 26*(2), 203–212. https://doi.org/10.2298/CICEQ190916039K

Kiranoudis, C. T., Tsami, E., & Maroulis, Z. B. (1997). Microwave vacuum drying kinetics of some fruits. *Drying Technology, 15*(10), 2421–2440.

Kwon, H. C., Won, I. H., & Lee, J. K. 2012. Electron heating mode transition induced by ultra-high frequency in atmospheric microplasmas for biomedical applications. *Applied Physical Letters, 100*, 183702.

Latorre, M. E., Bonelli, P. R., Rojas, A. M., & Gerschenson, L. N. (2012). Microwave inactivation of red beet (*Beta vulgaris* L. var. conditiva) peroxidase and polyphenoloxidase and the effect of radiation on vegetable tissue quality. *Journal of Food Engineering, 109*(4), 676–684. https://doi.org/10.1016/j.jfoodeng.2011.11.026

Lee, F. A. (1958). The Blanching Process. In *Advances in Food Research* (pp. 63–109). Academic Press.

Lee, H., Kim, J. E., Chung, M. S., & Min, S. C. (2015). Cold plasma treatment for the microbiological safety of cabbage, lettuce, and dried figs. *Food Microbiology, 51*, 74–80. https://doi.org/10.1016/j.fm.2015.05.004

Lin, C. C., & Li, C. F. (1971). Microwave sterilization of oranges in glass- pack. *Journal of Microwave Power, 6*(1), 45–47. https://doi.org/10.1080/00222739.1971.11688778

Linn, H., & Moller, M. (2003). Microwave heating. In *Proceedings of the Thermo Process Symposium*, Dusseldorf, Germany (pp. 16–21).

Liu, P., Mujumdar, A. S., Zhang, M., & Jiang, H. (2015). Comparison of Three Blanching Treatments on the Color and Anthocyanin Level of the Microwave-Assisted Spouted Bed Drying of Purple Flesh Sweet Potato. *Drying Technology, 33*(1), 66–71. https://doi.org/10.1080/07373937.2014.936558

Math, R., Nagender, A., Nayani, S., & Satyanarayana, A. (2014). Continuous microwave processing and preservation of acidic and non acidic juice blends. *IJAFST, 2*, 81–90.

Meredith, R. J. (1998). *Engineers' Handbook of Industrial Microwave Heating* (No. 25). IET.

Montero, M. L., Sablani, S., Tang, J., & Ross, C. F. (2020). Characterization of the sensory, chemical, and microbial quality of microwave-assisted, thermally pasteurized fried rice during storage. *Journal of Food Science, 85*(9), 2711–2719.

Morais, R. M. S. C. S. C., Morais, A. M. M. B. M. B., Dammak, I., et al.(2018). Functional dehydrated foods for health preservation. *Journal of Food Quality*, 1–29

Mujaffar, S., & Loy, A. L. (2016). Drying Kinetics of Microwave-Dried Vegetable Amaranth (Amaranthus dubius) Leaves. *Journal of Food Research, 5*(6), 33. https://doi.org/10.5539/jfr.v5n6p33

Mukhopadhyay, S., Ukuku, D. O., Juneja, V. K., Nayak, B., & Olanya, M. (2017). Principles of Food Preservation. *Microbial Control and Food Preservation,* 17–39. https://doi.org/10.1007/978-1-4939-7556-3_2

Nema, P. K., Mohapatra, D., Daniel, A., & Mishra, S. (2013). Modeling pulse microwave drying kinetics of ginger. *Journal of Food Research and Technology, 1*(2), 46–58.

Nguyen, T. V. L., Nguyen, P. B. D., Luu, X. C., Huynh, B. L., Krishnan, S., & Huynh, P. T. (2019). Kinetics of nutrient change and color retention during low-temperature microwave-assisted drying of bitter melon (*Momordica charantia* L.). *Journal of Food Processing and Preservation, 43*(12), 1–12. https://doi.org/10.1111/jfpp.14279

Niemira, B. A. (2012). Cold plasma decontamination of foods. *Annual Review of Food Science and Technology, 3,* 125–142

Ohlsson, T., & Bengtsson, N. (2000). Microwave technology and foods. *Advances in Food and Nutrition Research, 43,* 65140.

Okmen, Z. A., & Bayindirli, A. L. (1999). Effect of microwave processing on water soluble vitamins: Kinetic parameters. *International Journal of Food Properties, 2*(3), 255–264. https://doi.org/10.1080/10942919909524609

Onwude, D., Hashim, N., & Chen, G. (2016). Recent advances of novel thermal combined hot air drying of agricultural crops. *Trends in Food Science & Technology, 57,* 132–145

Orsat, V., Changrue, V., & Raghavan, G. S. V. (2006). Microwave drying of fruits and vegetables. *Stewart Post-Harvest Review, 6,* 4–9. https://doi.org/10.2212/spr.2006.6.4

Orsat, V., Raghavan, G. S. V., & Krishnaswamy, K. (2017). *Microwave technology for food processing: An overview of current and future applications.* In *The Microwave Processing of Foods: Second edition.* Elsevier Ltd. https://doi.org/10.1016/B978-0-08-100528-6.00005-X

Ozcan-Sinir, G., Ozkan-Karabacak, A., Tamer, C. E., & Copur, O. U. (2019). The effect of hot air, vacuum and microwave drying on drying characteristics, rehydration capacity, color, total phenolic content and antioxidant capacity of kumquat (Citrus japonica). *Food Science and Technology, 39*(2), 475–484. https://doi.org/10.1590/fst.34417

Patil, H., Shah, N. G., Hajare, S. N., Gautam, S., & Kumar, G. (2019). Combination of microwave and gamma irradiation for reduction of aflatoxin B1 and microbiological contamination in peanuts (*Arachis hypogaea* L.). *World Mycotoxin Journal, 12*(3), 269–280. https://doi.org/10.3920/WMJ2018.2384

Pérez-Tejeda, G., Vergara-Balderas, F. T., López-Malo, A., Rojas-Laguna, R., Abraham-Juárez, M. del R., & Sosa-Morales, M. E. (2016). Pasteurization treatments for tomato puree using conventional or microwave processes. *Journal of Microwave Power and Electromagnetic Energy, 50*(1), 35–42. https://doi.org/10.1080/08327823.2016.1157315

Picouet, P. A., Landl, A., Abadias, M., Castellari, M., & Viñas, I. (2009). Minimal processing of a Granny Smith apple purée by microwave heating. *Innovative Food Science and Emerging Technologies, 10*(4), 545–550. https://doi.org/10.1016/j.ifset.2009.05.007

Pina-Pérez, M. C., Benlloch-Tinoco, M., Rodrigo, D., & Martinez, A. (2014). Cronobacter sakazakii Inactivation by Microwave Processing. *Food and Bioprocess Technology, 7*(3), 821–828. https://doi.org/10.1007/s11947-013-1063-2

Pragalyaashree, M. M., Tiroutchelvame, D., & Gokularaman, S. (2019). Microwave Assisted Batch Sterilization of Apple Juice. *International Journal of Innovative Technology and Exploring Engineering, 9*(2), 1087–1091. https://doi.org/10.35940/ijitee.I3324.129219

Pucciarelli, A. B., & Benassi, F. O. (2005). Inactivation of *Salmonella enteritidis* on raw poultry using microwave heating. *Brazilian Archives of Biology and Technology, 48*(6), 939–945. https://doi.org/10.1590/S1516-89132005000800010

Ramesh, M. N., Wolf, W., Tevini, D., & Bognár, A. (2002). Microwave blanching of vegetables. *Journal of Food Science, 67*(1), 390–398. https://doi.org/10.1111/j.1365-2621.2002.tb11416.x

Rana, K. K., & Rana, S. (2014). *Microwave Reactors : A Brief Review on its Fundamental Aspects and Applications,* 1–21. https://doi.org/10.4236/oalib.1100686

Raner, K. D., Strauss, C. R., Vyskoc, F., & Mokbel, L. (1993). A comparison of reaction kinetics observed under microwave irradiation and conventional heating. *The Journal of Organic Chemistry, 58*(4), 950–953

Ren, G., & Chen, F. (1998). Drying of American ginseng (*Panax quinquefolium* roots by microwave-hot air combination. *Journal of Food Engineering, 35*(4), 433–443.

Rodriguez-Salinas, C., Salazar-Gonzalez, C., Lopez-Malo, A., & Sosa-Morales, M. E. (2011). Pasteurization of mango puree using microwaves. In *Proceedings of the 45th IMPI Microwave Power Symposium;* 2011 June 8–10; New Orleans, LA: International Microwave Power Institute.

Ruiz-Ojeda, L. M., & Peñas, F. J. (2013). Comparison study of conventional hot-water and microwave blanching on quality of green beans. *Innovative Food Science and Emerging Technologies, 20,* 191–197. https://doi.org/10.1016/j.ifset.2013.09.009

Salazar-Gonzalez, C., Sosa-Morales, M. E., Lopez-Malo, A., & San Martin-Gonzalez, M. F. (2011). Microwave heating as a potential treatment for guava nectar. In *Proceedings of the 45th IMPI Microwave Power Symposium;* 2011 June 8–10; New Orleans, LA: International Microwave Power Institute.

Samani, B. H., Khoshtaghaza, M. H., Lorigooini, Z., Minaei, S., & Zareiforoush, H. (2015). Analysis of the combinative effect of ultrasound and microwave power on *Saccharomyces cerevisiae* in orange juice processing. *Innovative Food Science and Emerging Technologies, 32,* 110–115. https://doi.org/10.1016/j.ifset.2015.09.015

Samani, B. H., Khoshtaghaza, M. H., Minaee, S., & Abbasi, S. (2015). Modeling the Simultaneous Effects of Microwave and Ultrasound Treatments on Sour Cherry Juice Using Response Surface Methodology, *17,* 837–846.

Sarah, M. (2018). Carotenoids preservation during sterilization of palm fruit using microwave irradiation. *ARPN Journal of Engineering and Applied Sciences, 13*(3), 1009–1014.

Schiffmann, R. F. (2001). Microwave processes for the food industry. In A. K. Datta & R. C. Anantheswaran (Eds.), *Handbook of Microwave Technology for Food Applications* (pp. 115–166). Marcel Dekker.

Schlegel, W. (1992). Commercial pasteurisation and sterilization of food products using microwave technology. Food Technology, 46, 62–63.

Scholtz, V., Pazlarová, J., Soušková, H., Khun, J., Julák, J. 2015. Nonthermal plasma – A tool for decontamination and disinfection. Biotechnol. Adv. 33, 1108–1119.

Severini, C., Baiano, A., De Pilli, T., Romaniello, R., & Derossi, A. (2004). Microwave blanching of sliced potatoes dipped in saline solutions to prevent enzymatic browning. *Journal of Food Biochemistry, 28*(1), 75–89. https://doi.org/10.1111/j.1745-4514.2004.tb00056.x

Severini, C., De Pilli, T., Baiano, A., Mastrocola, D., & Massini, R. (2001). Preventing enzymatic browning of potato by microwave blanching. *Sciences Des Aliments, 21*(2), 149–160. https://doi.org/10.3166/sda.21.149-160

Sharma, G. P., & Prasad, S. (2001). Drying of garlic (*Allium sativum*) cloves by microwave-hot air combination. *Journal of Food Engineering, 50*(2), 99–105. https://doi.org/10.1016/S0260-8774(00)00200-4

Siguemoto, É. S., Purgatto, E., Hassimotto, N. M. A., & Gut, J. A. W. (2019). Comparative evaluation of flavour and nutritional quality after conventional and microwave-assisted pasteurization of cloudy apple juice. *LWT, 111*(May), 853–860. https://doi.org/10.1016/j.lwt.2019.05.111

Smith, D. L., & Atungulu, G. G. (2018). Impact of drying deep beds of rice with microwave set at 915 MHz frequency on rice microbial community responses. *Cereal Chemistry, 95*(1), 130–140. https://doi.org/10.1002/cche.10018

Sorour, H., & El-Mesery, H. (2014). Effect of microwave and infrared radiation on drying of onion slices. *International Journal of Natural and Social Sciences, 2*, 119–130. https://doi.org/10.17221/310/2017-CJFS

Soysal, Y. (2004). Microwave drying characteristics of parsley. *Biosystems Engineering, 89*(2), 167–173. https://doi.org/10.1016/j.biosystemseng.2004.07.008

Stalam. (2015).www.stalam.com/en/rf-applications-for-industrial-bakeries/pasteurisation-equipment.html. Accessed in October, 2015.

Stratakos, A., Delgado-Pando, G., Linton, M., Patterson, M. F., & Koidis, A. (2016). Industrial scale microwave processing of tomato juice using a novel continuous microwave system. *Food Chemistry, 190*, 622628.

Sufer, O., Demir, H., & Sezer, S. (2017). Thin layer mathematical modelling of convective, vacuum and microwave drying of intact and brined onion slices. *Journal of Food Processing and Preservation, 41*, 1–13. https://doi.org/10.1111/jfpp.13239

Sutar, P. P., & Prasad, S. (2008, February). Microwave drying technology-recent developments and R&D needs in India. In *Proceedings of the 42nd ISAE Annual Convention* (pp. 1–3).

Thompson, J. S., & Thompson, A. (1990). In-home pasteurization of raw goat's milk by microwave treatment. *International Journal of Food Microbiology, 10*(1), 59–64.

Toshifuji, J., Katsumata, T., Takikawa, H., Sakakibara, T., & Shimizu, I. (2003). Cold arc-plasma jet under atmospheric pressure for surface modification. *Surface and Coating Technology, 171*, 302–306.

Uzel, R. A. (2018). Microwave-assisted green extraction technology for sustainable food processing. In *Emerging Microwave Technologies in Industrial, Agricultural, Medical and Food Processing*. London, UK: Intechopen.

Vagenas, G. K., & Marinos-kouris, D. (1991). The design and optimization of an industrial dryer for sultana raisins. *Drying Technology, 9*, 439–461. https://doi.org/10.1080/07373939108916675.

Vega-Mercado, H., & Marcela, M. (2001). Advances in dehydration of foods. *Journal of Food Engineering, 49*(4), 271–289.

Vergani, C., Ribeiro, D. G., Dovigo, L. N., Sanita, P. V., & Pavarina, A. C. (2011). Microwave assisted disinfection method in dentistry. In *Microwave Heating*. IntechOpen.

Villanueva, M. T. O., Marquina, A. D., Vargas, E. F., & Abellan, G. B. (2000). Modification of vitamins B1 and B2 by culinary process: Traditional systems and microwaves. Food Chemistry, 71(4), 417421.

Von Hippel, A. R. (1954). *Dielectrics Materials and Applications*. MIT Press.

Vriezinga, C. A., & Sanchez-Pedreno, S.(2002). Thermal runaway in microwave heating: A mathematical analysis. Applied Mathematics Modelling x, 26 (10), 1029–1038.

Wang, R., Zhang, M., Mujumdar, A. S., & Sun, J. C. (2009). Microwave freeze-drying characteristics and sensory quality of instant vegetable soup. *Drying Technology, 27*(9), 962–968. https://doi.org/10.1080/07373930902902040

Wiktor, A., Hrycak, B., Jasinski, M., Rybak, K., Kieliszek, M., Krasniewska, K., & Witrowa-Rajchert, D. (2020). Impact of atmospheric pressure microwave plasma treatment on quality of selected spices. *Applied Sciences (Switzerland), 10*(19). https://doi.org/10.3390/app10196815

Won, M. Y., Lee, S. J., & Min, S. C. (2017). Mandarin preservation by microwave-powered cold plasma treatment. *Innovative Food Science and Emerging Technologies, 39*, 25–32. https://doi.org/10.1016/j.ifset.2016.10.021

Xanthakis, E., Gogou, E., Taoukis, P., & Ahrné, L. (2018). Effect of microwave assisted blanching on the ascorbic acid oxidase inactivation and vitamin C degradation in frozen mangoes. *Innovative Food Science and Emerging Technologies, 48*, 248–257. https://doi.org/10.1016/j.ifset.2018.06.012

Yanqiu, M., Xinhuai, Z., Bingxin, L., Chenghai, L., & Xianzhe, Z. (2013). Influences of microwave vacuum puffing conditions on anthocyanin content of raspberry snack, *6*(3), 80–87. https://doi.org/10.3965/j.ijabe.20130603.0010

Yildiz, G., & Izli, G. (2019). Influence of microwave and microwave-convective drying on the drying kinetics and quality characteristics of pomelo. *Journal of Food Processing and Preservation, 43*(6), 1–11. https://doi.org/10.1111/jfpp.13812

Yongsawatdigul, J., & Gunasekaran, S. (1996). Microwave-vacuum drying of cranberries: Part I. Energy use and efficiency. *Journal of Food Processing and Preservation, 20*(2), 121–143. https://doi.org/10.1111/j.1745-4549.1996.tb00850.x

Zhang, M., Tang, J., Mujumdar, A. S., & Wang, S. (2006). Trends in microwave related drying of fruits and vegetables. *Trends in Food Science & Technology, 17*(10), 524–534

Zhou, L., Tey, C. Y., Bingol, G., & Bi, J. (2016). Effect of microwave treatment on enzyme inactivation and quality change of defatted avocado puree during storage. *Innovative Food Science and Emerging Technologies, 37*, 61–67. https://doi.org/10.1016/j.ifset.2016.08.002

Zielinska, M., Sadowski, P., & Błaszczak, W. (2015). Freezing/thawing and microwave-assisted drying of blueberries (*Vaccinium corymbosum* L.). *LWT – Food Science and Technology, 62*(1), 555–563. https://doi.org/10.1016/j.lwt.2014.08.002

11 Infrared Heating for Food Preservation

Sukumar Debnath and H. Umesh Hebbar

Corresponding author: *Sukumar Debnath*
Email: *debnath@cftri.res.in*

CONTENTS

11.1 INTRODUCTION

Infrared radiation (IR) is a form of electromagnetic waves having a wavelength range between UV and microwave (0.78–1000 µm) that does not require any medium for its emission. IR can be divided into three different categories, such as near infrared (NIR, 0.78–2 µm), mid-infrared (MIR, 2–4µm), and far infrared (FIR, 4–1000 µm; Jain and Pathare, 2004). The application of infrared heating for drying, blanching, pasteurization, disinfestation, enzyme inactivation, roasting, baking, and cooking of food materials is gaining popularity owing to its potential and several advantages such as faster heat treatment, higher heat transfer capacity, uniform heating, better process control, and higher retention of bioactive components as compared to traditional food-processing techniques.

In conventional heating, food material is heated by convection (surface), followed by conduction (inside), whereas in the case of IR heating, atoms or molecules experience vibration and rotational motion that leads to heat generation (Dagerskog and Osterstrom, 1979; Trivittayasil et al., 2011). Infrared heating depends on its power, the temperature of the commodity, wavelength, bandwidth, depth, water activity and moisture content of the sample (Abdul-kadir et al., 2000; Rifna et al., 2019). The literature reports revealed that drying time required for FIR and NIR are longer than that of MIR (Wang et al., 2014). The performance of blanching (dry) of red bell pepper (*Capsicum annuum* L.) slices using IR and microwave radiation in standardized processing conditions was compared with conventional water and steam blanching methods which substantiated that infrared (150°C) and microwave (17.5 W/g) resulted in higher retention of β-carotene (103.2% and 118.6%, respectively) in comparison to water (60.3%) and steam blanching (88.3%; Jeevitha et al., 2013). A study showed that the pretreatment of apple slices with potassium meta-bi-sulphate (KMS) and IR reduced the drying time by 23% and 17% in low-humidity air (LHA) and hot-air (HA) drying, respectively (Shewale et al., 2017). Vishwanathan et al., 2010 showed that HA-assisted IR drying of carrot and potato lessened the processing time by about 48% in comparison to HA drying alone. Studies have also revealed that IR-based blanching, as well as hybrid drying (IR-HA), could be used efficiently to produce good quality of bitter gourd chips as it retains higher quantity of ascorbic acid (93%) and charantin (74.8%) with respect to water and steam blanching (Nalawade et al., 2018). Kumar Praveen et al. (2005) explained combination mode of drying (HA as well as IR) of onion slices at 60°C and air velocity of 2 m/s and air temperature of 40°C that retained greater flavor and color in comparison to IR and HA drying alone. It was also investigated (Rajoriya et al., 2020) that FIR-assisted refractance window (RW) drying of apple slices reduced drying time by 50% and 69% in comparison to RW and HA, respectively, and saved time by 46% as compared to HA, with better retention of flavor and nutrients.

In this chapter, the principle of infrared heating, the interaction of IR with foods, its comparison with HA heating, and various factors affecting infrared heating are discussed. The infrared pretreatment on LHA drying, enzyme inactivation in green spices, dry blanching and hybrid drying of vegetables, the effects on microorganisms, infrared-assisted RW drying of fruits, cooking, freezing, sterilization, and the effect of infrared on the nutritional qualities of the product with respect to color, texture, sensory properties, microstructure, and the retention of bioactive properties are also discussed.

11.2 FUNDAMENTALS, PRINCIPLES, AND MECHANISM

The application of IR energy for food processing is known since the last few decades, although only recently it has gained popularity, and a number of applications have been reported. IR is a form of electromagnetic wave arising from the movement of electrons in atoms and molecules from a heat source and can travel without any medium (Dan et al., 2021). The mechanisms behind the energy absorption can be explained by the range of wavelength of the incident radiative energy, such as (a) changes in electronic state (wavelength, 0.2–0.7 µm, ultraviolet [UV] and visible rays), (b) changes in vibrational state (wavelength range 2.5–1000 µm, FIR), and (c) changes in rotational state (wavelengths > 1000 µm, microwaves; Decareau, 1985). Basically, infrared radiations are of three types (a) NIR (wavelength: 0.78–1.4 µm), (b) MIR (wavelength: 1.4–3 µm), and (c) FIR (wavelength: 3–1000 µm; Salam et al., 2019). The radiation (reflectivity, ϱ; absorptivity, α; and emissivity transmissivity, τ) falling on a specific food is transformed into heat and energy that could be absorbed, reflected, and/or transmitted (Figure 11.1).

$$\alpha + \rho + \tau = 1$$

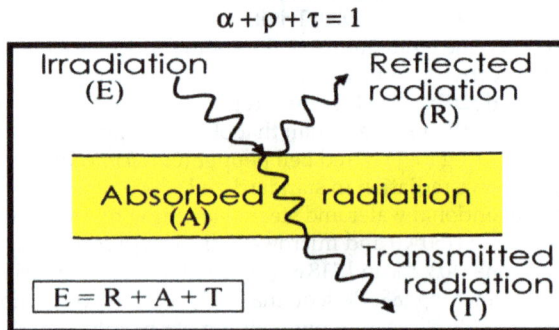

Figure 11.1 Extinction of radiation and total energy.

The amount of energy absorbed and the degree of heating are dependent on the food components that absorb radiation to different extents, while the temperature of the source depends on the wavelength of IR. However, the properties of electromagnetic radiation interacting with any material follow a set of rules known as radiation laws (Skjoldebrand, 2001; Krishnamurthy et al., 2008). These laws apply when a radiating body (black body radiator) absorbs all the radiant energy falling on it, attains equilibrium temperature, and emits the energy as fast as it absorbs. The energy coming out of an emitter consists of various wavelengths. The fraction of the radiation in each band depends on the temperature, as well as the emissivity of the emitter. The temperature of IR-heating elements governs the wavelength at which the maximum radiation occurs.

The electromagnetic radiation (EMR) produced by IR generates thermal movements of the molecules and the conversion efficiency depends on the frequency (energy) of the radiation that penetrates the food by a few millimeters at a higher wavelength. The radiation transmitted energy (whose wavelength is shorter than IR) causes electrochemical changes in radiation absorbing molecules, for example, chemical bonding, electronic excitation, dissipation of absorbed energy (less heat), and others. Several organic materials absorb IR at specific frequencies corresponding to the transport of internal molecules between the energy levels. This transition is expressed as rotational movement (1011–1013 Hz, wavelength 30 µm–1mm) and vibrational (stretching) movement of internal atomic bonds. Food materials absorb FIR energy most effectively through the mechanism of changes in molecular vibrational state. The water and organic compounds (proteins and starches) absorb FIR energy at different wavelengths (water: 3, 4.7, 6, and 15.3 µm; protein: 3–4 and 6–9 µm; fat: 3–4, 6, and 9–10 µm; sugar: 3 and 7–10 µm; Sakai and Hanzawa 1994; Salam et al., 2019; Yadav et al., 2020).

The mechanism of inactivation of microorganisms by infrared heating occurs due to DNA damage in addition to thermal effects (Hamanaka et al., 2000). The thermal inactivation can destroy in the order of protein > RNA > cell wall > DNA. The inhibition of microbial RNA polymerase (due to emission of generated infrared waves) binds the ribosomal subunits and slowly inhibits peptidyl transferase reactions. The IR resulted in total damage to cells by disintegrating the cell membrane and mesosomes of microorganisms (Rifna et al., 2019). The effectiveness of IR treatments on microorganism inactivation depends on several aspects, such as IR power, peak wavelength, temperature, depth and water activity of the food sample, and moisture content (Abdul-kadir et al., 2000).

In IR drying, heat is transferred to the food material in the form of radiant energy. Radiation drying involves the use of IR radiation generated by electricity or gas (using natural gas or propane). The higher absorption of incident IR and the coupling of absorbed energy with water in food are considered to improve the efficiency of drying. The IR source is enclosed in a chamber with a highly reflective surface within the enclosure to enhance energy efficiency (Tyagi et al., 2020). A continuous conveyorized (mesh: 4.5 mm, motor 0.18 kW) combined HA and infrared dryer (Figure 11.2) was developed at CSIR-CFTRI, Mysore for drying vegetables (Hebbar and Ramesh, 2002). This dryer consists of three heating chambers and has power input, 17 kW; quartz infrared (MIR) radiation, 2.4–3.0 µm; the size of drying chamber, 1.1 m × 0.9 m, a finned tube heater, 15 kW; and centrifugal blower, 156 m³/min, 3 HP · 960 rpm.

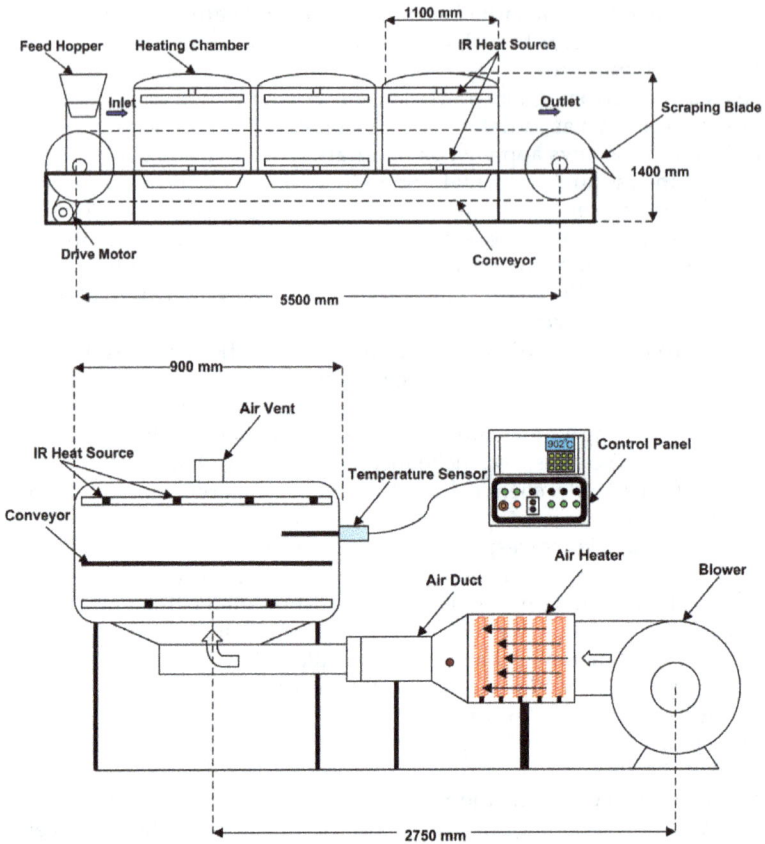

Figure 11.2 Combined infrared and HA dryer. Top: top view; bottom: side view. Reprinted from Hebbar et al., (2004).

11.3 ADVANTAGES AND DISADVANTAGES OF INFRARED HEATING

11.3.1 Advantages of Infrared Heating

The following points are the advantages of IR heating:

- Infrared heating is highly energy-efficient, lower water-consumption, environmentally friendly method.

- IR heating has a high heat-transfer rate or rapid heating (faster than convective or conduction heating), low heating times (e.g., baking etc.), low energy consumption (built-in thermostat with energy-saving function), and improved product quality (preserves vitamins, beta-carotene, and chlorophyll, nutrition, among others).

- It can be considered an alternative source of energy and heating as it is a contactless, clean (without greenhouse gas emissions), silent, safe, and chemical-free method.

- It is contemplated as an alternative source of heating and energy due to its higher thermal efficiency.

- IR does not cause environmental pollution, such as the production of volatile organic compounds, carbon monoxide, or nitrogen oxides, like fossil fuels.

- Infrared radiation can be used (as recommended by the U.S. Food and Drug Administration [FDA]) in food processing as it is nonionizing in nature.

- It is less costly to operate and more effective heating than traditional heaters.

11.3.2 Disadvantages of Infrared Heating

The disadvantages of IR heating follow:

- The delay of initial rising temperature in comparison to other technologies of electric heating.

- IR has limited penetration power so the size of food products should be considered accurately. The depth of its penetration into food is small.

- The long-term exposure to IR may cause tissue rupture.

11.4 INTERACTION OF IR WITH FOOD CONSTITUENTS

The interaction of light with food materials occurs in three different ways: at the surface of a material (regular reflection), through diffusion (light enters the material due to scattering) that undergoes absorption, and leaving the material (light scattering). When an infrared wave touches the surface of a food material, heat energy is released, which is independent of the surrounding temperature. A radiant electromagnetic energy imposes on a food surface induces changes in the electronic, rotational, and vibrational states of atoms and molecules, by which the IR is absorbed, reflected, or transmitted as shown in Figure 11.1. However, absorption intensities will be different at different wavelengths for various food materials. The design of an optimal infrared heating system for an effective interaction of IR with food constituents depends on its optical and physical properties (Halford, 1957). The IR absorption bands for chemical groups of related food components such as sugar, water (2.7–3.3 μm, contains hydroxyl O-H group), carbohydrate/sugar, lipid, protein (3.25–3.7 μm, contains aliphatic C-H bond), lipid (5.71–5.76 μm), carbonyl group (C=O, ester), protein (5.92 μm, carbonyl group C=O, amide), protein (2.83–3.33 μm, nitrogen–hydrogen group -NH-), unsaturated lipid (4.44–4.76 μm, carbon–carbon double bond C=C) during heating has been reported by Rosenthal (1992).

IR is absorbed by organic matter at separate frequencies corresponding to the transport of internal molecules between energy levels. The transition within the range of infrared energy is expressed as rotational (frequencies 1011–1013 Hz with a wavelength of 30 μm–1 mm) and vibrational (stretching) movement of internal atomic bonds. The water principally affects the absorption of incident radiation at all wavelengths. The amino acids, polypeptides, and proteins have two strong absorption bands (3–4 and 6–9 mm), while lipids and carbohydrates show three (3–4, 6, and 9–10 mm), and two strong absorption bands (3 and 7–10 mm), respectively. However, radiation-transmitted energy causes electron chemical changes in radiation-absorbing molecules, for example, chemical bonding, electronic excitation, and dissipation of absorbed energy at a shorter wavelength than IR. IR can be used to inhibit bacteria, spores, yeasts, and mold in liquid and solid food materials. The efficiency of infrared inhibition depends on the extent of infrared energy, food temperature, wavelength, wave width, food depth, time of exposure, moisture content, and more, and therefore, the total energy absorbed by microorganisms increases and thus increases microbial inhibition. The IR rays penetrate inside the food item, thus destroying the inhabiting pathogenic microorganisms (e.g., *A. niger*, *F. proliferatum*, etc.) inside as well as on the surface of the food product without any change in the quality attributes of foods, such as vitamins, proteins, and antioxidant activities. This is due to the fact that energy absorbed by the innate spores was greater at the selected wavelength, consequently leading to a greater rate of mortality (Salam et al., 2019). The specific wavelength denatures the protein in microorganisms, resulting in an increase in inhibition in comparison to the use of IR without determining a specific wavelength (Hamanaka et al., 2000).

11.5 EFFECTS OF IR

11.5.1 Effects on Microorganisms

The disinfestation of various fresh and dried fruits has been reported in the literature in order to inactivate (e.g., *Aspergillus, Claviceps, Alternaria, Fusarium,* and *Penicillium*) and degrade toxins (e.g., *ochratoxin zearalenone, aflatoxin, nivalenol,* and *fumonisin* release by various insects and microorganisms in foods). Various reports are available on the use of IR for decontamination of mycotoxin and the inactivation of fungi in the food and agriculture sectors (Ravindran and Jaiswal, 2019; Akhila et al., 2021). As discussed in an earlier section, infrared heating inhibits microbial RNA polymerase by binding with ribosomal subunits and thus inhibits the peptidyl transferase reactions. The food components absorb energy in the FIR region (31,000 mm) and involved in total damage of intracellular components (e.g., DNA, RNA, ribosomes, cell envelopes, proteins) of the cells by disintegrating cell membrane and mesosomes of microorganisms. Shirkole et al. (2020, 2021) reported the inactivation of *S. typhimurium* and *A. flavus* in paprika using short-time intensive microwave-infrared (MW-IR) radiation, and the results showed that spatial changes in temperature and water activity of paprika during this heating operation resulted in 8.849-log reduction of total plate count (*S. typhimurium*) and 7.372 log reduction of yeast and mold count (*A. flavus*) and proved a better method than steam-assisted sterilization in terms of quality degradation.

Microbial contamination from various sources (e.g., air and water, animals, insects, birds, rodents, rainfall, drought, temperature, unhygienic handling, harvesting, processing equipment and improper storage conditions) is one of the major reasons for the spoilage of food grains. Sirohi et al. (2021) reviewed the possibilities of controlling the microbial contamination of food grains using IR for the disinfestation of insects and mycotoxins and surface decontamination. Cai et al. (2020a) reported that FIR combined with CS@Fe_3O_4 nanoparticles thawing of red seabream (*Pagrus major*) fillets successfully delayed the increase of total volatile basic nitrogen (TVB-N) and pH and reduced the growth of various microorganisms such as *Pseudomonas*, and lactic acid bacteria (LAB). However, more research, including animal and epidemiological studies are needed to determine the potential toxicity of mycotoxin degradation products.

11.5.2 Effects on Food Enzymes

The IR heating has several advantages such as retention of water-soluble nutrients, solid-loss prevention, and no wastewater generation, among others. Jeevitha et al. (2013) investigated blanching of slices of red bell pepper (*Capsicum annuum* L.) containing various enzymes (peroxidase [POD], polyphenol oxidase [PPO]) using IR radiation, and compared its performance with conventional methods (water and steam blanching). The dry method (IR, 150°C) resulted in higher retention of β-carotene (103.2%) in comparison to water (60.3%) and steam blanching (88.3%). They also investigated the effect of IR-based blanching (dry) and a water and superheated steam enzyme inactivation process for evaluation of product quality of green bell pepper (*Capsicum annuum* L.), and they observed a firmer texture of IR-blanched slices in comparison to unprocessed as well as blanched by other methods. The retention of β-carotene (131.2%) and chlorophyll (136.4%) were found higher as compared to unprocessed bell pepper. Song et al. (2021) studied the effects of conveyor-belt catalytic infrared blanching (CBCIRB) on POD, the microstructure, and other physicochemical properties of sweet potato by changing the various parameters, such as irradiation distance, sample thickness, conveyor speed, and so on, and compared the effect of conventional hot-water blanching (HB) on similar material, and the result showed that shortening of the IR irradiation distance and reducing the slice thickness could accelerate the POD inactivation rate, maintained sample hardness in a better manner than HB, and concluded that CBCIRB technology could be applied to the blanching of fruits and vegetables.

Chen et al. (2018) investigated that a sequential infrared (IR, 552 power, 110s, 70 mm) blanching followed by HA drying is an energy-efficient sustainable technology. They developed a method for making healthy chip carrot snacks by IR blanching followed by HA drying at different temperatures and observed that IR blanching at 80°C reduced drying time by 45%, with the production of redder and crispier chips with less residual POD activity and a retention of 19.03% higher vitamin C (vit C) contents compared to an unblanched product. Nalawade et al. (2018) studied blanching (dry) (IR) and hybrid drying (HA as well as IR) of bitter gourd (*Momordica charantia* L.) slices and standardized degree of enzyme (POD as well as PPO) inactivation process and product quality. They also compared performance of IR blanching, water and steam blanching, and hybrid with HA drying. This study revealed that IR-blanched slices retained higher quantity of ascorbic acid (93%) and charantin (74.8%), in comparison to water and steam blanching. They also observed that IR blanching followed by hybrid drying required less (~66%) time than water-blanched HA drying and retained a higher quantity of charantin (68.4%), ascorbic acid (86.3%), and chlorophyll (42.9%) and consuming less energy (79%) than HA drying. This study concluded that the IR-based dry blanching and hybrid drying could be an alternative to wet blanching and HA drying for the production of a good-quality product.

11.5.3 Effects on Food Quality and Sensory Attributes

The application of electromagnetic radiation is gaining momentum in food processing, and this section addresses the effects of IR on quality characteristics and sensory attributes of foods. Hebbar et al. (2004) developed a combined (HA as well as infrared) unique conveyorized heating machine for the drying of vegetables that consists of three chambers fitted with MIR heaters for radiative heating. They investigated that this system performed at 80°C (air velocity: 1 m/s), reduced drying time (48%), and consumed less energy (63%) in a combination mode in comparison to HA drying alone. Wu et al. (2019a) developed a novel laboratory scale apparatus using infrared lamps (infrared freeze drying, IRFD) to replace the electric heating plate for the freeze drying of *Cordyceps militaris* and compared its performance with traditional freeze drying (TFD) at 40–70°C. These researchers found that IRFD could reduce drying time (7.21–17.78 %) and energy consumption (11.88–18.37%) and the higher retention of cordycepin, total phenolics, hydroxyl radical scavenging activity while reducing power, 3-octanone, 3-octanol, and 1,3-octadiene in comparison to TFD. Hnin et al. (2021) carried out drying experiments on edible rose flowers using IRFD with and without a pulsed spouted system (infrared pulse–spouted freeze drying, IRPSFD) and compared the results with conventional freeze drying (FD) process. They experienced that IRPSFD reduced processing time (8–30%), saved energy consumption (15–36%), and improved the overall quality of the dried products in comparison to FD process. Liu et al. (2020) carried out dehydration process for instant cream mushroom soup (CMS) by combining microwave pre-gelatinization (MPG) of wheat flour with IRFD, and compared its quality with freeze drying (FD). They observed that IRFD reduced drying time (23.81%) and energy consumption (28.43%) in comparison to FD and improved quality characteristics (flavor, viscosity of rehydrated soup, etc.) significantly. Ratseewo et al. (2020) applied FIR and HA drying to unpolished and polished rice (pigmented) varieties to study its quality characteristics, and the result showed that FIR increased total phenolic and flavonoid contents, anthocyanin, tocopherols, antioxidant capacity, quercetin, and gallic and ferulic acids, as it released bioactive compounds, although HA showed opposite results. However, in the case of polished rice, amino-acid content gets reduced in the FIR and HA process. This indicates that the polished rice provides lower nutritional quality with better sensory attributes in comparison to unpolished rice.

Ren et al. (2021) studied the combined effects of cutting (longitudinal, Lc, and transverse cut, Tc), hot-air (HA) drying, vacuum drying (VD), freeze-drying (FD), and catalytic-infrared drying (CID) on physiochemical properties of dried ginger. They found that Lc reduced drying time (15.12–29.6%) in comparison to Tc and improved physical properties, for example, the degree of shrinkage, hardness, and the rehydration rate of dried ginger. Among the various driers, CID exhibited minimum drying time (1.42–1.84 h) and retained most of the bioactive constituents such as gingerol, total phenolic content (TPC), total flavonoid content (TFC), volatile components, and others. Chen et al. (2015) investigated drying kinetics and quality characteristics of jujube slices using HA and short- and medium-wave IR radiation (SMIR) drying at 60–90°C. They observed that the HA and SMIR drying processes followed the logarithmic model and two-term model, respectively. The drying efficiency in SMIR drying was found two times higher than HA drying. Chao et al. (2022a, 2022b) evaluated effects of various drying methods (e.g., far-infrared radiation drying (FIRD), FD, VD, HA drying, etc.) on physicochemical properties, antioxidant activities, and antityrosinase capacity from hull and hull-less pumpkins (HLP) seed and revealed that in the case of HLP, FIRD produced highest total polyphenols content (37.11 ± 1.05 mg GAE/ g DW); FD showed better color and higher retention of β-carotene, ascorbic acid, and sugar content in comparison to HA drying; and VD possessed highest antityrosinase activity.

Guo et al. (2020) developed ultrasound-assisted IR (US-IR) dryer and investigated its effect on water diffusion kinetics of carrot slices during hybrid US-IR drying and compared quality parameters with IR-dried products. This experiment revealed that ultrasound improved the drying rate and rehydration ratio and reduced drying time in comparison to IR drying alone. Wu et al. (2021) investigated the combined effects of ultrasound (US, 80 W) with infrared (IR) treatments (900–1500 W) on the drying and quality characteristics and microstructure of jackfruit slices, and the result revealed that application of ultrasound reduced drying time (35–25%), preserved vit C, and increased TPC significantly, compared to IR treatment alone. Vishwanathan et al. (2010) studied HA-assisted IR drying of potato and carrot and concluded that combined effect of reduced drying time (48%) and improved quality of the products. The higher rehydration ratios with IR (2.9% and 5.6%) and combined mode (11.4% and 12.5%) dried potato and carrot indicated a better porous structure of the product due to rapid heating and quicker diffusion of water vapor in comparison to HA drying alone. The effective diffusivity values for IR and combined mode drying were higher compared to HA drying alone, which may be attributed to the rapid increase in product temperature due to vapor pressure difference between the surface and the inner layers.

11.6 APPLICATIONS OF IR HEATING IN PRESERVING FOODS

Several studies have been reported on application of IR in food products with respect to dehydration of vegetables, paddy, rice, roasting cereals, coffee and cocoa, biscuits and bread, thawing, surface pasteurization of bread and packaging materials, and more. Table 11.1 summarizes applications of IR on various type of food materials.

Table 11.1 Applications of IR on Various Food Products

Sl No.	Commodity/ species	Processing conditions	Application	References
1	Almonds	IR (5000 W/m², 11,000 W/m², 130–150°C,	Dry roasting reduced processing time produced and produced dry-roasted pasteurized almonds	Yang et al. (2010)
2	Apple slices	NIR (1.1–1.3 µm), 6 kW, 150°C, 200°C	Drying time decreased, retention of ascorbic acid and phenol content higher with IR treatment	Shewale et al. (2017)
3	Banana	Catalytic Infrared (CIR, 4600 W/m², 60–80°C)	Time of drying time reduced and color change increased with drying temperature	Pekke et al. (2013)
4	Bitter gourd	IR heater: NIR waves (1.1 THz), 150°C, 180°C, and 200°C)	IR blanching followed by hybrid drying with HA required lesser time and retain higher amounts of charantin, ascorbic acid and chlorophyll, and consume less energy	Nalawade et al., 2018
5	Button mushroom	Near infrared (150–375 W)	Drying time reduced whereas effective diffusivity coefficient and change of color enhanced	Salehi et al. (2016)
6	Carrot and potato	Quartz MIR (2.4–3.0 µm)	Combined mode of IR and hot air) drying reduced processing time, higher rehydration ratio lower browning index, and higher effective diffusivity	Vishwanathan et al. (2010)
7	Carrot	OR (IR, 62–125 W)	Time of drying time reduced, effective diffusivity coefficient enhanced, and rehydration capacity changed	Doymaz (2015)
8	Carrot	Catalytic IR (CIR, 60–80°C)	As the temperature increases, rehydration capacity reduced whereas effective diffusivity coefficient enhanced	Wu et al. (2014)
9	Carrot	IR (IR, 100–300 W)	Rate of drying and hardness enhanced, vit C reduced, reduction in shrinkage and color change observed	Guo et al. (2020)
10	Chinese yam	FIR (IR-HP, 500–2000 W, δ: 15 cm)	Time of drying was found to reduce whereas enhancement of shrinkage, hardness and colour observed	Song et al. (2018)
11	Citrus press-cake	Far infrared (FIR, 40 80°C, λ: 4–50 µm)	Time of drying reduced, later has no significant effect on total phenolic and flavonoid contents	Senevirathne et al. (2010)
12	Coffee (Turkish)	IR (600, 1200 W)	Roasting rate higher than moisture evaporation rate, lightness value of coffee decreased with increasing the power intensity	Yüksel et al. (2020)

(Continued)

Table 11.1 (Continued)

Sl No.	Commodity/ species	Processing conditions	Application	References
13	Cordyceps military	Intermediate-wave IR (IWIR, 40–70°C, δ: 14 cm, λmax: 3 μm)	Time of drying time and water activity lessened and total phenolic contents increased with drying temperature	Wu et al. (2019b)
14	Cordyceps military	IR (IR-FD, 40–70°C)	Time of drying and energy consumption lessened whereas total phenolic content increased first and then decreased with drying temperature	Wu et al. (2019a)
15	Eggplant	Infrared (IR-MW, 150–300 W)	Time of drying and shrinkage lessened, enhancement of rehydration capacity, effective diffusivity coefficient observed	Aydogdu et al. (2015)
16	Eggplant	Far infrared (FIR, 1000–2000 W, δ: 12 cm, λ: 9–9.45 μm)	Time of drying lessened whereas change in the total phenolic contents and color change observed	Jafari et al. (2020)
17	Grape and goji berries	Infrared (IR-VD, 55–75°C, λ: 1–30 μm)	Time of drying lessened	Zhang et al. (2020a)
18	Grapefruit	NIR (NIR-VD, 300–400 W)	Reduction of drying time as enhancement of effective diffusivity coefficient, colour change and shrinkage noticed	Salehi and Kashaninejad (2018a)
19	Grated carrot	Infrared (IR-HP, 45–50°C)	Reduction of drying time and energy consumption observed	Aktas et al. (2017)
20	Green Bell Pepper	NIR (1.1 THz, 0.26 kW/m²)	Texture of slices was firmer and higher retention of β-carotene as well as chlorophyll with of infrared treatment	Jeevitha et al. (2015)
21	Green pea	IR (IR-AD, 2–9 kW/m², λ: 0.6–2.5 μm)	Reduction in time of drying time and shrinkage noticed	Barzegar et al. (2015)
22	Green Pepper, carrot, cantaloupe	Infrared radiators (12 × 300 mm, 4–14 μm, 300 W)	Infrared-assisted thawing reduced thawing time, highly energy efficient, membrane permeabilities and thawing losses lower in the case of green pepper and carrot	Chen et al. (2022)
23	Green Tea	FIR-medium (three rounds) and FIR-heavy samples	Baking with far-infrared irradiated teas had distinct nutty aroma	Zhu et al. (2021)
24	Jujube	Short- and medium-wave IR (SMIR, 1125 W, 60–90°C, δ:11 cm)	Reduction of drying time, vit C content, and enhancement of effective diffusivity coefficient, color, and flavonoid contents observed	Chen et al. (2015)

25	Jujube powder	Short- and medium-wave IR (SMIR, 675–1575 W, 70–90°C, δ:11 cm)	Time of drying reduced	Bi et al. (2014)
26	Largemouth bass (*Micropterus salmoides*)	Ultrasonics (400 W and 40 kHz) combined with FIR (4–14 μm, 12 mm × 300 mm, 300 W) thawing	Improved fish quality during thawing with respect to protein stability and gelation property	Cai et al. (2020b)
27	Lemon	NIR (NIR-VD, 300–400 W)	Reduction of time of drying and enhancement of effective diffusivity coefficient, color, and shrinkage noticed	Salehi and Kashaninejad (2018b)
28	Longan	IR (IR-AD, 300–700 W, δ:20 cm)	Time of drying with enhancing of IR power noticed	Nuthong et al. (2011)
29	Longan	FIR (FIR-AD as well as FIR-HP, 250–450W, λ:7–1000 μm)	Reduction in time of drying time, shrinkage, hardness, water activity and enhancement of specific energy consumption color, rehydration capacity observed	Nathakaranakule et al. (2010)
30	Lotus root	IR (225 × 3 W)	Freeze-thaw pretreatment followed by infrared drying improved drying rate, reduced the total energy consumption, resulted in a more porous structure with low shrinkage rate of the product	Zhang et al. (2022)
31	Murta berry	IR (IR-AD, 400–800W)	Enhancement of rate of drying, effective diffusivity coefficient, total phenolic contents, and reduction of energy consumption with power noticed	Puente-Díaz et al. (2013)
32	Mushroom	Intermediate-wave IR (IWIR, 0.675 W/cm², 60–80°C)	Reduction of time of drying and change in hardness observed	Wang et al. (2014a)
33	Mushroom	Intermediate IR (IIR, 50–70°C, λ:2–4 μm)	Time of drying and rehydration capacity decreased whereas increase in change in color and hardness noticed	Qi et al. (2014)
34	Mushroom	IR (IR-AD, 0.22–0.49 W/cm²)	Reduction of consumption of energy noticed	Motevali et al. (2011)
35	Mushroom	FIR (FIR, 50–90°C, δ:15 cm)	Reduction of time of drying, energy consumption, and enhancement of effective diffusivity coefficient with temperature observed	Darvishi et al. (2013)

(Continued)

Table 11.1 (Continued)

Sl No.	Commodity/species	Processing conditions	Application	References
36	Noodle	IR (IR, 909–1673W, δ:20cm)	Drying time reduced	Basman and Yalcin (2011)
37	Onion	MIR (2.4–3.0 mm)	Lessened time of drying during combined effect of IR and hot air with better product color and pyruvic acid	Kumar Praveen et al. (2005)
38	Paddy	IR (IR-AD, 2–6 kW/m²)	Reduction of the time of drying and enhancement of specific energy consumption observed	Zare et al. (2014)
39	Paddy	FIR (FIR-AD, 0–2 kW/m²)	Drying time lessened	Nosrati et al. (2019)
40	Paddy	IR (IR-AD, 1000–1500W, 60–100°C)	The effective diffusivity coefficient and hardness enhanced with power	Bualuang et al. (2013)
41	Peach pomace	Intermediate-wave IR (IWIR, 60–80°C, δ:14 cm, λmax:3 μm)	Reduction of time of drying and enhancement of effective diffusivity coefficient noticed	Zhou et al. (2019)
42	Pea nut kernels	IR power (250–450 W), roasting time (10–30 min)	Roasting enhances flavor, color, texture and appearance, high roasting efficiency	Bagheri et al. (2019)
43	Pear	MIR (MIR-FD, 3–5.5 kW/ m², 40–70°C, λ: 2.4–3 μm)	Reduction in time of drying, specific energy consumption, total phenolic contents, rehydration ratio, and change in total antioxidant activity observed with temperature	Antal et al. (2017)
44	Peppermint leaves	IR (1.5–4.5 kW/m², δ:10–20 cm)	Reduction of the time of drying and enhancement of effective diffusivity coefficient observed with intensity and distance	Miraei et al. (2017)
45	Pomegranate aril	NIR (NIR-VD, 60–90°C, δ:5cm, λ: 280–6000 nm)	Reduction in time of drying and energy consumption and enhancement of effective diffusivity coefficient, shrinkage, color observed	Alaei et al. (2015)
46	Potato	NIR (104–167 W, δ:4 cm)	Reduction in time of drying, enhancement of effective diffusivity coefficient and change in rehydration capacity experienced	Doymaz (2012)
47	Potato	FIR (FIR-VD, 120–160°C, δ:0.08–0.16 m)	Reduction in time of drying with temperature and distance of radiation observed	Liu et al. (2013)

No.	Material	Conditions	Results	Reference
48	(a) Pumpkin (b) Seed-used Pumpkin slices	(a) NIR (NIR-VD, 204–272W) (b) FIR, 5–15 μm, 60°C	(a) Drying time decreased; effective diffusivity coefficient increased with infrared power (b) Drying rate enhanced, increased total carotenoids on pretreatments with ultrasound, freeze-thawing, and blanching	(a) Ghaboos et al. (2016) (b) Chao et al. (2022a, 2022b)
49	Red Bell Pepper	IR heater (0.26 kW/m²: NIR, 1.1 THz, 130–170°C	Dry blanching with IR (150°C) higher retention of β-carotene	Jeevitha et al. (2013)
50	Red seabream (*Pagrus major*) fillets	FIR (4–14 μm, 12 × 300 mm, 300 W)	Thawing with magnetic nanoparticles resulted desirable thermal stability, gelation property, and stable protein structure	Cao et al. (2018)
51	Rice	FIR (1–5 kW/m², 40°C, δ:20cm, λ:1.5–5.6 μm)	Reduction in time of drying, energy consumption, and enhancement of color and rehydration capacity revealed	Nachaisin et al. (2016)
52	Rice	IR (4685–5348 W/m², δ:5–10cm, λmax:3.1 μm)	Enhancement of the rate of drying observed	Pan et al. (2007)
53	Saffron	IR (60–110°C)	Reduction in the time of drying, enhancement of effective diffusivity coefficient noticed	Torki-Harchegani et al. (2017)
54	Shredded squid	IR (IR-AD, 50°C, δ:14cm, λ:2.5–3, 5–6 μm).	Enhancement of time of drying, specific energy consumption observed with wavelength	Wang et al. (2014)
55	Soybean grain	NIR (4–8 kW, λ:0.75–3 μm)	Improved in the rate of drying color observed	Dondee et al. (2011)
56	Sponge gourd	Short-wave and medium-wave IR (SMIR-AD, δ:8-16cm, λ:0.75–2, 2-4μm)	Enhancement of the time of drying, reduction in effective diffusivity coefficient observed in the given condition	Zhang et al. (2020b)
57	Strawberry	IR (100–300 W)	Reduction in drying time, improvement of total anthocyanin content, change in color, and total phenolic content observed	Adak et al. (2017)

11.7 CONCLUSION AND PERSPECTIVES

Several novel technologies such as UV radiation, gamma irradiation, electron beam, microwave, IR, and radiofrequency have been extensively studied for the inactivation of microorganisms and degradation of toxins in food. IR is an emerging technology that has many advantages like rapid heat transfer, effective surface decontamination, energy efficiency, and others that fulfill the contemporary consumer demands. IR heating is an environmentally friendly advanced thermal processing method that has been commercially used in food-processing industries for browning of bread and other baked products, drying, roasting, cooking, blanching, microbiological inhibition, decontamination of grains, and drying and peeling of fruit and vegetable. IR heating has limited penetration power, and thus, it is used for surface-heating applications. Therefore, a combination of IR heating with other modes of heating (such as HA, microwave, etc.) may be used for higher efficiency and efficient energy usage. Rapid heating, fast posttreatment cooling, less energy consumption (by up to 44%), and reduced processing time make IR treatment as an alternative to traditional decontamination and drying methods. IR heating can also be used to make powder of food material with good nutritional and sensory properties in shorter processing times but still limited in the food-processing industry due to higher initial investment.

REFERENCES

Abdul-kadir, R., Bargman, T.J., Rupnow, J.H. (2000). Effect of infrared heat processing on rehydration rate and cooking of *Phaseolus vulgaris* (var. Pinto), *Journal of Food Science*, **55(5)**, 1472–1473.

Adak, Nafiye, Heybeli, Nursel, Ertekin, Can (2017). Infrared drying of strawberry, *Food Chemistry*, **219**, 109–116.

Akhila, Plachikkattu Parambil, Sunooj, Kappat Valiyapeediyekkal, Aaliya, Basheer, Navaf, Muhammed, Sudheesh, Cherakkathodi, Sabu, Sarasan, Asidharan, Abhilash, S., Mir Shabir, Ahmad, George Johnsy, Khaneghah, Amin Mousavi (2021). Application of electromagnetic radiations for decontamination of fungi and mycotoxins in food products: A comprehensive review, *Trends in Food Science & Technology*, **114**, 399–409.

Aktas, M., Khanlari, A., Amini, A.S., Evik, S. (2017). Performance analysis of heat pump and infrared – heat pump drying of grated carrot using energy-exergy methodology, *Energy Conversion and Management*, **132**, 327–338.

Alaei, B., Amiri Chayjan, R. (2015). Drying characteristics of pomegranate arils under near infrared-vacuum conditions, *Journal of Food Processing and Preservation*, **39(5)**, 469–479.

Antal, T., Tarek-Tilisty´ak, J., Cziaky, Z., Sinka, L.T. (2017). Comparison of drying and quality characteristics of pear (*pyrus communis* L.) using mid-infrared-freeze drying and single stage of freeze drying, *International Journal of Food Engineering*, **13(4)**.

Aydogdu, A., Sumnu, G., Sahin, S. (2015). Effects of microwave-infrared combination drying on quality of eggplants. *Food and Bioprocess Technology*, **8(6)**, 1198–1210.

Bagheri, Hadi, Kashaninejad, Mahdi, Ziaiifar, Aman Mohammad, Aalami, Mehran (2019). Textural, color and sensory attributes of peanut kernels as affected by infrared roasting method, *Information Processing in Agriculture*, **6**, 255–264.

Barzegar, M., Zare, D., Stroshine, R.L. (2015). An integrated energy and quality approach to optimization of green peas drying in a hot air infrared-assisted vibratory bed dryer, *Journal of Food Engineering*, **166**, 302–315.

Basman, A., Yalcin, S. (2011). Quick-boiling noodle production by using infrared drying, *Journal of Food Engineering*, **106(3)**, 245–252.

Bi, J., Chen, Q., Zhou, Y., Liu, X., Wu, X., Chen, R. (2014). Optimization of short- and medium-wave infrared drying and quality evaluation of jujube powder, *Food and Bioprocess Technology*, **7(8)**, 2375–2387.

Bualuang, O., Tirawanichakul, Y., Tirawanichakul, S. (2013). Comparative study between hot air and infrared drying of parboiled rice: Kinetics and qualities aspects. *Journal of Food Processing and Preservation*, **37(6)**, 1119–1132.

Cai, Luyun, Dai, Yufei, Cao, Ailing, Cao, Minjie (2020a). The effects of CS@ Fe3O4 nanoparticles combined with microwave or far infrared thawing on microbial diversity of red seabream (*Pagrus major*) fillets based on high-throughput sequencing, *Food Microbiology*, **91**, 103511.

Cai, Luyun, Zhang, Wendi, Cao, Ailing, Cao, Minjie (2020b). Effects of different thawing methods on the quality of largemouth bass (Micropterus salmonides), *LWT – Food Science and Technology*, **120**, 108908.

Cao, Minjie, Cao, Ailing, Wang, Jing, Cai, Luyun, Regenstein, Joe, Ruan, Yunjie, Li, Xiuxia (2018). Effect of magnetic nanoparticles plus microwave or far-infrared thawing on protein conformation changes and moisture migration of red seabream (*Pagrus Major*) fillets, *Food Chemistry*, **266**, 498–507.

Chao, Erpeng, Li, Jinwei, Fan, Liuping (2022b). Enhancing drying efficiency and quality of seed-used pumpkin using ultrasound, freeze-thawing and blanching pretreatments, *Food Chemistry*, **384**, 132496.

Chao, Erpeng, Tian, Jianjun, Fan, Liuping, Zhang, Tao (2022a). Drying methods influence the physicochemical and functional properties of seed-used pumpkin, *Food Chemistry*, **369**, 130937.

Chen, Bing, Zhang, Min, Wang, Yuchuan, Devahastin, Sakamon, Yu Dongxing (2022). Comparative study of conventional and novel combined modes of microwave- and infrared-assisted thawing on quality of frozen green pepper, carrot and cantaloupe, *LWT – Food Science and Technology*, **154**, 112842.

Chen, Jing, Venkitasamy, Chandrasekar, Shen, Qun, McHugh, Tara, H., Zhang, Ruihong, Pan, Zhongli (2018). Development of healthy crispy carrot snacks using sequential infrared blanching and hot air drying method, *LWT – Food Science and Technology*, **97**, 469–475.

Chen, Qinqin, Bi, Jinfeng, Wu, Xinye, Yi, Jianyong, Zhou, Linyan, Zhou, Yuhan (2015). Drying kinetics and quality attributes of jujube (*Zizyphus jujuba* Miller) slices dried by hot-air and short- and medium-wave infrared radiation, *LWT – Food Science and Technology*, **64**, 759–766.

Dagerskog, M., Osterstrom, L. (1979). Infrared radiation for food processing. I. A study of the fundamental properties of infrared radiation, *LWT-Food Science and Technology*, **12(4)**, 237–242.

Dan, Huang, Pei, Yang, Xiaohong, Tang, Lei, Luo, Bengt, Sunden (2021). Application of infrared radiation in the drying of food products, *Trends in Food Science & Technology*, **110**, 765–777.

Darvishi, H., Najafi, G., Hosainpour, A., Khodaei, J., Aazdbakht, M. (2013). Farinfrared drying characteristics of mushroom slices, *Chemical Product and Process Modeling*, **8(2)**, 107–117.

Decareau, Robert V. (1985). *Microwaves in the Food Processing Industry*. Orlando: Academic Press.

Dondee, S., Meeso, N., Soponronnarit, S., Siriamornpun, S. (2011). Reducing cracking and breakage of soybean grains under combined near-infrared radiation and fluidized-bed drying, *Journal of Food Engineering*, **104(1)**, 6–13.

Doymaz, I. (2012). Infrared drying of sweet potato (*Ipomoea batatas* L.) slices. *Journal of Food Science and Technology*, **49(6)**, 760–766.

Doymaz, I. (2015). Infrared drying kinetics and quality characteristics of carrot slices, *Journal of Food Processing and Preservation*, **39(6)**, 2738–2745.

Ghaboos, S.H., Ardabili, S.M., Kashaninejad, M., Asadi, G., Aalami, M. (2016). Combined infrared-vacuum drying of pumpkin slices, *Journal of Food Science and Technology*, **53(5)**, 2380–2388.

Guo, Yiting, Wu, Bengang, Guo, Xiuyu, Ding, Fangfang, Pan, Zhongli, Ma, Haile (2020). Effects of power ultrasound enhancement on infrared drying of carrot slices: Moisture migration and quality characterizations, *LWT-Food Science and Technology*, **126**, 109312.

Halford, Ralph S. (1957). The influence of molecular environment on infrared spectra, *Annals of The New York Academy of Sciences*, 63–69.

Hamanaka, D., Dokan, S., Yasunaga, E., Kuroki, S., Uchino, T., Akimoto, K. (2000). The sterilization effects of infrared ray on the agricultural products spoirage microorganisms, *Bioproduction Environmental Sciences, ASAE Annual International Meeting, Technical Papers: Engineering Solutions for a New Century* – Milwaukee, WI., United States, pp. 971–979

Hebbar, U.H, Vishwanathan, K.H., Ramesh, M.N. (2004). Development of combined infrared and hot air dryer for vegetables, *Journal of Food Engineering*, **65**, 557–563.

Hebbar, U.H., Ramesh, M.N. (2002). Combined infrared and convective heating system for food processing. *Indian Patent application*, 336/DEL/02, Patent No. 222860.

Hnin, Kay Khaing, Zhang, Min, Ju, Ronghua, Wang, Bin (2021). A novel infrared pulse-spouted freeze drying on the drying kinetics, energy consumption and quality of edible rose flowers, *Lebensmittel-Wissenschaft und-Technologie*, **136(1)**, 110318.

Jafari, F., Movagharnejad, K., Sadeghi, E. (2020). Infrared drying effects on the quality of eggplant slices and process optimization using response surface methodology, *Food Chemistry*, **333**, 127423.

Jain, Dilip, Pathare, Pankaj, B. (2004). Selection and evaluation of thin layer drying models for infrared radiative and convective drying of onion slices, *Biosystems Engineering*, **89(3)**, 289–296.

Jeevitha, G.C., Anto, A., Chakkaravarthi, A., Hebbar, H.U. (2015). Application of electromagnetic radiations and superheated steam for enzyme inactivation in green bell pepper, *Journal of Food Processing and Preservation*, **39(6)**, 784–792.

Jeevitha, G.C., Hebbar, H. Umesh, Raghavarao, K.S.M.S. (2013). Electromagnetic radiation-based dry blanching of red bell peppers: A comparative study. *Journal of Food Process Engineering*, **36**, 663–674.

Krishnamurthy, K., Khurana, K., Soojin, J., Irudayaraj, J., Demirci, A. (2008). Infrared heating in food processing, an overview. *Comprehensive Reviews in Food Science and Food Safety*, **7**, 2–13.

Kumar Praveen, D.G., Umesh, H., Sukumar, D., Ramesh, M.N. (2005). Infrared and hot-air drying of onions, *Journal of Food Processing and Preservation*, **29(2)**, 132–150.

Liu, Wenchao, Zhang, Min, Adhikari, Benu, Chen, Jingjing (2020). A novel strategy for improving drying efficiency and quality of cream mushroom soup based on microwave pre-gelatinization and infrared freeze-drying, *Innovative Food Science and Emerging Technologies*, **66**, 102516.

Liu, Y., Zhu, W., Luo, L., Li, X., Yu, H. (2013). A mathematical model for vacuum farinfrared drying of potato slices, *Drying Technology*, 32(2), 180–189.

Miraei Ashtiani, S.-H., Salarikia, A., Golzarian, M.R. (2017). Analyzing drying characteristics and modeling of thin layers of peppermint leaves under hot-air and infrared treatments, *Information Processing in Agriculture*, **4(2)**, 128–139.

Motevali, A., Minaei, S., Khoshtaghaza, M.H., Amirnejat, H. (2011). Comparison of energy consumption and specific energy requirements of different methods for drying mushroom slices, *Energy*, **36(11)**, 6433–6441.

Nachaisin, M., Jamradloedluk, J., Niamnuy, C. (2016). Application of combined farinfrared radiation and air convection for drying of instant germinated Brown rice, *Journal of Food Process Engineering*, **39(3)**, 306–318.

Nalawade, S.A., Sinha, A., Hebbar, H.U. (2018). Infrared based dry blanching and hybrid drying of bitter gourd slices: Process efficiency evaluation, *Journal of Food Process Engineering*, **41(4)**, e12672, 1–11.

Nathakaranakule A., Jaiboon P., Soponronnarit S. (2010). Far-infrared radiation assisted drying of longan fruit, *Journal of Food Engineering*, **100(4)**, 662–668.

Nosrati, M., Zare, D., Singh, C.B., Stroshine, R.L. (2019). New approach in determination of moisture diffusivity for rough rice components in combined far-infrared drying by finite element method, *Drying Technology*, **38(13)**, 1721–1732.

Nuthong, P., Achariyaviriya, A., Namsanguan, K., Achariyaviriya, S. (2011). Kinetics and modeling of whole longan with combined infrared and hot air. *Journal of Food Engineering*, **102(3)**, 233–239.

Pan, Z., Khir, R., Bett-Garber, K.L., Champagne, E.T., Thompson, J.F., Salim, A., et al. (2007). Drying characteristics and quality of rough rice under infrared radiation heating, *Transactions of the ASABE*, **54(1)**, 203–210.

Pekke, M.A., Pan, Z.L., Atungulu, G.G., Smith, G., Thompson, J.F. (2013). Drying characteristics and quality of bananas under infrared radiation heating, *International Journal of Agricultural and Biological Engineering*, **6(3)**, 58–70.

Puente-Díaz, L., Ah-Hen, K., Vega-Galvez, A., Lemus-Mondaca, R., Scala, K.D. (2013). Combined infrared-convective drying of murta (*ugni molinae*Turcz) berries: Kinetic modeling and quality assessment, *Drying Technology*, **31(3)**, 329–338.

Qi, L.-L., Zhang, M., Mujumdar, A.S., Meng, X.-Y., Chen, H.-Z. (2014). Comparison of drying characteristics and quality of shiitake mushrooms (*lentinus edodes*) using different drying methods, *Drying Technology*, **32(15)**, 1751–1761.

Rajoriya, D., Shewale, S.R., Bhavya, M.L., Hebbar, H.U. (2020). Far infrared assisted refractance window drying of apple slices: Comparative study on flavour, nutrient retention and drying characteristics, *Innovative Food Science and Emerging Technologies*, **66**, 102530, 1–12.

Ratseewo, J., Meeso, N., Siriamornpun, S. (2020). Changes in amino acids and bioactive compounds of pigmented rice as affected by far-infrared radiation and hot air drying, *Food Chemistry*, **306**, 125644.

Ravindran, Rajeev, Jaiswal, Amit K. (2019). Wholesomeness and safety aspects of irradiated foods, *Food Chemistry*, **285(2)**, 363–368.

Ren, Zifei, Yu, Xiaojie, Yagoub, Abu ElGasim A., Fakayode, Olugbenga Abiola, Haile, Ma, Yanhui, Sun, Cunshan, Zhou (2021). Combinative effect of cutting orientation and drying techniques (hot air, vacuum, freeze and catalytic infrared drying) on the physicochemical properties of ginger (*Zingiber officinale* Roscoe), *LWT-Food Science and Technology*, **144**, 111238.

Rifna, E.J., Singh Sushil Kumar, Chakraborty, Snehasis, Dwivedi, Madhuresh (2019). Review: Effect of thermal and non-thermal techniques for microbial safety in food powder: Recent advances, *Food Research International*, **126**, 108654.

Rosenthal, I. (1992). *Electromagnetic Radiations in Food Science*. Berlin, Germany: Springer-Verlag.

Sakai, Noboru, Hanzawa, Tamotsu (1994). Applications and advances in far-infrared heating in Japan, *Trends in Food Science & Technology*, **5(11)**, 357–362.

Salam, A. Aboud, Ammar, B. Altemimi, Asaad, R.S. Al-Hilphy, Lee, Yi-Chen, Francesco, Cacciola (2019). Review a comprehensive review on infrared heating applications in food processing, *Molecules*, **24**, 4125.

Salehi, F., Kashaninejad, M. (2018a). Mass transfer and color changes kinetics of infrared-vacuum drying of grapefruit slices, *International Journal of Fruit Science*, **18(4)**, 394–409.

Salehi, F., Kashaninejad, M. (2018b). Modeling of moisture loss kinetics and color changes in the surface of lemon slice during the combined infrared-vacuum drying, *Information Processing in Agriculture*, **5(4)**, 516–523.

Salehi, F., Kashaninejad, M., Jafarianlari, A. (2016). Drying kinetics and characteristics of combined infrared-vacuum drying of button mushroom slices, *Heat and Mass Transfer*, **53(5)**, 1751–1759.

Senevirathne, M., Kim, S.-H., Kim, Y.-D., Oh, C.-K., Oh, M.-C., Ahn, C.-B., et al. (2010). Effect of far-infrared radiation drying of citrus press-cakes on free radical scavenging and antioxidant activities. *Journal of Food Engineering*, **97(2)**, 168–176.

Shewale, S.R., Hebbar, H.U. (2017). Effect of infrared pretreatment on low-humidity air drying of apple slices, *Drying Technology*, **35(4)**, 490–499.

Shirkole, S.S, Jayabalan, R., Sutar, P.P. (2020). Dry sterilization of paprika (*Capsicum annuum* L.) by short time-intensive microwave-infrared radiation: Establishment of process using glass transition, sorption, and quality degradation kinetic parameters, *Innovative Food Science and Emerging Technologies*, **62**, 102345.

Shirkole, Shivanand Shankarrao, Mujumdar, Arun Sadashiv, Jayabalan, Rasu, Sutar, Parag Prakash (2021). Dry pasteurization of paprika (*Capsicum annuum* L.) by short time intensive microwave-infrared radiation: Inactivation of *Salmonella Typhimurium* and *Aspergillus flavus* considering quality degradation kinetics, *Food Chemistry*, **338**, 128012.

Sirohi, Ranjna, Tarafdar, Ayon, Gaur, Vivek Kumar, Singh, Shikhangi, Raveendran, Sindhu, Rajasekharan, Reshmy, Madhavan, Aravind, Parameswaran, Binod, Kumar, Sunil, Pandey, Ashok (2021). Technologies for disinfection of food grains: Advances and way forward, *Food Research International*, **145**, 110396, 1–17.

Skjoldebrand, C. (2001). Infrared heating. In: Richardson, P. (Ed.), *Thermal Technologies in Food Processing*. Abington, England: Woodhead Publishing Limited, Abington Hall, pp. 208–228.

Song, X., Hu, H., Zhang, B. (2018). Drying characteristics of Chinese Yam (*Dioscorea opposita* Thunb) by far-infrared radiation and heat pump, *Journal of the Saudi Society of Agricultural Sciences*, **17(3)**, 290–296.

Song, Xiaoqian, Yu, Xiaojie, Zhou, Cunshan, Xu, Baoguo, Chen, Li, Yagoub, Abu ElGasim A., Emeka, Okonkwo Clinton, Wahia, Hafida (2021). Conveyor belt catalytic infrared as a novel apparatus for blanching processing applied to sweet potatoes in the industrial scale, *LWT-Food Science and Technology*, **149**, 111827, 1–11.

Torki-Harchegani, M., Ghanbarian, D., Maghsoodi, V., Moheb, A. (2017). Infrared thin layer drying of saffron (*Crocus sativus* L.) stigmas: Mass transfer parameters and quality assessment, *Chinese Journal of Chemical Engineering*, **25(4)**, 426–432.

Trivittayasil, Vipavee, Tanaka, Fumihiko, Uchino, Toshitaka (2011). Investigation of deactivation of mold conidia by infrared heating in a model-based approach, *Journal of Food Engineering*, **104(4)**, 565–570.

Tyagi, Leena, Sharma, G.P., Verma, R.C., Jain, S.K., Murdia, L.K., Mathur, S.M. (2020), Infrared heating in food processing: An overview, *International Journal of Chemical Studies*, **8(3)**, 327–336.

Vishwanathan, K.H., Hebbar, H. Umesh, Raghavarao, K.S.M.S (2010). Hot air assisted infrared drying of vegetables and its quality, *Food Science and Technology Research*, **16(5)**, 381–388.

Wang, L., Zhang, M., Fang, Z., Xu, B. (2014a). Application of intermediate-wave infrared drying in preparation of mushroom chewing tablets. *Drying Technology*, **32(15)**, 1820–1827.

Wang, Y., Zhang, M., Mujumdar, A.S., Chen, H. (2014). Drying and quality characteristics of shredded squid in an infrared-assisted convective dryer. *Drying Technology*, **32(15)**, 1828–1839.

Wu, B., Ma, H., Qu, W., Wang, B., Zhang, X., Wang, P., et al. (2014). Catalytic infrared and hot air dehydration of carrot slices, *Journal of Food Process Engineering*, **37(2)**, 111–121.

Wu, Bengang, Guo, Xiuyu, Guo, Yiting, Ma, Haile, Zhou, Cunshan (2021). Enhancing jackfruit infrared drying by combining ultrasound treatments: Effect on drying characteristics, quality properties and microstructure, *Food Chemistry*, **358(1)**, 129845, 1–7.

Wu, X.-f., Zhang, M., Bhandari, B. (2019a). A novel infrared freeze drying (IRFD) technology to lower the energy consumption and keep the quality of *Cordyceps militaris*, *Innovative Food Science and Emerging Technologies*, **54**, 34–42.

Wu, X.-f., Zhang, M., Li, Z. (2019b). Influence of infrared drying on the drying kinetics, bioactive compounds and flavor of *Cordyceps militaris*. *Lebensmittel-Wissenschaft & Technologie*, 111, 790–798.

Yadav, Garima, Gupta, Neeraj, Sood, Monika, Anjum, Nadira, Chib, Ankita (2020). Infrared heating and its application in food processing, *The Pharma Innovation Journal*, **9(2)**, 142–151.

Yang, Jihong, Bingol, Gokhan, Pan, Zhongli, Brandl, Maria T., McHugTara, H.h, Wang, Hua (2010). Infrared heating for dry-roasting and pasteurization of almonds, *Journal of Food Engineering*, **101**, 273–280.

Yüksel, Ayşe Nur, Barut, Kevser, Tuba, Özkara, Mustafa, Bayram (2020). The effects of roasting, milling, brewing and storage processes on the physico-chemical properties of Turkish coffee, *LWT – Food Science and Technology*, **131**, 109917.

Zare, D., Naderi, H., Ranjbaran, M. (2014). Energy and quality attributes of combined hot-air/infrared drying of paddy, *Drying Technology*, **33(5)**, 570–582.

Zhang, Long, Yu, Xiaojie, Mujumdar, Arun S., Zhou, Cunshan (2022). Effect of freeze-thaw pretreatment combined with variable temperature on infrared and convection drying of lotus root, *LWT – Food Science and Technology*, **154**, 112804.

Zhang, W.-P., Chen, C., Pan, Z., Xiao, H.-W., Xie, L., Gao, Z.-J., Zheng Zhi-An (2020a). Design and performance evaluation of a pilot-scale pulsed vacuum infrared drying (PVID) system for drying of berries, *Drying Technology*, **38(10)**, 1340–1355.

Zhang, Yue, Zhu, Guangfei, Li, Xingyi, Zhao, Yu, Lei, Dengwen, Ding, Guoqiang, Ambrose, Kingsly, Liu, Yanhong (2020b), Combined medium- and short-wave infrared and hot air impingement drying of sponge gourd (*Luffa cylindrical*) slices, *Journal of Food Engineering*, **284**, 110043.

Zhou, M., Li, C., Bi, J., Jin, X., Lyu, J., Li, X. (2019). Towards understanding the enhancement of moisture diffusion during intermediate-infrared drying of peach pomace based on the glass transition theory, *Innovative Food Science and Emerging Technologies*, **54**, 143–151.

Zhu, Yu-Meng, Dong, Jun-Jie, Liu, Jin-Hua, Zheng, Xin-Qiang, Lu, Jian-Liang, Liang, Yue-Rong (2021). Roasting process shaping the chemical profile of roasted green tea and the association with aroma features, *Food Chemistry*, **353(129428)**, 1–9.

12 Food Preservation by Coating Technology: Nanotechnology Approach

Madhu Kamle, Purnamsree Gogoi, Dipendra Kumar Mahato, Akansha Gupta, and Pradeep Kumar

Corresponding author:
Dr. Pradeep Kumar, Email: pkbiotech@gmail.com

CONTENTS

12.1 INTRODUCTION

Food preservation can be defined as a technique to maintain and protect foods from spoilage to enhance their shelf lives (Rahman 2020). *Food spoilage* refers a change that renders food undesirable for consumption and can be detected by the change in their color, flavor, texturem or smell. These changes are a result of contamination by microorganisms and insect infestations and by endogenous enzymes (Singh 2018). Physical and chemical modifications, like the tearing of body tissues or the oxidation of constituents of food, are the causes of food spoilage. The typical microorganisms responsible for food spoilage and foodborne illness are bacteria yeasts and molds. Insects and rodents also can cause a lot of damage by feeding on the product and by passing on pathogenic substances through their hair and droppings (Singh 2018; Rahman 2020).

Food production often occurs in a cyclical form, while consumption is considered more of a constant. Therefore, foods must be stored to ensure an abundant supply throughout the year. Fruits and vegetables have an abundant source of nutrients, vitamins and minerals. Food preservation is also required to prevent microbial growth and invasion and to control food quality such as taste, texture, aroma and color (Fellows 2009). A food preservation method is utilized to maintain and extend the shelf lives of foods and protect foods from any kind of spoilage. Drying, cooling, and freezing are a few conventional techniques used for food preservation (Ghio et al. 2000). However, recent studies and reports have highlighted that nanotechnology has become one of the suitable methods we can use for food preservation and processing of food. The new and emerging technology deals with nanoparticles (NPs), which have one or more dimensions to the order 100 nanometers or less, to create and use nanomaterials that should have novel functionalities and properties (Sridhar et al. 2021). Nanotechnology offers complete food solutions from food manufacturing and processing to packaging. Nanomaterials bring about a great difference not only in food quality and safety but also in the health benefits food delivers. Many organizations, researchers, and industries are coming up with novel techniques, methods, and products that have a direct application of nanotechnology in food science (Dasgupta et al., 2015). The concerns of consumers about the food quality and nutrition have led to an increased demand for exploring alternative ways to enhance food safety and quality and maintain food nutritious. The increasing demand for nanomaterials makes them popular in the food industry, and most important, many nanomaterials containing essential elements have been revealed to be nontoxic (Roselli et al. 2003). Apart from these, nanotechnology maintain versatility in terms of safety, toxicity and risk assessment in the fields of agriculture, environment and food (Kaphle et al. 2018). The excellent physico-chemical property and antimicrobial nature make nanomaterials more reliable, and hence, they are used against pathogens and in food safety and preservation (Das et al. 2019).

12.2 APPLICATION OF NANOCOATING IN FOOD PRESERVATION

Nanocoating is the technique of adding a thin layering (<100 nm) on to the food surface to improve its functionality and provide barriers against external environment. Conventional coatings had some disadvantages like improper adhesion between coating layer and substrate, less durability, less flexibility and others. The edible nanocoatings on manufactured food materials provide an exchange of gas and moisture and enhance colors, flavors, antioxidants properties and the shelf lives of foods (Weiss et al. 2006). The incorporation of nanoparticles increases the surface area per mass compared to large-sized particles, which makes them more biologically active and stable (Mahela et al. 2020). Nanotechnology can be utilized for the production of healthier food with low fat, sugar, and salt. Flores-López et al. (2016) found nanosized edible coating to be effective for preservation and shelf-life extensions of food. It was reported that the shelf life of tomatoes increased with bio-nanoencap-sulated quercetin, which could be applied in other vegetables and fruits as well (Yadav 2017). NPs such as SiO_2, carbon nanotubes and chitosan- and cellulose-based nanoparticles are filled in a polymeric matrix to make them lighter with improved thermal properties and lower gas permeability (Duncan 2011). Furthermore, Kulawik et al. (2020) reported the application of chitosan nanoparticles enhancing the shelf lives of foods without impacting the senso-rial properties. In addition, Morin-Crini et al. (2019) reported the utilization of chitin-based matrices for clarification, encapsulation and preservation of various fruit juices and other beverages. Bajpai et al. (2018) detailed the development of an 'intelligent' and 'smart' kind of nanocoated film that indicates whether any contamination has occurred during storage. Apart from this, Sridhar et al. (2021) mentioned that a combination of different nanomaterials, that is, nanocomposites, has the capability of performing efficient thermal and barrier properties at low cost.

12.2.1 Edible Nanocoating

Nanotechnology has enabled the development of nanocoatings that can be applied on various food surfaces as an alternative for natural protective waxy coatings to extend the storage lives of fresh fruits. These coatings control the moisture content and exchange of gases and maintain other characteristics like freshness, flavor, aroma, texture and nutritional value. The use of edible coat-ings also reduces synthetic packaging waste and protects the environment due to their biocompatibility and biodegradability (Mahajan et al. 2018). Research carried out by Li et al. (2021b) revealed that chitosan-based nanocoating can maintain available nutrients and minerals and control microbial invasion to extend the shelf life of fresh blueberry fruits. The various forms of nanocoating and their applicability in fruit preservation and postharvest management are reviewed in the following sections.

12.2.2 Nanoemulsion in Nanocoating

Nanoemulsions are kinetically stable colloidal systems composed of an oil phase distributed in water in such a way that each drop of oil is surrounded by a thin interfacial region of emulsifying molecules. Nanoemulsions have particle sizes ranging from 50 to 500 nanometers (Zambrano-Zaragoza et al. 2018). Nanoemulsions are divided into two categories based on their phases: (a) water/oil (w/o) and (b) oil/water (o/w). In most edible coatings, o/w nano-emulsions (Figure 12.1) are prioritized, because they allow the inclusion of a variety of lipophilic compounds with antioxidant and antibacterial proper-ties into a hydrophilic polymeric matrix (Zambrano-Zaragoza et al. 2018). Essential oils from plants (e.g., oregano, clove, mint), fatty acids, quinones, carotenoids, phytosterols and others are examples of lipophilic phytochemi-cals (Salvia-Trujillo et al. 2017). Furthermore, Bora et al. (2020) mentioned in their review that citrus essential oil (EO)–based nanoemulsions have the abil-ity to restrict the growth of bacteria, and hence, they can be utilized for the preservation of foodstuffs.

To improve the shelf lives of food products such as meat, dairy products and fresh and fresh-cut fruits, they can be coated with nanoemulsion coatings containing functional agents such as flavors and coloring components, enzymes, antioxidants, antimicrobials, and antibrowning agents (Aswathanarayan and Vittal 2019). Table 12.1 summarizes the microbial efficacy of lipophilic substances in a polymeric matrix and their applications in fruit preservation.

12.2.3 Polymeric NPs in Nanocoating

Zielińska et al. (2020) described polymeric NPs are solid colloidal particles with a size range of 100–1000 nm that are either packed with active chemicals or sur-face-adsorb onto the polymeric core. Nanospheres (produced by a dense poly-meric matrix), and nanocapsules are two forms of polymeric NPs (composed of an oil core surrounded by a polymeric membrane; Figure 12.2). According

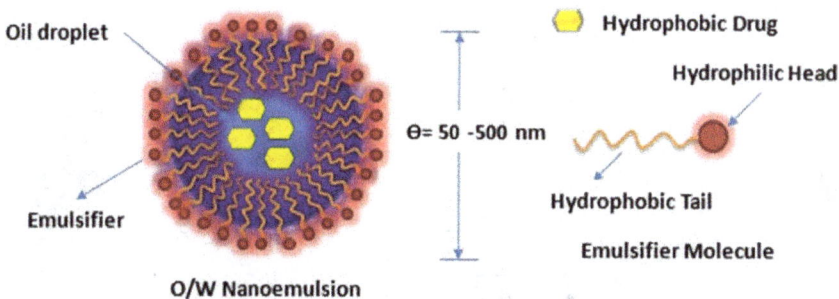

Figure 12.1 O/w nanoemulsion in edible coating.

Table 12.1 Nanoemulsions and Their Application as Coating in Fruits and Vegetables

Lipophilic or bioactive substance	Polymer matrix	Type of fruits	Application	Reference
Lemongrass essential oil	Sodium alginate	Fresh cut apple	The coating outperformed a typical emulsion in terms of antibacterial activity. Fresh cut apples' shelf life was extended by up to two weeks as a result of the edible coating.	Salvia-Trujillo et al. (2015)
Lemongrass oil	Chitosan	Grape berry	The use of nanocoating improves microbial safety against *S. typhimurium* and preserves grapes longer than conventional coating.	Oh et al. (2017)
Orange essential oil	Chitosan	Apple juices	Nanoemulsion coating displayed greater bactericidal activity against *E. coli* and represented an advantageous alternative to preserve fruit juices.	Bento et al. (2020)
Orange peel essential oil (OPEO)	Pectin	Freshly peeled orange	The result showed that the OPEO contained nanocoating extends the shelf life of orange without antibacterial and antifungal effects.	Radi et al. (2018)
Nutmeg seed oil	Chitosan	Strawberry	The nanocoated strawberry showed the best result suppressing and mold yeast growth during storage compared to control fruit.	Horison et al. (2019)
Pomegranate peel extract	Chitosan	Apricot fruit	Psychrophilic bacteria, yeasts, and mold were inhibited by nano chitosan-containing PPE throughout storage, suggesting that the nanoemulsion coating has a greater potential to safeguard the quality and extend the shelf life of apricots.	Gull et al. (2021)
Lemongrass essential oil (LG)	Alginate	'Rocha' Pear	Nanocoating decreased fruit color development and kept fruit firmness better than the control, suggesting that LG nanocoating might help preserve the Rocha pear.	Gago et al. (2020)
Lemon, mandarin, oregano, or clove essential oils	Modified chitosan	Arugula leaf (*Eruca sativa*)	Increase the antimicrobial activity of the essential oil and improvement of the homogeneity and stability of the emulsion	Sessa et al. (2015)
Carvacrol, bergamot, mandarin, and lemon essential oils	Chitosan	Green beans (*Phaseolus vulgaris*)	Increase the antimicrobial activity of essential oils	Severino et al. (2015)

Basil essential oil	Sodium alginate	Okra (*Abelmoschus esculentus*)	Increase the antimicrobial activity of essential oil	Gundewadi et al. (2018)
Cinnamon essential oil	Pullulan	Strawberry (*Fragaria* × *ananassa*)	Improve the distribution of oil in the matrix and increase its antimicrobial activity	(Chu et al. 2020)
Lemongrass essential oil	Carnauba wax	Plums (*Prunus salicina*)	Increase the antimicrobial activity of the essential oil and improve the homogeneity and stability of the emulsion	Kim et al. (2013)
Lemongrass essential oil	Carnauba wax	Grape berry (*Vitis labruscana* Bailey)	Increase the antimicrobial activity of essential oil	Kim et al. (2014)
Extract of tarbush	Candelilla wax	Fuji apple (*Malus domestica* 'Fuji')	Improved the wettability of the nanocoating on the Fuji apple surface	De León-Zapata et al. (2017)
Cellulose nanocrystal and oleic acid	Chitosan	Bartlett pears (*Pyrus communis*)	Increase coating stability at high humidity, adhesion on fruit surface and delayed ripening of pears	Deng et al. (2017)
Oleic acid and Carnauba wax	Carnauba wax	'Nova' mandarins (*Citrus reticulata*) and 'Unique' tangors (*C. reticulata, C. sinensis*)	Improve optical properties and emulsion stability	Miranda et al. (2020)
Citral	Sodium alginate	Fresh cut pineapples (*Ananas comosus*)	Improve the dispersion of the active compound in the matrix and increase its antimicrobial activity	Prakash et al. (2020)
Carnauba wax	Hydroxypropyl methylcellulose	'Redtainung' Papaya (*Carica papaya*)	Reduce gas permeability and moisture loss	Miranda et al. (2019)
Lemongrass essential oil	Sodium alginate	Fresh-cut Fuji apples (*Malus domestica* 'Fuji')	Improve the stability of the emulsion and increase the antimicrobial activity of the essential oil	Salvia-Trujillo et al. (2015)
Thymol	Quinoa protein/chitosan	Strawberry (*Fragaria ananassa*)	Increase the antimicrobial activity of the active compound and improve dispersion in the matrix	Robledo et al. (2018)

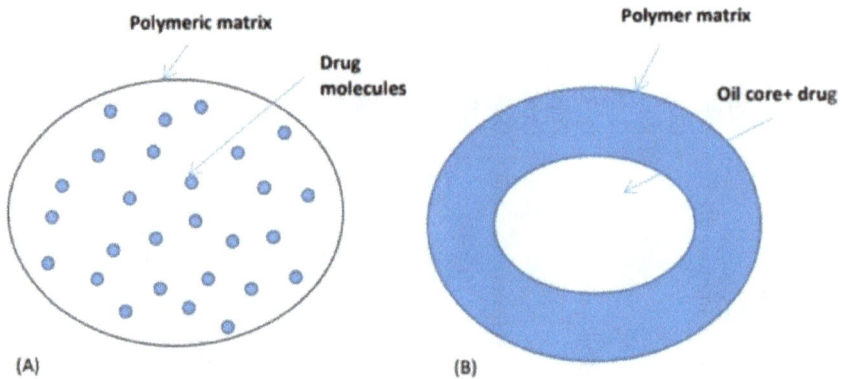

Figure 12.2 Schematic representation of (a) nanosphere and (b) nanocapsule.

to Baranwal et al. (2018), polymeric NPs have been thoroughly examined for their ability to deliver a wide range of antimicrobial drugs due to their various unique properties such as stable structure, restricted size distribution and zeta potential. In food packaging, nanopolymers are aiming to replace traditional polymers (Singh 2018). Although numerous biopolymers have been used for the preparation of polymeric NPs, chitosan NPs are gaining special interest for food preservation and packaging because of their biodegradability, nontoxicity, biocompatibility, film-forming capability, selective gas permeability and fungicidal and antimicrobial efficacy.

Correa-Pacheco et al. (2021) solved the economic losses in bell pepper caused by *Pectobacterium carotovorum* spoilage and developed a chitosan NP (CSNP) coating as an alternative to protect the food crop from *P. carotovorum*. Galindo-Pérez et al. (2015) investigated the role of xanthan gum coatings with nanoparticles (nanocapsules and nanospheres) on the production and oxidation of phenolic compounds generated by enzymatic activity (phenylalanine ammonia lyase [PAL] and polyphenol oxidase [PPO]) on fresh-cut red delicious apples stored for 21 days at 4 °C and reported that the best coating was the nanocapsules/xanthan gum combination because it decreased the initial respiration rate by 63% compared to controls. Esyanti et al. (2019) used chitosan-based NPs to extend the storage period of bananas and found that the rate of skin discoloration was reduced by 2–3 days when compared to the control. Piña-Barrera et al. (2019) extended the shelf life of grapes using a multisystem coating (NC-EOt-C) focusing on polymeric nanocapsules encapsulating EO of *Thymus vulgaris* L. (*Vitis vinifera* L). It was found that during storage, the grapes which have multisystem coating maintain their features like color, firmness and others, for a longer time duration. The EO of *Thymus vulgaris* contains various phytochemical constituent that possesses antimicrobial activities by restricting the biotic stress as well as antioxidant properties, which reduce the accumulation of free radicals in the fruits. In addition, the nanocapsules also increase the residence time of the volatile compound of the fruit by preventing its rapid evaporation (Piña-Barrera et al. 2019).

12.2.4 Solid Lipid NPs in Edible Nanocoating

Nanoencapsulation is a novel practice in the field of food safety and preservation. Because of its astounding development in recent years, it is gaining increased research attention for delivering necessary bioactive compounds to food products. The edible nanocoating component along with solid lipid NPs (SLNs), is considered one of the potent techniques in this new era. SLNs are composed of spherical solid lipid particles with diameters ranging from 50–1000 nm, within which the hydrophobic core is dispersed and subsequently consolidated by surfactants (Figure 12.3; Geszke-Moritz and Moritz 2016; Katouzian et al. 2017).

12.2.5 Nanotubes and Nanofibers in Edible Coating

Nanotubes and nanofibers have been used as novel and efficient materials for edible coatings. According to Zambrano-Zaragoza et al. (2018) milk protein nanotubes, particularly partially hydrolyzed-lactalbumin nanotubes with cavities of 8-nm diameter, have the ability to encapsulate diverse active substances in foods. Halloysite nanotubes in food packaging systems might be a potential way to distribute antimicrobials in a controlled manner, thereby increasing the quality, safety and shelf life of fresh produce throughout handling and distribution (Li et al. 2021a).

Nanofibers are fibrous scaffolds having diameter <100 nm or even <500 nm, and their potential applications in incorporation of active substances to maintain safety and preserve food quality have been explored (Mahela et al. 2020). The electrospinning technique provides functionalized polymeric nanofiber with different bioactive substances that have both antimicrobial and antioxidant characteristics (Zhang et al. 2020). EOs derived from plant extracts, which have antibacterial and food preservation properties due to a variety of active ingredients, such as terpenes, terpenoids, carotenoids, curcumins and so on are employed in nanofibers (Pandey et al. 2017). Figure 12.4 depicts the nanofibers coating embedded on food surfaces and diffusion of bioactive compounds from nanofibers.

In an experiment conducted by Aytac et al. (2017), the apples' surfaces were innoculated with *Botrytis cinerea* and *Penicillium expasum*, and then the fruit were coated by electrically spun zein nanotubes and stored for 15 days. The results revealed that microbial growth was prevented, and the fruits'

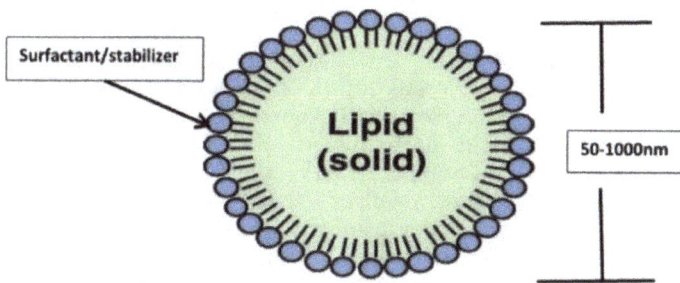

Figure 12.3 Showing the structure of SNP.

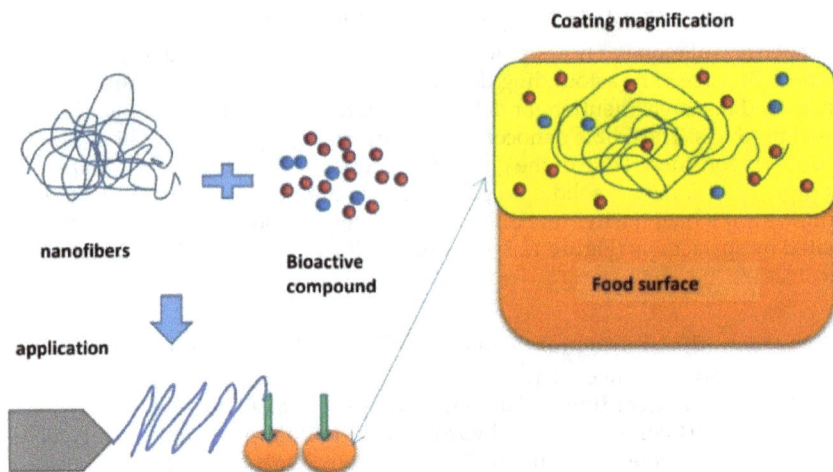

Figure 12.4 Nanofibers in an edible coating.

shelf life was increased. Further research carried out by Zhang et al. (2019) demonstrates that thymol-encapsulated nanofiber may suppress the growth of bacteria, fungus, and yeast, hence prolonging the shelf life of strawberry fruit. Zhai et al. (2020) investigated the properties of nanofibers derived from bacterial cellulose via HCl hydrolysis and their novel application in fresh-cut apples, resulting in improved properties such as delaying weight loss, increasing firmness and soluble solid content, lowering the browning index and titratable acidity, among others. Kwak et al. (2021) further applied carboxymethylated cellulose nanofibers (CM-CNFs) as an edible film on strawberry fruit by a dip-coating process and concluded that cellulose nanofiber is a promising nanomaterial that could be used for prolonged shelf life of the fruit.

12.2.6 Organic/Inorganic NPs in Edible Coating

A nanosystem consisting organic and inorganic material known as nanocomposites has been used for edible coating. These nanosystem blends improve the mechanical resistance, transparency, controlled release and gas barrier qualities of edible films and coatings. Inorganic substances that are often used include montmorillonite (MMT), nano SiOx, nano TiO$_2$ and nano ZnO, as well as AgNP (Zambrano-Zaragoza et al. 2018). Nano ZnO, an inorganic NP having multifunctional properties, is particularly known for antimicrobial action. ZnO has been designated as GRAS (generally recognized as safe) by the US Food and Drug Administration (Cotton et al. 2019). MMT contains silicate and has a multilayered form (width and length of 100–150 nm), but its thickness (only 1 nm) is the most significant factor in enhancing composite film mechanical characteristics (Zambrano-Zaragoza et al. 2018). Figure 12.5 explains a schematic description of coating of nanocomposites incorporated with polymer matrix on food surface, while Table 12.2 lists the antioxidant and antimicrobial responses of nanocomposites, including inorganic compounds and their role in fruit preservation.

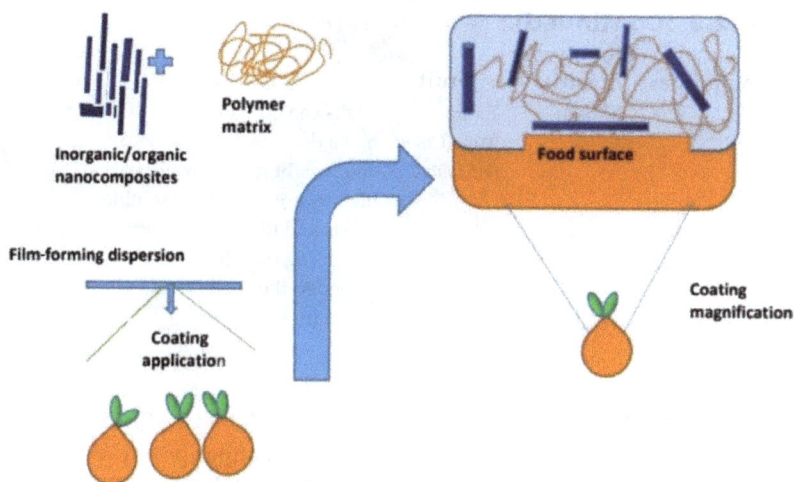

Figure 12.5 NPs in edible coating.

Table 12.2 Nanocomposites Containing Organic and Inorganic Components for Edible Coating of Fruits

Nanocomposites	Biopolymer matrix	Fruit	Finding	Reference
Nano-SiOx	Soy protein isolate	Apples	The inclusion of an edible coating to an SPI matrix reduced transpiration rate, preserved firmness, and extended shelf life of apple fruit.	Liu et al. (2017)
Montmorillonite (MMT)	*Oregano* essential oil	Fresh-cut papaya	MMT with oregano oil decreased weight loss while preserving the quality of papaya. Coated papaya has a slower microbiological growth rate and longer shelf life than untreated fruit.	Cortez-Vega et al. (2014)
MMT	Whey protein isolate (WPI) / Calcium caseinate	Strawberries	The edible coating inhibited mold development for at least 12 days and preserved the freshness of coated strawberries in postharvest handling.	Junqueira-Gonçalves et al. (2017)
ZnO	Chitosan/ Gram arabic (CH/GA)	Banana	The coating enhanced the banana's quality and shelf life. It permitted the long-term preservation of quality characteristics such as fruit firmness, weight loss, sugar reduction and titratable acidity.	La et al. (2021)

(Continued)

Table 12.2 (Continued)

Nanocomposites	Biopolymer matrix	Fruit	Finding	Reference
MMT	Chitosan	Tangerine fruit. (*Citrus reticulata*)	Fruits coated with CS/MMT had a decreased rate of decay and weight loss, as well as a higher content of total soluble solids and titratable acidity, indicating that the coating extended the shelf life of tangerines.	(Xu et al. 2018)

12.3 EDIBLE NANOCOATING FOR FRUITS AND VEGETABLES

The primary requisites for the application of edible coatings on any foodstuffs are to inhibit oxygen permeability (protection from oxidation), minimize the respiration rates as well as ethylene production, or prevent quick ripening and deterioration. Furthermore, edible coatings should be compatible with the food product, both organoleptically and functionally (Saha et al. 2017). After minimal processing such as washing, cutting, and peeling, fresh-cut fruit and vegetables are extremely prone to deteriorate owing to the commencement of enzymatic reactions, which further leads to surface browning, off-flavor, moisture loss and textural loss (Oms-Oliu et al. 2010). Among the various type of coatings, nanocoatings developed through nanoemulsion have been reported to provide the most promising results in terms of quality and functionality of fruit and vegetables (Salvia-Trujillo et al. 2017). The application of nanoemulsion-based coatings for fruits and vegetables is outlined in Table 12.3.

12.4 IMPACT OF NANOCOATED/NANO-FORMULATED FRUITS ON HUMAN HEALTH

Despite the tremendous potential and application of NPs in the preservation of fruits and other foodstuffs, there is great concern about human health. It is seen that NPs possess both direct and indirect impacts on the health of the population. Nanoparticles could be dangerous for both humans and the environment (Mahato et al. 2021). Berekaa (2015) raises his concern about toxicological aspects of nanoparticles in food, with emphasis on the risk assessment and safety issues. The author also indicated that there is an urgent need for a regulatory framework capable of managing any risks associated with the implementation of NPs in food technology. Cushen et al. (2012) documented that excessive intake of nanoemulsions might be detrimental as nanoemulsions are made up of surfactants and solvents. Mills and Hazafy (2009) reported that carbon nanotubes, which are commonly used in food packaging, can migrate into packaged food and may harm human health. Several nanocoatings are modified with chemical components that can trigger allergic reactions on the skin. When an NP interacts with biomolecules, it can cause harmful effects including the production of reactive oxygen species (ROS), resulting in mitochondrial damage and cell death (Hajipour et al. 2012). A few NPs, including nano-silver and titanium dioxide, used as a food additive have the potential to cause health problems. Investigation suggests that the NP can get stuck in the digestive

Table 12.3 Applications of Nanoemulsion-Based Coatings for Fruits and Vegetables

Core/functional material	Nanoemulsion technique	Application	Function	References
Lemongrass oil and carnauba wax	High-pressure homogenization/ Ultra-Turrax	Grape berry	Antimicrobial and texture enhancer	Kim et al. (2014)
Mandarin essential oil	High-pressure homogenization	Green bean	Antimicrobial activity	Donsì et al. (2015)
Oregano oil	Ultrasonication	Fresh lettuce	Antimicrobial	Bhargava et al. (2015)
α-Tocopherol	Ultra-Turrax	Fresh-cut apple	Antioxidant and texture enhancer	Zambrano-Zaragoza et al. (2014)
Candeuba wax and Xanthan gum	Ultra-Turrax	Guava	Texture enhancer	Zambrano-Zaragoza et al. (2013)
Essential oil (carvacrol) and chitosan	Ultra-Turrax/ stirring	Carrot slices	Antimicrobial and texture enhancer	Martínez-Hernández et al. (2017)
Lemongrass oil and Carnauba wax	High-pressure homogenization/ Ultra-Turrax	Apple	Antimicrobial and texture enhancer	Jo et al. (2014)
Lemongrass oil	Microfluidization	Fresh-cut apple	Antimicrobial activity	Salvia-Trujillo et al. (2015)
Peppermint oil	High-pressure homogenization	Fresh-cut apple	Antimicrobial activity	Liang et al. (2012)
Lemongrass oil and Carnauba wax	High-pressure homogenization	Plum	Antimicrobial and texture enhancer	Kim et al. (2013)
Lemongrass oil and chitosan	High-pressure homogenization/ Ultra-Turrax	Grape berry	Antimicrobial and texture enhancer	Oh et al. (2017)
Orange peel essential oil and pectin	Ultrasonication/ Stir/Ultra-Turrax	Orange slices	Antimicrobial and texture enhancer	Radi et al. (2018)
Essential oil (carvacrol)	High-pressure homogenization/ Ultra-Turrax	Zucchini	Antimicrobial and antioxidant	Donsì et al. (2014)
Essential oil (carvacrol) and chitosan	High-pressure homogenization/ Ultra-Turrax	Cucumber slices	Antimicrobial and antioxidant	Taştan et al. (2017)
Essential oil (carvacrol) and chitosan	High-pressure homogenization/ Ultra-Turrax	Shredded cabbage	Antimicrobial and antioxidant	Sow et al. (2017)
Mandarin essential oil	High-pressure homogenization	Green bean	Antimicrobial activity	Severino et al. (2014)

system, especially in the small intestine, possibly leading to leaky gut syndrome and autoimmune disease symptoms. However, NP toxicity is dependent on their characteristics and route of entry into the body, the quantity and the frequency of nanomaterial exposure, as well as individual sensitivity (Singh 2018). Aschberger et al. (2011) analyzed the oral transmission route and discovered evidence of toxicity when a higher amount of silver NP or nano-TiO$_2$ was applied to fruits. Cockburn et al. (2012) revealed that the small-sized engineered nanomaterials (ENMs) lead to the risk for bioaccumulation within body organs and tissues by triggering oxidation, irritation, genotoxic and toxicity of the cells. Furthermore, when human lung cells are exposed to NPs utilized as anti-caking agents, they can be lethal (Athinarayanan et al. 2014).

Although a nanomaterial is considered GRAS, more efforts and research are required to assess the threat of nano-counter components, as the different physical and chemical characteristics of nano-states differ significantly from those of microstates (Singh 2018). Human intake of NPs found in fresh produce and various food products are subjected to the digestion process in the gastrointestinal tract (GIT; McClements and Xiao 2017). However, there is currently limited research available on the metabolic activities of NPs after oral ingestion on human health through food consumption. Also, there is serious concern about nanoparticles designed to prolong the shelf life of fruit that could lead to the invasion of foreign substances into the bloodstream. It is clearly stated that physiochemical features like size, chemical composition, surface treatment and others are the key source of potential adverse health effects. At present, there is no clue and no clear information made about the consumption of nanocoated or nano-formulated fruits that are either safer or more dangerous to human health. Scientists and researchers have only evaluated the applicability of nanotechnology in fruit preservation and determined that, while consumers are likely to gain from this technique, advanced techniques to ensure and assess product safety and reliability may be required. However, while all the negative thinking and preconceptions about the possible risk of nanocoated fruits may not be real, it is morally imperative to bridge the information gap, uncover the truth and elucidate the process via further in-depth study and control measures of nanocoated fruits.

12.5 CONCLUSION AND FUTURE PERSPECTIVES
Nanotechnology is undoubtedly a huge breakthrough in the area of food science and technology and has versatile applications in food preservation and packaging, especially for products like fruits and vegetables on which edible nanocoatings are applied to improve antimicrobial and antioxidant properties. NPs have been used in the postharvest management of fruits to enhance their nutritional quality, flavor, color, fragrance and texture stability and control moisture loss and gas permeability, leading to shelf life extension. The current discoveries so far have mainly discussed the advantages of using edible nanocoatings of natural bioactive compounds over conventional synthetic plastic-based coatings to improve and ensure the overall quality, safety and the shelf life of fresh produce. Nonetheless, the majority of these studies were carried out on a laboratory level, and scale-up for commercial applications, as well as the potential impact of nanomaterials on human health, needs further exploration. Therefore, more extensive research and commercial-scale testing are required to provide more realistic information on applications of nanotechnology-based coatings. Furthermore, biologically active substances from plant extracts such as EOs, organic acids, and salts, reducing agents and others should be studied to develop efficient alternatives to synthetic preservatives and other active food additives.

Considering the potential toxicity of nanomaterials, further investigations are required to gain more knowledge about the toxicity and safety assessment of nanocoated foods. As each nanomaterial exhibits unique properties, toxicity will most likely be determined on a particular instance basis (Mahler et al. 2012). The gastrointestinal response of foods coated with nano edible coating should also be studied in the future. Nanotechnology in food science offers new opportunities for both government and the private sector. While processing food material, it must ensure consumers' acceptability, belief, satisfaction and positive response toward nanofoods. In the future, nanocoated fruits are expected to be commercially available to customers worldwide, and researchers will be able to address all the challenges related to this promising technology.

Acknowledgment: The authors are grateful to their respective authorities, departments, institutions, and university for their support and cooperation. Author (PK & MK) would like to thank DBT (BT/PR39789/NER/95/1664/2020), Government of India.

REFERENCES

Aschberger K, Micheletti C, Sokull-Klüttgen B, Christensen FM (2011) Analysis of currently available data for characterising the risk of engineered nano-materials to the environment and human health – lessons learned from four case studies. Environment International 37 (6):1143–1156

Aswathanarayan JB, Vittal RR (2019) Nanoemulsions and their potential applications in food industry. Frontiers in Sustainable Food Systems 3:95

Athinarayanan J, Periasamy VS, Alsaif MA, Al-Warthan AA, Alshatwi AA (2014) Presence of nanosilica (E551) in commercial food products: TNF-mediated oxidative stress and altered cell cycle progression in human lung fibroblast cells. Cell Biology Toxicology and Applied Pharmacology 30 (2):89–100

Aytac Z, Ipek S, Durgun E, Tekinay T, Uyar T (2017) Antibacterial electrospun zein nanofibrous web encapsulating thymol/cyclodextrin-inclusion complex for food packaging. Food Chemistry 233:117–124

Bajpai VK, Kamle M, Shukla S, Mahato DK, Chandra P, Hwang SK, Kumar P, Huh YS, Han Y-K (2018) Prospects of using nanotechnology for food preservation, safety, and security. Journal of Food Drug Analysis 26 (4):1201–1214

Baranwal A, Srivastava A, Kumar P, Bajpai VK, Maurya PK, Chandra P (2018) Prospects of nanostructure materials and their composites as antimicrobial agents. Frontiers in Microbiology 9:422

Bento R, Pagán E, Berdejo D, de Carvalho RJ, García-Embid S, Maggi F, Magnani M, de Souza EL, García-Gonzalo D, Pagán R (2020) Chitosan nanoemulsions of cold-pressed orange essential oil to preserve fruit juices. International Journal of Food Microbiology 331:108786

Berekaa MM (2015) Nanotechnology in food industry; advances in food processing, packaging and food safety. International Journal of Current Microbiology and Applied Sciences 4 (5):345–357

Bhargava K, Conti DS, da Rocha SRP, Zhang Y (2015) Application of an oregano oil nanoemulsion to the control of foodborne bacteria on fresh lettuce. Food Microbiology 47:69–73

Bora H, Kamle M, Mahato DK, Tiwari P, Kumar P (2020) Citrus essential oils (CEOs) and their applications in food: An overview. Plants 9 (3):357

Chu Y, Gao C, Liu X, Zhang N, Xu T, Feng X, Yang Y, Shen X, Tang X (2020) Improvement of storage quality of strawberries by pullulan coatings incorporated with cinnamon essential oil nanoemulsion. LWT 122:109054

Cockburn A, Bradford R, Buck N, Constable A, Edwards G, Haber B, Hepburn P, Howlett J, Kampers F, Klein C (2012) Approaches to the safety assessment of engineered nanomaterials (ENM) in food. Food Chemical Toxicology 50 (6):2224–2242

Correa-Pacheco ZN, Corona-Rangel ML, Bautista-Baños S, Ventura-Aguilar RI (2021) Application of natural-based nanocoatings for extending the shelf life of green bell pepper fruit. Journal of Food Science 86 (1):95–102

Cortez-Vega WR, Pizato S, de Souza JTA, Prentice C (2014) Using edible coatings from Whitemouth croaker (*Micropogonias furnieri*) protein isolate and organo-clay nanocomposite for improve the conservation properties of fresh-cut 'Formosa' papaya. Innovative Food Science Emerging Technologies 22:197–202

Cotton GC, Lagesse NR, Parke LS, Meledandri CJ (2019) Antibacterial Nanoparticles. Comprehensive Nanoscience and Nanotechnology (Second Edition). pp 65–82, Academic Press.

Cushen M, Kerry J, Morris M, Cruz-Romero M, Cummins E (2012) Nanotechnologies in the food industry – Recent developments, risks and regulation. Trends in Food Science Technology 24 (1):30–46

Das A, Kamle M, Bharti A, Kumar P (2019) Nanotechnology and it's applications in environmental remediation: An overview. Vegetos 32 (3):227–237

Dasgupta N, Ranjan S, Mundekkad D, Ramalingam C, Shanker R, Kumar A (2015) Nanotechnology in agrofood: From field to plate. Food Research International 69:381–400.

De León-Zapata MA, Pastrana-Castro L, Barbosa-Pereira L, Rua-Rodríguez ML, Saucedo S, Ventura-Sobrevilla JM, Salinas-Jasso TA, Rodríguez-Herrera R, Aguilar CN (2017) Nanocoating with extract of tarbush to retard Fuji apples senescence. Postharvest Biology and Technology 134:67–75

Deng Z, Jung J, Simonsen J, Wang Y, Zhao Y (2017) Cellulose nanocrystal reinforced chitosan coatings for improving the storability of postharvest pears under both ambient and cold storages. Journal of Food Sciences 82:453–462.

Donsì F, Cuomo A, Marchese E, Ferrari G (2014) Infusion of essential oils for food stabilization: Unraveling the role of nanoemulsion-based delivery systems on mass transfer and antimicrobial activity. Innovative Food Science Emerging Technologies 22:212–220

Donsì F, Marchese E, Maresca P, Pataro G, Vu KD, Salmieri S, Lacroix M, Ferrari G (2015) Green beans preservation by combination of a modified chitosan based-coating containing nanoemulsion of mandarin essential oil with high pressure or pulsed light processing. Postharvest Biology Technology 106:21–32

Duncan TV (2011) Applications of nanotechnology in food packaging and food safety: Barrier materials, antimicrobials and sensors. Journal of Colloid Interface Science 363 (1):1–24

Esyanti RR, Zaskia H, Amalia A Chitosan nanoparticle-based coating as postharvest technology in banana. In: Journal of Physics: Conference Series, 2019. vol 1. IOP Publishing, p 012109

Fellows PJ (2009) Food processing technology: Principles and practice. Elsevier.

Flores-López ML, Cerqueira MA, de Rodríguez DJ, Vicente AA (2016) Perspectives on utilization of edible coatings and nano-laminate coatings for extension of postharvest storage of fruits and vegetables. Food Engineering Reviews 8 (3):292–305

Gago C, Antão R, Dores C, Guerreiro A, Miguel MG, Faleiro ML, Figueiredo AC, Antunes MD (2020) The effect of nanocoatings enriched with essential oils on 'rocha' pear long storage. Foods 9 (2):240

Galindo-Pérez MJ, Quintanar-Guerrero D, Mercado-Silva E, Real-Sandoval SA, Zambrano-Zaragoza ML (2015) The effects of tocopherol nanocapsules/xanthan gum coatings on the preservation of fresh-cut apples: Evaluation of phenol metabolism. Food Bioprocess Technology 8 (8):1791–1799

Geszke-Moritz M, Moritz M (2016) Solid lipid nanoparticles as attractive drug vehicles: Composition, properties and therapeutic strategies. Materials Science Engineering: C 68:982–994

Ghio S, Barresi AA, Rovero G (2000) A comparison of evaporative and conventional freezing prior to freeze-drying of fruits and vegetables. Food Bioproducts Processing 78 (4):187–192

Gull A, Bhat N, Wani SM, Masoodi FA, Amin T, Ganai SA (2021) Shelf-life extension of apricot fruit by application of nanochitosan emulsion coatings containing pomegranate peel extract. Food Chemistry 349:129149

Gundewadi G, Rudra SG, Sarkar DJ, Singh D (2018) Nanoemulsion based alginate organic coating for shelf-life extension of okra. Food Packaging and Shelf Life 18:1–12

Hajipour MJ, Fromm KM, Ashkarran AA, de Aberasturi DJ, de Larramendi IR, Rojo T, Serpooshan V, Parak WJ, Mahmoudi M (2012) Antibacterial properties of nanoparticles. Trends in Biotechnology 30 (10):499–511

Horison R, Sulaiman FO, Alfredo D, Wardana AA (2019) Physical characteristics of nanoemulsion from chitosan/nutmeg seed oil and evaluation of its coating against microbial growth on strawberry. Food Research 3 (6):821–827

Jo W-S, Song H-Y, Song N-B, Lee J-H, Min SC, Song KB (2014) Quality and microbial safety of 'Fuji' apples coated with carnauba-shellac wax containing lemongrass oil. LWT-Food Science Technology 55 (2):490–497

Junqueira-Gonçalves MP, Salinas GE, Bruna JE, Niranjan KJ (2017) An assessment of lactobiopolymer-montmorillonite composites for dip coating applications on fresh strawberries. Journal of the Science of Food Agriculture 97 (6):1846–1853

Kaphle A, Navya P, Umapathi A, Daima HK (2018) Nanomaterials for agriculture, food and environment: Applications, toxicity and regulation. Environmental Chemistry Letters 16 (1):43–58

Katouzian I, Esfanjani AF, Jafari SM, Akhavan S (2017) Formulation and application of a new generation of lipid nano-carriers for the food bioactive ingredients. Trends in Food Science Technology 68:14–25

Kim IH, Lee H, Kim JE, Song KB, Lee YS, Chung DS, Min SC (2013) Plum coatings of lemongrass oil-incorporating carnauba wax-based nanoemulsion. Journal of Food Science 78 (10):E1551–E1559

Kim IH, Oh YA, Lee H, Song KB, Min SC (2014) Grape berry coatings of lemongrass oil-incorporating nanoemulsion. LWT-Food Science Technology 58 (1):1–10

Kulawik P, Jamróz E, Özogul F (2020) Chitosan role for shelf-life extension of seafood. Environmental Chemistry Letters 18 (1):61–74

Kwak H, Shin S, Kim J, Kim J, Lee D, Lee H, Lee EJ, Hyun J (2021) Protective coating of strawberries with cellulose nanofibers. Carbohydrate Polymers 258:117688

La DD, Nguyen-Tri P, Le KH, Nguyen PTM, Nguyen MD-B, Vo ATK, Nguyen MT, Chang SW, Tran LD, Chung WJ (2021) Effects of antibacterial ZnO nanoparticles on the performance of a chitosan/gum arabic edible coating for post-harvest banana preservation. Progress in Organic Coatings 151:106057

Li Q, Ren T, Perkins P, Hu X, Wang X (2021a) Applications of halloysite nanotubes in food packaging for improving film performance and food preservation. Food Control 124:107876

Li Y, Rokayya S, Jia F, Nie X, Xu J, Elhakem A, Almatrafi M, Benajiba N, Helal M (2021b) Shelf-life, quality, safety evaluations of blueberry fruits coated with chitosan nano-material films. Scientific Reports 11 (1):1–10

Liang R, Xu S, Shoemaker CF, Li Y, Zhong F, Huang Q (2012) Physical and antimicrobial properties of peppermint oil nanoemulsions. Journal of Agricultural Food Chemistry 60 (30):7548–7555

Liu R, Liu D, Liu Y, Song Y, Wu T, Zhang M (2017) Using soy protein SiOx nano-composite film coating to extend the shelf life of apple fruit. International Journal of Food Science Technology 52 (9):2018–2030

Mahajan BC, Tandon R, Kapoor S, Sidhu MK (2018) Natural coatings for shelf-life enhancement and quality maintenance of fresh fruits and vegetables – A review. Journal of Postharvest Technology 6 (1):12–26

Mahato DK, Mishra A, Kumar P (2021) Nanoencapsulation for agri-food application and associated health and environmental concerns. Frontiers in Nutrition 8:146

Mahela U, Rana DK, Joshi U, Tariyal YS (2020) Nano edible coatings and their applications in food preservation. Journal of Postharvest Technology 8 (4):52–63

Mahler GJ, Esch MB, Tako E, Southard TL, Archer SD, Glahn RP, Shuler ML (2012) Oral exposure to polystyrene nanoparticles affects iron absorption. Nature Nanotechnology 7 (4):264–271

Martínez-Hernández GB, Amodio ML, Colelli G (2017) Carvacrol-loaded chitosan nanoparticles maintain quality of fresh-cut carrots. Innovative Food Science Emerging Technologies 41:56–63

McClements DJ, Xiao H (2017) Is nano safe in foods? Establishing the factors impacting the gastrointestinal fate and toxicity of organic and inorganic food-grade nanoparticles. NPJ Science of Food 1 (1):1–13

Mills A, Hazafy D (2009) Nanocrystalline SnO2-based, UVB-activated, colourimetric oxygen indicator. Sensors Actuators B: Chemical 136 (2):344–349

Miranda M, Gozalbo AM, Sun X, Plotto A, Bai J, de Assis O, Ferreira M, Baldwin E (2019) Effect of mono and bilayer of carnauba wax based nano-emulsion and HPMC coatings on popst-harvest quality of 'redtainung' papaya. International Symposium on Agricultural & Food Processing Wastes Proceeding. 1:705–709

Miranda M, Sun X, Ference C, Plotto A, Bai J, Wood D, Assis OBG, Ferreira MD, Baldwin E (2020) Nano-and micro-carnauba wax emulsions versus shellac protective coatings on postharvest citrus quality. Journal of the American Society for Horticultural Science 1:1–10

Morin-Crini N, Lichtfouse E, Torri G, Crini G (2019) Applications of chitosan in food, pharmaceuticals, medicine, cosmetics, agriculture, textiles, pulp and paper, biotechnology, and environmental chemistry. Environmental Chemistry Letters 17 (4):1667–1692

Oh YA, Oh YJ, Song AY, Won JS, Song KB, Min SC (2017) Comparison of effectiveness of edible coatings using emulsions containing lemongrass oil of different size droplets on grape berry safety and preservation. LWT 75:742–750

Oms-Oliu G, Rojas-Graü MA, González LA, Varela P, Soliva-Fortuny R, Hernando MIH, Munuera IP, Fiszman S, Martín-Belloso O (2010) Recent approaches using chemical treatments to preserve quality of fresh-cut fruit: A review. Postharvest Biology Technology 57 (3):139–148

Pandey AK, Kumar P, Singh P, Tripathi NN, Bajpai VK (2017) Essential oils: Sources of antimicrobials and food preservatives. Frontiers in Microbiology 7:2161

Piña-Barrera AM, Álvarez-Román R, Báez-González JG, Amaya-Guerra CA, Rivas-Morales C, Gallardo-Rivera CT, Galindo-Rodríguez SA (2019) Application of a multisystem coating based on polymeric nanocapsules containing essential oil of *Thymus vulgaris* L. to increase the shelf life of table grapes (*Vitis vinifera* L.). IEEE Transactions on Nanobioscience 18 (4):549–557

Prakash A, Baskaran R, Vadivel V (2020) Citral nanoemulsion incorporated edible coating to extend the shelf life of fresh cut pineapples. LWT 118:108851

Radi M, Akhavan-Darabi S, Akhavan HR, Amiri S (2018) The use of orange peel essential oil microemulsion and nanoemulsion in pectin-based coating to extend the shelf life of fresh-cut orange. Journal of Food Processing Preservation 42 (2):e13441

Rahman MS (2020) Food preservation: An overview. In: Handbook of Food Preservation, pp 7–18. Boca Raton, London: CRC Press.

Robledo N, López L, Bunger A, Tapia C, Abugoch L (2018) Effects of antimicrobial edible coating of thymol nanoemulsion/quinoa protein/chitosan on the safety, sensorial properties, and quality of refrigerated strawberries (*Fragaria× ananassa*) under commercial storage environment. Food and Bioprocess Technology11:1566–1574

Roselli M, Finamore A, Garaguso I, Britti MS, Mengheri E (2003) Zinc oxide protects cultured enterocytes from the damage induced by *Escherichia coli*. The Journal of Nutrition 133 (12):4077–4082

Saha A, Tyagi S, Gupta RK, Tyagi YK (2017) Natural gums of plant origin as edible coatings for food industry applications. Critical Reviews in Biotechnology 37 (8):959–973

Salvia-Trujillo L, Rojas-Graü MA, Soliva-Fortuny R, Martín-Belloso O (2015) Use of antimicrobial nanoemulsions as edible coatings: Impact on safety and quality attributes of fresh-cut Fuji apples. Postharvest Biology Technology 105:8–16

Salvia-Trujillo L, Soliva-Fortuny R, Rojas-Graü MA, McClements DJ, Martín-Belloso O (2017) Edible nanoemulsions as carriers of active ingredients: A review. Annual Review of Food Science Technology 8:439–466

Sessa M, Ferrari G, Donsì F (2015) Novel edible coating containing essential oil nanoemulsions to prolong the shelf life of vegetable products. Chemical Engineering Transactions 43:55–60

Severino R, Ferrari G, Vu KD, Donsì F, Salmieri S, Lacroix M (2015) Antimicrobial effects of modified chitosan-based coating containing nanoemulsion of essential oils, modified atmosphere packaging and gamma irradiation against *Escherichia coli* O157: H7 and *Salmonella typhimurium* on green beans. Food Control 50:215–222

Severino R, Vu KD, Donsì F, Salmieri S, Ferrari G, Lacroix M (2014) Antibacterial and physical effects of modified chitosan based-coating containing nanoemulsion of mandarin essential oil and three non-thermal treatments against *Listeria innocua* in green beans. International Journal of Food Microbiology 191:82–88

Singh P (2018) Nanotechnology in food preservation. Food Science and Biotechnology 9:435–441

Sow LC, Tirtawinata F, Yang H, Shao Q, Wang S (2017) Carvacrol nanoemulsion combined with acid electrolysed water to inactivate bacteria, yeast in vitro and native microflora on shredded cabbages. Food Control 76:88–95

Sridhar A, Ponnuchamy M, Kumar PS, Kapoor A (2021) Food preservation techniques and nanotechnology for increased shelf life of fruits, vegetables, beverages and spices: A review. Environmental Chemistry Letters 19 (2):1715–1735

Taştan Ö, Pataro G, Donsì F, Ferrari G, Baysal T (2017) Decontamination of fresh-cut cucumber slices by a combination of a modified chitosan coating containing carvacrol nanoemulsions and pulsed light. International Journal of Food Microbiology 260:75–80

Weiss J, takhistov P, Mcclements DJ (2006) Functional materials in food nanotechnology. Journal of Food Science 71 (9):r107-r116

Xu D, Qin H, Ren D (2018) Prolonged preservation of tangerine fruits using chitosan/montmorillonite composite coating. Postharvest Biology Technology 143:50–57

Yadav SK (2017) Tissue science & engineering realizing the potential of nanotechnology for agriculture and food technology. Journal of Tissue Science and Engineering 8:8–11

Zambrano-Zaragoza ML, González-Reza R, Mendoza-Muñoz N, Miranda-Linares V, Bernal-Couoh TF, Mendoza-Elvira S, Quintanar-Guerrero D (2018) Nanosystems in edible coatings: A novel strategy for food preservation. International Journal of Molecular Sciences 19 (3):705

Zambrano-Zaragoza ML, Gutiérrez-Cortez E, Del Real A, González-Reza RM, Galindo-Pérez MJ, Quintanar-Guerrero D (2014) Fresh-cut Red Delicious apples coating using tocopherol/mucilage nanoemulsion: Effect of coating on polyphenol oxidase and pectin methylesterase activities. Food Research International 62:974–983

Zambrano-Zaragoza ML, Mercado-Silva E, Ramirez-Zamorano P, Cornejo-Villegas MA, Gutiérrez-Cortez E, Quintanar-Guerrero D (2013) Use of solid lipid nanoparticles (SLNs) in edible coatings to increase guava (*Psidium guajava* L.) shelf-life. Food Research International 51 (2):946–953

Zhai X, Lin D, Li W, Yang X (2020) Improved characterization of nanofibers from bacterial cellulose and its potential application in fresh-cut apples. International Journal of Biological Macromolecules 149:178–186

Zhang C, Li Y, Wang P, Zhang H (2020) Electrospinning of nanofibers: Potentials and perspectives for active food packaging. Comprehensive Reviews in Food Science Food Safety 19 (2):479–502

Zhang Y, Zhang Y, Zhu Z, Jiao X, Shang Y, Wen Y (2019) Encapsulation of thymol in biodegradable nanofiber via coaxial eletrospinning and applications in fruit preservation. Journal of Agricultural Food Chemistry 67 (6):1736–1741

Zielińska A, Carreiró F, Oliveira AM, Neves A, Pires B, Venkatesh DN, Durazzo A, Lucarini M, Eder P, Silva AM (2020) Polymeric nanoparticles: Production, characterization, toxicology and ecotoxicology. Molecules 25 (16):3731

13 Nanotechnology in Innovative Food Preservation and Packaging

Jayita Bandyopadhyay and Suprakas Sinha Ray

Corresponding author:
Dr. Suprakas Sinha Ray
Email: *ssinharay@uj.ac.za*

CONTENTS

13.1 INTRODUCTION

Strong economic growth and the rising population in developing countries will impose a huge burden on the supply and the demand of food in the coming decades. In spite of the huge demand, approximately 1.3 billion tons of food are lost per annum, and about 50% of it is wasted at the household level in Western developed countries.[1] Figure 13.1 shows an estimated waste percentages of milled and fresh produce across the food supply chain in Europe,[2] a significant portion of which occurs during consumption (households) followed by agricultural production. Fruits, vegetables, meat, and seafood are perishable unless preserved and packaged properly due to protein denaturation, lipid oxidation, and the growth of microorganisms, which leads to reduced shelf lives of such products. Such wastage of food may cause major food security concerns for a society and may lead to economic losses.

The typical food supply value chain is presented in Figure 13.2. Postharvest fresh produce or agricultural products are processed using common techniques like pasteurization, dehydration, and others prior to packaging. Referring to the food supply value chain demonstrated in Figure 13.2, packaging, particularly plastic packaging, plays a pivotal role in the food quality and safety and hence on the reduction in the food wastes during subsequent stages indicated in the food supply value chain. Ensuring packaging quality is vital for preventing possible contamination and resulting food spoilage. With global demand for fresh-cut and minimally processed food, innovative food packaging is required for the preservation of the food for a longer duration while retaining their quality.

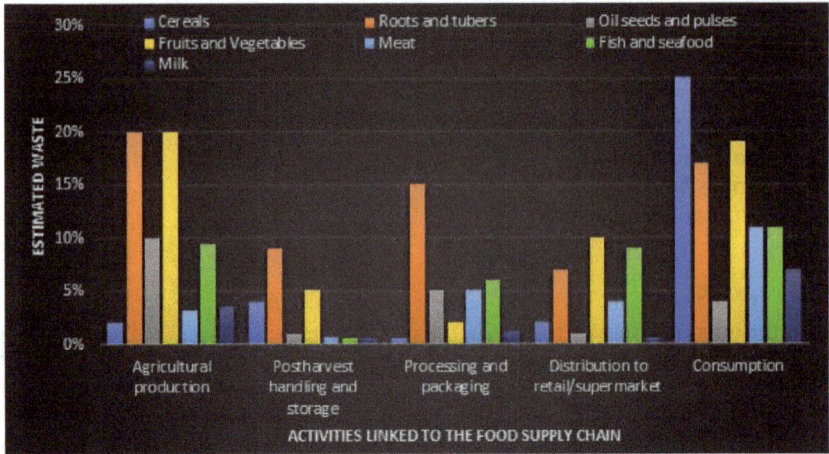

Figure 13.1 Estimated waste percentages of milled and fresh produce across the food supply chain in Europe.

The hanging lifestyle trends of consumers with having less time to prepare food has intrigued packaging industries, leading them to come up with innovative solutions.

Apart from the conventional function of product protection and acting as an inert barrier against the environment, modern-day packaging offer features like antimicrobial, oxygen and ethylene scavenging, carbon dioxide absorbers/emitters, aroma emitter, ripeness indicators, biosensors, and so on. Exposure to pulse electric field (PEF), microwave, high-pressure food preservation, and ozone and natural organic treatment are widely used to kill the pathogens and for preservation of foods.[3] Recent trend in polymer nanotechnology has shown prospects of emerging food packaging materials with nano-sensing that is capable of monitoring and tracing food condition during distribution and storage in addition to imparting improved antimicrobial, mechanical, and barrier properties. This chapter gives consideration to stakeholders' incentives to minimize food waste, the recent emerging trend of nano-enabled packaging intended for food packaging applications, safety regulations on the nanoparticle containing food packaging, and future prospects.

13.2 STAKEHOLDERS' INCENTIVES TO MINIMIZE FOOD WASTE

Different organizations define food loss and the food waste in different manners. Following the UN Food and Agriculture Organization (FAO), food loss refers to unintended loss of food during harvesting, post-harvest handling, processing, and distribution, while wastage is defined as loss of food at retail and consumption stages.[4] The World Resources Institute defines food loss as food that spoils before reaching the consumer mostly due to mishandling.[5] According to the U.S. Department of Agriculture (USDA) wastage of food that happens at any stage across the supply chain is considered food waste.[6] In the United States alone, 52.4 million tons of food ends up in landfill, and an additional 10.1 million tons remain unharvested at farms, totaling about 63 million tons of annual waste. USDA has provided a road map for minimizing food waste and identified

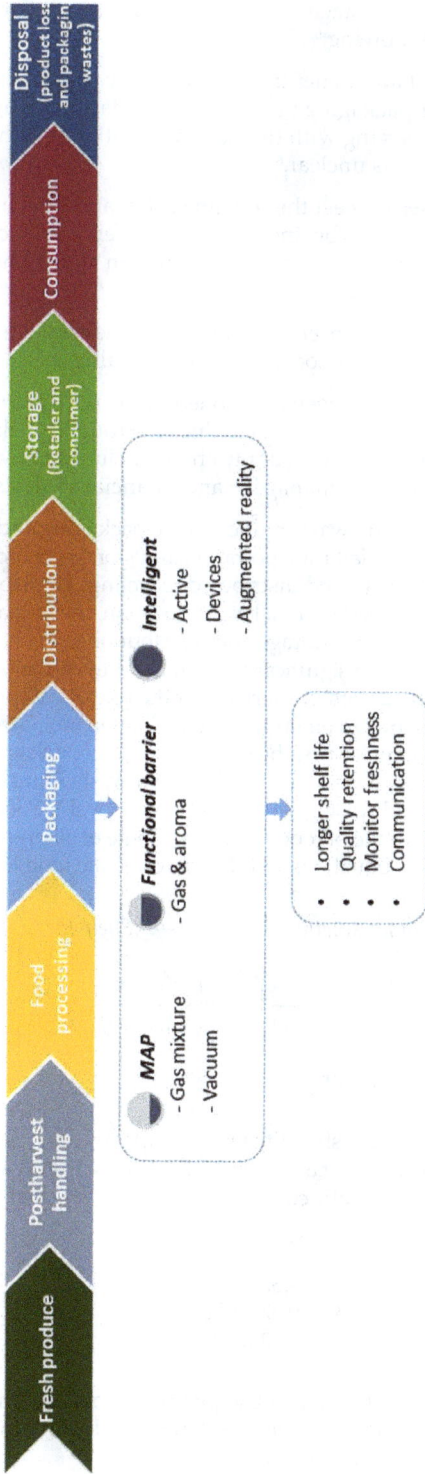

Figure 13.2 Importance of plastic packaging in the food supply value chain.

the barriers to prevent the wastage of food. Some hindrances for food waste prevention include the following:[6]

- **Misalignment of cost and benefit:** Businesses may find it hard to implement a new technology for packaging or standardized date labeling when consumers get more cost saving with the existing solution and the effect on the enhanced brand equity is unclear.[6]

- **Lack of social license:** To meet the consumer demand for variety and cosmetic perfection, sometimes businesses cannot offer cosmetically imperfect food, reduce portion sizes, or even invest in proven approaches such as cold chain and inventory management.[6]

- **Information gap:** There is uncertainty on where the wastage occurs, how much occurs, and what the associated economic value is.[6]

- **Organizational silos:** It has been mentioned in the road map to reduce U.S. food waste that the food waste can be reduced through collaboration between different stakeholders across the supply chain, including buyers, merchants, store managers, chefs, waste managers, and financial analysts.[6]

Innovations that can steer food waste reduction are packaging and labeling, information technology–enabled transportation and storage, coordinated logistics, value-added compost products, and distributed recycling. Therefore, improved packaging to keep the food fresher for a longer duration, the management of ethylene and microbes in food during storage and distribution, and changes in labeling practices are expected to play a significant role in the prevention of food waste. Brennan et al. studied consumers' concern about the role of packaging in food waste and their attitudes toward emerging packaging technologies.[7] The literature has shown that among many factors, the inconvenience of packaging design for decanting, damaged products, reduced product quality, shelf-life failure, and insufficient protection cause food wastage at the household level.[8,9] Food loss probability (FLP) decreases as the shelf life of food increases. Conte et al. proposed three different empirical equations, Equations 13.1–13.3, to correlate food loss and shelf life.[8]

$$First\ order\ kinetic: FLP = exp\left(-k_a.Shelf\ life\right) \tag{13.1}$$

$$Sigmoidal: FLP = \left(\frac{1}{1-exp(-1)}\right).\left(exp\left(\frac{1}{-k_b.Shelf\ life}\right)\right) - exp(-1) \tag{13.2}$$

$$Linear: FLP = k_c.Shelf\ life \tag{13.3}$$

Considering FLP attains zero if shelf life tends to infinity and vice versa in addition to an avoidable food waste of 8% by mass (i.e., FLP = 0.08 at shelf-life = 75), the authors have estimated the kinetic constants (k_a, k_b, and k_c) from the Equations 13.1–13.3 as follows:

$$k_a = 3.37\ 10^{-2}\ day^{-1}$$
$$k_b = 9.67\ 10^{-2}\ day^{-1}$$
$$k_c = 1.23\ day^{-1}$$

These kinetic constants can be used subsequently to correlate the FLP and the shelf life. There is a lack of data on the correlation of FLP and the shelf life, and more research are required in this direction.

Similarly, a handful of studies are available on the consumer attitude/perception toward emerging packaging technologies. A recent consumer study in the United Kingdom found that consumers recognize the benefits of packaging to keep food fresher and hygienic and protect it from damage.[10] However, the main concern is the impact of the packaging material on environment pollution. Several studies have claimed that the consumers' knowledge on emerging packaging technologies like active and intelligent packaging is insufficient.[11,12] Consumer perceptions of nanotechnologies are dependent on the description or information provided to the consumer about these technologies. Braska et al. has reported that despite consumers' acceptance of a higher price for active and intelligent packaging, the success of implementing such solutions requires raising awareness and promotion strategies. For instance, low-cost technologies like vacuum skin packaging (VSP) might be attractive to packaging industries, including the brand owners, as well as the consumers. Using food-saving technologies directly into the packaging system bypasses contamination concerns about sachets or pads and increases consumer confidence in active packaging techniques.[13]

It has been reported on many occasions that food packaging design, which can address food wastage at the household level includes protection (e.g., mechanical, physicochemical, reseal ability) facilitating handling (e.g., easy opening and closing features, providing correct quantity, and serving size), communication (e.g., freshness, storage, safety and sorting of household waste information), and sustainable packaging materials.[7,14] Generally, food packaging can be classified into four categories, namely, modified atmosphere packaging (MAP), improved, active, and intelligent, as shown in the Figure 13.3. Improved

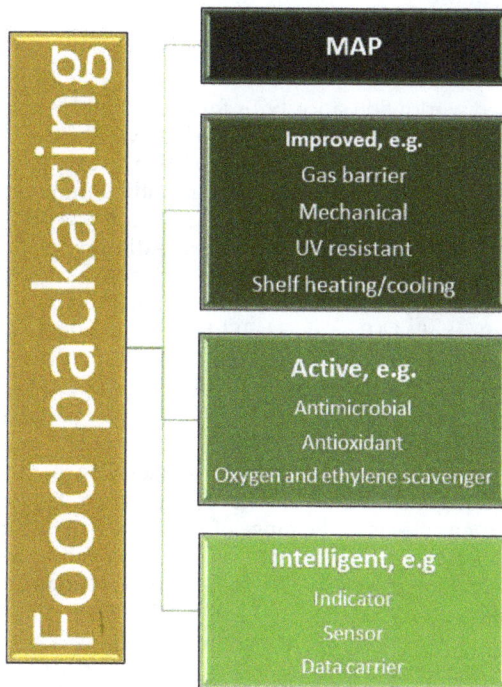

Figure 13.3 Types of food packaging.

properties, such as, gas barrier, mechanical, ultraviolet (UV) resistance, are desired for food packaging. Shelf heating and cooling do not aim to improve the shelf life but are related to the convenience, such as ready-to-eat meals. For example, calcium or magnesium oxide and water are commonly used for self-heating of packaged food by exothermic reactions.

Active packaging is intended to extend the shelf life of food by protecting against microbial growth, the prevention of oxidation of food, oxygen, and ethylene scavenging. Organic acids and their salts, plant extract and essential oils, bacteriocins, and metal oxide nanoparticles are commonly used in antimicrobial packaging. Additives like butylated hydroxytoluene, butylated hydroxyamino, gelatine, and carboxymethyl cellulose are incorporated in the packaging material to achieve antioxidant property. Common desiccants, for example, silica gel, calcium oxide, calcium sulphate, calcium chloride, potassium chloride, potassium carbonate, nanoclay, xylitol, sorbitol, and carboxymethyl cellulose, are either used in sachets, sheets, or pads to absorb moisture inside packages. Most oxygen scavengers are based on bags containing iron and different catalysts. Other nonmetallic oxygen scavengers include ascorbic acid, α-tocopherol, catechol, sorbose, lignin, gallic acid, enzymes, and others. Commonly used ethylene scavengers include natural nanoclays (e.g., bentonite, halloysite, zeolite, Japanese Oya stone, cristobalite, ceramics, etc.); regenerable solvents (e.g., propylene or hexylene glycol, phenymethylsilicon, squalene, etc.); catalytic oxidizer like potassium permanganate, electron-deficient diene or trienes (such as benzene, pyridine, tertazines); active charcoal; silica gel; kieselguhr; and aluminum oxide. Finally, intelligent packaging can be codified into three main categories: indicator, sensor, and the data carrier. MAP and nanotechnology-based improved, active, and intelligent packaging are discussed in the following sections.

13.3 MAP

MAP has been used worldwide since 1970s for preserving and extending the shelf life of fresh or minimally processed fruits and vegetables.[15] MAP can lower the respiration rate, delay ripening and discoloration, prevent odor buildup, and inhibit the growth of pathogens. Finnegan et al. studied the effect of intrinsic factors, such as physiological age, origin, and seasonality, as well as extrinsic factors like cut size, sharpness of the blade, and dipping treatment on the respiration rate of fresh-cut pineapple chunks.[16] Authors have found that the physiological age and origin are more important than the seasonality when determining the effect of MAP on the shelf life of fresh-cut pineapple chunks. MAP with lower oxygen levels (1–5%) can inhibit the growth of aerobic microorganisms, but the growth of anaerobic microorganisms still can lead to development of off-odor in fresh-cut melon or baby spinach, as reported by Amaro et al. and Tudela et al., respectively.[17,18] MAP with a mixture of low oxygen and 3% to 20% carbon dioxide is also used in many instances. The maximum level of carbon dioxide varies greatly from product to product; for instance, the carbon dioxide tolerance for apples is in the range of 2% while that for spinach and berries is 15%.[19] Mistriotis et al. reported that the high relative humidity within MAP can favor microbial growth, and hence, it should be managed carefully according to the packaged produce.[20]

The inherent gas permeability of the polymer used in packaging, the thickness of the packaging film or container, and the micropores/pinholes in the film influence the gaseous composition of MAP. A commonly used polymer for MAP includes low-density polyethylene (LDPE) and ethylene-vinyl acetate (EVA). Following the trend of sustainable and green packaging, the scope of using polylactic acid (PLA) for MAP has also been investigated.[20] Despite the fact that the MAP can preserve fruit and vegetables to a certain extent, innovations in complementary and augmented technologies are required to meet consumers' expectations for longer shelf life and quality retention. Unlike other MAP technologies, BreatheWay technology offers a balance of oxygen and carbon dioxide based on each product's unique respiration rate even when the temperature fluctuation occurs and does not require perforations in the packaging.[21] As an alternative to MAP, VSP used to package the raw meats is appealing to consumers as it provides adequate protection from the external environment with minimal packaging material.[7] VSP removes gases surrounding the product, which enables a reduction in microbiological growth and degradation rates, resulting in an extended shelf life of the product. Because of low-cost VSP is popular not only with consumers but also with supply chain industries.[22,23]

13.4 INNOVATION IN FOOD PACKAGING THROUGH NANOTECHNOLOGY APPROACH

Innovations in material science and nanotechnology is stipulating the advancement in the innovative and smart food packaging technologies. The major drawback of the plastic packaging materials is their inherent permeability to gases and other small molecules. Owing to the unique properties of nanostructures, the incorporation of nanostructures into packaging materials enhances performance and enriches the functionalities of the packaging materials. Nanotechnology is considered one of the cutting-edge technologies of the 21st century, and its applications in food packaging may enable overcoming the shortcomings of traditional preservation technologies.[24,25] Recent advancements on the nanotechnology-based food packaging are categorically discussed in the following sections.

13.4.1 Barrier Packaging Including Functional Barrier

Food packaging may contain a single polymer layer that prevents the permeation of O_2 and H_2O (Figure 13.4) or multilayered structure containing a nanocomposite layer sandwiched between polymers to enhance the barrier properties (Figure 13.4).[26] The dispersion and distribution of nanoparticles in the polymer nanocomposites provide an improved gas barrier that enhances the ability of the composite to maintain the stability of the gas composition inside the package required during the distribution and storage of the packaged food products. The high aspect ratio and absorption/adsorption capacity of nanofillers provide improved gas barrier properties in the polymer nanocomposite-based packaging materials. Owing to the impermeable nature of the inorganic nanofillers like nanoclay, the gas molecules need to follow a tortuous diffusion path through the space between the dispersed nanofillers in the polymer nanocomposite film/sheet. The Nielsen model (Equation 13.4) considers such lengthening effects of the diffusion of gases and allows theoretical prediction of the gas permeability in a polymer nanocomposite.[27]

$$\frac{P_s}{P_p} = \frac{1-\varphi}{1+\dfrac{L}{2W}\varphi}, \tag{13.4}$$

where P_s and P_p are the permeabilities of nanofiller and the polymer, respectively. φ indicates volume fraction of nanoscale filler, and L and W represent, respectively, the length and the width of the nanoparticles. Fredrickson and

Bicerano extended this model to forecast the effect of nanofillers dispersed in the dilute or semi-dilute composites as[28]

$$\frac{P_s}{P_p} = \frac{1-\varphi}{1+\mu\left(\dfrac{L}{2W}\right)^2 \varphi^2},$$

(13.5)

where $\mu = \pi^2/16 \ ln \ (L/2W)^2$.

The rate of permeation and the reaction rate can then be determined from Equations 13.6 and 13.7.[29]

$$Rate \ of \ permeation = \frac{P}{d} \cdot A \cdot \left(0.21 - p_{O_2}\right)$$

(13.6)

$$Reaction \ rate = k_{O_2} \cdot p_{O_2},$$

(13.7)

where P, d, A, p_{O_2}, and k_{O_2} represent, respectively, the oxygen permeability, the thickness of the film, the surface area, the partial pressure of oxygen, and the oxidation rate.

Continuous exchange of gases through the packaging material and the emission of moisture on the inner side of the package lead to food spoilage.[30] Improved gas barrier properties in polymer nanocomposite containing nanomaterials are a well-established field. For instance, Horst et al. achieved approximately 30% reduction in oxygen permeation in the LDPE and high-density polyethylene (HDPE) nanocomposite film composed of 15 wt% organically modified montmorillonite (MMT) nanoclay.[31] Maleic anhydride–grafted PE was used as a compatibilizer. Tajeddin et al. reported that carboxymethyl cellulose/polyvinyl alcohol–based control film, and the corresponding nanocomposite film containing 3 wt% nanoclay both experience moisture ingress, but there is approximately 52% reduction in moisture content inside the nanocomposite-based film when compared to the control film. Authors have investigated potential usage of the film for the walnut packaging;[32] however, since the film

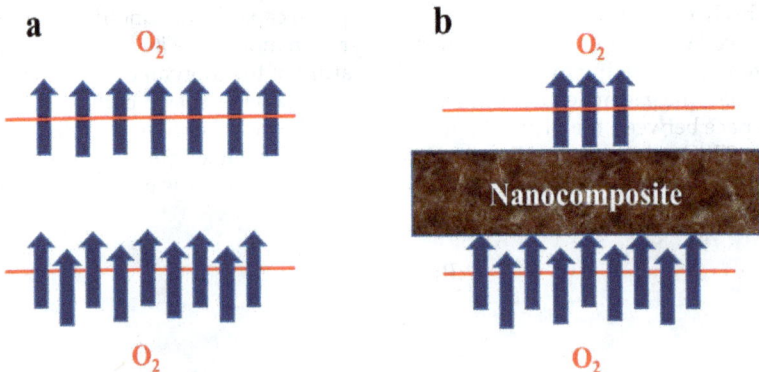

Figure 13.4 Gas barrier properties of packaging material before (a) and after (b) incorporation of polymer nanocomposite.[26]

was produced by solvent casting, despite improved properties, production at a commercial scale might be challenging.

Two-dimensional (2D) nanoclay platelets in the polymer nanocomposites exhibit many advantages with compared to the corresponding neat polymer. Nanocomposite-based packaging is lighter, stronger, and more heat-resistant and offers improved barrier against gases, moisture, volatiles materials, anti-microbial properties, and so forth.[33] Organically modified natural nanoclay at different modification levels, as well as synthetic nanoclay, and surface-treated synthetic nanoclay are also available from Council for Scientific and Industrial Research (CSIR), South Africa. The companies producing nano-reinforced food packaging materials include Colour Matrix Corporation, Mitsubishi Gas Chemical Company, Nanocore, Lanxess, and Honeywell.[34] We have developed a polyamide nanocomposite that offers approximately 40% higher oxygen barrier when compared with the neat polymer. Thus, by replacing a polyamide film with this polyamide nanocomposite may reduce the film thickness by 30%, resulting a 20% cost saving.

Metal oxide nanoparticles have attracted great research interest in food packaging due to their multifunctionalities, such as gas barrier, UV blocking, antimicrobial, and ethylene scavenging properties. The impermeable inorganic phase provides a physical barrier to diffusing gas molecules, creates tortu-ous path through the polymer, and hence reduces the gas permeability when incorporated in the polymer matrix. The effects of different nanofillers on the gas barrier properties of the nanocomposite films are tabulated in Table 13.1. Titanium dioxide (TiO_2),[35,36] silicon dioxide (SiO_2),[37-40] alumina (Al_2O_3),[41,4] and zinc oxide (ZnO)[43-45] are incorporated in the various polymer matrices includ-ing HDPE, LDPE, polypropylene (PP), polyvinyl alcohol (PVA) ethylene vinyl alcohol (EVOH), poly(ethylene terephthalate) (PET), and polylactide (PLA). As

Table 13.1 Effects of Different Nanofillers on the Gas Barrier Properties of the Nanocomposite Films

Nanoparticles	Matrix	Key properties	Reference
Nanoclay	Carboxymethyl cellulose and PVA film, solvent cast	High barrier against the diffusion of moisture vapor and oxygen and suitable for walnut packaging.	Tajeddin et al. (2019)
Organically modified nanoclay	LDPE, HDPE in the presence of PE grafted with maleic anhydride as compatibilizer	Oxygen permeation reduced by approximately 30% at a loading of 15 wt% nanoclay.	
TiO2 (Anatase, 40 μm)	LDPE, extruded film	Increased barrier properties of the packaging and decrease in ethylene generation for strawberries.	Li et al. (2017)
TiO2 (Rutile, 70–80 μm)	LDPE, blown film	The oxygen permeability of the nanocomposite decreased 16% which in turn reduced the rancidity of the packaged almonds by 78%.	Nasiri et al. (2012)

(Continued)

Table 13.1 (Continued)

Nanoparticles	Matrix	Key properties	Reference
SiO$_2$ (20–50 nm)	EVOH, blown film	The composite film exhibited 54% reduction in water permeability coefficient, 50% increase in mechanical strength, and improved heat resistance.	Liu et al. (2010)
SiO$_2$ (60 nm)	PLA	Oxygen and water vapor permeability reduced by 50%.	Bang and Kim (2012)
SiO$_2$ (50 nm)	PP surface treated with EVA, extruded film	The film showed 28% reduction in oxygen and 23.8% reduction in water vapor permeability in addition to 30% increase in tensile strength.	Li et al. (2016)
SiO$_2$ (<250 nm)	LDPE (titanate crosslinking), extruded film	Shelf life of packaged shrimp was enhanced by 33% due to the improved gas barrier properties.	Luo, Xu et al. (2015)
Al$_2$O$_3$ (25 nm)	PLA (coated with alginate and chitosan)	Water vapor transmission rate of PLA coated with alginate and chitosan decreased by 47% when compared with the neat polymer.	Hirvikorpi et al. (2011)
AlO$_x$ (25 nm)	PET and biaxially oriented PP, coated	AlO$_x$ coating improved the gas barrier properties of the PET film while biaxially oriented PP with coating produced inconsistent results.	Struller et al. (2014)
ZnO	PLA	18%, 17%, and 16% reduction in oxygen, carbon dioxide, and moisture vapor permeation, respectively, at a loading of 1 wt% ZnO. Further increase in ZnO content did not show any further reduction in the gas permeation.	Mara et al. (2016)
ZnO	PBAT	Lowest oxygen permeation was obtained at 10 wt% ZnO loading.	Venkatesan et al. (2017)
ZnO	LLDPE	17% and 12% reduction in oxygen and moisture vapor permeation, respectively, at a loading of 5 wt% ZnO.	Polat et al. (2018)
ZnO	PP	22% and 12% reduction in oxygen and moisture vapor permeation, respectively, at a loading of 5 wt% ZnO.	Polat et al. (2018)

mentioned in Table 13.1, the nanocomposite films possess improved gas barrier properties than the neat polymer. Pantani et al. also reported that the gas barrier property of PLA was increased due to the tortuosity created by the dispersed zinc oxide nanoparticles in PLA matrix.[46]

13.4.2 Active Packaging

Active packaging protects food using some mechanisms activated by internal or external factors. According to the regulation (EC) No 1935/2004 of the European Parliament and the council, "active food contact materials and articles are designed to deliberately incorporate 'active' components intended to be released into the food or to absorb substances from the food".[47] Active packaging comprised of nanoparticles shows antimicrobial properties due to the release of metal ions and reactive oxygen species (ROS).[48] As indicated in Figure 13.5, the microbial inactivation may occur via disruption of cell respiration due to mitochondrial damage, disruption of the cell membrane, the generation of ROS, and the interaction with thiol groups of DNA and sulfhydryl or phosphorous, proteins, and enzymes.[49] Hence, the controlled release of nanoparticles is highly desired to mitigate unforeseen health risks.

When exposed to UV light, TiO_2 nanoparticles exhibit photocatalytic antibacterial properties, generating ROS that destroys the microbes by peroxidation of lipids in the cell membrane.[50] This feature makes it attractive for decontaminating water and disinfecting food contact surfaces.[51] Hence, varieties of metal oxide nanoparticle containing polymer composite films are explored for food packaging applications. Table 13.2 summarizes exemplary food packaging materials composed of TiO_2 nanoparticles and shows the antimicrobial activity.[52–55] The microbial properties of LDPE and PLA film composed of dual filler (TiO_2 and the silver nanoparticles) have also been investigated, and the findings

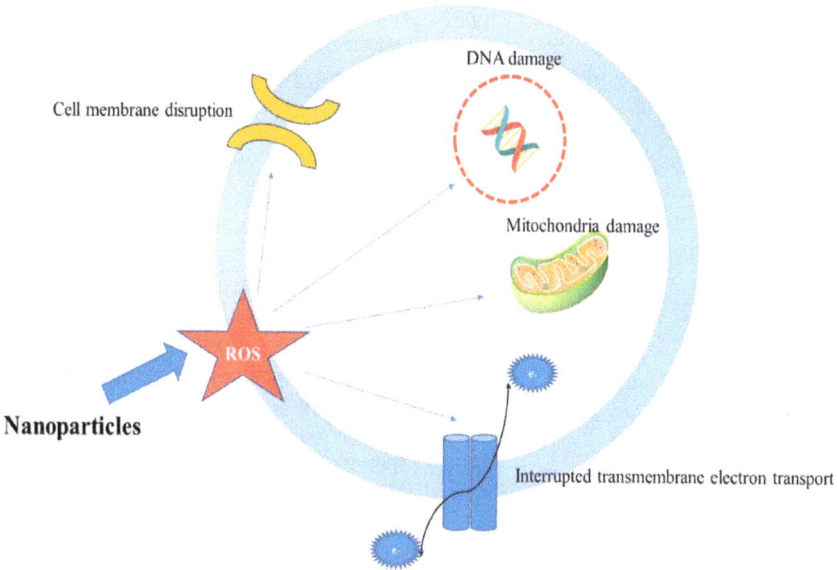

Figure 13.5 Mechanisms of microbial inactivation by nanoparticles.[26]

Table 13.2 Active Nanocomposites Containing Different Types of Nanoparticles

Nanoparticles	Matrix	Key properties	Reference
TiO_2 (anatase, 90 nm)	EVOH, compressed film	Self-sterilization, 2–5% TiO_2 can kill the Gram-positive and Gram-negative strains investigated.	Cerrada et al. (2008)
TiO_2 (anatase & rutile, 20–80 nm)	LDPE, blown film	Decrease of microorganisms e.g., *Pseudomonas* and *Rhodotorula* in packaged pears by 4 and 2 log10 CFU/mL, respectively.	Bodaghi et al. (2013)
TiO_2 (anatase, 25 nm)	HDPE (added CaCO3), blown film	Maintain the structure of packaged cheese by inhibiting the lactic acid bacteria and coliforms.	Gumiero et al. (2013)
TiO_2 (<100 nm)	LDPE, extruded film	Longer shelf life of packaged shrimp and increase in the sensory scores by 30.77%.	Luo, Qin, et al. (2015)
TiO_2 + Ag (nm range)	LDPE (added kaolin)	Decay of bayberries can be hindered by a combined effect of hot air treatment, inhibited respiration, ethylene production, yeasts, and molds, hindering the decay of bayberries.	Wang et al. (2010)
TiO2, TiO2 + Ag (<100 nm)	PLA, cast film	Extension of cheese shelf life of cheese to 25 days due to the antimicrobial activity of TiO_2 + Ag.	Li et al. (2018)
ZnO (200–400 nm)	PVC, coated film	Reduction in decay of cut apples due to inhibition of *E. coli* and *S. aureus*.	Li et al. (2011)
ZnO	PP, extruded film	Decreased *E. coli* numbers by 99.9% after 48 h. Resistance to photodegradation.	Silvestre et al. (2013)
ZnO	PLA	Water vapor barrier and antimicrobial activity	Pantani et al. (2013)
ZnO + Ag	LDPE	Reduced yeast and mold count in orange juice	Emamifar et al. (2010)
ZnO + Ag	LDPE, blown film	Reduced numbers of *Lactobacillus plantarum* in orange juice stored for 112 days.	Emamifar et al. (2011)
ZnO (silanized, 15–30 nm)	PLA, extruded film	Reduction of bacterial numbers by 3 log10 CFU/mL after 7 days; enhanced mechanical and anti-UV properties.	Pantani et al. (2013)
ZnO (silanized, 50 nm)	HDPE	Antimicrobial activity, decreasing *E. coli* and *S. aureus* by 97.7% and 99.9%, respectively.	Li & Li (2010)
ZnO + Al (7–9 nm, coating)	PLA, extruded film	Antibacterial activity against *E. coli*.	Valerini et al. (2018)

Nanoparticles	Matrix	Key properties	Reference
CuO (50 nm)	LDPE, extruded film	Antimicrobial properties by hindering the growth of coliform bacteria in cheese.	Beigmohammadi et al. (2016)
Ag	PE	High bactericide capacity against *Alicyclobacillus Acidoterrestris*.	Nobile et al. (2004)
Ag + TiO$_2$ + SiO$_2$ (40–60 nm)	LDPE, blown film	Packaging maintained the nutrient content and quality of mushrooms for 14 days due to reduced ethylene scavenging activity.	Donglu et al. (2016)
Ag + TiO$_2$ + Kaolin	PE	Slow ripening, firmer, less decay and browning in Chinese jujube.	Li et al. (2009_

are tabulated in Table 13.2.[56,57] Like TiO$_2$, ZnO show antibacterial performance when they come in contact with or penetrate the microbial cell. Typical examples of ZnO-based polymer nanocomposite films investigated for food packaging are summarized in Table 13.2.[46,58–62] Among many others, copper oxide (CuO) exhibits broad spectrum antimicrobial effects[63] due to the high redox potential of copper ions that disrupt the cellular components, which eventually results the cell death.[64] Nobile et al. have found that silver nanoparticles exhibit high bactericide capacity against *Alicyclobacillus acidoterrestris*.[65] Silver nanoparticles can get internalized within cells and then release silver ions (Ag$^+$) and subsequently disturb DNA replication and ATP production through the so-called Trojan horse mechanism.[66] Mixed filler systems like a combination of TiO$_2$, silver, and SiO$_2$ or nanoclay have proved to be more efficient in retarding the respiration rate and ethylene scavenging, as well as less decay and browning of fruits.[67,68]

Another emerging trend is incorporating doped nanoscale metal oxides in a polymer matrix to enhance the efficiency of the nanocomposites. Dopants are added at trace concentrations to alter the properties of a substance. For instance, in aluminum (Al)–doped ZnO-coated films, Al can replace Zn in the ZnO lattice or promote the formation of Al$_2$O$_3$ nanograins. As a result, the transparency, barrier, and antimicrobial properties of the films can be improved.[69] Similarly, in manganese (Mn)–doped ZnO[70] and Ta-doped ZnO,[71] resulting in enhanced antibacterial properties compared to pure ZnO. Figure 13.6 demonstrates the antibacterial activity of Ta-doped ZnO nanoparticles, and it has been reported that for *P. aeruginosa, E. coli*, and *S. aureus*, Ta-doped ZnO nanoparticles exhibit antimicrobial properties like ZnO without doping.[71] However, 1% Ta-doped ZnO nanoparticles exhibit slightly enhanced biocidal effect on *B. subtilis* probably due to its optimal photocatalytical activity. Emamifar et al. reported that the Ag doping onto ZnO can reduce the agglomeration of the ZnO nanoparticles while preparing nanocomposites.

13.4.3 Intelligent Packaging

According to Regulation (EC) No 1935/2004 of the European Parliament and the council on materials, intelligent food contact material is defined as the material or article that monitors the condition of packaged food or the environment surrounding the food.[18] As shown in Figure 13.3, intelligent packaging includes indicators, sensors, and data carriers. Indicators provide visual, qualitative, or semi-quantitative information about packed food through a change in color (or color intensity) or by diffusion of dye.[72] Typical

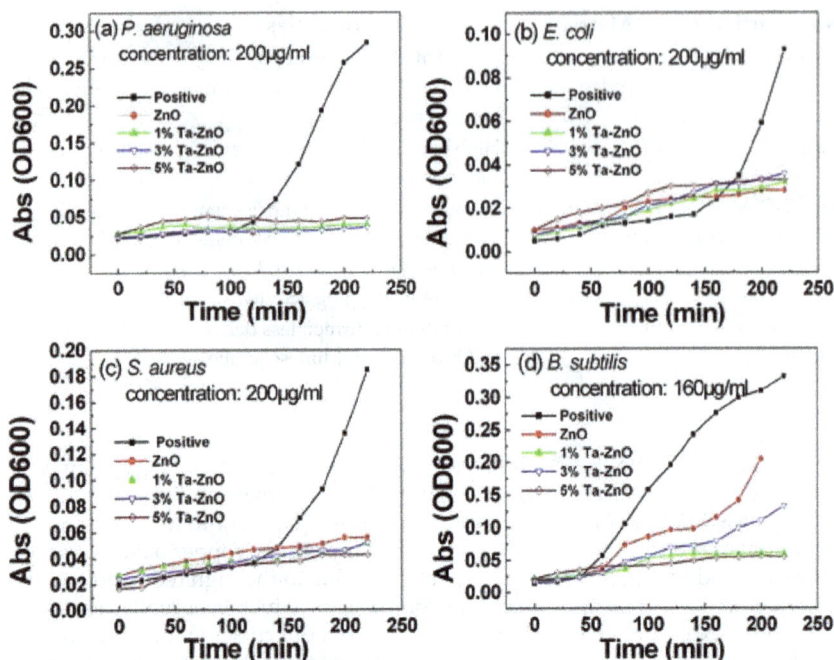

Figure 13.6 Growth curves of bacteria with time evolution in the presence of Ta-doped ZnO nanoparticles with different Ta contents under visible light: (a) *P. aeruginosa*, (b) *E. coli*, (c) *S. aureus*, and (d) *B. subtilis*.[71]

examples include time–temperature indicators (TTI), oxygen indicators, and freshness indicators.[73–75] Sensors can detect small molecules of pollutants, pathogens, and allergens in food.[76] Typical sensors used in food packaging is either a gas sensor (for detecting respiration products of nuts, metabolites of microorganism, gases in the headspace of the container) or a biosensor (for detecting pesticides in fruits and vegetables).[74,77] Data carriers are used for identification, automation, traceability, antitheft prevention or forgery protection across the supply chain. Typical examples of data carrier are radiofrequency identification (RFID) labels, bar codes, and quick response (QR) codes.[72,75]

13.4.3.1 TTI

Temperature is one key parameter that affects the shelf life of food and beverages. TTIs are simple, user-friendly, and inexpensive devices. There are three different types of TTIs available in the market:[78]

1. **Critical temperature indicator,** which shows if there is any temperature variation occurred than the permitted one

2. **Partial history indicator,** which indicates if there is any change in product quality due to the temperature alteration

3. **Full history indicator,** which records a temperature profile across the food supply chain

A broad range of materials are used in commercial TTIs; the functioning principle is based on the irreversible responses in the form of mechanical, chemical, electrochemical, molecular diffusion enzymatic, or microbiological changes when the temperature of the product changes than the permitted value. Schematic diagrams of different TTIs are presented in Figure 13.7.[79] CheckPoint® types M.L (enzymatic, color changes from green to orange to red), Fresh-Check® (polymeric, color changes from colorless to blue), On Vu™ (photochemical, color changes from dark blue to colorless), 3M MonitorMark™ (molecular diffusion, color changes from colorless to blue), Keep-it® (chemical, blue bar moves to the left), and time strip (diffusion bade, dye melts and migrates through the porous membrane of the indicator) are commercially available TTIs, having the price ranges between US$0.01 to US$0.15.[78]

Metal nanoparticles are also good candidates for TTI, since they exhibit irreversible color change when exposed to a particular temperature and time because of the change in their size, shape, and the surface morphology.[80] Lim et al. have developed a gelatin/gold nanoparticle–based thermal history indicator for low storage temperature food. This indicator exhibits a clear color signal after 6 h of exposure at 30°C.[81] The intensity of the color signal is proportional to the duration of exposure. Moreover, the color intensity of the gold nanoparticles (AuNPs) was maximal at a gelatin concentration of 2%. Wang et al. have developed a thermal history indicator using the localized surface plasmon resonance of gold nanoparticles synthesized in situ in alginate. Alginate facilitates the formation of solid hydrogel by adding divalent calcium ions.[82] With an increase in temperature and/or time, the viscosity of the hydrogel decreases, resulting in settling and agglomeration of the gold nanoparticles. As a result, a distinct change in color from gray to red can be observed at a high temperature (40°C). Newly developed colorimetric TTIs are made of plasticized polydiacetylene (PDA)/SiO_2 films. The film changes from purple to red within 4 days when stored below 35°C.[83] The monomer of PDA is pentacosadiynoic acid (PCDA), which comprises a hydrophilic carboxylic group and a hydrophobic long-chain hydrocarbon. The polymeric bilayers of PCDA show a reversible color transition when anchored onto the ZnO nanoparticles and an irreversible transition when PCDA forms vesicles.[84] Saenjaiban et al. have prepared a TTI based on PDA modified–silver nanoparticles (AgNP) and glycerol embedded in carboxymethyl cellulose.[85] The developed films change color from purplish-blue to purple and purple to reddish-purple over time at 35°C. AgNP improves the sensitivity and the temperature-induced color change of PDA. The good thermal conductivity of AgNP can increase the surface area of PDA by breaking up its bulky structure. PDA surface exposed to specific temperature is responsible for the observed color change. Drawbacks of TTIs are cost increase, potential undesirable migration of chemical components, and, hence, the safety of using TTIs, questionable accuracy, and reliability under uncontrolled conditions (e.g., impact, compression, and vibration), and legislative restrictions in Europe.[80] Therefore, future developments should be directed toward stable, nontoxic, and even edible material, and sensitivity of indicators, to real time–temperature history.

13.4.3.2 Freshness Indicators

Freshness indicators provide information on quality of the packaged food by detecting any biochemical changes or growth of microorganisms.[86] Figure 13.8 demonstrates a schematic diagram of the working principle of the freshness indicator.[87] The color change of such indicators occurs due to the release of some characteristic volatile from food. Certain metabolites such as carbon dioxide, ethanol, organic acids, glucose, ATP degradation products, volatile

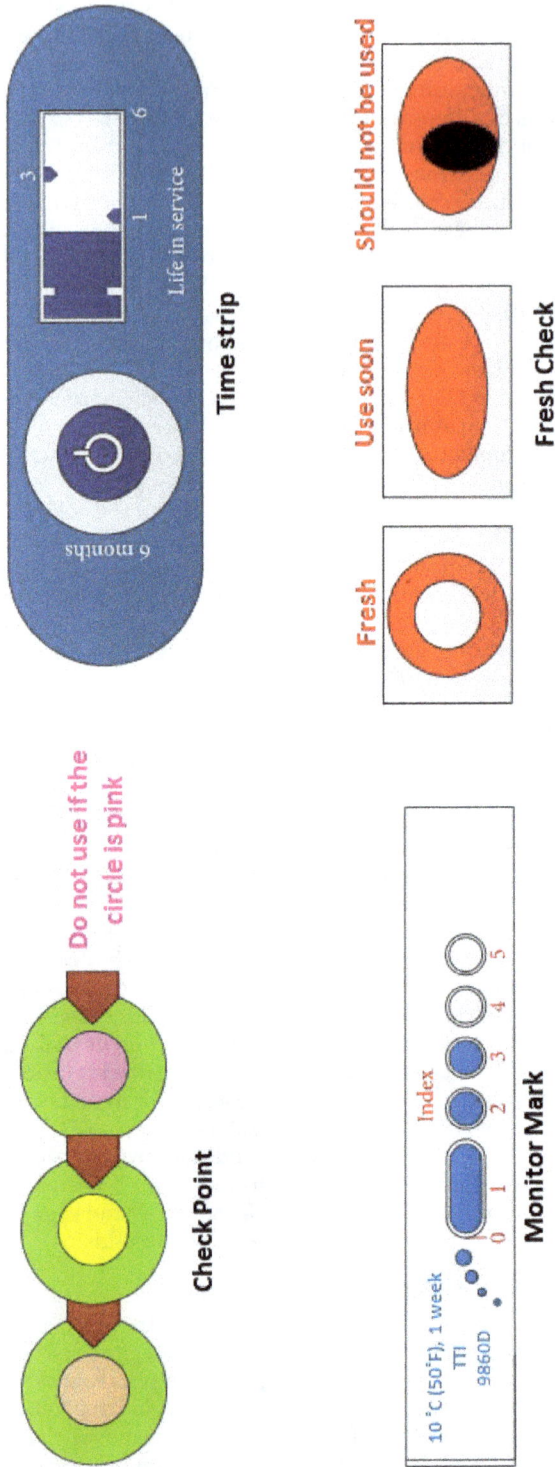

Figure 13.7 Schematic diagrams of different types of commercially available TTIs.[79]

nitrogen compounds, biogenic amines, and sulfuric compounds are considered quality indicators.[88] The reaction between these metabolites and indicators in the packaging system renders visual color change, indicating the quality of the products. Most commercial freshness sensor detecting spoilage are based on either time–temperature or pH sensing and uses pH indicator dyes. For instance, a pH indicator based on methyl red is used to detect the aldehyde emission as a marker for the ripening of the fruit. A label sensor, used as a sensor label for real-time monitoring of fresh meat of broiler chicken, is based on methyl red; a red/methyl cellulose membrane responds to the increase in pH due to volatile amines decomposition.[79] The indicator SensorQ™ developed by DSM NV can monitor the freshness of the fish.[79] RipeSense™ is the first intelligent sensor label that changes color to indicate the ripeness of the fruit as the aroma is released by the fruit.[79] Initial red color changes gradually to orange and finally to yellow, depending on the desired level of maturity for eating the fruit. Main component of ripeness indicator is pH indicator dyes and electrical circuits (e-noses sensor). In the case of MAP packaging, it is important to detect any leakage. Ageless eye, an oxygen-absorbing sachet, is an example of the commercial oxygen-sensitive MAP indicator.[89] Carbon dioxide–sensitive detection system is bit more complicated since its reliability is questionable as carbon dioxide gets dissolved or absorbed in the initial phase. Either a plastic optical fluorescent film or infrared technology is used in leak detection.

The ripening process of some fruits, such as tomatoes, bananas, and apples, is accelerated by the release of ethylene, a plant hormone.[89] Ethylene is released by fruits to trigger respiration so that energy is produced for internal biochemical processes resulting in changes in the flavor, texture, and nutrition of fruit. Scavenging and monitoring of ethylene thus provide freshness for a longer time. Commercially, potassium permanganate is widely used as an ethylene scavenger that oxidizes ethylene to ethylene glycol, which is further oxidized to manganese dioxide, carbon dioxide, and water. Low-temperature oxidation over a platinum catalyst on mesoporous silica can remove 50 ppm of ethylene at 0°C.[90] Ethylene sensors are fabricated using chemoresistance and electrochemical

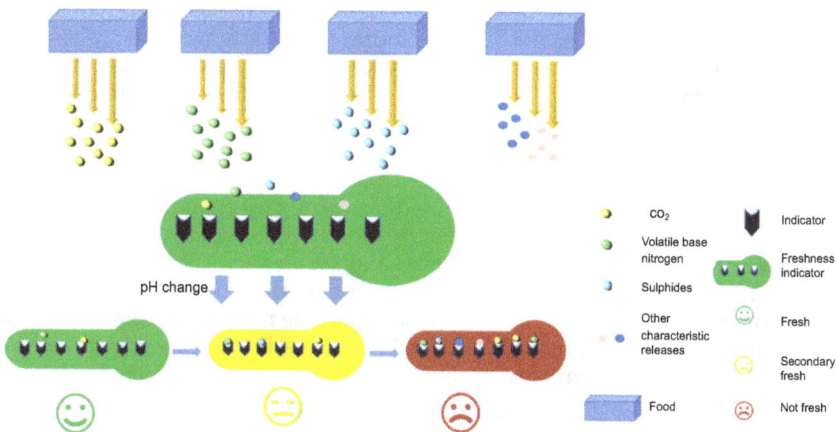

Figure 13.8 A schematic diagram of the working principle of the freshness indicator.[86]

methods. Eser et al. fabricated a chemo-resistive sensor by positioning a mixture of single-walled carbon nanotubes and copper (I) complex between gold electrodes,[91] and the resistance of the developed sensor changes due to the binding to ethylene. Agarwal et al. fabricated a tin-oxide nanoparticle–based capacitive sensor for ethylene detection by dip coating technique on a flexible polyimide substrate (Pyralux FR copper-clad laminate, DuPont).[92] Tin oxide nanoparticles are n-type material with electrons (e^-) as the majority carriers. When oxygen is absorbed, a depletion layer (depleted of e^-) consisting of O^- ionic species is formed in the tin oxide nanoparticles. As more oxygen is absorbed, the depletion region increases, and the thickness of the conduction region reduces. In presence of ethylene, the depleted electrons are replenished back to the tin oxide surface. As a result, the conduction region increases, and depletion region decreases.

Biosensors are devices capable of detecting pathogenic microbes via interaction with biological materials, for example, antibodies, enzymes, and genomes. It can be classified into three categories, namely, optical, electrochemical, and piezoelectric. Gold nanoparticles, graphene, carbon nanotubes, and photonic crystals are promising candidates for biosensors.[93] Biosensors are prepared using poly(dopamine)-modified magnetic nanoparticles coated with fourth-generation ethylenediamine and core polyamidoamine G-4 dendrimers and further decorated with platinum nanoparticles.[94] Yazdanparast et al. prepared poly(l-aspartic acid)/multiwalled carbon nanotube (MWCNT) film for the immobilization of xanthine oxidase on a glassy carbon electrode and applied this biosensor to determine the xanthine in fish at different storage time.[95] It is important to note that the different methodologies for synthesizing graphene and its derivatives can impart different properties and functionalities to graphene-based nanomaterials, which can affect the selectivity and accuracy of graphene-derived biosensors.[93]

All the sensing mechanisms have advantages and drawbacks.[89] For instance, pH sensors are cost-effectiveness, high sensitivity, and ease of use, but the main disadvantage is the chemical migration to food. Ripeness indicators are accurate and can be used in bulk packaging but are expensive. Leak detectors can be integrated into the packages, but it suffers problem with carbon dioxide absorption, low sensitivity, and spectral interference when an infrared technique is used. Gas detection sensors might be used in headspace; however, there is a possibility of the migration to food, and the sensitivity depends on the transducer and the quality of the antibody. Despite their significant potential, nanomaterial-based biosensors are still in their early stages. Ethylene sensors, based on electrical circuits, exhibit low sensitivity when exposed to temperature and humidity.

13.4.3.3 Data Carriers

Three main types of data carriers are electronic nose, electronic tongue, and RIFD tags. An electronic nose is composed of an array of sensors, an analogue-to-digital converter, and a computer equipped with recognition algorithm.[77] The difference between the electronic nose and the electronic tongue is that electronic nose interacts with the volatile compounds present in the container headspace while the electronic tongue reacts with the nonvolatile compounds dispersed in a liquid.[77] Although these devices are designed for smart packaging, these are expensive and complicated and require integration with the real packaging. In contrary, RFID tags use electromagnetic fields of radiofrequency to store and communicate product information in real time for their identification and traceability.[77] RFID tags consist of an integrated circuit connected to an antenna for the transmission of information stored on the chip to a reader. RIFD tags can be active (on-chip batteries for power) or passive (no internal power source).[89] The latest RFID tags allow the integration of other functions such as TTI or biosensors to monitor and communicate the thermal history of the product, as well as to provide information about product quality.[77]

13.5 PRODUCT-BASED PACKAGING REQUIREMENTS

Brand owners and the beverage packaging industries are seeking sustainable and cost-effective packaging solutions. They are focusing on the structural modification of packaging materials and the expansion of new active and/or intelligent systems. An improvement in packaging is expected to lengthen the shelf lives of beverages, such as milk, juice, wine, or beer. Oxidation poses a serious dilemma for producers of premium bag-in-box wines.[96] The oxidation of wine influences the organoleptic properties and color of the wine. Wine producers introduce sulfur dioxide (SO_2) in the form of sulfite in the must to prevent the oxidation of wine and, as a result, enhance its shelf life.[97] The main functions of SO_2 are to inhibit or kill yeast and unwanted bacteria and protect wine from oxidation. Sulfite attaches to sugar, acetaldehyde, and phenolic compounds like anthocyanin. Hence, the amount of free SO_2 may decrease before the expected shelf life, and the increase of SO_2 is not an option, as it will have a negative effect on its taste and color. Strawberry is one of the most favored fruits all over the world, is vulnerable to microbial contamination, and has a short shelf life, which is a major challenge to industrial production, distribution, and storage. Recent study has shown that an LDPE film containing nano-silver particles can improve the storage life and maintain the quality of fruit.[98] An edible coating on the cheese applied by a dipping method is commonly used for its longer shelf life. However, the challenges are their effect on the color and flavor of cheese, the difficulty of obtaining a uniform coating, and the price of the coating material. Nanocoating has been proposed as a technique for applying the edible coating, provided the safety of applying it on the food is evaluated carefully.[99]

13.6 SAFETY REGULATIONS ON NANOPARTICLE-CONTAINING FOOD PACKAGING

Despite of several advantages of nanoparticle/composites, public concern regarding the toxicity and the environmental effect of nanoparticles is quite substantial. The main concern stems from the lack of knowledge on the impact of nanosized materials on the consumer's health and the environment.[100] According to the relevant EU regulation, the overall migration limit from a plastic-based food packaging constituting foodstuffs is 10 mg dm^{-2} of the surface area of the packaging material.[101] However, in the case of 0.5- to 10-L food containers where the surface area in contact with foodstuffs cannot be determined accurately, or for the surface area of stoppers, caps, gaskets, or similar devices for sealing, the above limit can be extended to 60 mg of the constitutes released per kilogram of foodstuffs. The acute oral toxicity of bentonite in humans is very low. However, long-term occupational exposure to bentonite (e.g., for workers in mining or processing) can cause lung diseases.[102,103] Similarly, a very high dose of nano TiO_2 (10 mg m^{-3}) can cause lung tumors.[104] Echegoyen and Nerín reported that total silver migration varied between 1.66 and 31.46 ng/cm^2 from the commercial nano-silver-containing plastic food containers.[105] High doses of nano–zinc oxide (5000 mg.kg^{-1}) can affect zinc metabolism and biodistribution in mice.[106] Metal oxide nanoparticles (e.g., nanoTiO$_2$, nanosilica) can induce a cell membrane damage. It is apparent that high dose of exposure to nanoparticles imposes severe health risk. It will be beneficial to investigate product-based migration not only due the high dosage of exposure but also because of the effect of nanoparticles as a function of the frequency of exposure.

The most important regulations are to obtain food contact approval and environmental regulation. For end-of-life management of food and waste, industries in the United States follow the U.S. Environmental Protection Agency's Waste Reduction Model (WARM; version 13).[107] Industries in the European Union are required to make sure the environmental influence of their packaging operations accord with the Packaging and Packaging Waste Directive.[108] In the way forward, consumers should be more and better informed about the advantages and disadvantages of using nanotechnology in the food industry.

13.7 FUTURE PROSPECTS, HURDLES FOR COMMERCIAL TRANSLATION, AND A HOLISTIC APPROACH FOR MARKET IMPLEMENTATION

A brief statistical analysis of the different nanomaterials, their unique properties, and the different application areas of nano-related technologies mentioned based on the articles published within 2014–2018, conducted by Liu et al., gives readers an overview of the development of nano-related technologies in the shelf-life extension of fruits and vegetables.[25] A statistical analysis of different applications is presented in Figure 13.9. The figure shows that significant research has been conducted on microbial growth during fruit and vegetable storage followed by regulating the gas barrier properties.

A combination of nanotechnology with other fresh-keeping treatments has also been explored to achieve a synergistic effect in extending the shelf lives of fruits and vegetables. Fumigating *Pleurotus eryngii* mushroom with 0.3 μl L^{-1} 1-methylcyclopropene for 24 h and then packing it with nano-Ag PE bag significantly inhibits the activity of polyphenol oxidase and increase the activities of superoxide dismutase and catalase, which resulted in delayed ripening and prolonged shelf life of the mushroom.[109] The limitation of this method is a long fumigation time. Another important aspect is sustainable packaging. Sarfraz et al. have reported that the introduction of calcium carbonate and nanoclay in PE and silver in PP do not affect greatly on the mechanical properties, haze, pinhole, and smell after the first recycling step, whereas the incorporation of silver, zinc oxide, and nanoclay can have detrimental effect on the properties of the recycled material.[110] A systematic assessment on the recyclability of the nano-reinforced polymeric packaging material is required.

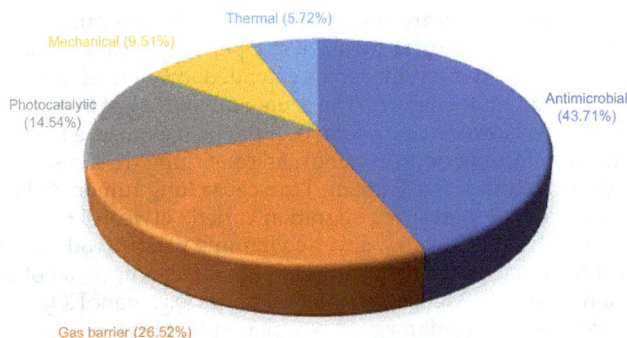

Thermal (5.72%)
Mechanical (9.51%)
Photocatalytic (14.54%)
Antimicrobial (43.71%)
Gas barrier (26.52%)

Figure 13.9 Statistical analysis of the different applications of nano-enabled packaging research between 2014 and 2018.[25]

The demand for innovative food packaging is evolving continuously. The first barrier to commercialization is the gap between science and industry.[111, 112] Academic research stops at the proof-of-concept stage. Researchers should consider the modifications required for the successive technology progression and manufacturing ability guided by the industry stakeholders. A second significant hurdle is the scale-up to mass production. A demonstration of pilot-scale production followed by testing the manufacturing ability and the validation of the value proposition by strategic collaborator can reduce the risk of adoption of the new technology. Requirement for the additional processing or equipment will increase the overall cost and will restrict the adoption of disruptive technologies by smaller companies. Finally, multiple roadblocks within the supply chain must be bridged for the successful translation of the emerging technologies to the commercial scale. Conflicting interests within the value chain and the supply chain can easily create boundaries to halt collaboration.

13.8 CONCLUSION AND PERSPECTIVES

There is a rising consumer demand for high-quality, minimally processed food preferably without synthetic preservatives while having longer shelf life. Packaging plays an important role in preserving food quality and ensuring food safety throughout the product lifetime. Efficiency, convenience, traceability, anticounterfeiting, and sustainability are the key drivers for innovative technologies suitable for food packaging. There is a knowledge gap on the correlation of FLP and shelf life, and therefore, more research is required in this direction. A variety of nanoparticles, such as the range of metal oxide nanoparticles, nanoclay, and others, are incorporated in the polymer matrix to achieve multifunctionalities, such as gas barrier, UV blocking, antimicrobial, and ethylene scavenging properties. The impermeable inorganic phase provides a physical barrier to diffusing molecules, creates tortuous paths, and, hence, reduces the gas permeability when incorporated into the polymer matrix. Active packaging exhibits antimicrobial activity and the use of mixed filler or doped nanofiller systems are emerging at a rapid pace. In addition to the improved or active functionality, it is important to monitor the condition of packaged food or the environment surrounding the food across the supply chain. As a result, several technologies have been developed on the indicators, sensors, and data carriers. Data carriers like RFID tags with indicators or sensors enables monitoring and communicating the thermal history of the product, as well as providing information about quality. Nanotechnology plays a key role in designing such innovative packaging materials, but it can impose health risks if people are exposed to high levels. The key barrier to commercialization is the gap between science and industry, consumer perception of pollution from plastic packaging, and the safety of using nanoparticles/nanomaterials. Nanotechnology-based innovative packaging can prove to be a blessing to modern society when developed sustainable packaging contemplates safety regulations. Consumers should be better informed about the advantages and disadvantages of using nanotechnology in the food industry. Moreover, multiple roadblocks within the supply chain must be overcome for the successful translation of the emerging technologies to the commercial scale.

REFERENCES

1. B. Schumann, M. Schmid. Packaging concepts for fresh and processed meat – Recent progress. Innovative Food Sci Emerg Technol, 47 (2018) pp. 88–100.

2. J. Gustavsson, C Cederberg, U. Sonesson, A. Emanuelsson. The methodology of the FAO study: "Global food losses and food waste – extent, causes and prevention" – FAO, 2011. SIK – The Swedish Institute for Food and Biotechnology report No. 857 (2013) pp. 1–70.

3. Nasrullah, M. Aslam, S. Yousaf, G. Atiq, R.K Iqbal. Conventional and modern method of preservation of foods. Food Sci Nutr Technol, 4 (2019) Article No. 00193.

4. Food loss and food waste. Food and Agriculture Organization of the United Nations. www.fao.org/foodloss-and-food-waste/en/.

5. B. Lipinski, C. Hanson, J. Lomax, L. Kitinoja, R. Waite, T. Searchinger. Reducing food loss and waste. Working Paper, Instalment 2 of "Creating a Sustainable Food Future". World Resources Institute (2013) pp. 1–40.

6. A roadmap to reduce U.S. food waste by 20 percent. ReFED Report (2016) pp. 1–96.

7. L. Brennan, S. Langley, K. Verghese, S. Lockrey, M. Ryder, C. Francis, N.T. Phan-Lee, A. Hill. The role of packaging in fighting food waste: A systematised review of consumer perception of packaging. J Cleaner Prod, 281 (2021) Article No. 125276.

8. A. Conte, G. Cappelletti, G. Nicoletti, C. Russo, M.A. Del Nobile. Environmental implications of food loss probability in packaging design. Food Res Int, 78 (2015) pp. 11–17.

9. F. Wikström, K. Verghese, R. Auras, A. Olsson, H. Williams, R. Wever, K. Grönman, M.K. Pettersen, H. Møller, R. Soukka. Packaging strategies that save food: A research agenda for 2030. J Ind Ecol, 23 (2018) pp. 532–540.

10. INCPEN, WRAP. Key Findings Report: UK Survey 2019 on Citizens' Attitudes & Behaviours Relating to Food Waste, Packaging and Plastic Packaging. WRAP, Banbury, UK (2019).

11. E. Crossin, K. Verghese, S. Lockrey. Review of emerging packaging technologies and trends for red meat. Meat and Livestock Australia Limited, (Australia, Sydney) (2015)

12. A. Barska, J. Wyrwa. Consumer perception of active and intelligent food packaging. Problems of Agricultural Economics, 4 (2016) pp. 138–159.

13. C.E. Realini, B. Marcos. Active and intelligent packaging systems for a modern society. Meat Sci, 98 (2014) pp. 404–419.

14. F. Wikström, H. Williams, K. Verghese, S. Clune. The influence of packaging attributes on consumer behaviour in food-packaging life cycle assessment studies – a neglected topic J Clean Prod, 73 (2014) pp. 100–108.

15. M.D. Wilson, R.A. Stanley, A. Eyles. Innovative process and technologies for modified atmosphere packaging of fresh and fresh-cut fruits and vegetables. Crit Rev Food Sci Nutr, 59 (2019) pp. 411–422.

16. E. Finnegan, P.V. Mahajan, M. O'Connell, G.A. Francis, D. O'Bierne. Modelling respiration in fresh-cut pineapple and prediction of gas permeability needs for optimal modified atmosphere packaging. Postharvest Biol Technol, 79 (2013) pp. 47–53.

17. A.L. Amaro, J.C. Beaulieu, C.C. Grimm, R.E. Stein, D.P.F. Almeida. Effect of oxygen on aroma volatiles and quality of fresh-cut cantaloupe and honeydew melons. Food Chem, 130 (2012) pp. 49–57.

18. J.A. Tudela, A. Marin, Y. Garrido, M. Cantwell, S. Medina-Martinez, M.I Gil. Off-odour development in modified atmosphere packaged baby spinach is an unresolved problem. Postharvest Biol Technol, 75 (2013) pp. 75–85.

19. A.A. Kader, D. Zagory, E.I Kerbel, C.Y. Wang. Modified atmosphere packaging of fruits and vegetables. Crit Rev Food Sci Nutr, 28 (1989) pp. 1–30.

20. A. Mistriotis, D. Briassoulis, A. Giannoulis, S. D'Aouino. Design of biodegradable bio-based equilibrium modified atmosphere packaging (EMAP) for fresh fruits and vegetables by using micro-perforated poly-lactic acid (PLA) films. Postharvest Biol Technol, 111 (2016) pp. 380–389.

21. BreatheWay is the better way. www.breatheway.com/how-it-works. Downloaded on October 13, 2021.

22. E. Crossin, K. Verghese, S. Lockrey Review of emerging packaging technologies and trends for red meat. Meat and Livestock Australia Limited, (Australia, Sydney) (2015)

23. E. Poyatos-Racionero, J.V. Ros-Lis, J.L. Vivancos, R. Martnez-Mez. Recent advances on intelligent packaging as tools to reduce food waste. J Clean Prod, 172 (2018) pp. 3398–3409.

24. L. Ma, M. Zhang, B. Bhandari, Z. Gao. Recent developments in novel shelf life extension technologies of fresh-cut fruits and vegetables. Trends Food Sci Technol, 64 (2017) pp. 23–38.

25. W. Liu, M. Zhang, B. Bhandari. Nanotechnology – A shelf life extension strategy for fruits and vegetables. Critic Rev Food Sci Technol, 60 (2020) pp. 1706–1721.

26. M.A. Emamhadi, M. Sarafraz, M. Akbari, V.N. Thai, Y. Fakhri, N.T.T. Linh, A.M. Khaneghah. Nanomaterials for food packaging applications: A systematic review. Food Chem Toxicology, 146 (2020) Article No. 111825.

27. L.E. Nielsen. Models for the permeability of filled polymer systems. J Macromol Sci, Part A 1 (1967) pp. 929–942.

28. G.H. Fredrickson, J. Bicerano. Barrier properties of oriented disk composites. J Chem Phys, 110 (1999) pp. 2181–2188.

29. S. Park, Y. Jeon, T. Han, S. Kim, Y. Gwon, J Kim. Nanoscale manufacturing as an enabling strategy for the design of smart food packaging systems. Food Pack Shelf life, 26 (2020) Article No. 100570.

30. N. Bumbudsanpharoke, S. Ko. Nano-food packaging: An overview of market, migration research, and safety regulations. J Food Sci, 80 (2015) pp. 910–923.

31. M.F. Horst, L.M. Quinzani, M.D. Failla. Rheological and barrier properties of nanocomposites of HDPE and exfoliated montmorillonite. Thermoplast Compos Mater 27 (2014) pp. 106–125.

32. B. Tajeddin, N. Ramedani, H.J.P.J. Mirzaei. Preparation and characterization of a bionanopolymer film for walnut packaging. Polyolefins J, 6 (2019) pp. 159–167.

33. C. Vasile. Polymeric nanocomposites and nanocoatings for food packaging: A Review. Materials, 11 (2018) Article No. 1834.

34. E. Kny. Polymer nanocomposite materials used for food packaging. In Ecosustainble Polymer Nanomaterials for Food Packaging. Innovative Solutions, Characterisation Needs, Safety and Environmental Issues; Silvestre, C., Cimmino, S., Eds.; CRC Press Taylor & Francis Group: Boca Raton, FL, USA, 2013; Chapter 13; pp. 337–375, ISBN 9781138034266.

35. D. Li, Q. Ye, L. Jiang, Z. Luo. Effects of nano-TiO2-LDPE packaging on postharvest quality and antioxidant capacity of strawberry (*Fragaria ananassa* Duch.) stored at refrigeration temperature. J Sci Food Agri, 97 (2017) pp. 1116–1123.

36. A. Nasiri., M.S.-N. Shariaty-Niasar, Z. Akbari. Synthesis of LDPE/ nano TiO2 nanocomposite for packaging applications. Int J Nanosci Nanotechnol, 8 (2012) pp. 165–170.

37. Y. Liu, Y. Liu, S. Wei. Processing technologies of EVOH/nano-SiO2 high-barrier packaging composites. Proceedings of the 17th IAPRI world conference on packaging. (2010) pp. 269–274.

38. G. Bang, S.W. Kim. Biodegradable poly(lactic acid)-based hybrid coating materials for food packaging films with gas barrier properties. J Ind Eng Chem, 18 (2012) pp. 1063–1068.
39. D. Li, J. Zhang, W. Xu, Y. Fu, Y. Effect of SiO2/EVA on the mechanical properties, permeability, and residual solvent of polypropylene packaging films. Polym Compo, 37 (2016) pp. 101–107.
40. Z. Luo, Y. Xu, Q. Ye. Effect of nano-SiO2-LDPE packaging on biochemical, sensory, and microbiological quality of Pacific white shrimp *Penaeus vannamei* during chilled storage. Fisheries Sci, 81 (2015) pp. 983–993.
41. T. Hirvikorpi, M. Vähä-Nissi, A. HarlinM. Salomäki, S. Areva, J.T. Korhonen. Enhanced water vapor barrier properties for biopolymer films by polyelectrolyte multilayer and atomic layer deposited Al_2O_3 double-coating. Appl Surf Sci, 257 (2011) pp. 9451–9454.
42. C.F. Struller, P.J. Kelly, N.J. Copeland. Aluminum oxide barrier coatings on polymer films for food packaging applications. Surf Coat Technol, 241 (2014) pp. 130–137.
43. A. Marra, C. Silvestre, D. Duraccio, S. Cimmino. Polylactic acid/zinc oxide biocomposite films for food packaging application. Int J Biol Macromol, 88 (2016) pp. 254–262.
44. R. Venkatesan, N. Rajeswari. ZnO/PBAT nanocomposite films: Investigation on the mechanical and biological activity for food packaging. Polym Adv Technol, 28 (2017) pp. 20–27.
45. S. Polat, H. Fenercioğlu, M. Güçlü. Effects of metal nanoparticles on the physical and migration properties of low density polyethylene films. J Food Eng, 229 (2018) pp. 32–42.
46. R. Pantani, G. Gorrasi, G. Vigliotta, M. Murariu, P. Dubois. PLA-ZnO nanocomposite films: Water vapor barrier properties and specific end-use characteristics Eur Polym J, 49 (2013) pp. 3471–3482.
47. Regulation (EC) No 1935/2004 of the European Parliament and the council on materials and articles intended to come into contact with food and repealing Directives 80/590/EEC and 89/109/EEC, Official Journal of the European Union, L338 (2004) pp. 4–17.
48. T.V.Duncan. Applications of nanotechnology in food packaging and food safety: Barrier materials, antimicrobials and sensors. J Colloid Interface Sci, 363 (2011) pp. 1–24.
49. P. Kanmani, J.-W. Rhim. Nano and nanocomposite antimicrobial materials for food packaging applications. Future Med, 6 (2014) pp. 15–23.
50. G. Carré, E. Hamon, S. Ennahar, M. Estner, M.C. Lett, P. Horvatovich. TiO2 photocatalysis damages lipids and proteins in *Escherichia coli*. Applied and Environmental Microbiology, 80 (2014) pp. 2573–2581.
51. A. Llorens, E. Lloret, P.A. Picouet, R. Trbojevich, A. Fernandez. Metallicbased micro and nanocomposites in food contact materials and active food packaging. Trends Food Sci Technol, 24 (2012) pp. 19–29.
52. M.L. Cerrada, C. Serrano, M. Sánchez-Chaves, M. Fernández-García, F. Fernández-Martín, A. de Andrés, A. Self-sterilized EVOH-TiO2 nanocomposites: Interface effects on biocidal properties. Adv Func Mater, 18 (2008) pp. 1949–1960.
53. H Bodaghi, Y. Mostofi, A. Oromiehie, Z. Zamani, B. Ghanbarzadeh, C. Costa. Evaluation of the photocatalytic antimicrobial effects of a TiO2 nanocomposite food packaging film by in vitro and in vivo tests. LWT – Food Sci Technol, 50 (2013) pp. 702–706.

54. M. Gumiero, D. Peressini, A. Pizzariello, A. Sensidoni, L. Iacumin, G. Comi, G. Effect of TiO2 photocatalytic activity in a HDPE-based food packaging on the structural and microbiological stability of a short-ripened cheese. Food Chem, 138 (2013) pp. 1633–1640.

55. Z. Luo, Y. Qin, Q. Ye. Effect of nano-TiO2-LDPE packaging on microbiological and physicochemical quality of Pacific white shrimp during chilled storage. Int J Food Sci Technol, 50, (2015) pp. 1567–1573.

56. K. Wang, P. Jin, H. Shang, H. Li, F. Xu, Q. Hu, Q. A combination of hot air treatment and nano-packing reduces fruit decay and maintains quality in postharvest Chinese bayberries. J Sci Food Agri, 90 (2010) pp. 2427–2432.

57. W. Li, L. Li, H. Zhang, M. Yuan, Y. Qin. Evaluation of PLA nanocomposite films on physicochemical and microbiological properties of refrigerated cottage cheese. J Food Process Preservation, 42 (2018) Article No. e13362.

58. X. Li, W. Li, Y. Xing, Y. Jiang, Y. Ding, P. Zhang. Effects of nano-ZnO power-coated PVC film on the physiological properties and microbiological changes of fresh-cut" Fuji" apple. Adv Mater Res, 152 (2011) pp. 450–453.

59. C. Silvestre, S. Cimmino, M. Pezzuto, A. Marra, V. Ambrogi, J. Dexpert-Ghys. Preparation and characterization of isotactic polypropylene/zinc oxide microcomposites with antibacterial activity. Polym J, 45 (2013) pp. 938–945.

60. A. Emamifar, M. Kadivar, M. Shahedi, S. Soleimanian-Zad. Evaluation of nanocomposite packaging containing Ag and ZnO on shelf life of fresh orange juice. Innovat Food Sci Emerg Technol, 11 (2010) pp. 742–748.

61. A. Emamifar, M. Kadivar, M. Shahedi, S. Soleimanian-Zad. Effect of nanocomposite packaging containing Ag and ZnO on inactivation of Lactobacillus plantarum in orange juice. Food Contr. 22 (2011) pp. 408–413.

62. S.C. Li, Y.N. Li, Mechanical and antibacterial properties of modified nano-ZnO/high-density polyethylene composite films with a low doped content of nano- ZnO. J Appl Polym Sci, 116 (2010) pp. 2965–2969.

63. F. Beigmohammadi, S.H. Peighambardoust, J. Hesari, S. Azadmard-Damirchi, S.J. Peighambardoust, N.K. Khosrowshahi. Antibacterial properties of LDPE nanocomposite films in packaging of UF cheese. LWT – Food Sci Technol, 65 (2016) pp. 106–111.

64. L. Nan, W. Yang, Y. Liu, H. Xu, Y. Li, M. Lu. Antibacterial mechanism of copper-bearing antibacterial stainless steel against E. coli. J Material Sci Technol, 24 (2008) pp. 197–201.

65. M.D. Nobile, M. Cannarsi, C. Altieri, M. Sinigaglia, P. Favia, G. Iacoviello, R. D'agostino. Effect of Ag-containing nano-composite active packaging system on survival of Alicyclobacillus acidoterrestris. J Food Sci, 69 (2004) pp. E379–E383.

66. L.K. Limbach, P. Wick, P. Manser, R.N. Grass, A. Bruinink, W.J.J.E.s. Stark. Exposure of engineered nanoparticles to human lung epithelial cells: Influence of chemical composition and catalytic activity on oxidative stress. Environ Sci Technol, 41 (2007) pp. 4158–4163.

67. F. Donglu, Y. Wenjian, B.M. Kimatu, A.M. Mariga, Z. Liyan, A. Xinxin. Effect of nanocomposite-based packaging on storage stability of mushrooms (Flammulina velutipes). Innovat Food Sci Emerg Technol, 33, (2016) pp. 489–497.

68. H. Li, F. Li, L. Wang, J. Sheng, Z. Xin, L. Zhao, H. Xiao, Y. Zheng, Q. Hu. Effect of nano-packing on preservation quality of Chinese jujube (Ziziphus jujuba Mill. var. inermis (Bunge) Rehd). Food Chem, 114 (2009) pp. 547–552.

69. D. Valerini, L. Tammaro, F. Di Benedetto, G. Vigliotta, L. Capodieci, R. Terzi. Aluminum-doped zinc oxide coatings on polylactic acid films for antimicrobial food packaging. Thin Solid Films, 645 (2018) pp. 187–192.

70. K. Rekha, M. Nirmala, M.G. NairA. Anukaliani, A. Structural, optical, photocatalytic and antibacterial activity of zinc oxide and manganese doped zinc oxide nanoparticles. Physica B: Condensed Matter, 405 (2010) pp. 3180–3185.

71. B.L. Guo, P. Han, L.C. Guo, Y.Q. Cao, A.D. Li, J.Z. Kong, J.-Z. The antibacterial activity of Ta-doped ZnO nanoparticles. Nanoscale Res Lett, 10 (2015) Article No. 336.

72. P.R. Salgado, L.D. Giorgio, Y.S. Musso, A.N. Mauri. Recent developments in smart food packaging focused on biobased and biodegradable polymers. Front Sustain Food Syst, 5 (2021) Article No. 630393.

73. S.A. Hogan, J.P. Kerry. Chapter 3: Smart packaging of meat and poultry products. Smart Packag. Technol. Fast Mov. Consum. Goods, John Wiley & Sons, Ltd. ISBN: 978-0-470-02802-5 (2008) pp. 33–54.

74. S.Y. Lee, S.J. Lee, D.S. Choi, S.J. Hur. Current topics in active and intelligent food packaging for preservation of fresh foods. J Sci Food Agric, 95 (2015) pp. 2799–2810.

75. M. Ghaani, C.A. Cozzolino, G. Castelli, S. Farris. An overview of the intelligent packaging technologies in the food sector. Trends Food Sci Technol, 51 (2016) pp. 1–11.

76. M. Ramos, A. Valdés, A.C. Mellinas, M.C. Garrigós. New trends in beverage packaging systems: A review. Beverages, 1 (2015) pp. 248–272.

77. K.B. Biji, C.N. Ravishankar, C.O. Mohan, T.K.S. Gopal. Smart packaging systems for food applications: A review. J Food Sci Technol, 52 (2015) pp. 6125–6135.

78. P Müller, M. Schmid. Intelligent packaging in the food sector: A brief overview. Foods, 8 (2019) pp. 1–12.

79. G. Fuertes, I Soto, R. Carrasco, M. Vargas, J. Sabattin, C. Lagos. Intelligent packaging systems: Sensors and nanosensors to monitor food quality and safety. J Sensors, 2016 (2016) Article No. 4046061.

80. J-W. Han, L. Ruiz-Garcia, J-P. Qian, X-T. Yang. Food packaging: A comprehensive review and future trends. Comp Rev Food Sci Food Safety, 17 (2018) pp. 860–877.

81. S. Lim, S. Gunasekaran, J.Y. Imm. Gelatin-templated gold nanoparticles as novel time-temperature indicator. Journal of Food Science, 77 (2012) pp. N45–N49.

82. Y.C. Wang, L. Lu, S. Gunasekaran. Biopolymer/gold nanoparticles composite plasmonic thermal history indicator to monitor quality and safety of perishable bioproducts. Biosensors and Bioelectronics, 92 (2017) pp. 109–116.

83. A. Nopwinyuwong, T. Kaisone, P. Hanthanon, C. Nandhivajrin, W. Boonsupthip, C. Pechyen, P. Suppakul. Effects of nanoparticle concentration and plasticizer type on colourimetric behavior of polydiacetylene/silica nanocomposite as time-temperature indicator. Energy Procedia, 56 (2014) pp. 423–430.

84. N. Traiphol, M. Rungruangviriya, R. Potai, R. Traiphol. Stable polydiacetylene/ZnO nanocomposites with two-steps reversible and irreversible thermochromism: The influence of strong surface anchoring. J Colloid Interface Sci, 356 (2011) pp. 481–489.

85. A. Saenjaiban, T. Singtisan, P. Suppakul, K. Jantanasakulwong, W. Punyodom, P. Rachtanapun. Novel colour change film as a time – temperature indicator using polydiacetylene/silver nanoparticles embedded in carboxymethyl cellulose. Polymers, 12 (2020) Article No. 2306.

86. B. Kuswandi, C. Maryska, A. Abdullah, L.Y. Heng. Real time on package freshness indicator for guavas packaging. J Food Meas Charact 7 (2013) pp. 29–39.
87. P. Shao, L. Liu, J. Yu, Y. Lin, H. Gao, H. Chen, P. Sun. An overview of intelligent freshness indicator packaging for food quality and safety monitoring. Trends Food Sci Technol, https://doi.org/10.1016/j.tifs.2021.10.012.
88. Z. Fang, Y. Zhao, R.D. Warner, S.K. Johnson. Active and intelligent packaging in meat industry. Trends Food Sci Technol, 61 (2017) pp. 60–71.
89. A.U. Alam, P. Rathi, H. Beshai, G.K. Sarabha, M.J. Deen. Fruit quality monitoring with smart packaging. Sensors, 21 (2021) Article No. 1509.
90. W. Weber, S. Luzi, M. Karlsson, M. Fussenegger. A novel hybrid dual-channel catalytic-biological sensor system for assessment of fruit quality. J Biotechnol 139 (2009) pp. 314–317.
91. B. Esser, J.M. Schnorr, T.M. Swager. Selective detection of ethylene gas using carbon nanotube-based devices: Utility in determination of fruit ripeness. Angew Chem Int Ed, 51 (2012) pp. 5752–5756.
92. M. Agarwal, M.D. Balachandran, S. Shrestha, K. Varahramyan. SnO2 nanoparticle-based passive capacitive sensor for ethylene detection. J Nanomater, 2012 (2012) Articl No. 145406.
93. F.T.T. Cavalcante, I.R. de A. Falcão, J.E. da S. Souza, T.G. Rocha, I.G. de Sousa, A.L.G. Cavalcante, A.L.B. de Oliveira, M.C.M. de Sousa, J.C.S. dos Santos. Designing of nanomaterials-based enzymatic biosensors: Synthesis, properties, and applications. Electrochem, 2 (2021) pp. 149–184.
94. B. Borisova, A. Sánchez, S. Jiménez-Falcao, M. Martín, P.Salazar, C. Parrado, J.M. Pingarrón, R. Villalonga. Reduced graphene oxide-carboxymethylcellulose layered with platinum nanoparticles/PAMAM dendrimer/magnetic nanoparticles hybrids. Application to the preparation of enzyme electrochemical biosensors. Sens. Actuators B Chem, 232 (2016) pp. 84–90.
95. S. Yazdanparast, A. Benvidi, S. Abbasi, M. Rezaeinasab. Enzyme-based ultrasensitive electrochemical biosensor using poly(Laspartic acid)/MWCNT bio-nanocomposite for xanthine detection: A meat freshness marker. Microchem J (2019) Article No. 149.
96. A. Pretorius. Bubble trouble: How premium is that box, really? www.wineland.co.za/bubble-trouble-how-premium-is-that-box-really/ (downloaded on October 16, 2021)
97. P. Shea, J-C Vidal, S. Vialis. The Measurement of Total Oxygen in Filled BIB Wine. Performance BIB Meetings in Bordeaux, France, 2010.
98. N.V. Motlagh, M. Alibadi, E. Rahmani, S. Ghorbanpour. The effect of nano-silver packaging on quality maintenance of fresh strawberry. Int Scholarly Scientific Innovat, 14 (2020) pp. 124–128.
99. M.J. Costa, L.C. Maciel, J.A. Teixeira, A.A. Vicente, M.A. Cerqueira. Use of edible films and coatings in cheese preservation: Opportunities and challenges. Foof Res Int, 107 (2018) pp. 84–92.
100. Q. Chaudhry, M. Scotter, J. Blackburn, B. Ross, A. Boxall, L. Castle, R. Aitken, R. Watkins. Applications and implications of nanotechnologies for the food sector. Food Additives Contaminants, 25 (2008) pp. 241–258.
101. Corrigendum to Commission Directive 90/128/EEC of 23 February 1990 relating to plastics materials and articles intended to come into contact with foodstuffs (Official Journal of the European Communities No. L 75 of 21 March 1990).
102. J. Bandyopadhyay, S Sinha Ray. Are nanoclay-containing polymer composites safe for food packaging applications? – An overview. J Appl Polym Sci, 136 (2019) Article No. 47214.

103. A. Bratovcic. Nanomaterials in food processing and packaging, its toxicity and food labeling. Acta Sci Nutri Health, 4 (2020) pp. 07–13.
104. C. Sharma, R. Dhiman, N. Rokana, H. Panwar. Nanotechnology: An untapped resource for food packaging. Front Microbiol, 8 (2017) Article No. 1735.
105. Y. Echegoyen, C. Nerín. Nanoparticle release from nano-silver antimicrobial food containers. Food Chem Toxicol, 62 (2013) pp. 16–22.
106. C. Wang, J. Lu, L. Zhou, J. Li, J. Xu, W. Li. Effects of long-term exposure to zinc oxide nanoparticles on development, zinc metabolism and biodistribution of minerals (Zn, Fe, Cu, Mn) in mice. PloS One, 11 (2016) Article No. e0164434.
107. US EPA (U.S. Environmental Protection Agency) (2015). Documentation for Greenhouse Gas Emission and Energy Factors Used in the Waste Reduction Model (WARM), version 13.https://archive.epa.gov/ epawaste/ conserve/tools/warm/pdfs/WARM_Documentation.pdf
108. European Commission (2018). Packaging and packaging waste. https:// single-market-economy.ec.europa.eu/single-market/european-standards/ harmonised-standards/packaging-and-packaging-waste_en.
109. F. Xu, Y. Liu, X. Shan, S. Wang. Evaluation of 1-methylcyclopropene (1-MCP) treatment combined with nano-packaging on quality of pleurotus eryngii. J Food Sci Technol 55 (2018) pp. 4424–4431.
110. J. Sarfraz, T. Gulin-Sarfraz, J. Nilsen-Nygaard, M.K. Pettersen. Nanocomposites for food packaging applications: An overview. Nanomaterials, 10 (2021) Articla No. 11.
111. B.G. Werner, J.L. Koontz, J.M. Goddard. Hurdles to commercial translation of next generation active food packaging technologies. Current Opinion food Sci, 16 (2017) pp. 40–48.
112. S. Tiekstra, A. Dopico-Parada, H. Koivula, J. Lahti, M. Buntinx. Holistic approach to a successful market implementation of active and intelligent food packaging. Foods, 10 (2021) Article No. 465.

14 Laws and Regulations for Emerging Food-Processing Technologies

Pradip Chakraborty, Avik Mukherjee, and Santosh Kumar

Corresponding author: *Pradip Chakraborty*
Email: *pradipchakraborty91@yahoo.com*

CONTENTS

14.1 INTRODUCTION

A lot of innovations have taken place all over the world, particularly in the area of food-processing technology. Food processing is now not limited to the conventional processes like freezing, canning, drying, smoking, and curing only. Due to the increasing demand for high-quality processed foods with fresh-like characteristics, nonthermal processing is being introduced. The retention of nutritive value and sensory quality of food is the main concern along with the elimination or reduction of microbes from the food products. Nonthermal technologies, such as pulsed light (PL) technology, irradiation, ultraviolet (UV) light radiation, ultrasound processing, and high-pressure processing (HPP), are some of novel technologies that are being used nowadays. Nonthermal technologies have some limitations, especially in terms of microbial safety, and thus, combination treatments with conventional methods have been found to be more effective (Jadhav et al., 2021). Since food products and technologies are monitored by the regulatory authorities of countries, these novel technologies require approval by the concerned regulatory authority. As per section 22(4) of the Food Safety and Standards Act, 2006 (India), *novel food* means an article of food for which standards have not been specified but it is not unsafe. Per approval of nonspecified food and food ingredients regulations of the Food Safety and Standards Authority of India (FSSAI), specific approval is also required for novel food or food ingredients or processed with the use of novel technology. The FSSAI has yet to specify these nonthermal technologies except irradiated food. Hence, specific approval is required from the FSSAI to use these nonthermal technologies except irradiation.

Therefore, innovative food products and new technologies can be commercialized only when regulatory approval is obtained from the competent authority. Among these nonthermal technologies, irradiated food regulations have been notified by the FSSAI. Per the Food Safety and Standards (Prohibition and

DOI: 10.1201/9781003147978-14

Restriction on Sales) Regulations, 2011, no person shall sell or expose for sale, or distribute, or offer for sale, or dispatch, or deliver to any person for the purpose of sale of any food product, which does not conform to the standards of quality as provided in the Food Safety and Standards Act, 2006 and rules and regulations made there under. While managing innovation and technology in the food industry, particularly in India, these things should be kept in mind. Otherwise, the entire investment in innovation and technology may be wasted. Novel technology shall be supported by enough safety studies to ensure the safety of the ingredient and the product. Some of the nonthermal technologies like irradiation, PL technology, HPP, and ultrasound treatment are gaining momentum as a result of continuous research in these fields (Jadhav et al., 2021; Oms-Oliu et al., 2008). In this chapter, regulatory requirements and safety aspects of these novel technologies are described.

14.2 FOOD-PROCESSING LAWS AND REGULATIONS

Regulatory approval for application of a novel, emerging food-processing technology, such as irradiation technology, HPP, PL technology, ultrasound processing, and others, depends on the concerned country and its food legislation(s) and the types of products. These technologies are at various stages of research and development, and although they have shown great promise as nonthermal, least invasive food-processing technologies, they still need substantial research and exploration before commercialization. Regulations on the applications of these technologies in food processing are yet to be finalized, and many developed countries like the US, certain EU countries, and even developing countries like India are still in the process of evaluating risk to public health, nutritional disadvantage, and sustainability of such applications in food processing. Also, the commercialization of these technologies by the food-processing/manufacturing industries depends primarily on relative advantage of the new technology, its ease of adoption versus its complexity compared to the existing technologies, and consumer perception of the technology (Priyadarshini et al., 2019). Food safety management tools, for example, Hazard Analysis Critical Control Point (HACCP) and Good Manufacturing Practices (GMPs), however, must be implemented during the application of any of these food-processing technologies (Featherstone, 2015). The regulatory aspect of commonly used novel technologies are discussed in the following sections.

14.2.1 Irradiation Technology

Food irradiation is a process in which food commodities are exposed to controlled doses of energy of ionizing radiation such as gamma rays to achieve certain technological objectives. The major applications of food irradiation include preventing or eliminating microbial contamination and sprouting inhibition of spices and cereals. Food irradiation results in killing of parasites, pathogens, and other microorganisms, which lead to an extended shelf life of food. India exports irradiated spices and dry ingredients to several countries. Indian mangos are treated with gamma radiation to control insects and pests. In fact, it has become mandatory to treat mangoes with gamma radiation before export to the US. Even fruits and vegetables are subjected to gamma radiation before being exported to New Zealand from Australia. Since the food is irradiated by gamma radiation, some consumers have confusion about the residual dose of gamma radiation. This is a misconception and irradiation has been proved to be a safe technology used for making safe food. In India, dose has been prescribed by Bhaba Atomic Research Centre, Trombay, Mumbai, Maharastra by taking the utmost care on consumer safety. Conventional methods such as freezing, heating, smoking, and chemical preservatives have some limitations, particularly nutritional loss and food intoxication due to the production of toxins by microbes. Irradiated food preserves its nutritional, physicochemical, and organoleptic properties.

Several safety studies have been conducted nationally and internationally to assess the safety of food processed by irradiation technology (Ehlermann, 2014). It has been found that food processed by irradiation technology presents no toxicological, nutritional, or microbiological problems. The Codex Alimentarius Commission (CAC), a joint body set up by the Food and Agricultural Association (FAO) and World Health Organization (WHO), prescribed a general standard for irradiated foods. Several scientific bodies and associations like the American Medical Association, the American Dietetic Association, and the Institute of Food Technologists have also endorsed the safety of irradiated foods. FSSAI identified regulations for irradiated food in its regulation no. 2.13 of Food Products and Food Additives regulations, 2011, which has been subsequently amended. As per the amended regulations, minimum and maximum doses of radiation have been specified for onions, spices, potatoes, rice, wheat, maida, atta, semolina (sooji or rawa), pulses, mango, raisins, ginger, garlic and shallots (small onions), figs and dried dates, and meat and meat products. The dose of food irradiation must not exceed the limits specified in the regulations. In the amended regulations, 2016, classes of food, the purpose of radiation, and dose limits have been specified for allied products, namely, food additives, nutraceuticals, dietary supplements, health foods, and packaging materials. Routine quantitative dosimetry records of food irradiation operation shall be kept as per Department of Atomic Energy (Control of Irradiation of Food) Rules, 1991.

Food irradiation facilities must approve and licensed under the Atomic Energy (Radiation processing of food and allied products) Rules, 2012 before it is used for the treatment of food. However, a separate license has to be obtained from the FSSAI following due procedure. The FSSAI has already notified a few facilities in India after consultation with the Bhaba Atomic Research Centre, Trombay, Mumbai, India. These FSSAI-notified facilities shall comply with the conditions for approval, operation, license, and process control prescribed under the Atomic Energy (Radiation processing of food and allied products) Rules, 2012. No food or irradiated food shall be allowed to leave the irradiation facility unless it has been irradiated in accordance with the provision of the Department of Atomic Energy (Radiation processing of food and allied products) Rules, 2012, and the competent authority shall issue a certificate of irradiation indicating the dose and the purpose of the irradiation. The dose limit and source of radiation at specific conditions prescribed for each category or type of food under the Atomic Energy (Radiation processing of food and allied products) Rules, 2012 must be followed. Irradiated foods shall not be re-irradiated without permission from the Licensing Authority, and food, which has been irradiation, shall be identified so that it is not re-irradiation. The irradiation shall be carried out by personnel having the minimum qualification and training as prescribed for the purpose under Atomic Energy (Radiation processing of food and allied products) Rules, 2012. Irradiated food shall be offered for sale only in prepackaged condition. Any treatment of food shall be recorded by an officer authorized by the competent authority as specified under the Department of Atomic Energy (Radiation processing of food and allied products) Rules, 2012.

The record for irradiated food shall contain the following information:

- Name of the article

- License no.

- Name, address and other details of licensee
- Purpose of irradiation
- Source of irradiation
- Date of irradiation
- Dose of irradiation
- Serial number of batch
- The nature, quality of food to be irradiated and the Batch number
- Quantity of food irradiated
- Physical appearance of article; before and after irradiation
- Type of packaging used during the irradiation treatment and for packing the irradiated food

The label of an irradiated food shall carry a written statement indicating the radiation treatment adjacent to the name of the food, and shall bear the "Radura" logo in green color and declaration, namely, the following:

- PROCESSED BY IRRADIATION
- Radura Logo
- Name of the Product
- Purpose of Radiation Processing
- Operating License No
- Batch Identification No (as provided by facility)
- Date of Processing

14.2.2 HPP

HPP is an innovative novel food preservation technology, and it is also called high hydrostatic pressure processing, ultra-HPP, Pascalization, or cold pasteurization. HPP applies high pressure to solid or liquid foods to improve safety, quality, and organoleptic properties. Thermal processing carries demerits like vitamin destruction or flavor changes. These constraints are overcome in HPP, and it is also environmentally friendly process and can be operated at room temperature by using electrical energy. It requires less processing time and produces no waste. It is carried out at room temperature using pressure between 400–600 MPa (Hogan et al., 2005). Pressurized foods were first commercialized in Japan in 1992. HPP has several advantages over conventional food preservation techniques. It uses a small amount of electrical energy for eliminating spoilage microorganisms at room temperature. It preserves the natural flavor of the product. The transmission of pressure in the food occurs uniformly irrespective of its geometry, chemical composition, and packaging material. The process assures health benefits to consumers, particularly those who prefer raw seafood. It acts instantaneously unlike conventional thermal processing and requires less time. The temperature of the product in the pressure chamber can rise by 3–6°C for every megapascal increase in pressure, depending on the composition of the food product (Hogan et al., 2005).

According to the FSSAI Regulations, food processed by using novel technology is also considered novel food. Manufacture or import of novel food cannot be done without the prior approval of the FSSAI. Approval is based on adequate scientific risk analysis studies, and scientific evidence about the safety of the product/ingredient is most important. Approval can be acquired through an application together with necessary documents and details as specified by the FSSAI, and a fee of Rs 50,000/per product/ingredient. As per novel food regulations, novel technology means a food/ingredient is obtained by using new technology with ab innovative engineering process, in which the process may give rise to a significant change in the composition or structure or size of the food or food ingredients, which may alter the nutritional value, metabolism, or level of undesirable substances. The labeling of a novel food shall be

- in accordance with the specific labeling requirements, if any, or

- specific to claims relating to the novel product or

- as per category notified by the FSSAI in the specific regulations.

The Food Safety Authority of Ireland have issued guidelines in 2015 on HPP of foods, according to which all food business operators proposing to use HPP as part of their production process must document and implement a procedure based on the principles of HACCP for all their products as part of their Food Safety Management System (FSMS). They should document how the chosen HPP parameters of pressure and holding time were validated and verified to ensure food safety during shelf life. Food business operators should use accredited laboratories, and relevant microbiological parameters are to be examined.

If the food business operator (FBO) plan to use HPP, it should consider the following:

- The intended purpose of high-pressure treatment

- The sensitivity of microorganisms to HPP

- The characteristics of the food

- The storage condition of the final product

- The potential for the introduction of food safety hazards due to shelf-life extension resulting from HPP

HPP provides a unique opportunity for food business operators to develop a new generation of value-added food products having superior quality and extended shelf life. HPP is a very promising technology for ready-to-eat (RTE) meats because there are few barriers to approval by regulatory authorities. No special labeling is required because no chemicals are added. There are no changes in sensory quality such as flavors, texture, and so on, and no changes in consumer acceptability in RTE meats that are pressure treated at 600 megapascals at 20°C for 180 s. An HPP treatment of 600 megapascals for 10 min at 30°C extend the shelf life of food, including cooked ham, dry cured ham, and marinated beef loins (Gracia et al., 2020). An HPP treatment of 500 megapascals extends the shelf life of cooked pork ham and raw smoked pork up to 8 weeks (Pietrzak et al., 2007).

14.2.3 PL Technology

PL is an emerging technology for various operations in the food industry, and it is also known as pulsed UV light or pulsed white light. This technology involves the application of high intensity and short duration (100–400 microseconds) pulses of a continuous broad-spectrum light in which 40% of the light emitted corresponds to the UV region (Palmieri & Cacace, 2005). These pulses have high peak energy, producing a light intensity per time unit, which is approximately 20,000 times greater than sunlight at sea level. The lethal effect of PL technology depends on the type of microbes as well as the absorption parameter of the liquid food. PL technology has been found to be effective in sterilizing food-packaging materials. Although PL intensity lasts for only a second, it is 20,000 times brighter than sunlight without any thermal effect. The xenon flash lamp used in PL treatment is eco-friendly. However, the packaging materials subjected to PL treatment should be chemically stable (Oms-Oliu et al., 2008). The PL treatment is a very recent development that can have several applications in the food industry as a novel non thermal method of preservation. The effectiveness of the PL treatment depends on the food to be processed, the type of microbial contamination, and the microbial load. PL treatment alone has a few limitations, and thus, combining PL technology with other processing techniques can improve food preservation by reducing effects on food quality. Some microbial species are found to be resistant to PL technology. This technology has immense potential in reducing peanut allergy. It is a novel technology that is less commercialized to date, and further research is required in this field for effective commercialization.

The application of pulsed light technology in foods was approved by US Food and Drug Administration (USFDA) in 1996 for the use of up to 12 joules/cm^2 for decontamination of food or food contact surfaces (Mandal et al., 2020). The French company Claranor has commercialized specific solutions for caps, cups, and trays, which are difficult to decontaminate using conventional technologies. Since 2000, several industrial experiments in food preservation have been carried out in North America and Europe. PL pasteurized coffee and cooking syrups have been commercialized by Taylerson's Malmesbury Syrups (UK). Xenon Corporation, USA, and the Dole Company, USA, have commercialized mushrooms enriched with vitamin D by means of PL treatment (Mandal et al., 2020).

However, this novel technology has not been approved by the Indian regulator, FSSAI, to date. Since it is a novel technology considered by the FSSAI, specific approval is required from the FSSAI by submitting an application for the approval of nonspecified food and food ingredients, along with the details of the new technology, safety information (documents on risk assessment or toxicity studies), a history of consumption of food product/food ingredient with an established history of safe use in at least two countries with well-established regulatory status, and the quantity of consumption and duration of consumption in years. Until and unless the application is approved by the FSSAI, no FBO can process any food by using the novel technology. In view of the limitations of the traditional thermal food preservation techniques and increase in consumer awareness, PL technology will find more applications in the coming days.

14.2.4 UV Light Treatment

UV light holds considerable promise in food processing as an alternative to traditional thermal processing. Its commercial application includes pasteurization of juices and milk, the treatment of meat, the treatment of food contact surfaces, and the extension of shelf life of fresh produce like fruits and vegetables (Chawla et al., 2021; Pinheiro et al., 2015). UV has been proved to have antibacterial and antifungal properties and active against viruses and molds. At certain wavelengths, UV destroys and damages the DNA of different types of microbes; however, UV radiation does not denature proteins which is the major advantage for UV-treated food products. UV radiation is considered to be nonionizing radiation that has germicidal properties at wavelengths in the range of 200–280 nm (Shahi et al., 2021). UV light radiation is a cost-effective tool and requires low energy consumption in comparison to thermal processing that makes it suitable and affordable for industrial point of view. The USFDA and US Department of Agriculture (USDA) have approved UV light as an alternative treatment to the thermal pasteurization of fresh fruit juice products in 2000. The definition of *pasteurization* is now revised and includes any process, treatment, or combination thereof, that is applied to food to reduce the most organisms of public health significance. UV treatment is a new technology that is used as an alternative to thermal processing to produce microbiologically safe foods with better retention of nutritional quality, extended shelf life, and health benefits. The USFDA considers UV light as radiation used to treat food as a food additive, which is literally not added to food. In 2000, the USFDA amended the food additive regulations to provide for the safe use of UV radiation at 253.7 nm to reduce human pathogens and other microorganisms in juice products. The amount of UV radiation necessary for human pathogen reduction depends on the type of juice, initial microbial load, and the design of the irradiation system. The USFDA did not specify a minimum or maximum UV dose by regulation (Koutchma, 2022). This should be achieved for individual usage situation in a manner consistent with GMPs. The levels of UV irradiation applied to juice will be limited by possible alteration in quality, nutritional and organoleptic characteristics of the juice.

In the EU and five countries, namely, Great Britain, Canada, Australia, New Zealand, and China, UV-treated food products belong to the category of novel food. Novel foods and novel ingredients are regulated in varying manners in each country and require approval by the regulatory body before being used. The FSSAI regulates and defines novel foods as an article of food for which standards have not been specified but is not unsafe, provided that such food does not contain any of the foods and ingredients prohibited under the Food Safety and Standards Act, 2006 and the rules and regulations made there under. Not only the novel food or ingredients but even the novel technology requires approval by the FSSAI. In 2013, based on the application and supporting documents, the Scientific Committee of the FSSAI accorded product approval to UV-treated raw milk. The fast spread of UV technology in food processing dictates a need for the acceleration of regulatory approval the harmonization and globalization of the regulations. There are several UV-treated products such as juice products, raw and pasteurized milk, mushroom, bread, and baker's yeast that have received approval from some regulators (Koutchma, 2009). UV light treatment is considered environmentally friendly emergent nonthermal technology for decontamination of juice and dairy products. It is less costly and has efficiency in the maintenance of quality parameters. However, further research is needed for the selection of the lowest possible radiation dose to minimize the loss of health-promoting compounds with a maximum reduction of pathogen loads in fruit juice and beverages.

14.2.5 Ultrasound Processing

Sound is a mechanical wave that propagates through a transmission medium such as solid, liquid, or gas but not through vacuum. A region of increased pressure on a sound wave is called a compression and that of decreased pressure is called a rarefaction. Humans are capable of hearing sounds between 20 Hz and 20 kHz, and sounds that have frequencies above this range are called ultrasound, whereas below this range, sounds are called infrasound. Ultrasound was first used for clinical purposes and now has widespread applications, including food processing and preservation (Gallo et al., 2018). Although in a relatively infant stage, ultrasound technology is capable of replacing traditional processing technologies like freezing, thawing, filtration, cutting, meat tenderization, and others. When combined with other technologies, ultrasound has the potential to effectively control microorganisms and ensure food safety. Vitamins, antioxidants, polyphenols, and other sensorial characteristics are affected due to a high-temperature treatment. Studies have shown that ultrasound or ultrasonic technology can be utilized in preservation, pasteurization, bioactive extraction, and the enhancement of shelf life by deactivating microorganisms at mild temperatures. It is a relatively simple power-saving emerging technology. Ultrasound can be classified as high-power ultrasound and low-power ultrasound. High-power ultrasound is in the band of 20–100 kHz and is also known as power ultrasound, which has the ability to cause cavitation, that is, the implosion of gas bubbles (Yao et al., 2020). Since power ultrasound generates pressure, temperature, and shear in the medium through which it propagates, it is capable of changing the physical and chemical properties of the material and inactivation of microbes, enzymes, degassing, deforming, and so on (Yao et al., 2020). Ultrasound technology is used in combination with other preservation techniques, and these are known as

- manosonication – combination of ultrasound and pressure.

- thermosonication – combination of ultrasound and heat.

- manothermosonication – combination of ultrasound, pressure, and heat.

The global market of ultrasonicated foods was valued at USD93.7 million in 2017 and is projected to reach USD134.6 million by 2022, growing at a rate of 7.5% from 2017. An increase in the demand for processed food products has led FBOs to focus on novel technologies for better quality and extended shelf life. Innovation and research on nonthermal processing technology have been increasing, particularly ultrasonication for the multifunctionalities in food processing. Europe is one of the leading suppliers of ultrasound-processed and -packaged food. In Europe, Germany dominated ultrasound-processed food market in 2017. The demand for ultrasonic-processed food markets is also high in Japan and China. Other countries in the Asia Pacific region, such as Indonesia, Malaysia, Thailand, Vietnam, and South Korea, are favorable markets for ultrasound-processed food products. The FSSAI consider ultrasound technology as novel technology. There is no regulation for ultrasound-processed food products notified by the FSSAI to date, and specific approval is to be obtained from the FSSAI before using this novel technology.

14.2.6 Other Emerging Processing Technologies

For example, the Federal Grain Inspection Service (FGIS) under the USDA-FGIS, the US Environmental Protection Agency (US-EPA), and the US Food and Drug Administration (US-FDA) are exploring regulatory oversight, treatment efficacy, product labeling, and residue determination for direct application of cold plasma on whole wheat (Laroque et al., 2022). The reactive species produced in cold plasma discharge is responsible for antimicrobial activity against foodborne pathogens, ensuring safety of the processed food; however, they can potentially alter the morphology and regulatory functions of biological cells and therefore need further examination before commercial applications. Thus, the precision and consistency required for confirming safety and regulatory compliance, and the resulting accompanying high cost often slow down or discourage commercial applications of these emerging food-processing technologies (Priyadarshini et al., 2019). Ozonation in food processing in the US, for example, started its journey as an effective sanitizer for beverage bottles in 1982, but it took almost another 20 years to approve direct food application of ozone in fruits, vegetables, beverages, meat, and poultry products (Sarron et al., 2021).

14.3 CONSUMER PERCEPTION OF NOVEL FOOD-PROCESSING TECHNOLOGIES

Consumers' perception on novel, emerging food-processing technologies is one of the important determinants for their societal, regulatory, and commercial acceptance and success. The popularity of organic foods and the unpopularity of irradiated foods are examples of such implications of consumer perception. Factors affecting consumers' perception of newly emerging food-processing technologies include sociodemographic attributes, education, availability of information, and reliability of sources of such information (Priyadarshini et al., 2019). Nowadays, consumers are preferring sustainable food-processing technologies, for example, consumers have positive perceptions of HPP and pulse lectric field (PEF) compared to traditional processing technologies like pasteurization and drying. Consumers in many countries have been found to be willing to pay a premium for their sustainable food choices. Besides safety and health benefit(s), consumers' demand for transparency is also driving food industries to disclose and/or advertise sustainable, novel food processing while including warning declarations for perceived harmful processes on the label. Thus, consumers' awareness and, consequently, their demands have forced legislators, retailers, technology developers, and food manufacturers to give due diligence to consumers' perception, sometimes even when it is not based on a sound technical understanding of the process. Since 2015, ISO14001 has specified guidelines to achieve environmental management in an organization, including the food-processing/handling industries, to achieve sustainable food processing catering to the increasing consumers' demands. Often, such consumers' perceptions and demands drive regulatory agencies to bring about necessary regulations, for example, mandatory display of "Radura" symbol for irradiated foods.

14.4 GLOBAL HARMONIZATION OF REGULATIONS ON NOVEL FOOD-PROCESSING TECHNOLOGIES

In this day and age of global food businesses and an intercontinental food chain, the harmonization of regulations applicable on food-processing technologies is a major challenge. With food being part and parcel of history, culture, and heritage of a society, achieving global harmony in regulations applicable on food-processing technologies is complex and challenging. Thus, the harmonization of the regulations in different countries/different parts of the globe is a major priority for the concerned authorities and stakeholders. There are a number of internationally recognized institutions/organizations such as CAC and the International Organization for Standardization (ISO) that thrive to achieve global harmony and consistency in food regulations (Barbosa-Cánovas et al., 2022). Even within a country, harmonization efforts are needed to bring about consistency among regulations mandated on food-processing technologies practiced by communities of different ethnicities and religious beliefs. In the US, the Food Safety Modernization Act, 2011 was implemented during the last decade to simplify enactments of food legislatures, particularly for applications on novel food-processing technologies and newly developed food ingredients. The EU, on the other hand, has had Regulation EC 258/97 since 1997 to regulate novel food processing and novel food ingredients. In 2018, the regulation was amended as Regulation (EU) 2015/2283 to expand novel food categories and make the process of bringing novel foods and food-processing technologies to the EU market in compliance with the safety norms simpler. The Global Harmonization Initiative, established by the Institute of Food Technologists and the European Federation of Food Science and Technology, is an effective international nonprofit network of individual scientists and scientific organizations working together to promote the harmonization of global food regulations and legislation. The initiative has been making significant progress in eliminating trade barriers, reducing unnecessary destruction of safe food within and across international borders, and minimizing inconsistencies among food safety regulations and legislation.

14.5 CONCLUSION

Nonthermal and noninvasive food-processing technologies are becoming increasingly relevant, and concerned legislation is also being updated accordingly to ensure food safety. Among these novel food-processing technologies, irradiation and HPP are highly regulated in various countries throughout the world, whereas research efforts are continuing to develop regulations on the applications of cold plasma, PL, and pulsed electric field technologies in food processing. Complexity and varieties of food matrices, vastly diverse consumer preference, and the high cost of research of novel food-processing technologies are primary challenges against achieving uniform legislations throughout the globe. However, food safety management system tools like HACCP and GMPs are universal requirements in all food-processing/handling businesses. Guidelines for the application of such tools, as published by the CAC, ISO, and others, are also globally acceptable and are routinely updated to cater to the emerging needs of regulating novel food-processing technologies and food ingredients. The harmonization of food regulations has been a recent and ongoing effort to minimize inconsistencies among food legislations of different countries and barriers to global food trade and ensure safe foods to consumers everywhere.

REFERENCES

Barbosa-Cánovas, G. V., Bermúdez-Aguirre, D., Gonçalves Franco, B., Candoğan, K., & Shin, G. Y. (2022). Chapter 12 – Novel food processing technologies and regulatory hurdles. In A. Martinović, S. Oh, & H. Lelieveld (Eds.), *Ensuring Global Food Safety (Second Edition)*, (pp. 221–228). Academic Press.

Chawla, A., Lobacz, A., Tarapata, J., & Zulewska, J. (2021). UV Light Application as a Mean for Disinfection Applied in the Dairy Industry. *Applied Sciences*, 11(16), 7285.

Ehlermann, D. A. E. (2014). Safety of food and beverages: Safety of irradiated foods. In Y. Motarjemi (Ed.), *Encyclopedia of Food Safety* (pp. 447–452). Academic Press.

Featherstone, S. (2015). 9 – Hazard analysis and critical control point (HACCP) systems in food canning. In S. Featherstone (Ed.), *A Complete Course in Canning and Related Processes (Fourteenth Edition)* (pp. 215–234). Woodhead Publishing.

Gallo, M., Ferrara, L., & Naviglio, D. (2018). Application of ultrasound in food science and technology: A perspective. *Foods (Basel, Switzerland)*, 7(10), 164.

Gracia, M. J., Lázaro, R., Pérez-Arquillué, C., Pagán, R., Ramos, S., Garcia, J. L., & Bayarri, S. (2020). High-pressure processing (HPP) of raw and dry-cured ham from experimentally infected pigs as a potential tool for the risk control of Toxoplasma gondii. *Innovative Food Science & Emerging Technologies*, 61, 102315.

Hogan, E., Kelly, A. L., & Sun, D.-W. (2005). 1 – High pressure processing of foods: An overview. In D.-W. Sun (Ed.), *Emerging Technologies for Food Processing*, (pp. 3–32). Academic Press.

Jadhav, H. B., Annapure, U. S., & Deshmukh, R. R. (2021). Non-thermal technologies for food processing. *Frontiers in Nutrition, 8.*

Koutchma, T. (2009). Advances in ultraviolet light technology for non-thermal processing of liquid foods. *Food and Bioprocess Technology, 2*(2), 138–155.

Koutchma, T. (2022). Chapter 4 – Validation of light-based processes. In T. Koutchma (Ed.), *Validation of Food Preservation Processes Based on Novel Technologies* (pp. 113–167): Academic Press.

Laroque, D. A., Seó, S. T., Valencia, G. A., Laurindo, J. B., & Carciofi, B. A. M. (2022). Cold plasma in food processing: Design, mechanisms, and application. *Journal of Food Engineering, 312,* 110748.

Mandal, R., Mohammadi, X., Wiktor, A., Singh, A., & Pratap Singh, A. (2020). Applications of pulsed light decontamination technology in food processing: An overview. *Applied Sciences, 10*(10), 3606.

Oms-Oliu, G., Martín-Belloso, O., & Soliva-Fortuny, R. (2008). Pulsed light treatments for food preservation. A review. *Food and Bioprocess Technology, 3*(1), 13.

Palmieri, L., & Cacace, D. (2005). 11 – High intensity pulsed light technology. In D.-W. Sun (Ed.), *Emerging Technologies for Food Processing* (pp. 279–306). Academic Press.

Pietrzak, D., Fonberg-Broczek, M., Mucka, A., & Windyga, B. (2007). Effects of high pressure treatment on the quality of cooked pork ham prepared with different levels of curing ingredients. *High Pressure Research, 27*(1), 27–31.

Pinheiro, J., Alegria, C., Abreu, M., Gonçalves, E. M., & Silva, C. L. M. (2015). Use of UV-C postharvest treatment for extending fresh whole tomato (Solanum lycopersicum, cv. Zinac) shelf-life. *Journal of Food Science and Technology, 52*(8), 5066–5074.

Priyadarshini, A., Rajauria, G., O'Donnell, C. P., & Tiwari, B. K. (2019). Emerging food processing technologies and factors impacting their industrial adoption. *Critical Reviews in Food Science and Nutrition, 59*(19), 3082–3101.

Sarron, E., Gadonna-Widehem, P., & Aussenac, T. (2021). Ozone treatments for preserving fresh vegetables quality: A critical review. *Foods (Basel, Switzerland), 10*(3), 605.

Shahi, S., Khorvash, R., Goli, M., Ranjbaran, S. M., Najarian, A., & Mohammadi Nafchi, A. (2021). Review of proposed different irradiation methods to inactivate food-processing viruses and microorganisms. *Food Science & Nutrition, 9*(10), 5883–5896.

Yao, Y., Pan, Y., & Liu, S. (2020). Power ultrasound and its applications: A state-of-the-art review. *Ultrasonics Sonochemistry, 62,* 104722.

Index

Note: Page numbers in *italics* refer to figures, those in **bold** refer to tables.

For Product Safety Concerns and Information please contact our EU
representative GPSR@taylorandfrancis.com
Taylor & Francis Verlag GmbH, Kaufingerstraße 24, 80331 München, Germany